Fluid Mechanics and Hydraulics

Fluid Mechanics and Hydraulics

Fourth Edition

Ranald V. Giles

Late Professor of Civil Engineering
Drexel Institute of Technology

Jack B. Evett, PhD

Professor of Civil Engineering
The University of North Carolina at Charlotte

Cheng Liu

Professor of Civil Engineering Technology
The University of North Carolina at Charlotte

Schaum's Outline Series

New York Chicago San Francisco Athens London Madrid
Mexico City Milan New Delhi Singapore Sydney Toronto

Ranald V. Giles was formerly a Professor of Civil Engineering at Drexel Institute of Technology in Philadelphia, PA. He was the author of the first and second editions.

Jack B. Evett holds BS and MS degrees from the University of South Carolina and a PhD from Texas A&M University. A registered professional engineer and land surveyor, he is a professor of civil engineering at The University of North Carolina at Charlotte, where he was formerly associate dean of engineering. He is author/coauthor of eleven books, including *Fundamentals of Fluid Mechanics* and *2500 Solved Problems in Fluid Mechanics and Hydraulics*.

Cheng Liu holds a BSCE from National Taiwan University and an MSCE from West Virginia University. He is a registered professional engineer and a professor of civil engineering technology at The University of North Carolina at Charlotte. He is coauthor of seven books, including *Fundamentals of Fluid Mechanics* and *2500 Solved Problems in Fluid Mechanics and Hydraulics*.

5 6 7 8 9 10 QVS/QVS 21 20 19 18 17

ISBN 978-0-07-183145-1
MHID 0-07-183145-2

e-ISBN 978-0-07-183084-3 (basic e-book)
e-MHID 0-07-183084-7

e-ISBN 978-0-07-183146-8 (enhanced e-book)
e-MHID 0-07-183146-0

Library of Congress Control Number: 2013946497

Preface

This book is designed primarily to supplement standard textbooks in fluid mechanics and hydraulics. It is based on the authors' conviction that clarification and understanding of the basic principles of any branch of mechanics can be accomplished best by means of numerous illustrative problems.

Previous editions of this book have been very favorably received. This third edition contains two new chapters—one on fluid statics, the other on flow of compressible fluids. Additionally, many chapters have been revised and expanded to keep pace with the most recent concepts, methods, and terminology. Another very important feature of this new edition is the use of the International System of Units (SI). Precisely half of all problems that involve units of measure utilize SI units, the other half employing the British Engineering System.

The subject matter is divided into chapters covering duly recognized areas of theory and study. Each chapter begins with statements of pertinent definitions, principles, and theorems together with illustrative and descriptive material. This material is followed by graded sets of solved and supplementary problems. The solved problems illustrate and amplify the theory, present methods of analysis, provide practical examples, and bring into sharp focus those fine points which enable the student to apply the basic principles correctly and confidently. Free-body analysis, vector diagrams, the principles of work and energy and of impulse-momentum, and Newton's laws of motion are utilized throughout the book. Efforts have been made to present original problems developed by the authors during many years of teaching the subject. Numerous proofs of theorems and derivations of formulas are included among the solved problems. The large number of supplementary problems serve as a complete review of the material of each chapter.

In addition to its use by engineering students of fluid mechanics and hydraulics, this book should be of considerable value as a reference for practicing engineers. They will find well-detailed solutions to many practical problems and can refer to the summary of the theory when the need arises. Also, the book should serve individuals who must review the subject for licensing examinations or other reasons.

We hope you will enjoy using this book and that it will help a great deal in your study of fluid mechanics and hydraulics. We would be pleased to receive your comments, suggestions, and/or criticisms.

Jack B. Evett
Cheng Liu

Contents

Symbols and Abbreviations

The following tabulation lists the letter symbols used in this book. Because the alphabet is limited, it is impossible to avoid using the same letter to represent more than one concept, Since each symbol is defined when it is first used, no confusion should result.

a	acceleration, area
A	area
b	weir length, width of water surface, bed width of open channel
c	coefficient of discharge, celerity of pressure wave (acoustic velocity)
c_c	coefficient of contraction
c_v	coefficient of velocity
C	coefficient (Chezy), constant of integration
CB	center of buoyancy
CG	center of gravity
C_p	center of pressure, power coefficient for propellers
C_D	coefficient of drag
C_F	thrust coefficient for propellers
C_L	coefficient of lift
C_T	torque coefficient for propellers
C_1	Hazen-Williams coefficient
cfs	cubic feet per second
CP	center of pressure
d, D	diameter
D_1	unit diameter
e	efficiency
E	bulk modulus of elasticity, specific energy
f	friction factor (Darcy) for pipe flow
F	force, thrust
F_B	buoyant force
FE	pressure energy
Fr	Froude number
g	gravitational acceleration ($= 32.2\ \text{ft/sec}^2 = 9.81\ \text{m/s}^2$)
gpm	gallons per minute
h	head, height or depth, pressure head
H	total head (energy)
H_L, h_L	lost head (sometimes LH)
hp	horsepower = 0.746 kW
I	moment of inertia
I_{xy}	product of inertia
J	joule
k	ratio of specific heats, isentropic (adiabatic) exponent, von Karman constant
K	discharge factors for trapezoidal channels, lost head factor for enlargements, any constant
K_c	lost head factor for contractions
KE	kinetic energy
l	mixing length
L	length
L_E	equivalent length
m	roughness factor in Bazin formula, weir factor for dams
mc	metacenter
M	mass, molecular weight
$\overline{MB}$	distance from CB to mc
n	roughness coefficient, exponent, roughness factor in Kutter's and Manning's formulas
N	rotational speed
N_s	specific speed
N_u	unit speed
N_M	Mach number
p	pressure, wetted perimeter
p'	pressure
P	power
Pa	pascal
PE	potential energy
P_u	unit power
psf	lb/ft^2
psia	lb/in^2, absolute
psig	lb/in^2, gage
q	unit flow
Q	volume rate of flow
Q_u	unit discharge

r	any radius
r_o	radius of pipe
R	gas constant, hydraulic radius
Re	Reynolds number
S	slope of hydraulic grade line, slope of energy line
S_0	slope of channel bed
sp gr	specific gravity
t	time, thickness, viscosity in Saybolt seconds
T	temperature, torque, time
u	peripheral velocity of rotating element
u, v, w	components of velocity in X, Y, and Z directions
v	volume, local velocity, relative velocity in hydraulic machines
v_s	specific volume $(= 1/\gamma)$
v_*	shear velocity
V	average velocity
V_c	critical velocity
V_d	volume of fluid displaced
W	weight, weight flow
We	Weber number
x	distance
y	depth, distance
y_c	critical depth
y_N	normal depth
Y	expansion factor for compressible flow
z	elevation (head)
Z	height of weir crest above channel bottom

α (alpha)	angle, kinetic energy correction factor
β (beta)	angle, momentum correction factor
γ (gamma)	specific (or unit) weight
δ (delta)	boundary layer thickness
Δ (delta)	flow correction term
ϵ (epsilon)	surface roughness
η (eta)	eddy viscosity
θ (theta)	any angle
μ (mu)	absolute viscosity
ν (nu)	kinematic viscosity
π (pi)	dimensionless parameter
ρ (rho)	density
σ (sigma)	surface tension, intensity of tensile stress
τ (tau)	shear stress
ϕ (phi)	speed factor, velocity potential, ratio
ψ (psi)	stream function
ω (omega)	angular velocity

Conversion Factors

1 cubic foot = 7.48 U.S. gallons = 28.32 liters
1 U.S. gallon = 8.338 pounds of water at 60°F
1 cubic foot per second = 0.646 million gallons per day
= 448.8 gallons per minute
1 pound-second per square foot (μ) = 478.7 poises
1 square foot per second (ν) = 0.0929 square meter per second
1 horsepower = 550 foot-pounds per second = 0.746 kilowatt
30 inches of mercury = 34 feet of water = 14.7 pounds per square inch
762 millimeters of mercury = 10.4 meters of water = 101.3 kilopascals

Parameter	British Engineering System to International System	International System to British Engineering System
Length	1 in = 0.0254 m 1 ft = 0.3048 m	1 m = 39.37 in 1 m = 3.281 ft
Mass	1 slug = 14.59 kg	1 kg = 0.06854 slug
Force	1 lb = 4.448 N	1 N = 0.2248 lb
Time	1 sec = 1 s	1 s = 1 sec
Specific (or unit) weight	1 lb/ft^3 = 157.1 N/m^3	1 N/m^3 = 0.006366 lb/ft^3
Mass density	1 slug/ft^3 = 515.2 kg/m^3	1 kg/m^3 = 0.001941 slug/ft^3
Specific gravity	Same dimensionless value in both systems	Same dimensionless value in both systems
Dynamic viscosity	1 lb-sec/ft^2 = 47.88 N·s/m^2	1 N·s/m^2 = 0.02089 lb-sec/ft^2
Kinematic viscosity	1 ft^2/sec = 0.09290 m^2/s	1 m^2/s = 10.76 ft^2/sec
Pressure	1 lb/ft^2 = 47.88 Pa 1 lb/in^2 = 6.895 kPa	1 Pa = 0.02089 lb/ft^2 1 kPa = 0.1450 lb/in^2
Surface tension	1 lb/ft = 14.59 N/m	1 N/m = 0.06853 lb/ft

Chapter 1

Properties of Fluids

FLUID MECHANICS AND HYDRAULICS

Fluid mechanics and hydraulics represent that branch of applied mechanics that deals with the behavior of fluids at rest and in motion. In the development of the principles of fluid mechanics, some fluid properties play principal roles, others only minor roles or no roles at all. In fluid statics, specific weight (or unit weight) is the important property, whereas in fluid flow, density and viscosity are predominant properties. Where appreciable compressibility occurs, principles of thermodynamics must be considered. Vapor pressure becomes important when negative pressures (gage) are involved, and surface tension affects static and flow conditions in small passages.

DEFINITION OF A FLUID

Fluids are substances that are capable of flowing and conform to the shape of containing vessels. When in equilibrium, fluids cannot sustain tangential or shear forces. All fluids have some degree of compressibility and offer little resistance to change of form.

Fluids can be classified as liquids or gases. The chief differences between liquids and gases are (*a*) liquids are practically incompressible whereas gases are compressible and usually must be so treated and (*b*) liquids occupy definite volumes and have free surfaces whereas a given mass of gas expands until it occupies all portions of any containing vessel.

BRITISH ENGINEERING (OR FPS) SYSTEM OF UNITS

In this system the fundamental mechanical dimensions are *length*, *force*, and *time*. The corresponding fundamental units are the foot (ft) of length, pound (lb) of force (or pound weight), and second (sec) of time. All other units can be derived from these. Thus unit volume is the ft^3, unit acceleration is the ft/sec^2, unit work is the ft-lb, and unit pressure is the lb/ft^2.

The unit for mass in this system, the *slug*, is derived from the fundamental units as follows. For a freely falling body in vacuum, the acceleration is that of gravity ($g = 32.2\ ft/sec^2$ at sea level), and the only force acting is its weight. From Newton's second law,

$$\text{force in pounds} = \text{mass in slugs} \times \text{acceleration in ft/sec}^2$$

Then $$\text{weight in pounds} = \text{mass in slugs} \times g(32.2\ \text{ft/sec}^2)$$

or $$\text{mass } M \text{ in slugs} = \frac{\text{weight } W \text{ in pounds}}{g(32.2\ \text{ft/sec}^2)} \tag{1}$$

By equation (*1*), slug = $lb\text{-}sec^2/ft$.

The temperature unit of the British system is the degree Fahrenheit (°F) or, on the absolute scale, the degree Rankine (°R).

INTERNATIONAL SYSTEM OF UNITS (SI)

In the SI, the fundamental mechanical dimensions are *length*, *mass* (unlike the British system), and *time*. The corresponding fundamental units are meter (m), kilogram (kg), and second (s). In terms of these, unit volume is the m^3, unit acceleration the m/s^2, and unit (mass) density the kg/m^3.

The SI unit of force, the newton (N), is derived via Newton's second law:

$$\text{force in N} = (\text{mass in kg}) \times (\text{acceleration in m/s}^2) \tag{2}$$

Thus, $1\ \text{N} = 1\ \text{kg} \cdot \text{m/s}^2$. Along with the newton are derived the joule (J) of work, where $1\ \text{J} = 1\ \text{N} \cdot \text{m}$, and the pascal (Pa) of pressure or stress, where $1\ \text{Pa} = 1\ \text{N/m}^2$.

In the SI, temperatures are usually reported in degrees Celsius (°C); the unit of absolute temperature is the kelvin (K).

SPECIFIC OR UNIT WEIGHT

The specific (or unit) weight γ of a substance is the weight of a unit volume of the substance. For liquids, γ may be taken as constant for practical changes of pressure. The specific weight of water for ordinary temperature variations is 62.4 lb/ft³, or 9.79 kN/m³. See Appendix, Table 1, for additional values.

The specific weight of a gas can be calculated using its *equation of state*,

$$\frac{pv}{T} = R \tag{3}$$

where pressure p is absolute pressure, v is the volume per unit weight, temperature T is the absolute temperature, and R is the *gas constant* of that particular species:

$$R = \frac{R_0}{Mg} = \frac{\text{universal gas constant}}{\text{molar weight}} \tag{4}$$

Since $\gamma = 1/v$, equation (*3*) can be written

$$\gamma = \frac{p}{RT} \tag{5}$$

MASS DENSITY OF A BODY ρ (rho) = mass per unit volume = γ/g.

In the British Engineering system of units, the mass density of water is $62.4/32.2 = 1.94\ \text{slugs/ft}^3$. In the International system, the density of water is $1000\ \text{kg/m}^3$ at 4°C. See Appendix, Table 1.

SPECIFIC GRAVITY OF A BODY

The specific gravity of a body is the dimensionless ratio of the weight of the body to the weight of an equal volume of a substance taken as a standard. Solids and liquids are referred to water (at 68°F = 20°C) as standard, while gases are often referred to air free of carbon dioxide or hydrogen (at 32°F = 0°C and 1 atmosphere = 14.7 lb/in² = 101.3 kPa pressure) as standard. For example,

$$\begin{aligned}\text{specific gravity of a substance} &= \frac{\text{weight of substance}}{\text{weight of equal volume of water}} \\ &= \frac{\text{specific weight of substance}}{\text{specific weight of water}} \\ &= \frac{\text{density of substance}}{\text{density of water}}\end{aligned} \tag{6}$$

Thus if the specific gravity of a given oil is 0.750, its specific weight is $(0.750)(62.4\ \text{lb/ft}^3) = 46.8\ \text{lb/ft}^3$, or $(0.750)(9.79\ \text{kN/m}^3) = 7.34\ \text{kN/m}^3$. Specific gravities are tabulated in the Appendix, Table 2.

VISCOSITY OF A FLUID

The viscosity of a fluid is that property which determines the amount of its resistance to a shearing force. Viscosity is due primarily to interaction between fluid molecules.

Referring to Fig. 1-1, consider two large, parallel plates a small distance y apart, the space between the plates being filled with a fluid. To keep the upper plate moving at constant velocity U, it is found that a constant force F must be applied. Thus there must exist a viscous interaction between plate and fluid, manifested as a drag on the former and a shear force on the latter. The fluid in contact with the upper plate will adhere to it and will move at velocity U, and the fluid in contact with the fixed plate will have velocity zero. If distance y and velocity U are not too great, the velocity profile will be a straight line. Experiments have shown that shear force F varies with the area of the plate A, with velocity U, and inversely with distance y. Since by similar triangles, $U/y = dV/dy$, we have

$$F \propto \left(\frac{AU}{y} = A\frac{dV}{dy}\right) \quad \text{or} \quad \left(\frac{F}{A} = \tau\right) \propto \frac{dV}{dy}$$

where $\tau = F/A$ = shear stress. If a proportionality constant μ (mu), called the *absolute (dynamic) viscosity*, is introduced,

$$\tau = \mu\frac{dV}{dy} \quad \text{or} \quad \mu = \frac{\tau}{dV/dy} = \frac{\text{shear stress}}{\text{rate of shear strain}} \tag{7}$$

It follows that the units of μ are Pa · s or $\frac{\text{lb-sec}}{\text{ft}^2}$. Fluids for which the proportionality of equation (7) holds are called *Newtonian fluids* (see Problem 1.10).

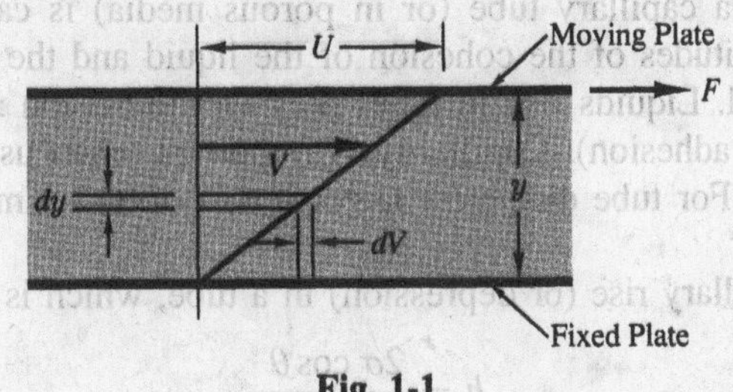

Fig. 1-1

Another viscosity coefficient, the *coefficient of kinematic viscosity*, is defined as

$$\text{kinematic viscosity } \nu \text{ (nu)} = \frac{\text{absolute viscosity } \mu}{\text{mass density } \rho}$$

or

$$\nu = \frac{\mu}{\rho} = \frac{\mu}{\gamma/g} = \frac{\mu g}{\gamma} \tag{8}$$

The units of ν are $\frac{\text{m}^2}{\text{s}}$ or $\frac{\text{ft}^2}{\text{sec}}$.

Viscosities are reported in older handbooks in poises or stokeses (cgs units) and on occasion in Saybolt seconds, from viscosimeter measurements. Conversions to the fps system are illustrated in Problems 1.7 through 1.9. A few values of viscosities are given in Tables 1 and 2 of the Appendix.

Viscosities of liquids decrease with temperature increases but are not affected appreciably by pressure changes. The absolute viscosity of gases increases with increase in temperature but is not appreciably changed by changes in pressure. Since the specific weight of gases changes with pressure changes (temperature constant), the kinematic viscosity varies inversely as the pressure.

VAPOR PRESSURE

When evaporation takes place within an enclosed space, the partial pressure created by the vapor molecules is called vapor pressure. Vapor pressures depend upon temperature and increase with it. See Table 1 in the Appendix for values for water.

SURFACE TENSION

A molecule in the interior of a liquid is under attractive forces in all directions, and the vector sum of these forces is zero. But a molecule at the surface of a liquid is acted on by a net inward cohesive force that is perpendicular to the surface. Hence it requires work to move molecules to the surface against this opposing force, and surface molecules have more energy than interior ones.

The surface tension σ (sigma) of a liquid is the work that must be done to bring enough molecules from inside the liquid to the surface to form one new unit area of that surface (J/m^2 or ft-lb/ft^2). Equivalently, the energized surface molecules act as though they compose a stretched sheet, and

$$\sigma = \Delta F / \Delta L \tag{9}$$

where ΔF is the elastic force transverse to any length element ΔL in the surface. Definition (*9*) gives the units N/m or lb/ft. The value of surface tension of water with air is 0.0756 N/m at 0°C, or 0.00518 lb/ft at 32°F. Table 1C gives values of surface tension for other temperatures.

CAPILLARITY

Rise or fall of liquid in a capillary tube (or in porous media) is caused by surface tension and depends on the relative magnitudes of the cohesion of the liquid and the adhesion of the liquid to the walls of the containing vessel. Liquids rise in tubes they wet (adhesion > cohesion) and fall in tubes they do not wet (cohesion > adhesion). Capillarity is important when using tubes smaller than about $\frac{3}{8}$ inch (10 mm) in diameter. For tube diameters larger than $\frac{1}{2}$ in (12 mm), capillary effects are negligible.

Figure 1-2 illustrates capillary rise (or depression) in a tube, which is given approximately by

$$h = \frac{2\sigma \cos\theta}{\gamma r} \tag{10}$$

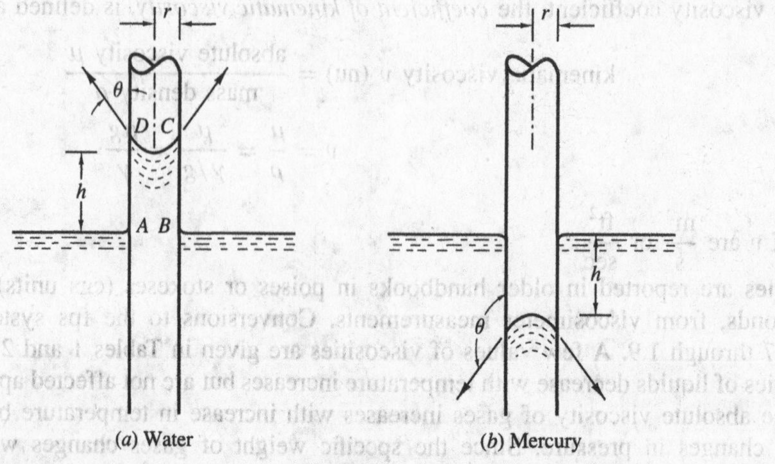

(*a*) Water (*b*) Mercury

Fig. 1-2

where

h = height of capillary rise (or depression)
σ = surface tension
θ = wetting angle (see Fig. 1-2)
γ = specific weight of liquid
r = radius of tube

If the tube is clean, θ is 0° for water and about 140° for mercury.

BULK MODULUS OF ELASTICITY (E)

The bulk modulus of elasticity (E) expresses the compressibility of a fluid. It is the ratio of the change in unit pressure to the corresponding volume change per unit of volume.

$$E = \frac{dp}{-dv/v} \tag{11}$$

Because a pressure increase, dp, results in a decrease in fractional volume, dv/v, the minus is inserted to render E positive. Clearly, the units of E are those of pressure—Pa or lb/in^2.

ISOTHERMAL CONDITIONS

For a fixed temperature, the ideal gas law, equation (3) or (5), becomes

$$p_1 v_1 = p_2 v_2 \qquad \text{and} \qquad \frac{\gamma_1}{\gamma_2} = \frac{p_1}{p_2} = \text{constant} \tag{12}$$

Also,

$$\text{bulk modulus } E = p \tag{13}$$

ADIABATIC OR ISENTROPIC CONDITIONS

If no heat is exchanged between the gas and its container, equations (12) and (13) are replaced by

$$p_1 v_1^k = p_2 v_2^k \qquad \text{or} \qquad \left(\frac{\gamma_1}{\gamma_2}\right)^k = \frac{p_1}{p_2} = \text{constant} \tag{14}$$

Also,

$$\frac{T_2}{T_1} = \left(\frac{p_2}{p_1}\right)^{(k-1)/k} \tag{15}$$

and

$$\text{bulk modulus } E = kp \tag{16}$$

Here k is the ratio of the specific heat at constant pressure to the specific heat at constant volume.

PRESSURE DISTURBANCES

Pressure disturbances imposed on a fluid move in waves, at speed

$$c = \sqrt{E/\rho} \tag{17}$$

For gases, the acoustic velocity is

$$c = \sqrt{kp/\rho} = \sqrt{kgRT} \tag{18}$$

Solved Problems

1.1. Calculate the specific weight γ, specific volume v_s, and density ρ of methane at 100°F and 120 psi absolute.

Solution:

From Table 1A in the Appendix, $R = 96.3$ ft/°R.

$$\text{specific weight } \gamma = \frac{p}{RT} = \frac{120 \times 144}{(96.3)(460 + 100)} = 0.320 \text{ lb/ft}^3$$

$$\text{density } \rho = \frac{\gamma}{g} = \frac{0.320}{32.2} = 0.00994 \text{ slug/ft}^3$$

$$\text{specific volume } v_s = \frac{1}{\rho} = \frac{1}{0.00994} = 101 \text{ ft}^3/\text{slug}$$

1.2. If 6 m³ of oil weighs 47 kN, calculate its specific weight γ, density ρ, and specific gravity.

Solution:

$$\text{specific weight } \gamma = \frac{47 \text{ kN}}{6 \text{ m}^3} = 7.833 \text{ kN/m}^3$$

$$\text{density } \rho = \frac{\gamma}{g} = \frac{7833 \text{ N/m}^3}{9.81 \text{ m/s}^2} = 798 \text{ kg/m}^3$$

$$\text{specific gravity } = \frac{\gamma_{\text{oil}}}{\gamma_{\text{water}}} = \frac{7.833 \text{ kN/m}^3}{9.79 \text{ kN/m}^3} = 0.800$$

1.3. At 90°F and 30.0 psi absolute the volume per unit weight of a certain gas was 11.4 ft³/lb. Determine its gas constant R and the density ρ.

Solution:

Since $\gamma = \dfrac{p}{RT}$,

$$R = \frac{p}{\gamma T} = \frac{pv}{T} = \frac{(30.0 \times 144)(11.4)}{460 + 90} = 89.5 \text{ ft/°R}$$

$$\text{density } \rho = \frac{\gamma}{g} = \frac{1/v}{g} = \frac{1}{vg} = \frac{1}{11.4 \times 32.2} = 0.00272 \text{ slug/ft}^3$$

1.4. (*a*) Find the change in volume of 1.00 ft³ of water at 80°F when subjected to a pressure increase of 300 psi.

(*b*) From the following test data determine the bulk modulus of elasticity of water: at 500 psi the volume was 1.000 ft³, and at 3500 psi the volume was 0.990 ft³.

Solution:

(*a*) From Table 1C in the Appendix, E at 80°F is 325,000 psi. Using formula (*11*),

$$dv = -\frac{v\,dp}{E} = -\frac{1.00 \times 300}{325{,}000} = -0.00092 \text{ ft}^3$$

(*b*)

$$E = -\frac{dp}{dv/v} = -\frac{3500 - 500}{(0.990 - 1.000)/1.000} = 3 \times 10^5 \text{ psi}$$

1.5. At a great depth in the ocean, the pressure is 80 MPa. Assume that specific weight at the surface is 10 kN/m^3 and the average bulk modulus of elasticity is 2.340 GPa. Find: (*a*) the change in specific volume between the surface and that great depth, (*b*) the specific volume at that depth, and (*c*) the specific weight at that depth.

Solution:

(*a*) $$(v_s)_1 = \frac{1}{\rho_1} = \frac{g}{\gamma_1} = \frac{9.81}{10 \times 10^3} = 9.81 \times 10^{-4}\ \text{m}^3/\text{kg}$$

$$E = \frac{dp}{-dv_s/v_s}$$

$$2.340 \times 10^9 = \frac{(80 \times 10^6) - 0}{dv_s/(9.81 \times 10^{-4})}$$

$$dv_s = -0.335 \times 10^{-4}\ \text{m}^3/\text{kg}$$

(*b*) $$(v_s)_2 = (v_s)_1 + dv_s = (9.81 - 0.335) \times 10^{-4} = 9.475 \times 10^{-4}\ \text{m}^3/\text{kg}$$

(*c*) $$\gamma_2 = g/(v_s)_2 = 9.81/(9.475 \times 10^{-4}) = 10.35\ \text{kN/m}^3$$

1.6. A cylinder contains 12.5 ft^3 of air at 120°F and 40 psi absolute. The air is compressed to 2.50 ft^3. (*a*) Assuming isothermal conditions, what is the pressure at the new volume, and what is the bulk modulus of elasticity? (*b*) Assuming adiabatic conditions, what is the final pressure and temperature, and what is the bulk modulus of elasticity?

Solution:

(*a*) For isothermal conditions, $p_1 v_1 = p_2 v_2$

Then $(40 \times 144)(12.5) = (p_2 \times 144)(2.50)$ and $p_2 = 200$ psi absolute

The bulk modulus $E = p = 200$ psi.

(*b*) For adiabatic conditions, $p_1 v_1^k = p_2 v_2^k$, and Table 1A in the Appendix gives $k = 1.40$.

Then $(40 \times 144)(12.5)^{1.40} = (p_2 \times 144)(2.50)^{1.40}$ and $p_2 = 381$ psi absolute

The final temperature is obtained by using equation (*15*):

$$\frac{T_2}{T_1} = \left(\frac{p_2}{p_1}\right)^{(k-1)/k}, \qquad \frac{T_2}{460 + 120} = \left(\frac{381}{40}\right)^{0.40/1.40}, \qquad T_2 = 1104°R = 644°F$$

The bulk modulus $E = kp = 1.40 \times 381 = 533$ psi.

1.7. From the International Critical Tables, the viscosity of water at 20°C (68°F) is 1.008 cp (centipoises). (*a*) Compute the absolute viscosity in lb-sec/ft^2. (*b*) If the specific gravity at 20°C is 0.998, compute the kinematic viscosity in ft^2/sec.

Solution:

Using 1 poise = 1 dyne-sec/cm^2, 1 lb = 444,800 dynes, and 1 ft = 30.48 cm, we obtain

$$1\frac{\text{lb-sec}}{\text{ft}^2} = \frac{444{,}800\ \text{dyne-sec}}{(30.48\ \text{cm})^2} = 478.8\ \text{poises}$$

(*a*) $$\mu = \frac{1.008 \times 10^{-2}\ \text{poise}}{(478.8\ \text{poise})/(\text{lb-sec/ft}^2)} = 2.11 \times 10^{-5}\frac{\text{lb-sec}}{\text{ft}^2}$$

(*b*) $\nu = \dfrac{\mu}{\rho} = \dfrac{\mu}{(\text{sp gr})\rho_{\text{water}}}$

$$= \frac{2.11 \times 10^{-5}\ \text{lb-sec/ft}^2}{(0.998)\left(1.94\ \text{slugs/ft}^3\right)} = 1.090 \times 10^{-5}\frac{\text{ft-lb-sec}}{\text{slug}} = 1.090 \times 10^{-5}\ \text{ft}^2/\text{sec}$$

1.8. Convert 15.14 poises to ft²/sec if the liquid has specific gravity 0.964.

Solution:

From Problem 1.7, the overall conversion factor is

$$\frac{1}{(478.8)(1.94)} = 0.001077\ \left(\text{ft}^2/\text{sec}\right)/\text{poise}$$

Thus, $\nu = (15.14/0.964)(0.001077) = 0.0169\ \text{ft}^2/\text{sec}$.

1.9. Convert a kinematic viscosity of 510 Saybolt seconds at 60°F to ft²/sec.

Solution:

Absolute and kinematic viscosities are converted according to

(*a*) For $t \le 100$, μ in poises $= (0.00226t - 1.95/t) \times$ sp gr
For $t > 100$, μ in poises $= (0.00220t - 1.35/t) \times$ sp gr

(*b*) For $t \le 100$, ν in stokeses $= (0.00226t - 1.95/t)$
For $t > 100$, ν in stokeses $= (0.00220t - 1.35/t)$

where t = Saybolt seconds. To convert stokeses $\left(\text{cm}^2/\text{sec}\right)$ to ft²/sec units, divide by $(30.48)^2$ or 929.

Using group (*b*), and since $t > 100$, $\nu = \left(0.00220 \times 510 - \dfrac{1.35}{510}\right)\left(\dfrac{1}{929}\right) = 0.001205\ \text{ft}^2/\text{sec}$.

1.10. Discuss the shear characteristics of the fluids for which the curves have been drawn in Fig. 1-3.

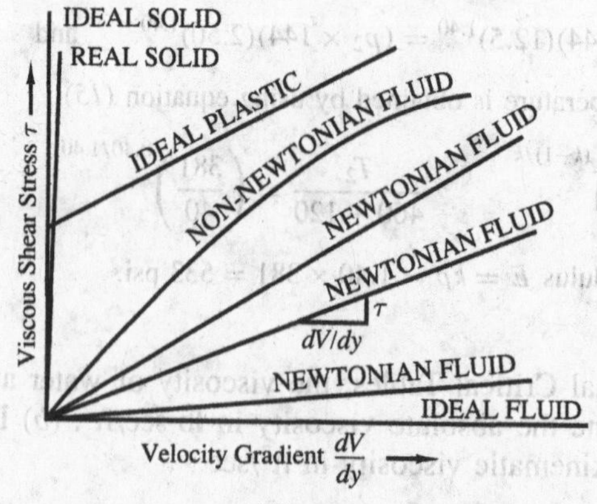

Fig. 1-3

Solution:

(*a*) The Newtonian fluids behave according to the law $\tau = \mu(dV/dy)$, or the shear stress is proportional to the velocity gradient or rate of shearing strain. Thus for these fluids the plot of shear stress against velocity gradient is a straight line passing through the origin. The slope of the line determines the viscosity.

(*b*) For the "ideal" fluid, the resistance to shearing deformation is zero, and hence the plot coincides with the x axis. Although no ideal fluids exist, in certain analyses the assumption of an ideal fluid is useful and justified.

(*c*) For the "ideal" or elastic solid, no deformation will occur under any loading condition, and the plot coincides with the y axis. Real solids have some deformation, and within the proportional limit (Hooke's law) the plot is a straight line that is almost vertical.

(*d*) Non-Newtonian fluids deform in such a way that shear stress is not proportional to rate of shearing deformation, except perhaps at very low shear stresses. The deformation of these fluids might be classified as plastic.

(*e*) The "ideal" plastic material could sustain a certain amount of shear stress without deformation, and thereafter it would deform in proportion to the shear stress.

1.11. Refer to Fig.1-4. A fluid has absolute viscosity 0.0010 lb-sec/ft^2 and specific gravity 0.913. Calculate the velocity gradient and the intensity of shear stress at the boundary and at points 1 in, 2 in, and 3 in from the boundary, assuming (*a*) a straight-line velocity distribution and (*b*) a parabolic velocity distribution. The parabola in the sketch has its vertex at A. Origin is at B.

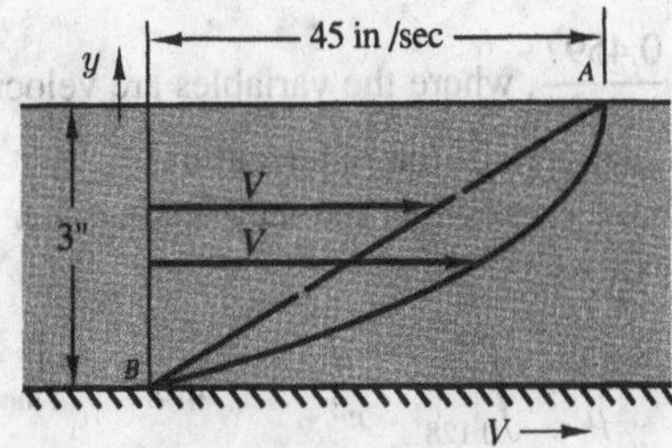

Fig. 1-4

Solution:

(*a*) For the straight-line assumption, the relation between velocity and distance y is $V = 15y$. Then $dV = 15dy$, or the velocity gradient is $dV/dy = 15$.

For $y = 0$, $V = 0$, $dV/dy = 15\ \text{sec}^{-1}$ and

$$\tau = \mu(dV/dy) = 0.0010 \times 15 = 0.015\ \text{lb/ft}^2$$

Similarly, for other values of y we also obtain $\tau = 0.015\ \text{lb/ft}^2$.

(*b*) The equation of the parabola must satisfy the condition that the velocity is zero at the boundary B. The equation of the parabola is $V = 45 - 5(3 - y)^2$. Then $dV/dy = 10(3 - y)$, and tabulation of results yields the following:

y	V	dV/dy	$\tau = 0.0010(dV/dy)$
0	0	30	0.030 lb/ft^2
1	25	20	0.020 lb/ft^2
2	40	10	0.010 lb/ft^2
3	45	0	0

It will be observed that where the velocity gradient is zero (which occurs at the centerline of a pipe flowing under pressure, as will be seen later) the shear stress is also zero.

Note that the units of velocity gradient are sec^{-1}, and therefore the product $\mu(dV/dy) = (\text{lb-sec/ft}^2)(\text{sec}^{-1}) = \text{lb/ft}^2$, the correct dimensions of shear stress τ.

1.12. A cylinder of 0.122-m radius rotates concentrically inside a fixed cylinder of 0.128-m radius. Both cylinders are 0.305 m long. Determine the viscosity of the liquid that fills the space between the cylinders if a torque of 0.881 N · m is required to maintain an angular velocity of 60 revolutions per minute.

Mathcad

Solution:

(*a*) The torque is transmitted through the fluid layers to the outer cylinder. Since the gap between the cylinders is small, the calculation can be made without integration.

$$\text{tangential velocity of the inner cylinder} = r\omega = (0.122 \text{ m})(2\pi \text{ rad/s}) = 0.767 \text{ m/s}$$

For the small space between cylinders, the velocity gradient can be assumed to be a straight line, and the mean radius can be used. Then $dV/dy = (0.767 \text{ m})/(0.128 - 0.122) = 127.8 \text{ s}^{-1}$.

$$\text{torque applied} = \text{torque resisting}$$

$$0.881 = \tau(\text{area})(\text{arm}) = \tau(2\pi \times 0.125 \times 0.305)(0.125) \quad \text{and} \quad \tau = 29.4 \text{ Pa}$$

$$\text{Then } \mu = \frac{\tau}{dV/dy} = \frac{29.4}{127.8} = 0.230 \text{ Pa} \cdot \text{s}.$$

(*b*) The more exact mathematical approach uses calculus, as follows.

As before, $0.881 = \tau(2\pi r \times 0.305)r$, from which $\tau = 0.4597/r^2$.

Now $\dfrac{dV}{dy} = \dfrac{\tau}{\mu} = \dfrac{0.4597}{\mu r^2}$, where the variables are velocity V and radius r. The velocity is 0.767 m/s at the inner radius and zero at the outer radius.

Rearranging the above expression and substituting $-dr$ for dy (the minus sign indicates that r decreases as V increases), we obtain

$$\int_{V_{\text{outer}}}^{V_{\text{inner}}} dV = \frac{0.4597}{\mu} \int_{0.128}^{0.122} \frac{-dr}{r^2} \quad \text{and} \quad V_{\text{inner}} - V_{\text{outer}} = \frac{0.4597}{\mu} \left[\frac{1}{r}\right]_{0.128}^{0.122}$$

Then $(0.767 - 0) = \left(\dfrac{0.4597}{\mu}\right)\left(\dfrac{1}{0.122} - \dfrac{1}{0.128}\right)$, from which $\mu = 0.230 \text{ Pa} \cdot \text{s}$.

1.13. Develop the expression for the relation between the gage pressure p inside a droplet of liquid and the surface tension σ.

Mathcad

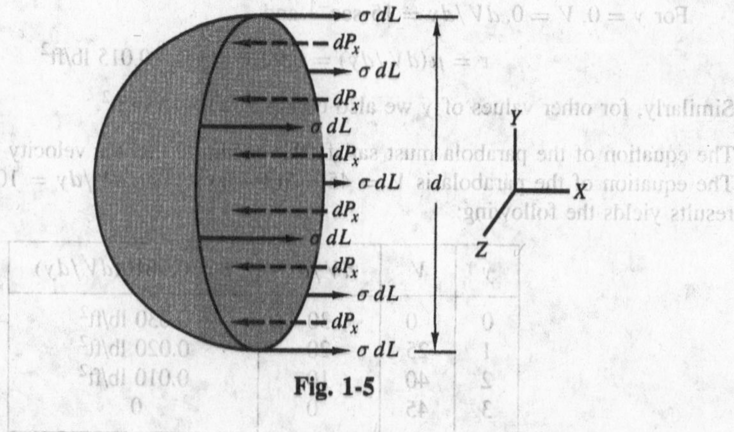

Fig. 1-5

Solution:

The surface tension in the surface of a small drop of liquid causes the pressure inside the drop to be greater than the pressure outside.

Figure 1-5 shows the forces that cause equilibrium in the X direction of half of a small drop of diameter d. The forces $\sigma\, dL$ are due to surface tension around the perimeter, and the forces dP_x are the X components of the $p\, dA$ forces (see Chapter 2). Then, from $\Sigma X = 0$,

sum of forces to the right = sum of forces to the left

$$\sigma \int dL = \int dP_x$$

surface tension × perimeter = pressure × projected area

$$\sigma(\pi d) = p(\pi d^2/4)$$

or $p = 4\sigma/d$.

It should be observed that the smaller the droplet, the greater the pressure.

1.14. A small drop of water at 80°F is in contact with the air and has diameter 0.0200 in. If the pressure within the droplet is 0.082 psi greater than the atmosphere, what is the value of the surface tension?

Solution:

By Problem 1.13, $\sigma = \frac{1}{4}pd = \frac{1}{4}\left[(0.082 \times 144)\text{lb/ft}^2\right] \times [(0.0200/12)\text{ft}] = 0.00492 \text{ lb/ft}$.

1.15. A needle 35 mm long rests on a water surface at 20°C. What force over and above the needle's weight is required to lift the needle from contact with the water surface?

Solution:

From Table 1C,

$$\sigma = 0.0728 \text{ N/m}$$

$$\sigma = F/L$$

$$0.0728 = F/(2 \times 0.035)$$

$$F = 0.00510 \text{ N}$$

1.16. Derive equation (*10*) for calculating the height to which a liquid will rise in a capillary tube exposed to the atmosphere.

Solution:

The rise in the tube can be approximated by considering the mass of liquid $ABCD$ in Fig. 1-2(*a*) as a free body. Since ΣY must equal 0, we obtain

components of force due to surface tension (up) − weight of volume $ABCD$ (down)
+ pressure force on AB (up) − pressure force on CD (down) = 0

or

$$+\left(\sigma \int dL\right)\cos\theta - \gamma\left(\pi r^2 \times h\right) + p(\text{area } AB) - p(\text{area } CD) = 0$$

It can be seen that the pressure at levels AB and CD are both atmospheric. Thus the last two terms on the left-hand side of the equation cancel, and, since $\sigma \int dL = \sigma(2\pi r)$, we obtain

$$h = \frac{2\sigma\cos\theta}{\gamma r}$$

1.17. Calculate the approximate depression of mercury at 20°C in a capillary tube of radius 1.5 mm. Surface tension (σ) for mercury is 0.514 N/m at 20°, and its specific weight is 133.1 kN/m³.

Solution:

$$h = \frac{2\sigma \cos\theta}{\gamma r} = \frac{(2)(0.514)(\cos 140°)}{(133.1 \times 10^3)(1.5 \times 10^{-3})} = -0.00394 \text{ m}, \text{ or } -3.94 \text{ mm}$$

1.18. Estimate the height to which water at 70°F will rise in a capillary tube of diameter 0.120 in.

Solution:

From Table 1C, $\sigma = 0.00497$ lb/ft.

$$h = \frac{2\sigma \cos\theta}{\gamma r} = \frac{(2)(0.00497)(\cos 0°)}{(62.4)(0.060/12)} = 0.0319 \text{ ft} = 0.382 \text{ in}$$

Supplementary Problems

1.19. If the density of a liquid is 835 kg/m^3, find its specific weight and specific gravity. *Ans.* 8.20 kN/m^3, 0.837

1.20. Check the values of the density and specific weight of air at 80°F shown in Table 1B.

1.21. Check the values of the specific weights of carbon dioxide and nitrogen in Table 1A.

1.22. At what pressure will air at 49°C weigh 18.7 N/m^3? *Ans.* 176 kPa

1.23. Two cubic feet of air at atmospheric pressure is compressed to 0.50 ft^3. For isothermal conditions, what is the final pressure? *Ans.* 58.8 psia

1.24. In the preceding problem, what would be the final pressure if no heat were lost during compression? *Ans.* 102 psia

1.25. Determine the absolute viscosity of mercury in N · s/m^2 if the viscosity in poises is 0.0158. *Ans.* 1.58×10^{-3} N · s/m^2

1.26. If an oil has an absolute viscosity of 510 poises, what is its viscosity in the fps system? *Ans.* 1.07 lb-sec/ft^2

1.27. What are the absolute and kinematic viscosities in fps units of an oil having a Saybolt viscosity of 155 seconds, if the specific gravity of the oil is 0.932? *Ans.* 646×10^{-6}, 358×10^{-6}

1.28. Two large plane surfaces are 1 in apart, and the space between them is filled with a liquid of absolute viscosity 0.0200 lb-sec/ft^2. Assuming the velocity gradient to be a straight line, what force is required to pull a very thin plate of 4.00 ft^2 area at a constant speed of 1.00 ft/sec if the plate is $\frac{1}{3}$ in from one of the surfaces? *Ans.* 4.32 lb

1.29. What force is necessary to lift a thin wire ring 45 mm in diameter from a water surface at 20°C? Neglect the weight of the wire. *Ans.* 0.0206 N

1.30. What is the smallest diameter glass tube that will keep the capillary height-change of water at 20°C less than 0.9 mm? *Ans.* 33.1 mm

1.31. Find the change in volume of 10.00000 ft^3 of water at 80°F when subjected to a pressure increase of 500 psi. Water's bulk modulus of elasticity at this temperature is approximately 325,000 psi. *Ans.* −0.0154 ft^3

1.32. Approximately what pressure must be applied to water to reduce its volume by 1.25% if the bulk modulus of elasticity is 2.19 GPa? *Ans.* 0.0274 GPa

Chapter 2

Fluid Statics

ON

essure refers to the effects of a force acting against and distributed over a surface. The erted by a solid, liquid, or gas. Often, the force causing a pressure is simply the weight

very important factor in many fluid mechanics and hydraulics problems. As will be ntly in this chapter, pressure exerted by a fluid varies directly with depth. Hence, the ttom of a dam is considerably greater than that near the top of the dam, and enormous on a submarine at the ocean's bottom. Needless to say, pressure effects such as these account in designing structures such as dams and submarines.

RE

is transmitted with equal intensity in all directions and acts normal to any plane. In l plane the pressure intensities in a liquid are equal. Measurements of unit pressures by using various forms of gages. Unless otherwise stated, gage or relative pres- throughout this book. Gage pressures represent values above or below atmospheric

OR PRESSURE is expressed as force divided by area. In general,

$$p = \frac{dF}{dA}$$

Pressure is commonly given in units of lb/ft^2 (psf), lb/in^2 (psi), or Pa (N/m^2), depending on units of force and area.

For conditions where force F is uniformly distributed over an area, we have

$$p = \frac{F}{A}$$

DIFFERENCE IN PRESSURE

Difference in pressure between any two points at different levels in a liquid is given by

$$p_2 - p_1 = \gamma(h_2 - h_1) \qquad (1)$$

when γ = unit weight of the liquid and $h_2 - h_1$ = difference in elevation.

If point 1 is in the free surface of the liquid and h is positive downward, the above equation becomes

$$p = \gamma h \quad \text{(gage)} \qquad (2)$$

These equations are applicable as long as γ is constant (or varies so slightly with h as to cause no significant error in the result).

PRESSURE HEAD h

Pressure head h represents the height of a column of homogeneous fluid that will produce a given intensity of pressure. Then

$$h = \frac{p}{\gamma} \tag{3}$$

PRESSURE VARIATIONS IN A COMPRESSIBLE FLUID

Pressure variations in a compressible fluid are usually very small because of the small unit weights and the small differences of elevation being considered in hydraulic calculations. Where such differences must be recognized for small changes in elevation dh, the law of pressure variation can be written

$$dp = -\gamma\, dh \tag{4}$$

The negative sign indicates that the pressure decreases as the altitude increases, with h positive upward. For applications, see Problems 2.23 through 2.25.

VACUUM AND ATMOSPHERIC PRESSURE

In the context of pressure, the term *vacuum* is used to refer to a space that has a pressure less than atmospheric pressure. Atmospheric pressure refers, of course, to the prevailing pressure in the air around us. It varies somewhat with changing weather conditions, and it decreases with increasing altitude. At sea level, average atmospheric pressure is 14.7 psi, 101.3 kPa, 29.9 in (760 mm) of mercury, or 1 atmosphere. This is commonly referred to as "standard atmospheric pressure."

A vacuum is quantified in terms of how much its pressure is below atmospheric pressure. For example, if air is pumped out of a pressure vessel until the internal pressure is 10.0 psi, and if atmospheric pressure is standard (i.e., 14.7 psi), the pressure in the vessel could be indicated as a vacuum of 14.7 − 10.0, or 4.7 psi.

ABSOLUTE AND GAGE PRESSURE

Pressure measurements are generally indicated as being either *absolute pressure* or *gage pressure*. Absolute pressure uses absolute zero, which is the lowest possible pressure and the pressure that would exist in a perfect vacuum, as its base (i.e., zero reading). Gage pressure is measured with atmospheric pressure as its base. Thus if a fluid pressure is 5.5 kPa above standard atmospheric pressure (101.3 kPa), its gage pressure would be 5.5 kPa and its absolute pressure 5.5 + 101.3, or 106.8 kPa. Sometimes absolute and gage pressures are indicated by appending "a" or "g," respectively, to the pressure units (e.g., psia or psig). If no a or g (or other indication) is given, pressure is usually gage pressure.

BAROMETERS

A barometer is a device for measuring atmospheric pressure. A simple barometer consists of a tube more than 30 in (762 mm) long inserted in an open container of mercury with a closed tube end at the top and an open tube end at the bottom and with mercury extending from the container up into the tube. Mercury rises in the tube to a height of approximately 30 in (762 mm) at sea level. Inasmuch as the tube is longer than 30 in, there will be a vacuum (pressure near absolute zero) above the mercury in the tube. The only pressure causing the mercury to rise in the tube is that of the atmosphere; and, of course, the amount the mercury rises varies with the applied atmospheric pressure.

If atmospheric pressure is 14.7 psi, the actual height to which mercury will rise can be computed using equation (*3*).

$$h = \frac{(14.7\ \text{lb/in}^2)[(144\ \text{in}^2)/(1\ \text{ft}^2)]}{(13.6)(62.4\ \text{lb/ft}^3)} = 2.49\ \text{ft, or } 29.9\ \text{in}$$

The level of mercury will rise and fall as atmospheric pressure changes; direct reading of the mercury level gives prevailing atmospheric pressure as a pressure head (of mercury), which can be converted to pressure, if desired, by using equation (*2*).

PIEZOMETERS AND MANOMETERS

Although a barometer can be used to measure atmospheric pressure, it is often necessary to measure pressures of other fluids. There are a number of ways to accomplish this task. For liquids, a tube may be attached to the wall of the container (or conduit) in which the liquid resides so liquid can rise in the tube. By determining the height to which liquid rises and using equation (*2*), the pressure of the liquid in the container (or conduit) can be determined. Such a device is known as a *piezometer*. To avoid capillary effects, a piezometer tube's diameter should be about $\frac{1}{2}$ in (13 mm) or greater. For applications, see Problems 2.11 and 2.12.

A somewhat more complicated device for measuring fluid pressure consists of a bent tube (or tubes) containing one or more liquids of different specific gravities. Such a device is known as a *manometer*. In using a manometer, generally a known pressure (which may be atmospheric) is applied to one end of the manometer tube and the unknown pressure to be determined is applied to the other end. In some cases, however, the difference between the pressures at the ends of the manometer tube is desired rather than the actual pressure at either end. A manometer to determine this pressure difference is known as a *differential manometer*. The liquids in a manometer will rise or fall as the pressure at either end (or both ends) of the tube changes.

The pressure in a container (or conduit) using a manometer is determined by transforming heights of liquids within the manometer tube to pressures, using equation (*2*). The general procedure in calculation is to start at one end of the manometer tube and proceed from each fluid level to the next, adding or subtracting pressures as the elevation decreases or increases, respectively. Each pressure is determined from equation (*2*), using the appropriate specific gravities of the liquids in the manometer. For applications, see Problems 2.14 through 2.22.

Solved Problems

2.1. Show that the pressure intensity at a point is of equal magnitude in all directions.

Solution:

Consider a small triangular prism of liquid at rest (Fig. 2-1) acted upon by the fluid around it. The values of average unit pressures on the three surfaces are p_1, p_2, and p_3. In the z direction the forces are equal and opposite and cancel each other.

Summing forces in the x and y directions, we obtain

$$\Sigma X = 0, \quad F_2 - F_3 \sin\theta = 0$$

or

$$p_2(dy\,dz) - p_3(ds\,dz)\sin\theta = 0$$

$$\Sigma Y = 0, \quad F_1 - F_3\cos\theta - dW = 0$$

or

$$p_1(dx\,dz) - p_3(ds\,dz)\cos\theta - \gamma\left(\tfrac{1}{2}\,dx\,dy\,dz\right) = 0$$

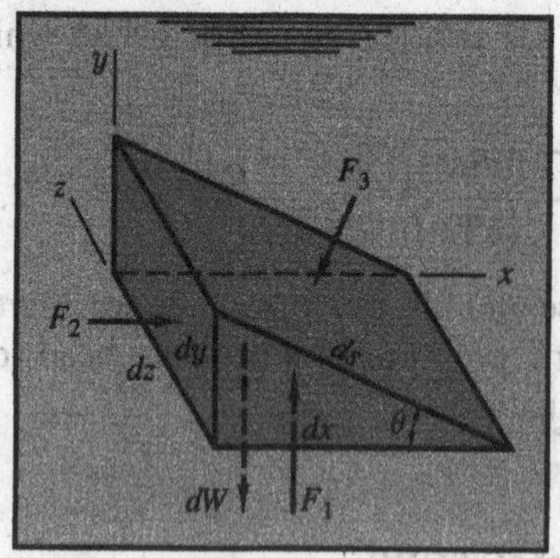

Fig. 2-1

Since $dy = ds \sin\theta$ and $dx = ds\cos\theta$, the equations reduce to

$$p_2\,dy\,dz - p_3\,dy\,dz = 0 \quad \text{or} \quad p_2 = p_3 \tag{A}$$

and

$$p_1\,dx\,dz - p_3\,dx\,dz - \gamma\left(\tfrac{1}{2}\,dx\,dy\,dz\right) = 0 \quad \text{or } p_1 - p_3 - \gamma\left(\tfrac{1}{2}dy\right) = 0 \tag{B}$$

As the triangular prism approaches a point, dy approaches zero as a limit, and the average pressures become uniform or even "point" pressures. Then putting $dy = 0$ in equation (B), we obtain $p_1 = p_3$ and hence $p_1 = p_2 = p_3$.

2.2. Derive the expression $p_2 - p_1 = \gamma(h_2 - h_1)$.

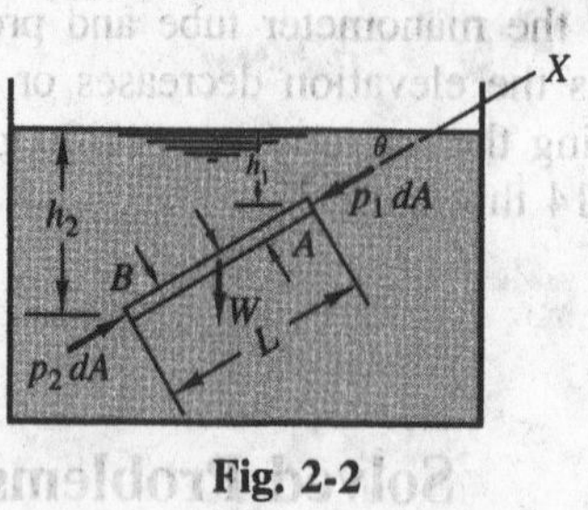

Fig. 2-2

Solution:

Consider a portion AB of the liquid in Fig. 2-2 as a free body of cross-sectional area dA, held in equilibrium by its own weight and the effects of other particles of liquid on the body AB.

At A the force acting is $p_1 dA$; at B it is $p_2 dA$. The weight of the free body AB is $W = \gamma v = \gamma L\,dA$. The other forces acting on the free body AB are normal to its sides, and only a few of these are shown in the diagram. In taking $\Sigma X = 0$, such normal forces will not enter into the equation. Therefore,

$$p_2\,dA - p_1\,dA - \gamma L\,dA\sin\theta = 0$$

Since $L\sin\theta = h_2 - h_1$, the above equation reduces to $p_2 - p_1 = \gamma(h_2 - h_1)$.

2.3. Determine the pressure in psi at a depth of 20.0 ft below the free surface of a body of water.

Solution:

Using an average value of 62.4 lb/ft³ for γ,

$$p = \frac{\gamma h}{144} = \frac{62.4 \text{ lb/ft}^3 \times 20.0 \text{ ft}}{144 \text{ in}^2/\text{ft}^2} = 8.67 \text{ psig}$$

2.4. Determine the pressure at a depth of 9.00 m in oil of specific gravity 0.750.

Solution:

$$p = \gamma h = (9.79 \times 0.750)(9.00) = 66.1 \text{ kPa}$$

2.5. Find the absolute pressure in psi in Problem 2.3 when the barometer reads 29.90 in of mercury, sp gr 13.57.

Solution:

$$\text{absolute pressure} = \text{atmospheric pressure} + \text{pressure due to 20.0 ft water}$$

$$= \frac{(13.57 \times 62.4)(29.90/12)}{144} + \frac{62.4 \times 20.0}{144} = 23.32 \text{ psia}$$

2.6. What depth of oil, sp gr 0.750, will produce a pressure of 40.0 psi? What depth of water?

Solution:

$$h_{\text{oil}} = \frac{p}{\gamma_{\text{oil}}} = \frac{40.0 \times 144}{0.750 \times 62.4} = 123 \text{ ft}, \qquad h_{\text{water}} = \frac{p}{\gamma_{\text{water}}} = \frac{40.0 \times 144}{62.4} = 92.3 \text{ ft}$$

2.7. Find the pressure at the bottom of a tank containing glycerin under pressure as shown in Fig. 2-3.

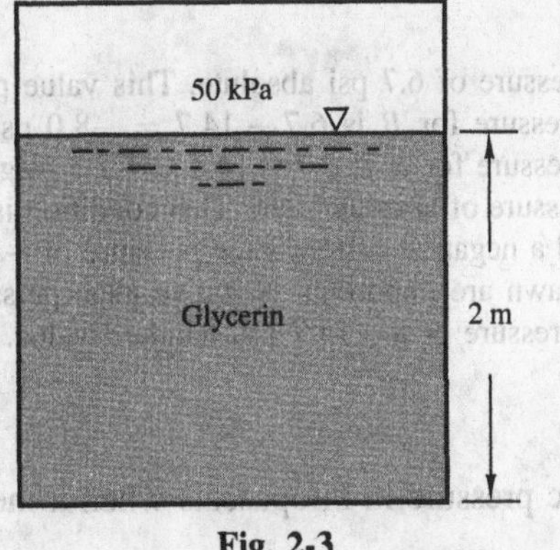

Fig. 2-3

Solution:

$$\text{pressure at bottom} = 50 + \gamma h$$

$$\text{pressure at bottom} = 50 + (12.34)(2) = 74.68 \text{ kPa}$$

2.8. (*a*) Convert a pressure head of 4.60 m of water to meters of oil, sp gr 0.750.

(*b*) Convert a pressure head of 24 in of mercury to feet of oil, sp gr 0.750.

Solution:

$$(a)\quad h_{oil} = \frac{h_{water}}{\text{sp gr oil}} = \frac{4.60}{0.750} = 6.13\text{ m} \qquad (b)\quad h_{oil} = \frac{h_{water}}{\text{sp gr oil}} = \frac{13.57 \times 24/12}{0.750} = 36.2\text{ ft}$$

2.9. Prepare a chart so that gage and absolute pressures can be compared readily, with limitations noted.

Solution:

Let A in Fig. 2-4 represent an absolute pressure of 54.7 psi. The gage pressure will depend upon the atmospheric pressure at the moment. If such pressure is standard for sea level (14.7 psi), then the gage pressure for A is $54.7 - 14.7 = 40.0$ psi. If the current barometer reads a pressure equivalent to 14.4 psi, then the gage pressure will be $54.7 - 14.4 = 40.3$ psi gage.

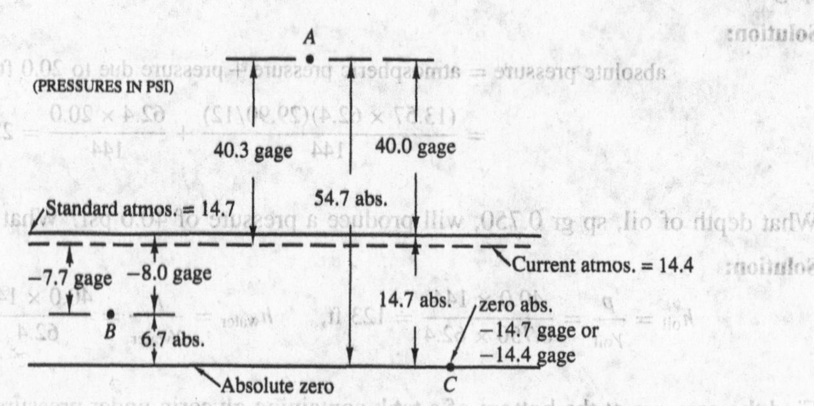

Fig. 2-4

Let B represent a pressure of 6.7 psi absolute. This value plots graphically as less than the standard 14.7 psi, and the gage pressure for B is $6.7 - 14.7 = -8.0$ psig. If the current atmospheric pressure is 14.4 psi, then the gage pressure for B is $6.7 - 14.4 = -7.7$ psig.

Let C represent a pressure of absolute zero. This condition is equivalent to a negative "standard" gage pressure of -14.7 psi and a negative current gage pressure of -14.4 psi.

Conclusions to be drawn are important. Negative gage pressures cannot exceed a theoretical limit of the current atmospheric pressure or a -14.7 psi standard value. Absolute pressures cannot have negative values assigned to them.

2.10. What is the atmospheric pressure in kilopascals when a mercury barometer reads 742 mm?

Solution:

$$p = \gamma h = (133.1)(742/1000) = 98.8\text{ kPa}$$

2.11. For the pressure vessel containing glycerin, with piezometer attached, shown in Fig. 2-5, find the pressure at point A.

Solution:

$$\text{pressure at } A = \gamma h = (78.5)(40.8/12) = 267\text{ psf, or } 1.85\text{ psi}$$

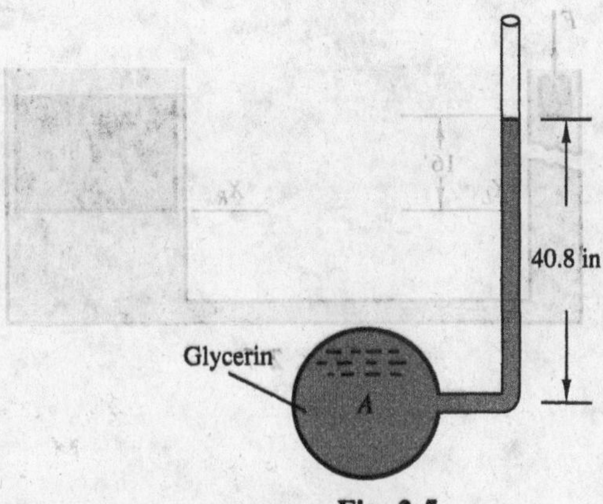

Fig. 2-5

2.12. For the open tank, with piezometers attached on the side, containing two different immiscible liquids as shown in Fig. 2-6, find (*a*) the elevation of the liquid surface in piezometer *A*, (*b*) the elevation of the liquid surface in piezometer *B*, and (*c*) the total pressure at the bottom of the tank.

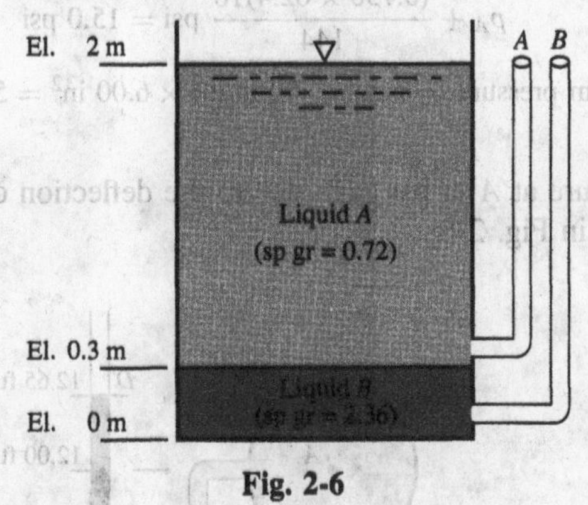

Fig. 2-6

Solution:

(*a*) Liquid *A* will simply rise in piezometer *A* to the same elevation as liquid *A* in the tank (i.e., to elevation 2 m).

(*b*) Liquid *B* will rise in piezometer *B* to elevation 0.3 m (as a result of the pressure exerted by liquid *B*) plus an additional amount, h_A, as a result of the overlying pressure of liquid *A*, p_A.

$$p_A = \gamma h = (0.72 \times 9.79)(1.7) = 11.98 \text{ kPa}$$

$$h_A = p/\gamma = 11.98/(2.36 \times 9.79) = 0.519 \text{ m}$$

Liquid *B* will rise in piezometer *B* to elevation $0.3 + 0.519 = 0.819$ m.

(*c*) Pressure at bottom = $(0.72 \times 9.79)(1.7) + (2.36 \times 9.79)(0.3) = 18.9$ kPa

2.13. In Fig. 2-7 the areas of the plunger *A* and cylinder *B* are 6.00 and 600 in^2, respectively, and the weight of *B* is 9000 lb. The vessel and the connecting passages are filled with oil of specific gravity 0.750. What force *F* is required for equilibrium, neglecting the weight of *A*?

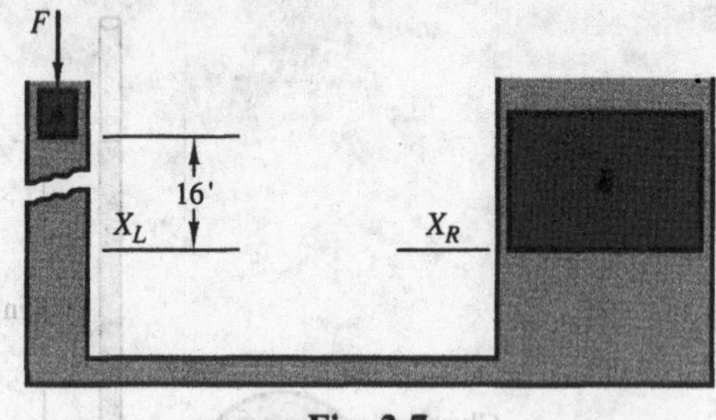

Fig. 2-7

Solution:

First determine the unit pressure acting on the plunger A. Since X_L and X_R are at the same level in the same liquid,

$$\text{pressure at } X_L \text{ in psi} = \text{pressure at } X_R \text{ in psi}$$

or
$$\text{pressure under } A + \text{pressure due to 16 ft oil} = \frac{\text{weight of } B}{\text{area of } B}$$

Substituting,
$$p_A + \frac{\gamma h}{144} = \frac{9000 \text{ lb}}{600 \text{ in}^2}$$

$$p_A + \frac{(0.750 \times 62.4)16}{144} \text{ psi} = 15.0 \text{ psi} \qquad \text{and} \qquad p_A = 9.80 \text{ psi}$$

Force F = uniform pressure × area = 9.80 lb/in^2 × 6.00 in^2 = 58.8 lb.

2.14. Determine the pressure at A in psi gage due to the deflection of the mercury, sp gr 13.57, in the U-tube gage shown in Fig. 2-8.

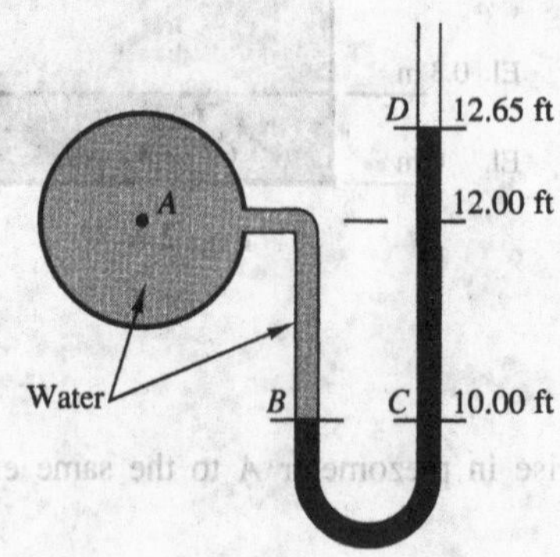

Fig. 2-8

Solution:

B and C are at the same level in the same liquid, mercury; hence we equate the pressures at B and C in *psf gage*.

$$\text{pressure at } B = \text{pressure at } C$$

$$p_A + \gamma h(\text{for water}) = p_D + \gamma h \text{ (for mercury)}$$

$$p_A + 62.4(12.00 - 10.00) = 0 + (13.57 \times 62.4)(12.65 - 10.00)$$

Solving, $p_A = 2119$ psf and $p_A = 2119/144 = 14.7$ psi gage.

An alternative solution using pressure heads in feet of water usually requires less arithmetic, as follows:

$$\text{pressure head at } B = \text{pressure head at } C$$

$$p_A/\gamma + 2.00 \text{ ft water} = 2.65 \times 13.57 \text{ ft water}$$

Solving, $p_A/\gamma = 34.0$ ft water, and $p_A = (62.4 \times 34.0)/144 = 14.7$ psi gage, as before.

2.15. A manometer is attached to a tank containing three different fluids, as shown in Fig. 2-9. Find the difference in elevation of the mercury column in the manometer (i.e., y in Fig. 2-9).

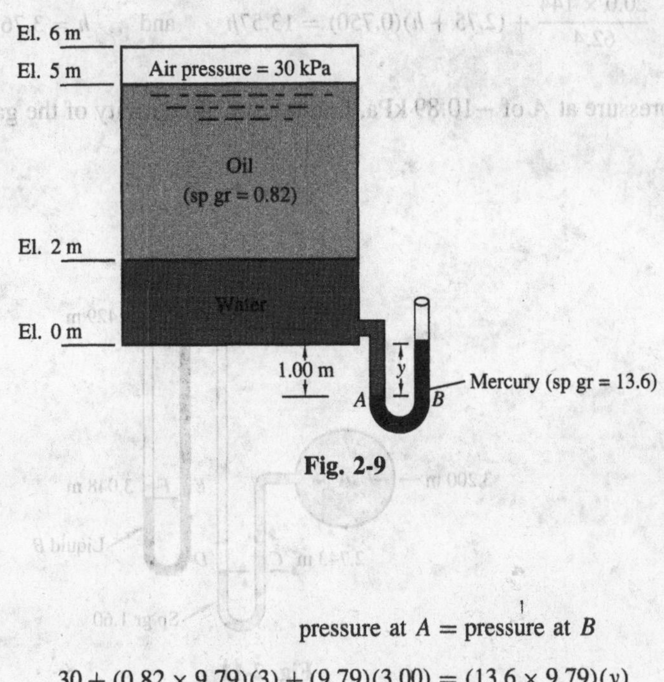

Fig. 2-9

Solution:

$$\text{pressure at } A = \text{pressure at } B$$

$$30 + (0.82 \times 9.79)(3) + (9.79)(3.00) = (13.6 \times 9.79)(y)$$

$$y = 0.627 \text{ m}$$

2.16. Oil of specific gravity 0.750 flows through the nozzle shown in Fig. 2-10 and deflects the mercury in the U-tube gage. Determine the value of h if the pressure at A is 20.0 psi.

Mathcad

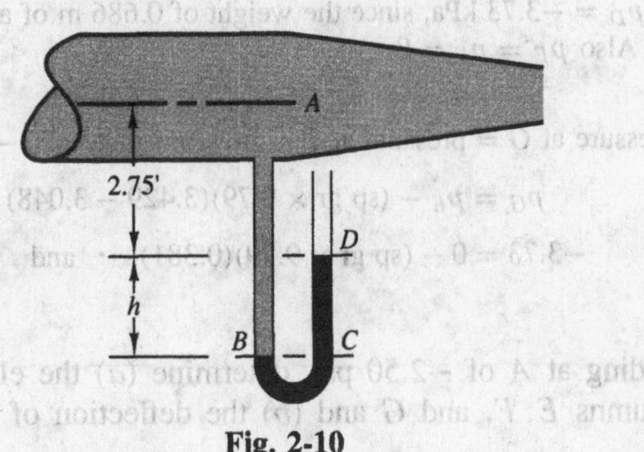

Fig. 2-10

Solution:

$$\text{pressure at } B = \text{pressure at } C$$

or, using psi units, $$p'_A + \frac{\gamma h}{144}\text{(oil)} = p'_D + \frac{\gamma h}{144}\text{(mercury)}$$

$$20.0 + \frac{(0.750 \times 62.4)(2.75 + h)}{144} = \frac{(13.57 \times 62.4)h}{144} \qquad \text{and} \qquad h = 3.76 \text{ ft}$$

Another method:

Using convenient feet of water units,

$$\text{pressure head at } B = \text{pressure head at } C$$

$$\frac{20.0 \times 144}{62.4} + (2.75 + h)(0.750) = 13.57h \qquad \text{and} \qquad h = 3.76 \text{ ft, as before}$$

2.17. For a gage pressure at A of -10.89 kPa, find the specific gravity of the gage liquid B in Fig. 2-11.

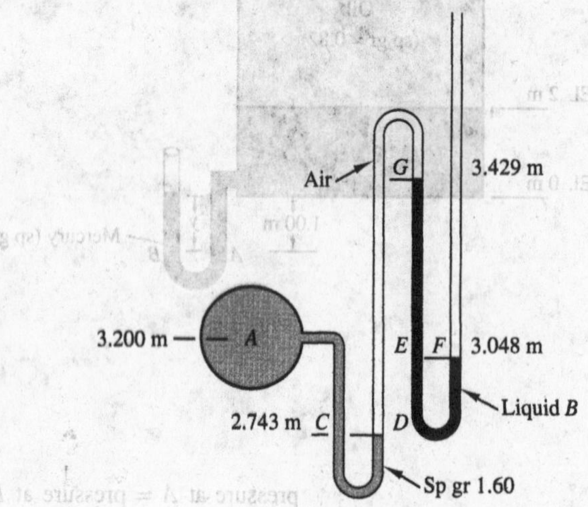

Fig. 2-11

Solution:

$$\text{pressure at } C = \text{pressure at } D$$

$$p_A + \gamma h = p_D$$

$$-10.89 + (1.60 \times 9.79)(3.200 - 2.743) = -3.73 \text{ kPa}$$

Now $p_G = p_D = -3.73$ kPa, since the weight of 0.686 m of air can be neglected without introducing significant error. Also $p_E = p_F = 0$.

Thus,

$$\text{pressure at } G = \text{pressure at } E - \text{pressure of } (3.429 - 3.048) \text{ m of gage liquid}$$

or $$p_G = p_E - (\text{sp gr} \times 9.79)(3.429 - 3.048)$$

$$-3.73 = 0 - (\text{sp gr} \times 9.79)(0.381) \qquad \text{and} \qquad \text{sp gr} = 1.00$$

2.18. For a gage reading at A of -2.50 psi, determine (*a*) the elevations of the liquids in the open piezometer columns E, F, and G and (*b*) the deflection of the mercury in the U-tube gage in Fig. 2-12.

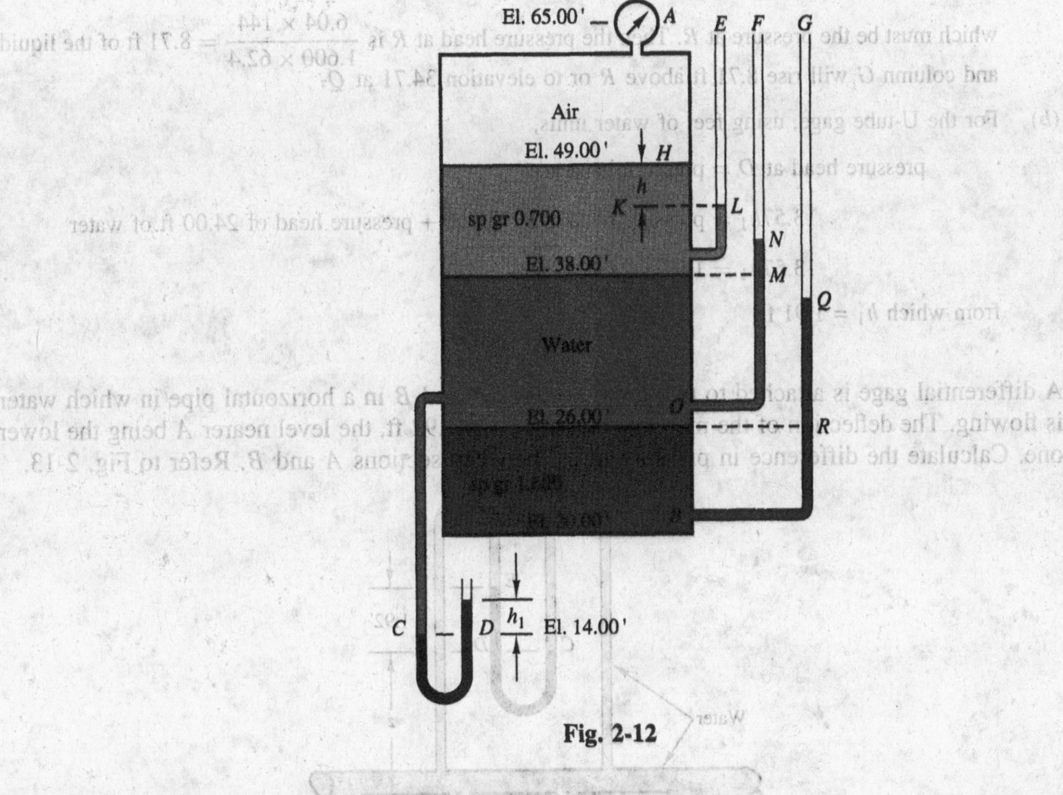

Fig. 2-12

Solution:

(*a*) Since the unit weight of the air (about 0.08 lb/ft^3) is very small compared with that of the liquids, the pressure at elevation 49.00 may be considered to be −2.50 psi without introducing significant error in the calculations.

For column E:

The elevation at *L* being assumed as shown, we have, in psf gage,

$$p_K = p_L$$

Then
$$p_H + \gamma h = 0$$

or $\quad -2.50 \times 144 + (0.700 \times 62.4)h = 0 \quad$ and $\quad h = 8.24$ ft.

Hence the elevation at *L* is 49.00 − 8.24 = 40.76 ft.

For column F:

pressure at El. 38.00 = pressure at El. 49.00 + pressure of 11 ft of liquid of sp gr 0.700

$$= -2.50 + \frac{(0.700 \times 62.4)(49.00 - 38.00)}{144} = 0.837 \text{ psi}$$

which must equal the pressure at *M*. Thus the pressure head at *M* is $\dfrac{0.837 \times 144}{62.4} = 1.93$ ft of water, and column *F* will rise 1.93 ft above *M* or to elevation 39.93 at *N*.

For column G:

pressure at El. 26.00 = pressure at El. 38.00 + pressure of 12 ft of water

or $$p_O = 0.837 + \frac{62.4 \times 12}{144} = 6.04 \text{ psi}$$

which must be the pressure at R. Then the pressure head at R is $\dfrac{6.04 \times 144}{1.600 \times 62.4} = 8.71$ ft of the liquid, and column G will rise 8.71 ft above R or to elevation 34.71 at Q.

(*b*) For the U-tube gage, using feet of water units,

$$\text{pressure head at } D = \text{pressure head at } C$$

$$13.57h_1 = \text{pressure head at El. 38.00} + \text{pressure head of 24.00 ft of water}$$

$$13.57h_1 = 1.93 + 24.00$$

from which $h_1 = 1.91$ ft.

2.19. A differential gage is attached to two cross sections A and B in a horizontal pipe in which water is flowing. The deflection of the mercury in the gage is 1.92 ft, the level nearer A being the lower one. Calculate the difference in pressure in psi between sections A and B. Refer to Fig. 2-13.

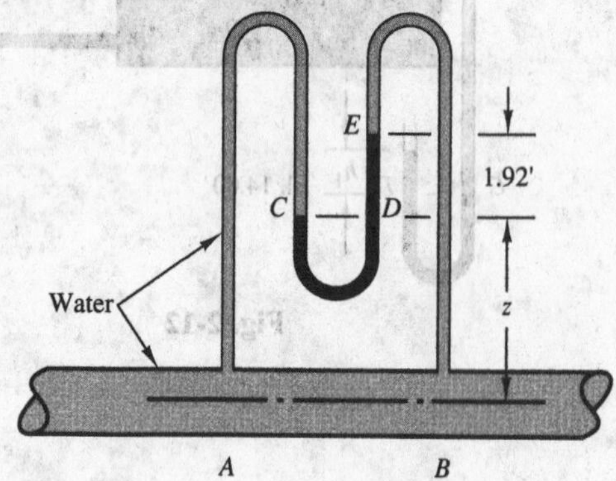

Fig. 2-13

Solution:

Note: A sketch will be of assistance in clarifying the analysis of all problems as well as in reducing mistakes. Even a single-line diagram will help.

$$\text{pressure head at } C = \text{pressure head at } D$$

or, using ft of water units, $p_A/\gamma - z = [(p_B/\gamma) - (z + 1.92)] + (13.57)(1.92)$

Then $p_A/\gamma - p_B/\gamma = \text{difference in pressure heads} = (1.92)(13.57 - 1) = 24.1$ ft water

and $p_A - p_B = (24.1)(62.4/144) = 10.44$ psi.

If $(p_A - p_B)$ had been negative, the proper interpretation of the sign would be that the pressure at B was greater than the pressure at A by 10.44 psi.

Differential gages should have the air extracted from all tubing before readings are taken.

2.20. The loss through a device X is to be measured by a differential gage using oil of specific gravity 0.750 as the gage fluid. The flowing liquid has sp gr 1.50. Find the change in pressure head between A and B for the deflection of the oil shown in Fig. 2-14.

Solution:

$$\text{pressure at } C \text{ in psf} = \text{pressure at } D \text{ in psf}$$

$$p_B - (1.50 \times 62.4)(2.00) - (0.750 \times 62.4)(3.00) = p_A - (1.50 \times 62.4)(11.00)$$

Then $p_A - p_B = 702$ psf, and difference in pressure heads $= \dfrac{702}{\gamma} = \dfrac{702}{1.50 \times 62.4} = 7.50$ ft of liquid.

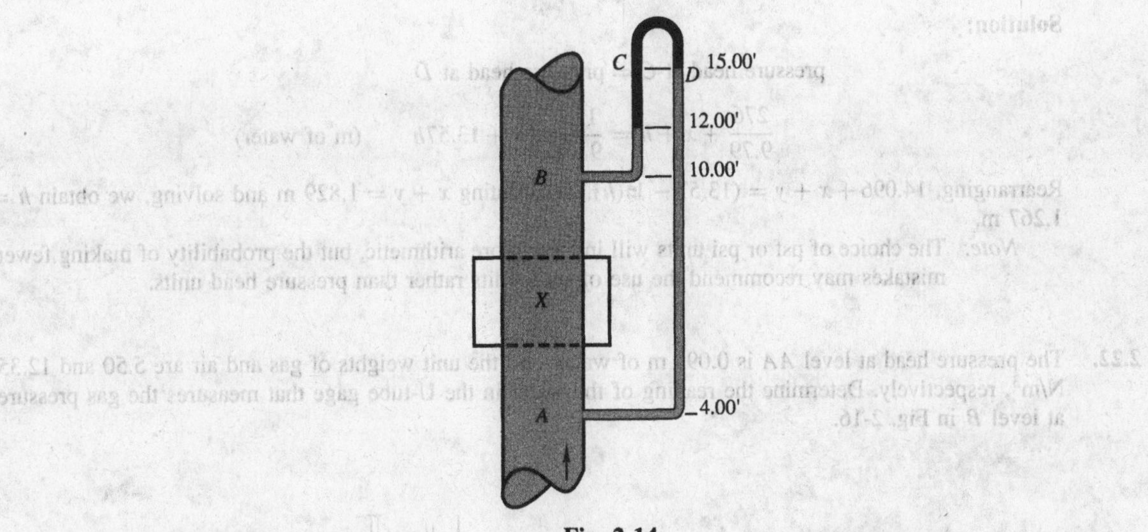

Fig. 2-14

Another method:

Using feet of liquid (sp gr 1.50) as the units,

$$\text{pressure head at } C = \text{pressure head at } D$$

$$\frac{p_B}{\gamma} - 2.00 - \frac{0.750 \times 3.00}{1.50} = \frac{p_A}{\gamma} - 11.00$$

Then $p_A/\gamma - p_B/\gamma$ = difference in pressure heads = 7.50 ft of the liquid, as before.

2.21. Vessels A and B contain water under pressures of 276 kPa and 138 kPa, respectively. What is the deflection of the mercury in the differential gage in Fig. 2-15?

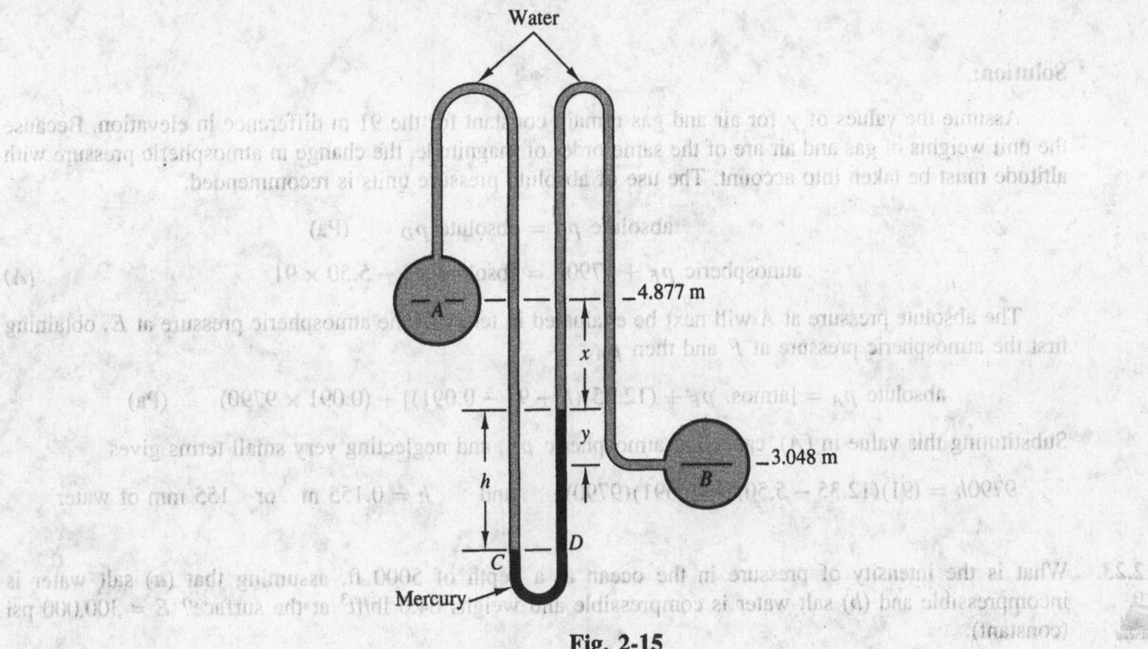

Fig. 2-15

Solution:

$$\text{pressure head at } C = \text{pressure head at } D$$

$$\frac{276}{9.79} + x + h = \frac{138}{9.79} - y + 13.57h \qquad \text{(m of water)}$$

Rearranging, $14.096 + x + y = (13.57 - 1)(h)$. Substituting $x + y = 1.829$ m and solving, we obtain $h = 1.267$ m.

Note: The choice of psf or psi units will involve more arithmetic, but the probability of making fewer mistakes may recommend the use of such units rather than pressure head units.

2.22. The pressure head at level AA is 0.091 m of water, and the unit weights of gas and air are 5.50 and 12.35 N/m^3, respectively. Determine the reading of the water in the U-tube gage that measures the gas pressure at level B in Fig. 2-16.

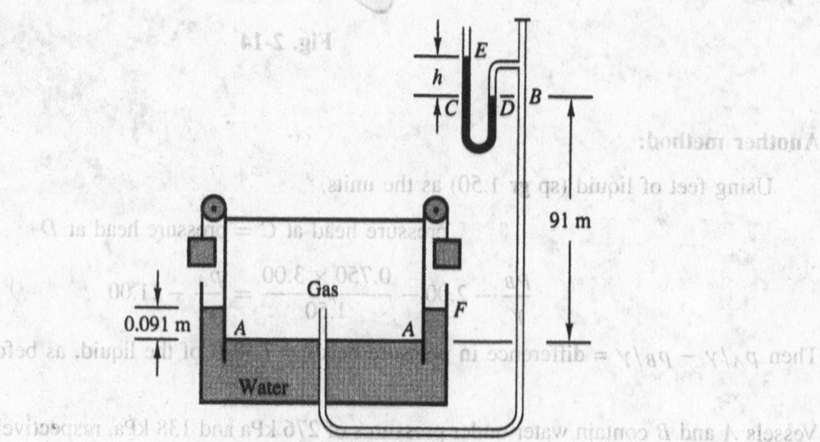

Fig. 2-16

Solution:

Assume the values of γ for air and gas remain constant for the 91 m difference in elevation. Because the unit weights of gas and air are of the same order of magnitude, the change in atmospheric pressure with altitude must be taken into account. The use of absolute pressure units is recommended.

$$\text{absolute } p_C = \text{absolute } p_D \qquad \text{(Pa)}$$

$$\text{atmospheric } p_E + 9790h = \text{absolute } p_A - 5.50 \times 91 \qquad (A)$$

The absolute pressure at A will next be evaluated in terms of the atmospheric pressure at E, obtaining first the atmospheric pressure at F and then p_A.

$$\text{absolute } p_A = [\text{atmos. } p_E + (12.35)(h + 91 - 0.091)] + (0.091 \times 9790) \qquad \text{(Pa)}$$

Substituting this value in (A), canceling atmospheric p_E, and neglecting very small terms gives

$$9790h = (91)(12.35 - 5.50) + (0.091)(9790) \quad \text{and} \quad h = 0.155 \text{ m} \quad \text{or} \quad 155 \text{ mm of water}$$

2.23. What is the intensity of pressure in the ocean at a depth of 5000 ft, assuming that (*a*) salt water is incompressible and (*b*) salt water is compressible and weighs 64.0 lb/ft^3 at the surface? E = 300,000 psi (constant).

Solution:

(*a*) Intensity of pressure $p = \gamma h = (64.0)(5000) = 320{,}000$ psf gage.

(*b*) Since a given mass does not change its weight as it is compressed, $dW = 0$; then

$$dW = d(\gamma v) = \gamma dv + v d\gamma = 0 \qquad \text{or} \qquad dv/v = -d\gamma/\gamma \qquad (A)$$

From equations (*4*) and (*11*), the latter from Chapter 1, $dp = -\gamma\, dh$ and $dv/v = -dp/E$. Substituting in (*A*),

$$dp/E = d\gamma/\gamma \qquad (B)$$

Integrating, $p = E \ln\gamma + C$. At the surface, $p = p_0$, $\gamma = \gamma_0$; then $C = p_0 - E\ln\gamma_0$ and

$$p = E\ln\gamma + p_0 - E\ln\gamma_0 \qquad \text{or} \qquad p - p_0 = E\ln(\gamma/\gamma_0) \qquad (C)$$

Putting $dp = -\gamma\, dh$ in (*B*), $\dfrac{-\gamma\, dh}{E} = \dfrac{d\gamma}{\gamma}$ or $dh = -\dfrac{E d\gamma}{\gamma^2}$. Integrating,

$$h = E/\gamma + C_1 \qquad (D)$$

At the surface, $h = 0$, $\gamma = \gamma_0$; then $C_1 = -E/\gamma_0$, $h = (E/\gamma) - (E/\gamma_0)$, and hence

$$\gamma = \frac{\gamma_0 E}{\gamma_0 h + E} = \frac{(64.0)(300{,}000 \times 144)}{(64.0)(-5000) + (300{,}000 \times 144)} = 64.5\ \text{lb/ft}^3$$

remembering that h is positive upwards and using psf units for E. From (*C*),

$$p = (300{,}000 \times 144)\ln(64.5/64.0) = 336{,}000 \text{ psf gage}$$

2.24. Compute the barometric pressure at an altitude of 1200 m if the pressure at sea level is 101.4 kPa. Assume isothermal conditions at 20° C.

Solution:

The unit weight of air at 20° C is $\gamma = \dfrac{p}{(29.3)(273+20)}$. Then from equation (*4*),

$$dp = -\gamma dh = -\frac{p}{(29.3)(293)}dh \qquad \text{or} \qquad \frac{dp}{p} = -0.0001165\, dh \qquad (A)$$

Integrating (*A*), $\ln p = -0.0001165h + C$, where C is the constant of integration.
To evaluate C: when $h = 0$, $p = 101.4$ kPa absolute. Hence $C = \ln 101.4$, and

$$\ln p = -0.0001165h + \ln 101.4 \qquad \text{or} \qquad 0.0001165h = \ln(101.4/p) \qquad (B)$$

Transforming (*B*) to $\log_{10}$,

$$2.3026\log_{10}(101.4/p) = (0.0001165)(1200)$$

$$\log_{10}(101.4/p) = 0.0607, \qquad 101.4/p = \text{antilog } 0.0607 = 1.150$$

from which $p = \dfrac{101.4}{1.150}$ kPa or $p = 88.2$ kPa.

2.25. Derive a general expression for the relation between pressure and elevation for isothermal conditions, using $dp = -\gamma dh$.

Solution:

For isothermal conditions, the equation $\dfrac{p}{\gamma T} = \dfrac{p_0}{\gamma_0 T_0}$ becomes $\dfrac{p}{\gamma} = \dfrac{p_0}{\gamma_0}$ or $\gamma = \gamma_0\dfrac{p}{p_0}$.

Then $dh = -\dfrac{dp}{\gamma} = -\dfrac{p_0}{\gamma_0} \times \dfrac{dp}{p}$. Integrating, $\displaystyle\int_{h_0}^{h} dh = -\frac{p_0}{\gamma_0}\int_{p_0}^{p} \frac{dp}{p}$ and

$$h - h_0 = -\frac{p_0}{\gamma_0}(\ln p - \ln p_0) = +\frac{p_0}{\gamma_0}(\ln p_0 - \ln p) = \frac{p_0}{\gamma_0}\ln\frac{p_0}{p}$$

Actually, the temperature of the atmosphere decreases with altitude. Hence an exact solution requires knowledge of the temperature variations with altitude and the use of the gas law $p/\gamma T$ = a constant.

2.26. Find the pressure difference between A and B for the setup shown in Fig. 2-17.

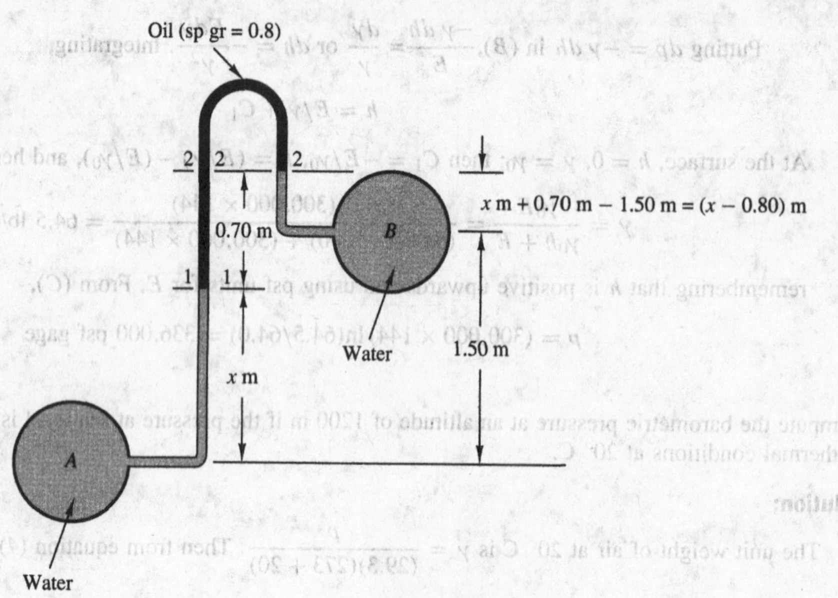

Fig. 2-17

Solution:

$$p_A - 9.79x - (0.8 \times 9.79)(0.70) + (9.79)(x - 0.80) = p_B$$

$$p_A - 9.79x - 5.482 + 9.79x - 7.832 = p_B$$

$$p_A - p_B = 13.3 \text{ kPa}$$

2.27. A differential manometer is attached to two tanks as shown in Fig. 2-18. Calculate the pressure difference between chamber A and chamber B.

Solution:

$$\gamma_{\text{mercury}} = 132.8 \text{ kN/m}^3 \qquad \gamma_{\text{SAE 30 oil}} = 8.996 \text{ kN/m}^3, \qquad \gamma_{\text{carbon tetrachloride}} = 15.57 \text{ kN/m}^3$$

$$p_A + (8.996)(1.1) + (132.8)(0.3) - (15.57)(0.8) = p_B$$

$$p_A - p_B = -37.28 \text{ kPa (i.e., } p_B > p_A)$$

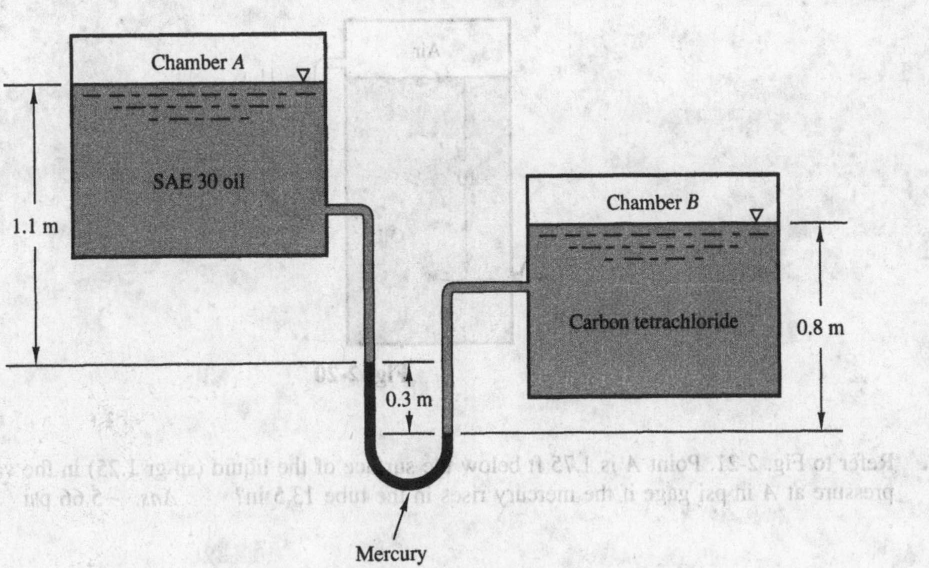

Fig. 2-18

Supplementary Problems

2.28. A glass U-tube open to the atmosphere at both ends is shown in Fig. 2-19. If the U-tube contains oil and water as shown, determine the specific gravity of the oil. *Ans.* 0.86

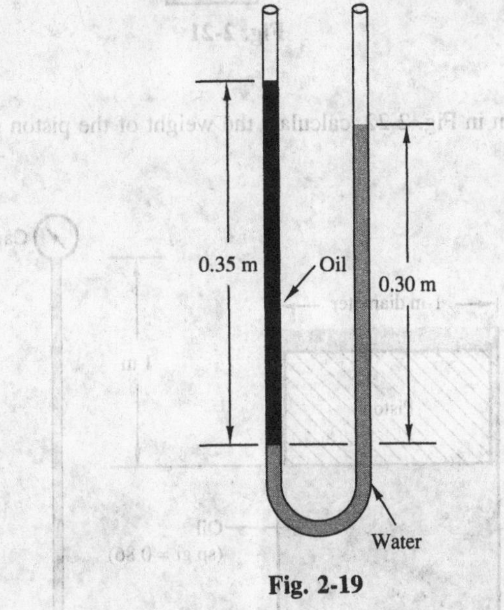

Fig. 2-19

2.29. The tank in Fig. 2-20 contains oil of specific gravity 0.750. Determine the reading of gage A in psi. *Ans.* -1.16 psi.

2.30. A closed tank contains 0.610 m of mercury, 1.524 m of water, 2.438 m of oil of specific gravity 0.750, and an air space above the oil. If the pressure at the bottom of the tank is 276 kPa gage, what should be the reading of the gage in the top of the tank? *Ans.* 161 kPa

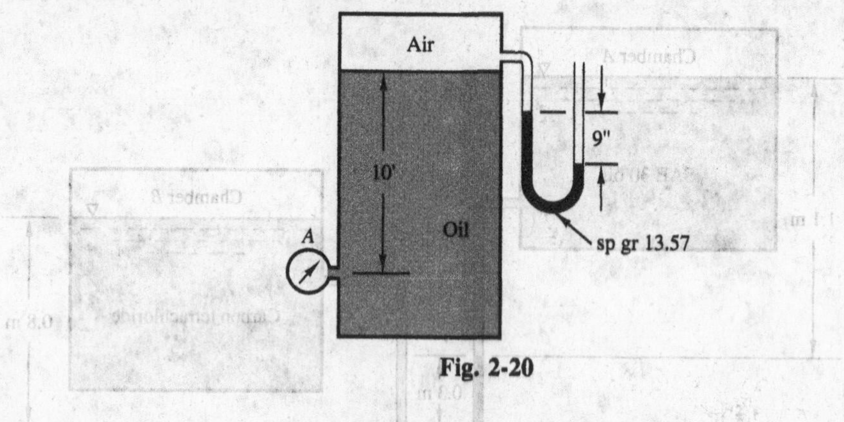

Fig. 2-20

2.31. Refer to Fig. 2-21. Point *A* is 1.75 ft below the surface of the liquid (sp gr 1.25) in the vessel. What is the pressure at *A* in psi gage if the mercury rises in the tube 13.5 in? *Ans.* −5.66 psi

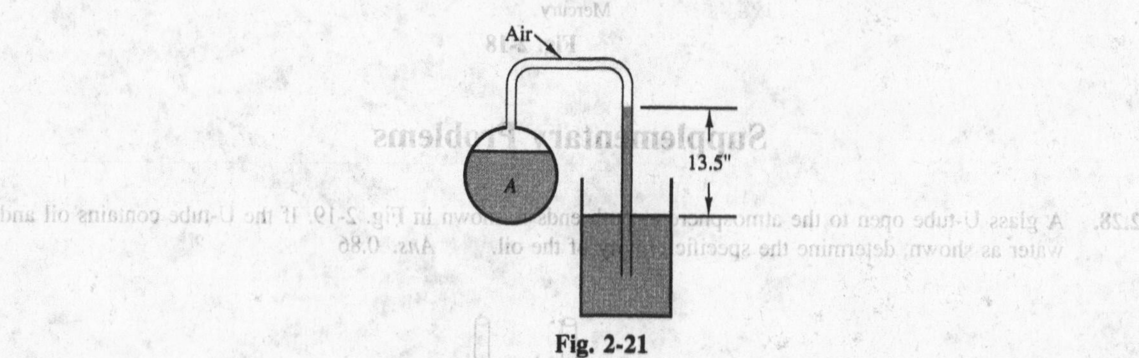

Fig. 2-21

2.32. For the configuration shown in Fig. 2-22, calculate the weight of the piston if the gage pressure reading is 70 kPa. *Ans.* 61.6 kN

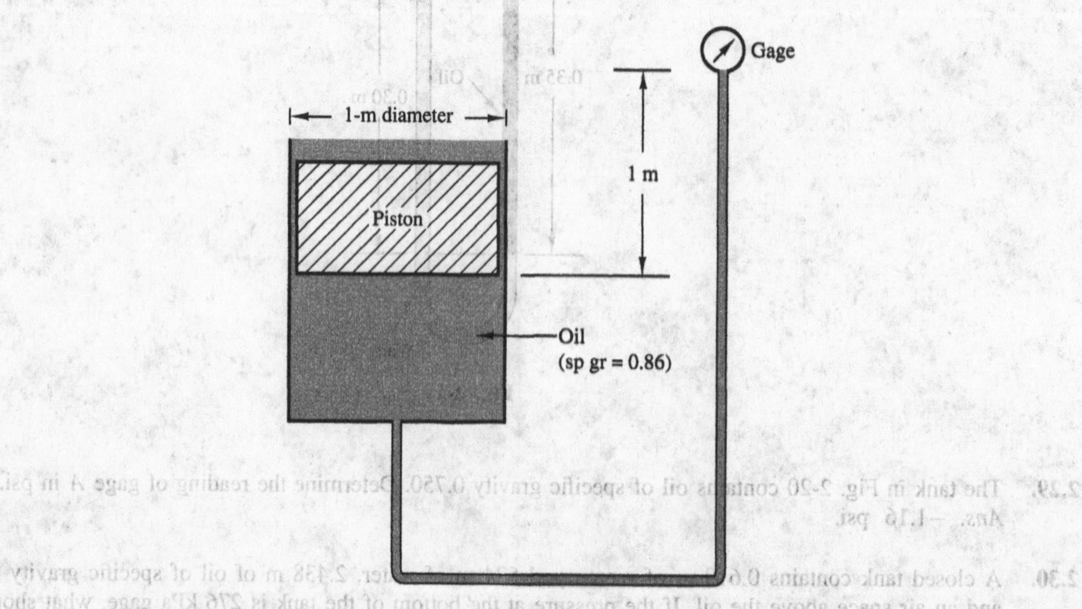

Fig. 2-22

2.33. Refer to Fig. 2-23. Neglecting friction between piston *A* and the gas tank, find the gage reading at *B* in ft of water. Assume gas and air to be of constant specific weight and equal to 0.0351 and 0.0750 lb/ft^3, respectively. *Ans.* 1.76 ft of water

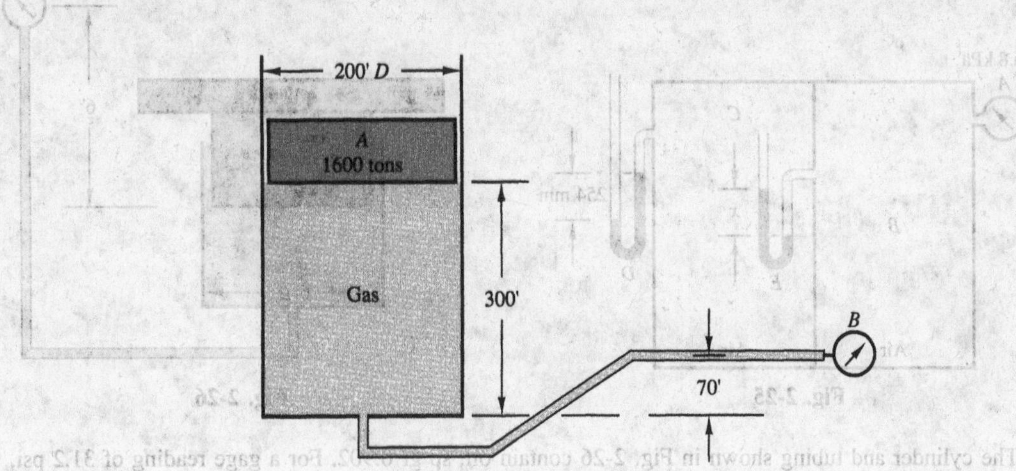

Fig. 2-23

2.34. Tanks *A* and *B*, containing oil and glycerin of specific gravities 0.780 and 1.25, respectively, are connected by a differential gage. The mercury in the gage is at elevation 1.60 on the *A* side and at elevation 1.10 on the *B* side. If the elevation of the surface of the glycerin in tank *B* is 21.10, at what elevation is the surface of the oil in tank *A*? *Ans.* elevation 24.90

2.35. Vessel *A*, at elevation 2.438 m, contains water under 103.4 kPa pressure. Vessel *B*, at elevation 3.658 m, contains a liquid under 68.95 kPa pressure. If the deflection of the differential gage is 305 mm of mercury, with the lower level on the *A* side at elevation 0.305 m, determine the specific gravity of the liquid in vessel *B*. *Ans.* 0.500

2.36. In the left-hand tank in Fig. 2-24, the air pressure is −9 in of mercury. Determine the elevation of the gage liquid in the right-hand column at *A*. *Ans.* elevation 86.7 ft

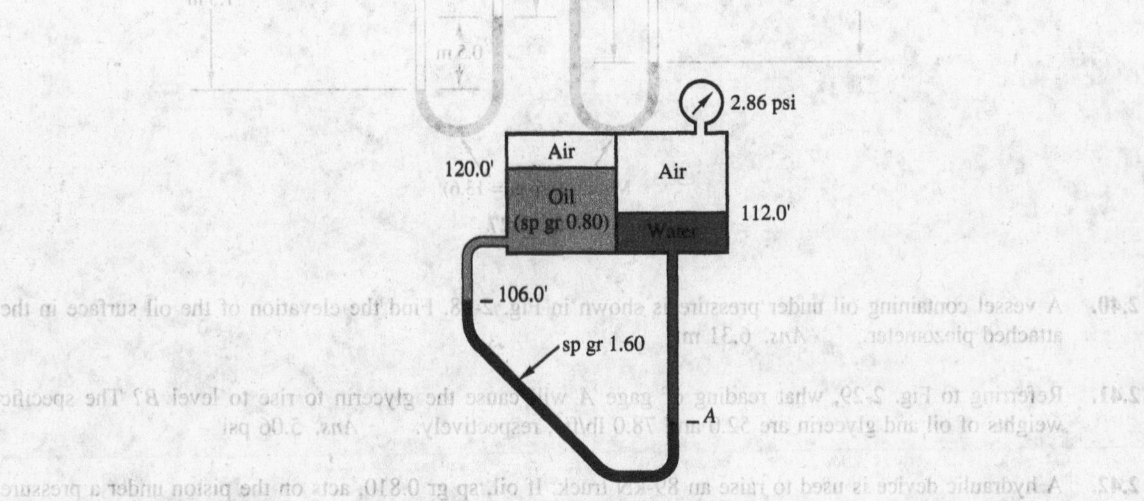

Fig. 2-24

2.37. Compartments B and C in Fig. 2-25 are closed and filled with air. The barometer reads 99.98 kPa. When gages A and D read as indicated, what should be the value of x for gage E (mercury in each U-tube gage)? *Ans.* 1.82 m

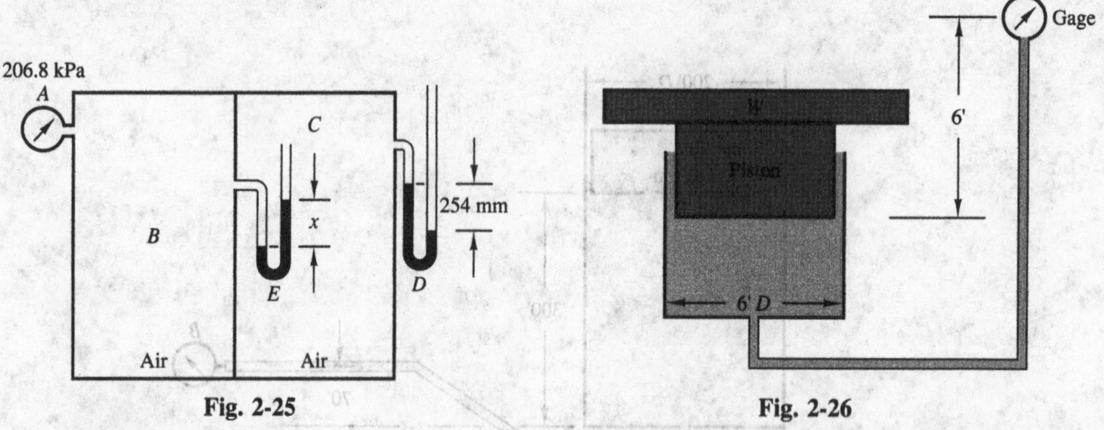

Fig. 2-25

Fig. 2-26

2.38. The cylinder and tubing shown in Fig. 2-26 contain oil, sp gr 0.902. For a gage reading of 31.2 psi, what is the total weight of piston and weight W? *Ans.* 136,800 lb

2.39. Determine the pressure difference between pipes A and B for the differential manometer shown in Fig. 2-27. *Ans.* 124 kPa

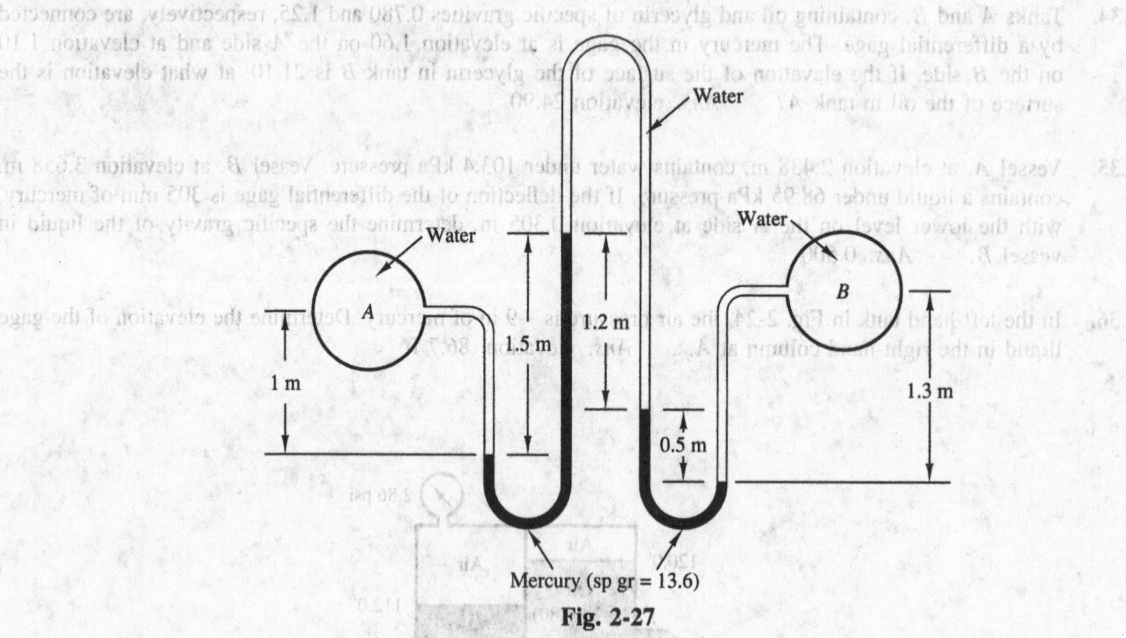

Fig. 2-27

2.40. A vessel containing oil under pressure is shown in Fig. 2-28. Find the elevation of the oil surface in the attached piezometer. *Ans.* 6.31 m

2.41. Referring to Fig. 2-29, what reading of gage A will cause the glycerin to rise to level B? The specific weights of oil and glycerin are 52.0 and 78.0 lb/ft^3, respectively. *Ans.* 5.06 psi

2.42. A hydraulic device is used to raise an 89-kN truck. If oil, sp gr 0.810, acts on the piston under a pressure of 1.22 MPa, what diameter is required? *Ans.* 305 mm

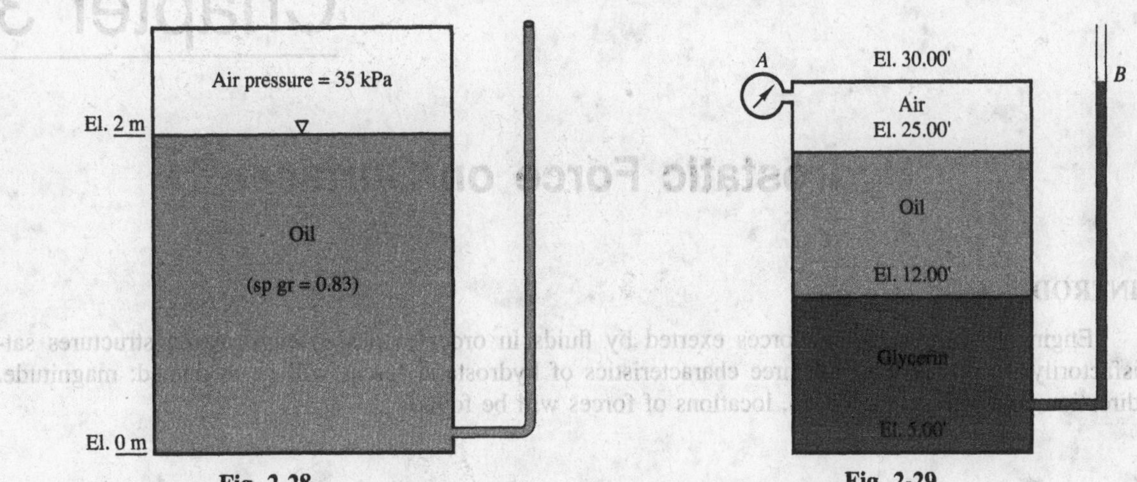

Fig. 2-28

Fig. 2-29

2.43. If the specific weight of glycerin is 79.2 lb/ft^3, what suction pressure is required to raise the glycerin 9 in vertically in a $\frac{1}{2}$-in-diameter tube? *Ans.* −0.412 psi

2.44. What is the pressure inside a raindrop that is 0.06 in in diameter when the temperature is 70°F? *Ans.* 0.0276 psig

2.45. The liquid surface in a piezometer attached to a conduit stands 1.0 m above point A, as shown in Fig. 2-30. Find the pressure at point A if the liquid is (*a*) water and (*b*) mercury. *Ans.* (*a*) 9.79 kPa, (*b*) 133 kPa

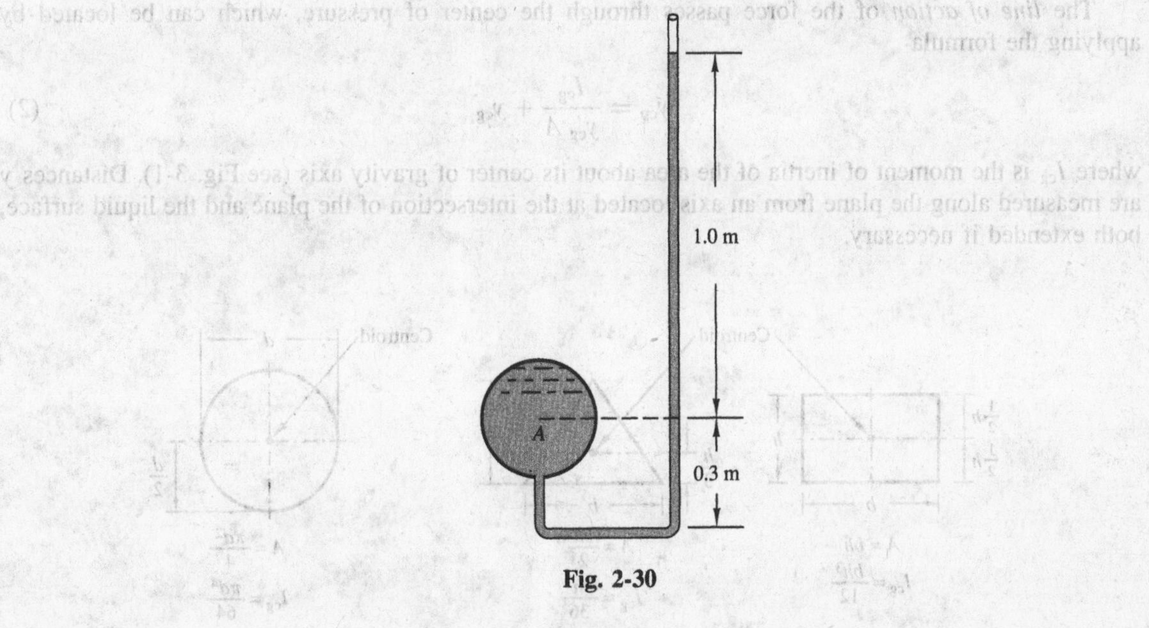

Fig. 2-30

Chapter 3

Hydrostatic Force on Surfaces

INTRODUCTION

Engineers must calculate forces exerted by fluids in order to design constraining structures satisfactorily. In this chapter all three characteristics of hydrostatic forces will be evaluated: magnitude, direction, and sense. In addition, locations of forces will be found.

FORCE EXERTED BY A LIQUID ON A PLANE AREA

The force F exerted by a liquid on a plane area A is equal to the product of the specific weight γ of the liquid, depth of the center of gravity of the area h_{cg}, and the area.

The equation is $$F = \gamma h_{cg} A \qquad (1)$$

typical units being $$\text{lb} = \frac{\text{lb}}{\text{ft}^3} \times \text{ft} \times \text{ft}^2 \qquad \text{or} \qquad \text{N} = \frac{\text{N}}{\text{m}^3} \times \text{m} \times \text{m}^2$$

Note that the product of specific weight and depth of the center of gravity of the area yields the intensity of pressure at the area's center of gravity.

The *line of action* of the force passes through the center of pressure, which can be located by applying the formula

$$y_{cp} = \frac{I_{cg}}{y_{cg} A} + y_{cg} \qquad (2)$$

where I_{cg} is the moment of inertia of the area about its center of gravity axis (see Fig. 3-1). Distances y are measured along the plane from an axis located at the intersection of the plane and the liquid surface, both extended if necessary.

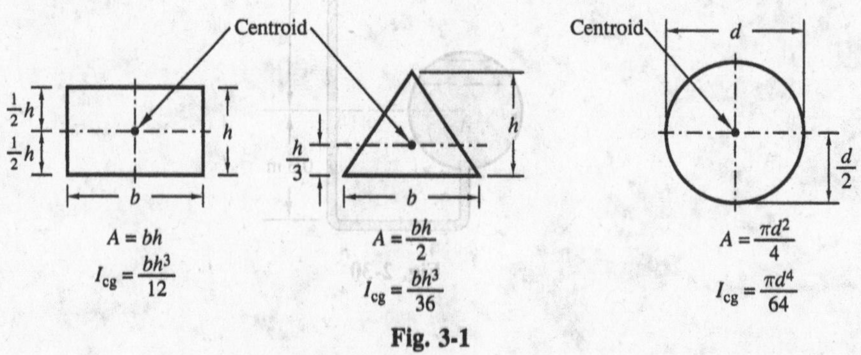

Fig. 3-1

FORCE EXERTED BY A LIQUID ON A CURVED SURFACE

The *horizontal component* of the hydrostatic force on a curved surface is equal to the normal force on the vertical projection of the surface. The component acts through the center of pressure for the vertical projection.

The *vertical component* of the hydrostatic force on a curved surface is equal to the weight of the volume of liquid above the area, real or imaginary. The force passes through the center of gravity of the volume.

HOOP OR CIRCUMFERENTIAL TENSION

Hoop tension or circumferential tension is created in the walls of a cylinder subjected to internal pressure. For thin-walled cylinders ($t < 0.1d$),

$$\text{intensity of stress } \sigma = \frac{\text{pressure } p \times \text{radius } r}{\text{thickness } t} \tag{3}$$

typical units being lb/in^2 (psi) or Pa.

LONGITUDINAL STRESS IN THIN-WALLED CYLINDERS

Longitudinal stress in thin-walled cylinders closed at the ends is equal to half the hoop tension.

HYDROSTATIC FORCES ON DAMS

Large hydrostatic forces to which dams are subjected tend to cause a dam to (1) slide horizontally along its base and (2) overturn about its downstream edge (which is known as the *toe* of the dam). Another factor that may affect dam stability is hydrostatic uplift along the bottom of the dam, caused by water seeping under the dam. Checks for dam stability are made by finding (1) the factor of safety against sliding, (2) the factor of safety against overturning, and (3) the pressure intensity on the base of the dam.

The factor of safety against sliding is determined by dividing sliding resistance by sliding force. The factor of safety against overturning is computed by dividing the total righting (resisting) moment by the total overturning moment, all moments being taken about the toe of the dam. Pressure intensity on the base of the dam can be calculated using the flexural formula

$$p = F/A \pm M_y\, x/I_y \pm M_x\, y/I_x \tag{4}$$

where p = pressure intensity
F = total vertical load
A = area of base of dam
M_x, M_y = total moment about the x and y axes, respectively
I_x, I_y = moment of inertia about the x and y axes, respectively
x, y = distance from centroid to the point at which pressure intensity is computed along the x and y axes, respectively

Equation (*4*) gives the pressure distribution along the dam's base if the resultant reaction on the base acts within its middle third.

Solved Problems

3.1. (*a*) Develop the equation for the hydrostatic force acting on a plane area, and (*b*) locate the force.

Solution:

(*a*) Let trace AB represent any plane area acted upon by a fluid and making an angle θ with the horizontal, as shown in Fig. 3-2. Consider an element of area such that every particle is the same distance h

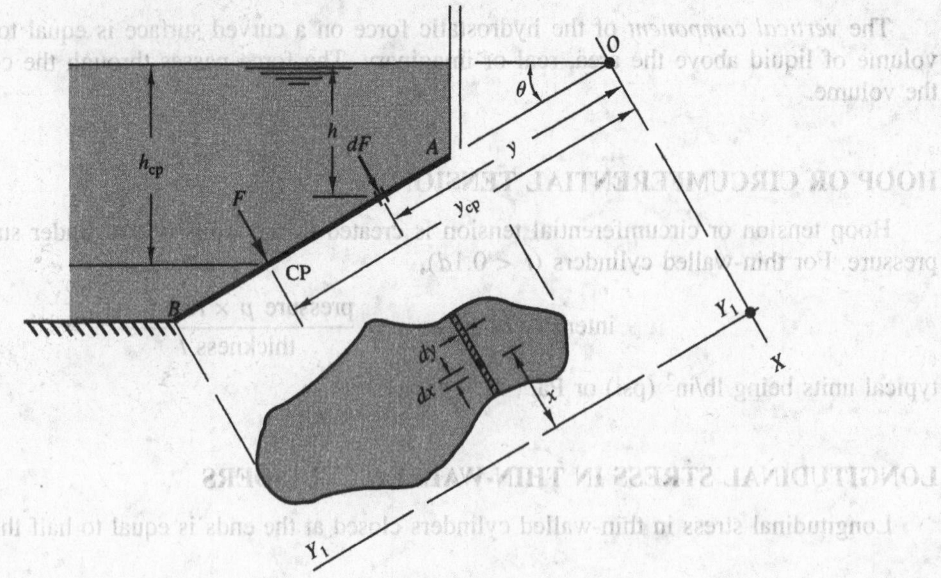

Fig. 3-2

below the surface of the liquid. The strip shown crosshatched is such an area (dA), and the pressure is *uniform* over this area. Then the force acting on the area dA is equal to the uniform intensity of pressure p times the area dA, or

$$dF = p\,dA = \gamma h\,dA$$

Summing all the forces acting on the area and considering that $h = y\sin\theta$,

$$F = \int \gamma h\,dA = \int \gamma(y\sin\theta)\,dA$$

$$= (\gamma\sin\theta)\int y\,dA = (\gamma\sin\theta)y_{cg}A$$

where γ and θ are constants and, from statics, $\int y\,dA = y_{cg}A$. Since $h_{cg} = y_{cg}\sin\theta$,

$$F = \gamma h_{cg} A \qquad (1)$$

(b) To locate this force F, proceed as in static mechanics by taking moments. Axis O is chosen as the intersection of the plane area and the water surface, both extended if necessary. All distances y are measured from this axis, and the distance to the resultant force is called y_{cp}, which is the distance to the center of pressure. Since the sum of the moments of all the forces about axis O = the moment of the resultant force, we obtain

$$\int (dF \times y) = F \times y_{cp}$$

But $dF = \gamma h\,dA = \gamma(y\sin\theta)dA$ and $F = (\gamma\sin\theta)(y_{cg}A)$. Then

$$(\gamma\sin\theta)\int y^2\,dA = (\gamma\sin\theta)(y_{cg}A)y_{cp}$$

Since $\int y^2\,dA$ is the moment of inertia of the plane area about axis O,

$$\frac{I_O}{y_{cg}A} = y_{cg}$$

In more convenient form, from the parallel axis theorem,

$$y_{cp} = \frac{I_{cg} + Ay_{cg}^2}{y_{cg}A} = \frac{I_{cg}}{y_{cg}A} + y_{cg} \tag{2}$$

Note that the position of the center of pressure is always *below* the center of gravity of the area, or $y_{cp} - y_{cg}$ is always positive because I_{cg} is always positive.

3.2. Locate the lateral position of the center of pressure. Refer to Fig. 3-2.

Solution:

While in general the lateral position of the center of pressure is not required to solve most engineering problems concerning hydrostatic forces, occasionally this information may be needed. Using the sketch in the preceding problem, area dA is chosen as $(dx\,dy)$ so that the moment arm x is properly used. Taking moments about any nonintersecting axis Y_1Y_1,

$$Fx_{cp} = \int (dF\,x)$$

Using values derived in Problem 3.1,

$$(\gamma h_{cg}A)x_{cp} = \int p(dx\,dy)x = \int \gamma h(dx\,dy)x$$

or

$$(\gamma \sin\theta)(y_{cg}A)x_{cp} = (\gamma \sin\theta)\int xy(dx\,dy) \tag{3}$$

since $h = y\sin\theta$. The integral represents the product of inertia of the plane area about the X and Y axes chosen, designated by I_{xy}. Then

$$x_{cp} = \frac{I_{xy}}{y_{cg}A} = \frac{(I_{xy})_{cg}}{y_{cg}A} + x_{cg} \tag{4}$$

Should *either* of the centroidal axes be an axis of symmetry of the plane area, I_{xy} becomes zero, and the lateral position of the center of pressure lies on the Y axis, which passes through the center of gravity (not shown in the figure). Note that the product of inertia about the center of gravity axes, $(I_{xy})_{cg}$, may be positive or negative, so that the lateral position of the center of pressure may lie on *either* side of the centroidal y axis.

3.3. Determine the resultant force F due to water acting on the 3 m by 6 m rectangular area AB shown in Fig. 3-3.

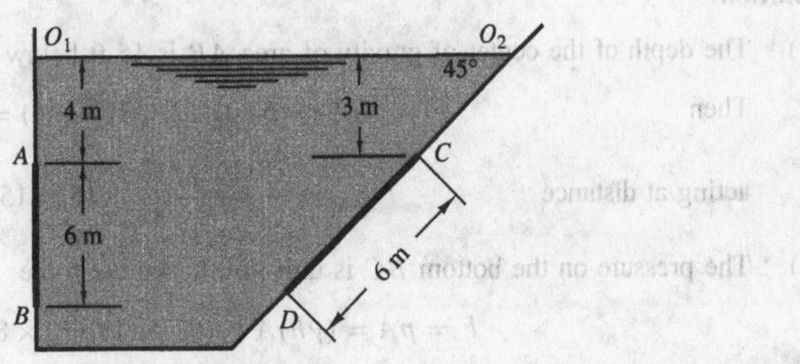

Fig. 3-3

Solution:

$$F = \gamma h_{cg} A = (9.79) \times (4 + 3) \times (6 \times 3) = 1234 \text{ kN}$$

This resultant force acts at the center of pressure which is at a distance y_{cp} from axis O_1 and

$$y_{cp} = \frac{I_{cg}}{y_{cg} A} + y_{cg} = \frac{(3)(6^3)/12}{(7)(3 \times 6)} + 7 = 7.43 \text{ m from } O_1$$

3.4. Determine the resultant force due to water acting on the 4 m by 6 m triangular area CD shown in Fig. 3-3. The apex of the triangle is at C.

Solution:

$$F_{CD} = (9.79)\left[3 + \left(\frac{2}{3} \times \sin 45° \times 6\right)\right]\left(\frac{1}{2} \times 4 \times 6\right) = 685 \text{ kN}$$

This force acts at a distance y_{cp} from axis O_2 and is measured along the plane of the area CD.

$$y_{cp} = \frac{(4)(6^3)/36}{(5.83/\sin 45°)(\frac{1}{2} \times 4 \times 6)} + \frac{5.83}{\sin 45°} = 8.49 \text{ m from axis } O_2$$

3.5. Water rises to level E in the pipe attached to tank $ABCD$ in Fig. 3-4. Neglecting the weight of the tank and riser pipe, (*a*) determine and locate the resultant force acting on area AB, which is 8 ft wide; (*b*) compute the total force on the bottom of the tank; and (*c*) compare the total weight of the water with the result in (*b*) and explain the difference.

Mathcad

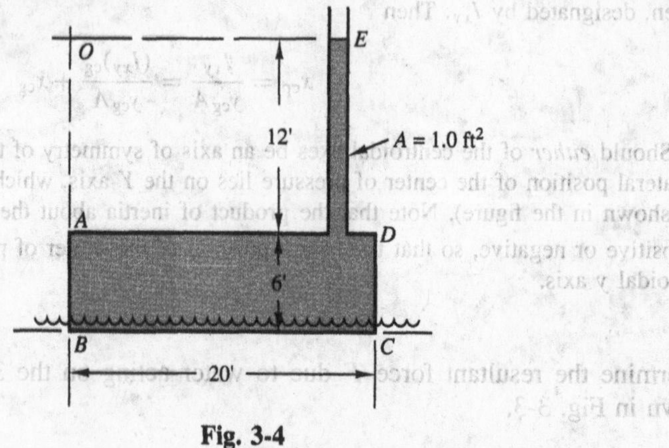

Fig. 3-4

Solution:

(*a*) The depth of the center of gravity of area AB is 15 ft below the free surface of the water at E.

Then $$F = \gamma h A = (62.4)(12 + 3)(6 \times 8) = 44{,}900 \text{ lb}$$

acting at distance $$y_{cp} = \frac{(8)(6^3)/12}{(15)(6 \times 8)} + 15 = 15.20 \text{ ft from } O$$

(*b*) The pressure on the bottom BC is uniform; hence the force

$$F = pA = (\gamma h)A = (62.4)(18)(20 \times 8) = 179{,}700 \text{ lb}$$

(*c*) The total weight of the water is $W = (62.4)[(20 \times 6 \times 8) + (12 \times 1)] = 60{,}700$ lb.

A free body of the lower part of the tank (cut by a horizontal plane just above level BC) will indicate a downward force on area BC of 179,700 lb, vertical tension in the walls of the tank, and the reaction of the supporting plane. The reaction must equal the total weight of water or 60,700 lb. The tension in the walls of the tank is caused by the upward force on the top AD of the tank, which is

$$F_{AD} = (\gamma h)A = (62.4)(12)(160 - 1) = 119{,}000 \text{ lb upward}$$

An apparent paradox is thus clarified since, for the free body considered, the sum of the vertical forces is zero, i.e.,

$$179{,}700 - 60{,}700 - 119{,}000 = 0$$

and hence the condition for equilibrium is satisfied.

3.6. Gate AB in Fig. 3-5(*a*) is 4 ft wide and is hinged at A. Gage G reads -2.17 psi, and oil of specific gravity 0.750 is in the right-hand tank. What horizontal force must be applied at B for equilibrium of gate AB?

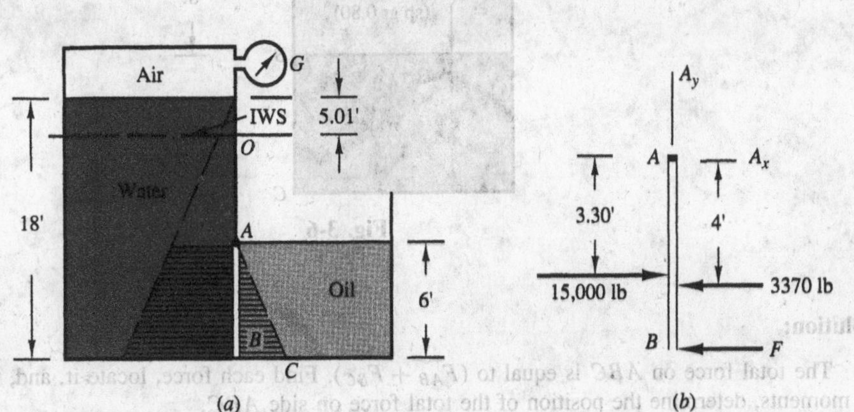

Fig. 3-5

Solution:

The forces acting on the gate due to the liquids must be evaluated and located. For the right-hand side,

$$F_{\text{oil}} = \gamma h_{cg} A = (0.750 \times 62.4)(3)(6 \times 4) = 3370 \text{ lb to the left}$$

acting
$$y_{cp} = \frac{(4)\left(6^3\right)/12}{(3)(4 \times 6)} + 3 = 4.00 \text{ ft from } A$$

It should be noted that the pressure intensity acting on the right-hand side of rectangle AB varies linearly from zero gage to a value due to 6 ft of oil ($p = \gamma h$ is a linear equation). Loading diagram ABC indicates this fact. For a rectangular area only, the center of gravity of this loading diagram coincides with the center of pressure. The center of gravity is located $\left(\frac{2}{3}\right)(6) = 4$ ft from A, as above.

For the left-hand side, it is necessary to convert the negative pressure due to the air to its equivalent in feet of the liquid, water.

$$h = -\frac{p}{\gamma} = -\frac{2.17 \times 144 \text{ lb/ft}^2}{62.4 \text{ lb/ft}^3} = -5.01 \text{ ft}$$

This negative pressure head is equivalent to having 5.01 ft less of water above level A. It is convenient and useful to employ an imaginary water surface (IWS) 5.01 ft below the real surface and solve the problem by direct use of basic equations. Thus,

$$F_{\text{water}} = (62.4)(6.99 + 3)(6 \times 4) = 15{,}000 \text{ lb acting to the right at the center of pressure}$$

For the submerged rectangular area, $y_{cp} = \dfrac{(4)(6^3)/12}{(9.99)(6\times 4)} + 9.99 = 10.29$ ft from O, or the center of pressure is $(10.29 - 6.99) = 3.30$ ft from A.

In Fig. 3-5(b), the free-body diagram of gate AB shows the forces acting. The sum of the moments about A must equal zero. Taking clockwise as plus,

$$+3370 \times 4 + 6F - 15{,}000 \times 3.30 = 0 \quad \text{and} \quad F = 6000 \text{ lb to the left}$$

3.7. The tank in Fig. 3-6 contains oil and water. Find the resultant force on side ABC, which is 4 ft wide.

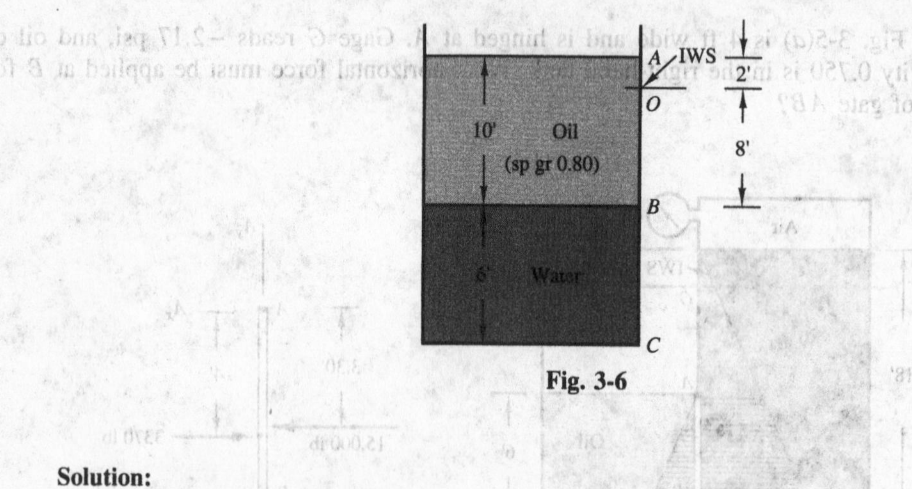

Fig. 3-6

Solution:

The total force on ABC is equal to $(F_{AB} + F_{BC})$. Find each force, locate it, and, using the principle of moments, determine the position of the total force on side ABC.

(a) $F_{AB} = (0.800 \times 62.4)(5)(10 \times 4) = 9980$ lb acting at a point $\left(\frac{2}{3}\right)(10)$ ft from A or 6.67 ft down. The same distance can be obtained by formula as follows:

$$y_{cp} = \frac{(4)\left(10^3\right)/12}{(5)(4\times 10)} + 5 = 6.67 \text{ ft from } A$$

(b) Water is acting on area BC, and any superimposed liquid can be converted into an equivalent depth of water. Employ an imaginary water surface (IWS) for this second calculation, locating the IWS by changing 10 ft of oil to $0.800 \times 10 = 8$ ft of water. Then

$$F_{BC} = (62.4)(8+3)(6\times 4) = 16{,}470 \text{ lb acting at the center of pressure}$$

$$y_{cp} = \frac{(4)\left(6^3\right)/12}{(11)(4\times 6)} + 11 = 11.27 \text{ ft from } O \quad \text{or} \quad (2 + 11.27) = 13.27 \text{ ft from } A$$

The total resultant force $= 9980 + 16{,}470 = 26{,}450$ lb acting at the center of pressure for the entire area. The moment of this total force = the sum of the moments of its two parts. Using A as a convenient axis,

$$26{,}450\, Y_{cp} = (9980)(6.67) + (16{,}470)(13.27) \quad \text{and} \quad Y_{cp} = 10.78 \text{ ft from } A$$

Other methods of attack may be employed, but it is believed that the method illustrated will greatly reduce mistakes in judgment and calculation.

3.8. In Fig. 3-7, gate ABC is hinged at B and is 4 m long. Neglecting the weight of the gate, determine the unbalanced moment due to the water acting on the gate.

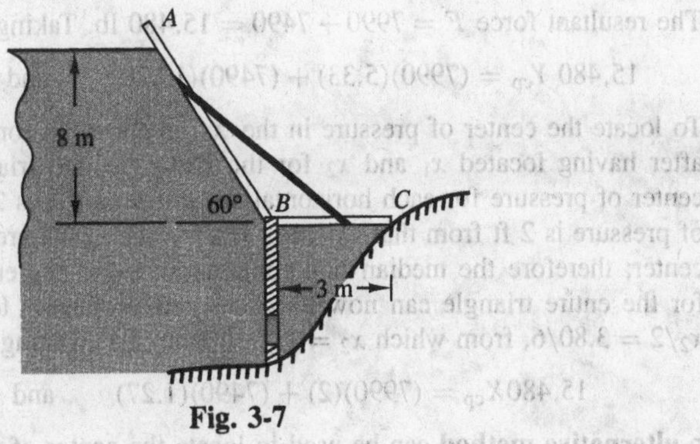

Fig. 3-7

Solution:

$F_{AB} = (9.79)(4)(9.24 \times 4) = 1447$ kN, acting $\left(\frac{2}{3}\right)(9.24) = 6.16$ m from A.
$F_{BC} = (9.79)(8)(3 \times 4) = 940$ kN, acting at the center of gravity of BC since the pressure on BC is uniform. Taking moments about B (clockwise plus),

$$\text{unbalanced moment} = +(1447 \times 3.08) - (940 \times 1.50)$$

$$= +3047 \text{ kN} \cdot \text{m clockwise}$$

3.9. Determine the resultant force due to the water acting on the vertical area shown in Fig. 3-8(*a*), and locate the center of pressure in the x and y directions.

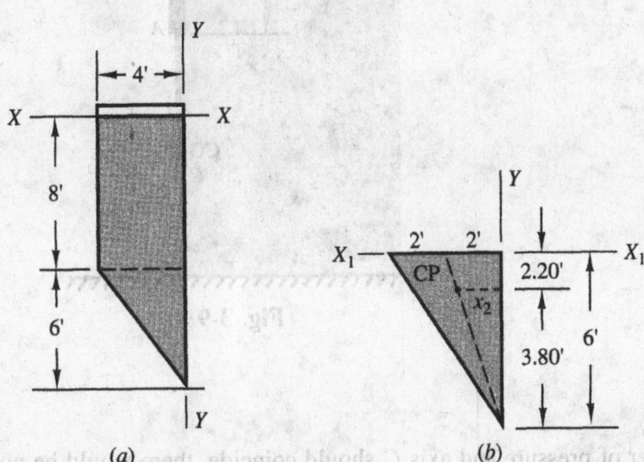

Fig. 3-8

Solution:

Divide the area into a rectangle and a triangle. The total force acting is equal to force F_1 acting on the rectangle plus force F_2 acting on the triangle.

(*a*) $F_1 = (62.4)(4)(8 \times 4) = 7990$ lb acting $\left(\frac{2}{3}\right)(8) = 5.33$ ft below surface XX.

$$F_2 = (62.4)(10)\left(\tfrac{1}{2} \times 6 \times 4\right) = 7490 \text{ lb at } y_{cp} = \frac{(4)\left(6^3\right)/36}{(10)\left(\frac{1}{2} \times 4 \times 6\right)} + 10 = 10.20 \text{ ft below } XX.$$

The resultant force $F = 7990 + 7490 = 15{,}480$ lb. Taking moments about axis XX,

$$15{,}480\, Y_{cp} = (7990)(5.33) + (7490)(10.20) \quad \text{and} \quad Y_{cp} = 7.69 \text{ ft below surface } XX$$

(b) To locate the center of pressure in the X direction (seldom required), use the principle of moments after having located x_1 and x_2 for the rectangle and triangle, respectively. For the rectangle, the center of pressure for each horizontal strip of area dA is 2 ft from the YY axis; therefore its center of pressure is 2 ft from that axis. For the triangle, each area dA has its center of pressure at its own center; therefore the median line contains all these centers of pressure, and the center of pressure for the entire triangle can now be calculated. Referring to Fig. 3-8(b) and using similar triangles, $x_2/2 = 3.80/6$, from which $x_2 = 1.27$ ft from YY. Taking moments,

$$15{,}480 X_{cp} = (7990)(2) + (7490)(1.27) \quad \text{and} \quad X_{cp} = 1.65 \text{ ft from axis } YY$$

An **alternative method** can be used to locate the center of pressure. Instead of dividing the area into two parts, calculate the center of gravity position for the *entire* area. Using the parallel-axis theorem, determine the moment of inertia and the product of inertia of the entire area about these center of gravity axes. The values of y_{cp} and x_{cp} are then calculated by formulas (2) and (4), Problems 3.1 and 3.2. Generally this alternative method has no particular advantage and may involve more arithmetic.

3.10. The 2-m-diameter gate AB in Fig. 3-9 swings about a horizontal pivot C located 40 mm below the center of gravity. To what depth h can the water rise without causing an unbalanced clockwise moment about pivot C?

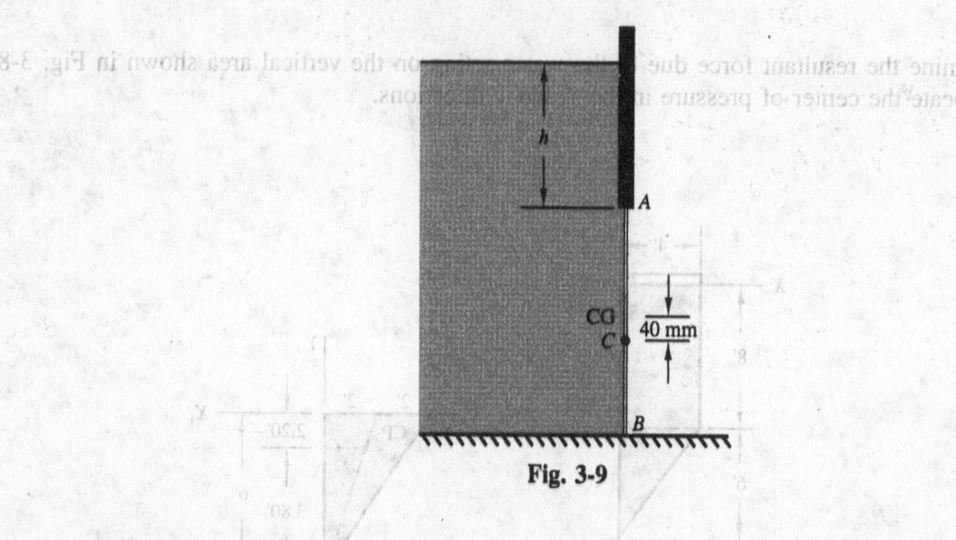

Fig. 3-9

Solution:

If the center of pressure and axis C should coincide, there would be no unbalanced moment acting on the gate. Evaluating the center of pressure distance,

$$y_{cp} = \frac{I_{cg}}{y_{cg} A} + y_{cg} = \frac{\pi d^4/64}{y_{cg}(\pi d^2/4)} + y_{cg}$$

Then
$$y_{cp} - y_{cg} = \frac{\pi 2^4/64}{(h+1)(\pi 2^2/4)} = \frac{40}{1000} \text{ m (given)}$$

from which $h = 5.25$ m above A.

3.11. Determine and locate the components of the force due to the water acting on curved area AB in Fig. 3-10, per meter of its length.

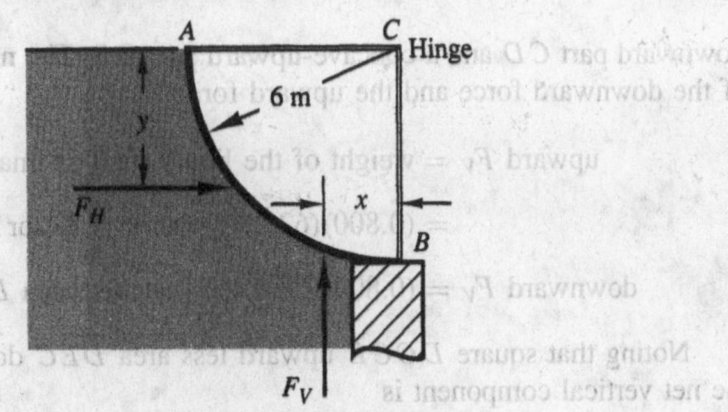

Fig. 3-10

Solution:

$$F_H = \text{force on vertical projection } CB = \gamma\, h_{cg} A_{CB}$$
$$= (9.79)(3)(6 \times 1) = 176 \text{ kN} \quad \text{acting } \left(\tfrac{2}{3}\right)(6) = 4 \text{ m from } C$$

$$F_V = \text{weight of water above area } AB = (9.79)\left(\pi 6^2/4 \times 1\right) = 277 \text{ kN}$$

acting through the center of gravity of the volume of liquid. The center of gravity of a quadrant of a circle is located at a distance $(4/3) \times (r/\pi)$ from either mutually perpendicular radius. Thus

$$x_{cp} = (4/3) \times (6/\pi) = 2.55 \text{ m to the left of line } BC$$

Note: Each force dP acts normal to curve AB and would therefore pass through hinge C upon being extended. The total force should also pass through C. To confirm this statement, take moments of the components about C, as follows.

$$\Sigma M_c = -(176 \times 4) + (277 \times 2.55) \cong 0 \qquad \text{(satisfied)}$$

3.12. The 6-ft-diameter cylinder in Fig. 3-11 weighs 5000 lb and is 5 ft long. Determine the reactions at A and B, neglecting friction.

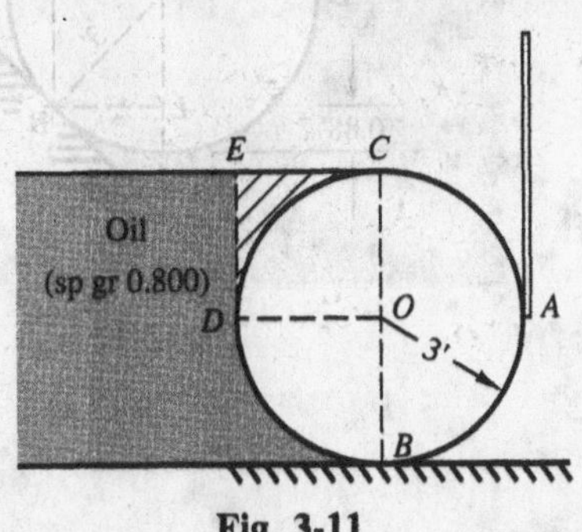

Fig. 3-11

Solution:

(*a*) The reaction at A is due to the horizontal component of the liquid force acting on the cylinder or

$$F_H = (0.800 \times 62.4)(3)(6 \times 5) = 4490 \text{ lb}$$

to the right. Hence the reaction at A must be 4490 lb to the left.

(*b*) The reaction at B is the algebraic sum of the weight of the cylinder and the net vertical component of the force due to the liquid. The curved surface CDB acted upon by the liquid consists of a concave-downward part CD and a concave-upward part DB. The net vertical component is the algebraic sum of the downward force and the upward force.

$$\text{upward } F_V = \text{weight of the liquid (real or imaginary) above curve } DB$$
$$= (0.800)(62.4)(5)\ (\text{area of sector } DOB + \text{area of square } DOCE)$$
$$\text{downward } F_V = (0.800)(62.4)(5)\ (\text{hatched area } DEC)$$

Noting that square $DOCE$ upward less area DEC downward equals quadrant of circle DOC, the net vertical component is

$$\text{net } F_V = (0.800)(62.4)(5)\ (\text{sectors } DOB + DOC) \text{ upward}$$
$$= (0.800)(62.4)(5)\left(\tfrac{1}{2}\pi 3^2\right) = 3530 \text{ lb upward}$$

Finally, $\Sigma Y = 0, \quad 5000 - 3530 - B = 0, \quad \text{and} \quad B = 1470 \text{ lb upward}$

In this particular problem, the upward component (buoyant force) equals the weight of the displaced liquid to the left of the vertical plane COB.

3.13. Referring to Fig. 3-12, determine the horizontal and vertical forces due to the water acting on the 6-ft-diameter cylinder per foot of its length.

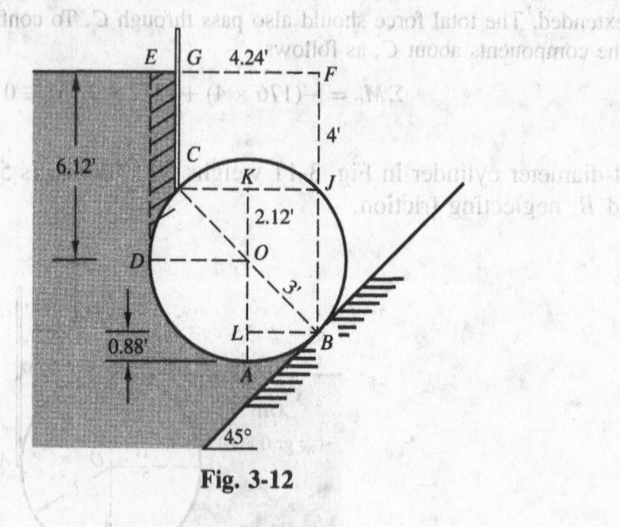

Fig. 3-12

Solution:

(*a*) Net F_H = force on CDA − force on AB. Using the vertical projection CDA and of AB,

$$F_H(CDA) = (62.4)(4 + 2.56)(5.12 \times 1) = 2090 \text{ lb to the right}$$
$$F_H(AB) = (62.4)(4 + 4.68)(0.88 \times 1) = 477 \text{ lb to the left}$$

Net $F_H = 2090 - 477 = 1613$ lb to the right.

(*b*) Net F_V = upward force on DAB − downward force on DC
= weight of (volume $DABFED$ − volume $DCGED$).

The hatched area (volume) is contained in each of the above volumes, one force being upward and the other downward. Thus they cancel, and

$$\text{net } F_V = \text{weight of volume } DABFGCD$$

Dividing this volume into convenient geometric shapes,

$$\text{net } F_V = \text{weight of (rectangle } GFJC + \text{triangle } CJB + \text{semicircle } CDAB)$$

$$= 62.4\left[(4 \times 4.24) + \left(\tfrac{1}{2} \times 4.24 \times 4.24\right) + \left(\tfrac{1}{2}\pi 3^2\right)\right](1) = 2500 \text{ lb upward}$$

If it is desired to locate this resultant vertical component, the principle of moments is employed. Each part of the 2500-lb resultant acts through the center of gravity of the volume it represents. By static mechanics, the centers of gravity are found and the moment equation is written (see Problems 3.7 and 3.9).

3.14. In Fig. 3-13, an 8-m-diameter cylinder plugs a rectangular hole in a tank that is 3 m long. With what force is the cylinder pressed against the bottom of the tank due to the 9 m of water?

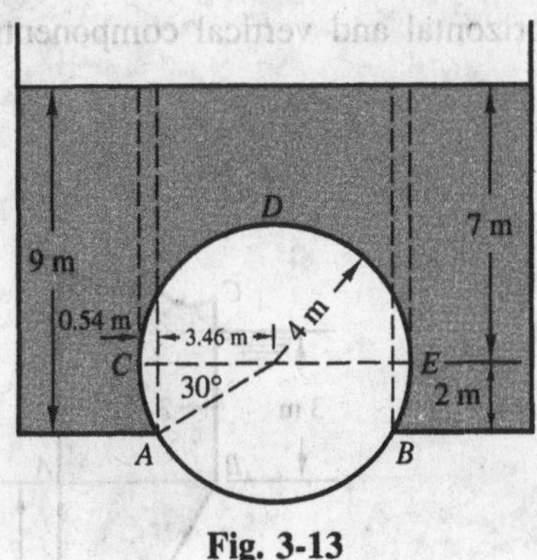

Fig. 3-13

Solution:

$$\text{net } F_V = \text{downward force on } CDE - \text{upward force on } CA \text{ and } BE$$

$$= 9.79 \times 3\left\{\left[(7 \times 8) - \left(\tfrac{1}{2}\pi 4^2\right)\right] - 2\left[(7 \times 0.54) + \left(\tfrac{1}{12}\pi 4^2\right) - \left(\tfrac{1}{2} \times 2 \times 3.46\right)\right]\right\}$$

$$= 642 \text{ kN downward}$$

3.15. In Fig. 3-14, the 8-ft-diameter cylinder weighs 500 lb and rests on the bottom of a tank that is 3 ft long. Water and oil are poured into the left- and right-hand portions of the tank to depths

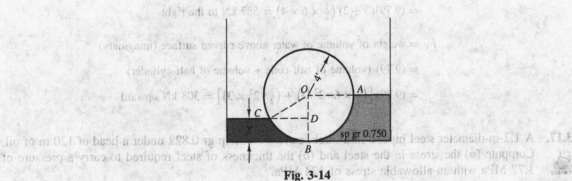

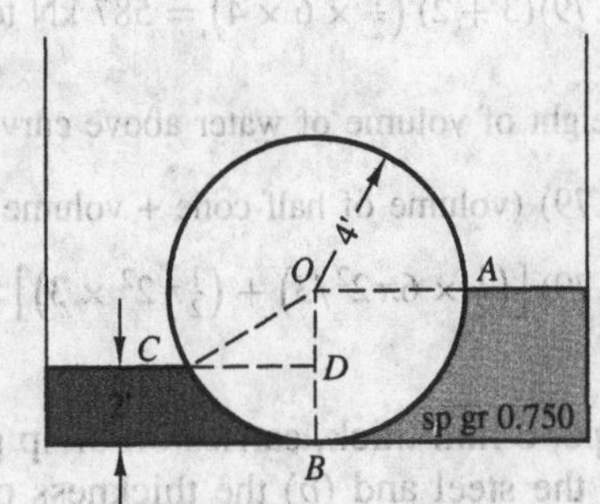

Fig. 3-14

of 2 and 4 ft, respectively. Find the magnitudes of the horizontal and vertical components of the force that will keep the cylinder touching the tank at B.

Solution:

$$\text{net } F_H = \text{component on } AB \text{ to left} - \text{component on } CB \text{ to right}$$

$$= [0.750 \times 62.4 \times 2(4 \times 3)] - [62.4 \times 1(2 \times 3)] = 749 \text{ lb to left}$$

$$\text{net } F_V = \text{component upward on } AB + \text{component upward on } CB$$

$$= \text{weight of quadrant of oil} + \text{weight of (sector} - \text{triangle) of water}$$

$$= \left(0.750 \times 62.4 \times 3 \times \tfrac{1}{4}\pi 4^2\right) + \left\{62.4 \times 3\left[\tfrac{1}{6}\pi 4^2 - \left(\tfrac{1}{2} \times 2\sqrt{12}\right)\right]\right\} = 2680 \text{ lb upward}$$

The components to hold the cylinder in place are 749 lb to the right and 2180 lb downward.

3.16. The half-conical buttress ABE shown in Fig. 3-15 is used to support a half-cylindrical tower $ABCD$. Calculate the horizontal and vertical components of the force due to water acting on buttress ABE.

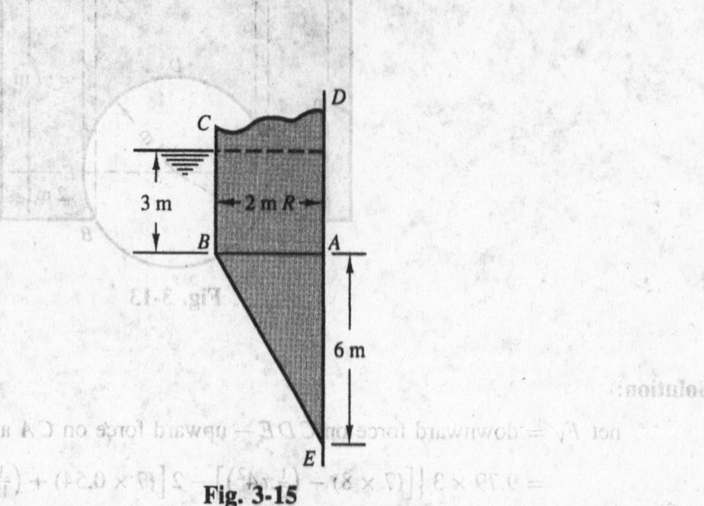

Fig. 3-15

Solution:

$$F_H = \text{force on vertical projection of half-cone}$$

$$= (9.79)(3+2)\left(\tfrac{1}{2} \times 6 \times 4\right) = 587 \text{ kN to the right}$$

$$F_V = \text{weight of volume of water above curved surface (imaginary)}$$

$$= (9.79)\ (\text{volume of half-cone + volume of half-cylinder})$$

$$= (9.79)\left[\left(\tfrac{1}{2} \times 6\pi 2^2/3\right) + \left(\tfrac{1}{2}\pi 2^2 \times 3\right)\right] = 308 \text{ kN upward}$$

3.17. A 1.2-m-diameter steel pipe, 6 mm thick, carries oil of sp gr 0.822 under a head of 120 m of oil. Compute (a) the stress in the steel and (b) the thickness of steel required to carry a pressure of 1.72 MPa with an allowable stress of 124 MPa.

Mathcad

Solution:

(*a*)
$$\sigma \text{ (stress in kPa)} = \frac{p \text{ (pressure in kPa)} \times r \text{ (radius in m)}}{t \text{ (thickness in m)}}$$
$$= \frac{(0.822 \times 9.79 \times 120)(1.2/2)}{6/1000} = 96{,}600 \text{ kPa, or } 96.6 \text{ MPa}$$

(*b*) $\sigma = pr/t$, $124 = 1.72 \times 0.6/t$, $t = 0.0083 \text{ m} = 8.3 \text{ mm}$

3.18. A wooden storage vat, 20 ft in outside diameter, is filled with 24 ft of brine, sp gr 1.06. The wood staves are bound by flat steel bands, 2 in wide by $\frac{1}{4}$ in thick, whose allowable stress is 16,000 psi. What is the spacing of the bands near the bottom of the vat, neglecting any initial stress? Refer to Fig. 3-16.

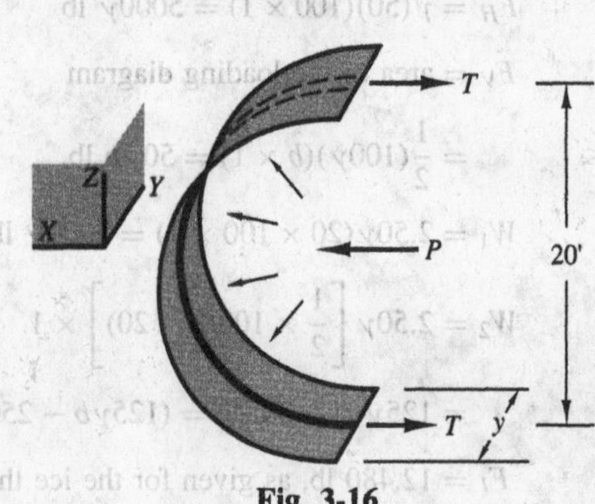

Fig. 3-16

Solution:

Force P represents the sum of all horizontal components of small forces dP acting on length y of the vat, and forces T represent the total tension carried in a band loaded by the same length y. Since the sum of the forces in the X direction must be zero, $2T$ (lb) $- P$ (lb) $= 0$, or

$$(2)\text{ (area steel} \times \text{stress in steel)} = p \times Z \text{ projection of semicylinder}$$

Then $(2)\left(2 \times \frac{1}{4}\right)(16{,}000) = (1.06 \times 62.4 \times 24/144)(20 \times 12y)$

and $y = 6.05$ in spacing of bands

3.19. Refer to Fig. 3-17. What is the minimum width b for the base of a dam 100 ft high if upward pressure beneath the dam is assumed to vary uniformly from full hydrostatic head at the heel to zero at the toe, and also assuming an ice thrust F_I of 12,480 lb per linear foot of dam at the top? For this study make the resultant of the reacting forces cut the base at the downstream edge of the middle third of the base (at O) and take the weight of the masonry as 2.50γ.

Solution:

Shown in the diagram are the H and V components of the reaction of the foundation, acting through O. Consider a length of 1 ft of dam, and evaluate all the forces in terms of γ and b, as follows:

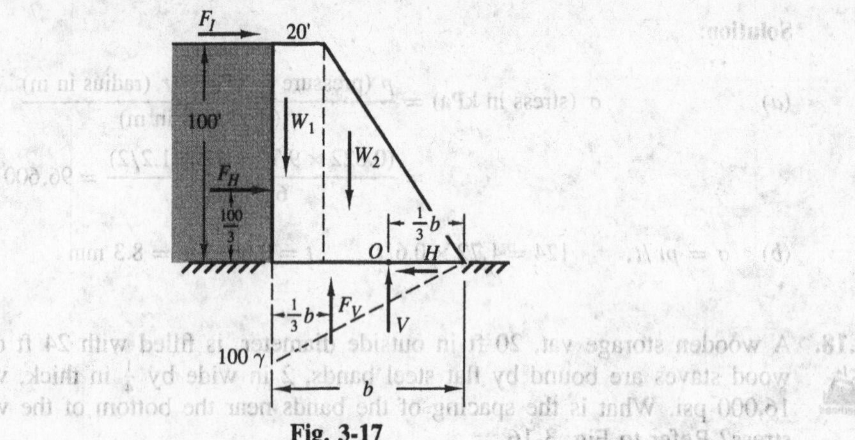

Fig. 3-17

$$F_H = \gamma(50)(100 \times 1) = 5000\gamma \text{ lb}$$

$$F_V = \text{area of the loading diagram}$$

$$= \frac{1}{2}(100\gamma)(b \times 1) = 50\gamma b \text{ lb}$$

$$W_1 = 2.50\gamma(20 \times 100 \times 1) = 5000\gamma \text{ lb}$$

$$W_2 = 2.50\gamma\left[\frac{1}{2} \times 100(b-20)\right] \times 1$$

$$= 125\gamma(b-20) \text{ lb} = (125\gamma b - 2500\gamma) \text{ lb}$$

$$F_I = 12{,}480 \text{ lb, as given for the ice thrust}$$

To find the value of b for equilibrium, take moments of these forces about axis O. Considering clockwise moments positive,

$$5000\gamma\left(\frac{100}{3}\right)+50\gamma b\left(\frac{b}{3}\right)-5000\gamma\left(\frac{2}{3}b-10\right)-(125\gamma b-2500\gamma)\left[\frac{2}{3}(b-20)-\frac{b}{3}\right]+12{,}480(100)=0$$

Simplifying and solving, $3b^2 + 100b - 24{,}400 = 0$ and $b = 75$ ft wide.

3.20. A concrete dam retaining 6 m of water is shown in Fig. 3-18(a). The unit weight of the concrete is 23.5 kN/m^3. The foundation soil is impermeable. Determine (a) the factor of safety against sliding, (b) the factor of safety against overturning, and (c) the pressure intensity on the base of the dam. The coefficient of friction between the base of the dam and the foundation soil is 0.48.

Solution:

$$F_H = \gamma h_{cg} A = (9.79)(3)(6 \times 1) = 176.2 \text{ kN}$$

$$F_V = 0$$

Refer to Fig. 3-18(b).

$$\text{weight of part 1 of dam} = (1)\,[(2)(7)/2](23.5) = 164.5 \text{ kN}$$

$$\text{weight of part 2 of dam} = (1)(2)(7)(23.5) = 329.0 \text{ kN}$$

Total weight of dam = 164.5 + 329.0 = 493.5 kN.

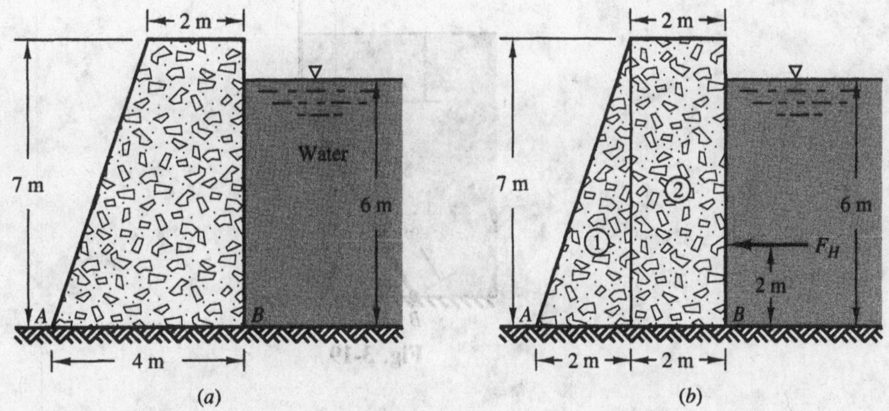

Fig. 3-18

(a) $$FS_{\text{sliding}} = \frac{\text{sliding resistance}}{\text{sliding force}} = \frac{(0.48)(493.5)}{176.2} = 1.34$$

(b) $$FS_{\text{overturning}} = \frac{\text{total righting moment}}{\text{total overturning moment}} = \frac{(164.5)(1.333) + (329.0)(3.000)}{(176.2)(2)} = 3.42$$

(c) Resultant (R) on base $= \sqrt{(164.5 + 329.0)^2 + 176.2^2} = 524$ kN. Let $\bar{x}$ be the distance from A to the point where R intersects the base of the dam.

$$\bar{x} = \frac{\Sigma M_A}{R_y} = \frac{[(164.5)(1.333) + (329.0)(3.000)] - [(176.2)(2)]}{493.5} = 1.730 \text{ m}$$

$$\text{eccentricity} = \frac{4}{2} - 1.730 = 0.270 \text{ m} < \frac{4}{6} = 0.667 \text{ m}$$

Therefore, the resultant lies within the middle third of the base.

$$p = F/A \pm M_y x/I_y \pm M_x y/I_x$$

$$= \frac{493.5}{(4)(1)} \pm \frac{[(493.5)(0.270)](2)}{(1)(4)^3/12} \pm 0$$

$$p_A = 123.4 + 50.0 = 173.4 \text{ kPa} \qquad p_B = 123.4 - 50.0 = 73.4 \text{ kPa}$$

Supplementary Problems

3.21. For an 8-ft length of gate AB in Fig. 3-19, find the compression in strut CD due to water pressure (B, C, and D are pins). *Ans.* 15,850 lb

3.22. A 3.7-m high by 1.5-m wide rectangular gate AB is vertical and is hinged at a point 150 mm below its center of gravity. The total depth of water is 6.1 m. What horizontal force F must be applied at the bottom of the gate for equilibrium? *Ans.* 15 kN

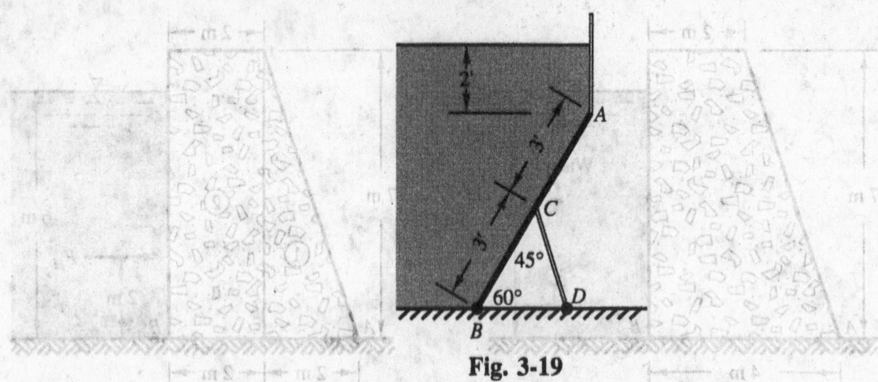

Fig. 3-19

3.23. Find dimension z so that the total stress in rod BD in Fig. 3-20 will be not more than 18,000 lb, using a 4-ft length perpendicular to the paper and considering BD pinned at each end. *Ans.* 5.87 ft

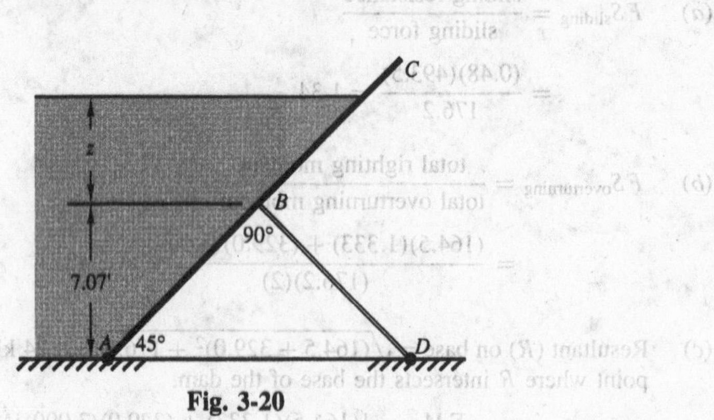

Fig. 3-20

3.24. A dam 20 m long retains 7 m of water, as shown in Fig. 3-21. Find the total resultant force acting on the dam and the location of the center of pressure. *Ans.* 5541 kN, 4.667 m below water surface.

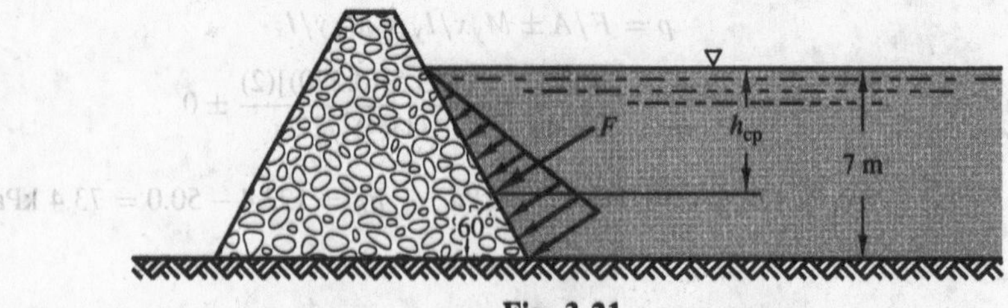

Fig. 3-21

3.25. Oil of specific gravity 0.800 acts on a vertical triangular area whose apex is in the oil surface. The triangle is 9 ft high and 12 ft wide. A vertical rectangular area 8 ft high is attached to the 12-ft base of the triangle and is acted upon by water. Find the magnitude and position of the resultant force on the entire area. *Ans.* 83,300 lb; 12.18 ft down

3.26. In Fig. 3-22, gate AB is hinged at B and is 1.2 m wide. What vertical force, applied at the center of gravity of the 20-kN gate, will keep it in equilibrium? *Ans.* 54 kN

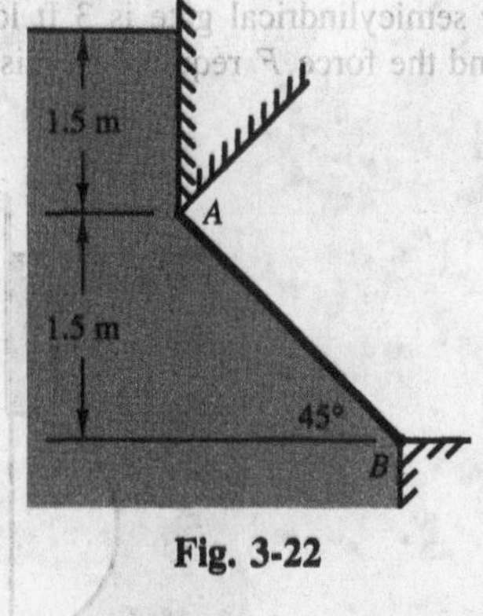

Fig. 3-22

3.27. A tank is 20 ft long and of cross section shown in Fig. 3-23. Water is at level AE. Find (*a*) the total force acting on side BC and (*b*) the total force acting on end $ABCDE$ in magnitude and position.
Ans. 200,000 lb; 98,000 lb at 11.17 ft

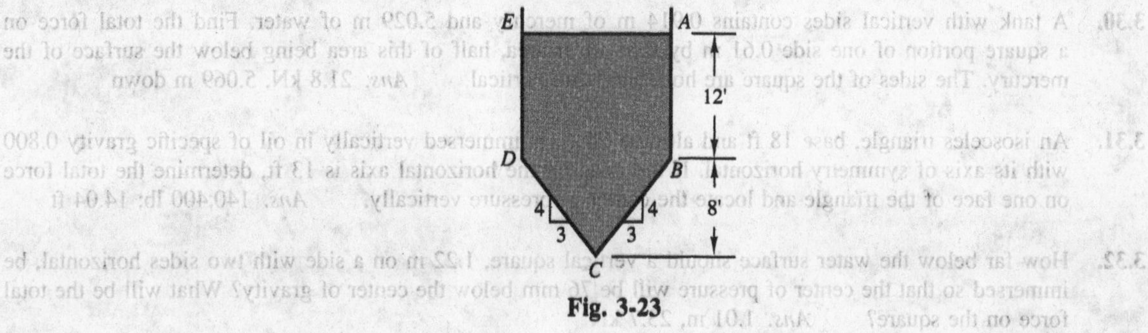

Fig. 3-23

3.28. Given the vertical rectangular gate with water on one side, shown in Fig. 3-24, find the total resultant force acting on the gate and the location of the center of pressure.
Ans. 84.59 kN, 3.633 m below water surface

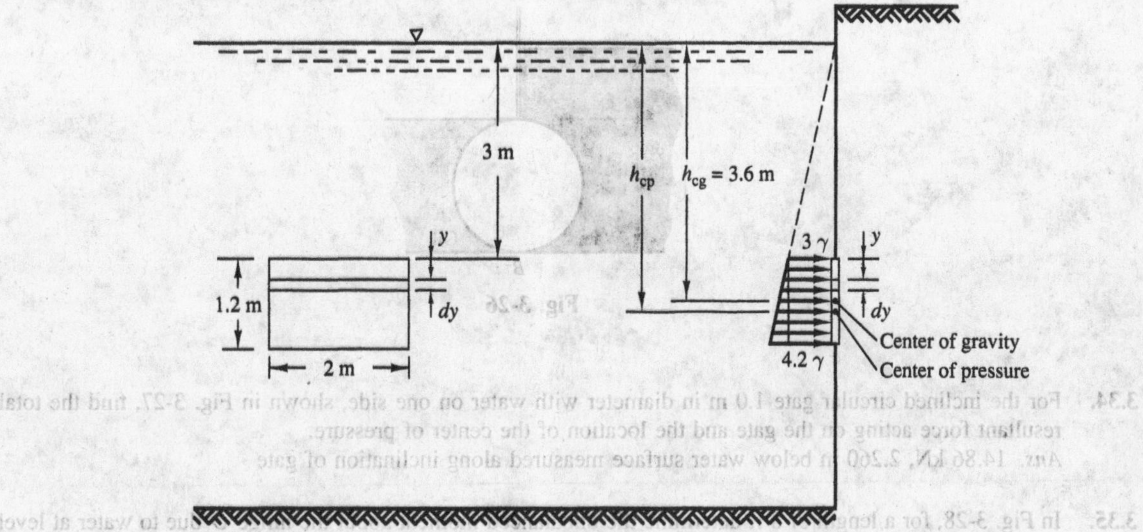

Fig. 3-24

3.29. In Fig. 3-25, the 4-ft-diameter semicylindrical gate is 3 ft long. If the coefficient of friction between the gate and its guides is 0.100, find the force F required to raise the 1000-lb gate. *Ans.* 347 lb

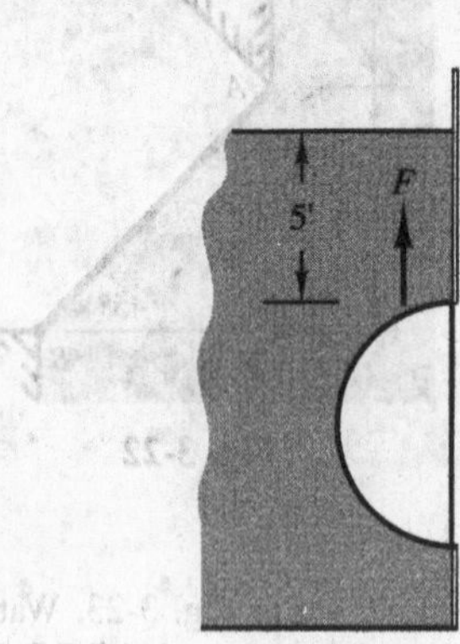

Fig. 3-25

3.30. A tank with vertical sides contains 0.914 m of mercury and 5.029 m of water. Find the total force on a square portion of one side 0.61 m by 0.61 m in area, half of this area being below the surface of the mercury. The sides of the square are horizontal and vertical. *Ans.* 21.8 kN, 5.069 m down

3.31. An isosceles triangle, base 18 ft and altitude 24 ft, is immersed vertically in oil of specific gravity 0.800 with its axis of symmetry horizontal. If the head on the horizontal axis is 13 ft, determine the total force on one face of the triangle and locate the center of pressure vertically. *Ans.* 140,400 lb; 14.04 ft

3.32. How far below the water surface should a vertical square, 1.22 m on a side with two sides horizontal, be immersed so that the center of pressure will be 76 mm below the center of gravity? What will be the total force on the square? *Ans.* 1.01 m, 23.7 kN

3.33. In Fig. 3-26, the 4-ft-diameter cylinder, 4 ft long, is acted upon by water on the left and oil of sp gr 0.800 on the right. Determine (*a*) the normal force at B if the cylinder weighs 4000 lb and (*b*) the horizontal force due to oil and water if the oil level drops 1 ft. *Ans.* 1180 lb, 3100 lb to right

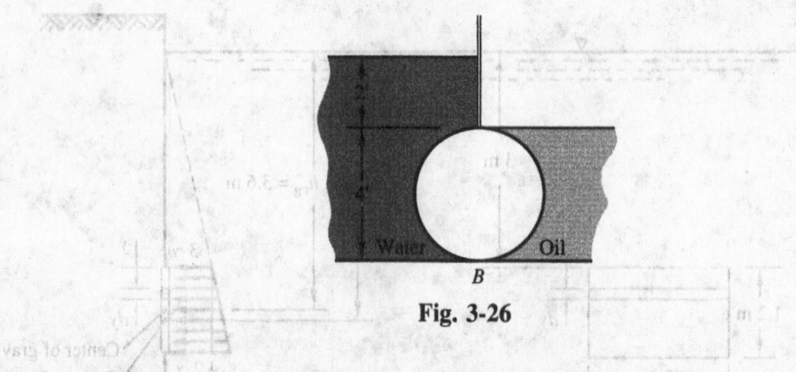

Fig. 3-26

3.34. For the inclined circular gate 1.0 m in diameter with water on one side, shown in Fig. 3-27, find the total resultant force acting on the gate and the location of the center of pressure.
Ans. 14.86 kN, 2.260 m below water surface measured along inclination of gate

3.35. In Fig. 3-28, for a length of 8 ft determine the unbalanced moment about the hinge O due to water at level A. *Ans.* 18,000 ft-lb clockwise

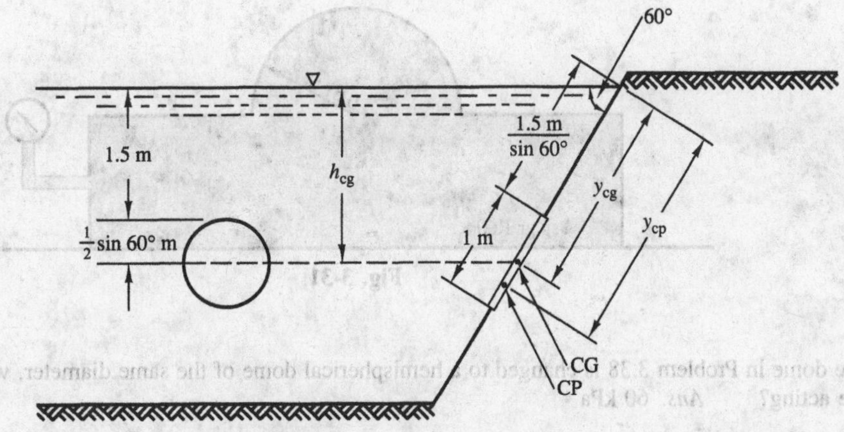

Fig. 3-27

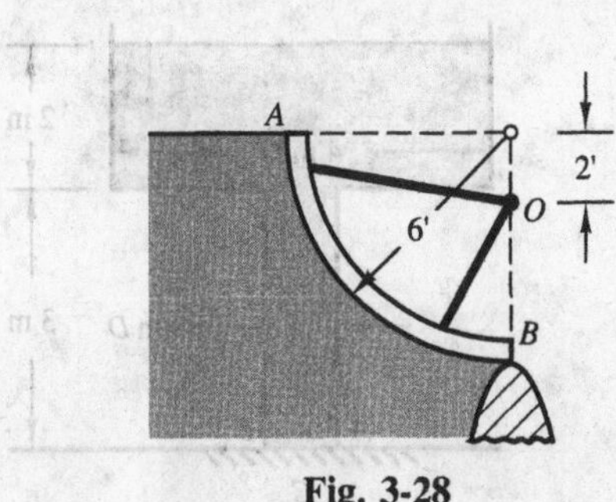

Fig. 3-28

3.36. The tank whose cross section is shown in Fig. 3-29 is 1.2 m long and full of water under pressure. Find the components of the force required to keep the cylinder in position, neglecting the weight of the cylinder. *Ans.* 14 kN down, 20 kN to left

3.37. Determine, per foot of length, the horizontal and vertical components of water pressure acting on the Tainter-type gate shown in Fig. 3-30. *Ans.* 3120 and 1130 lb

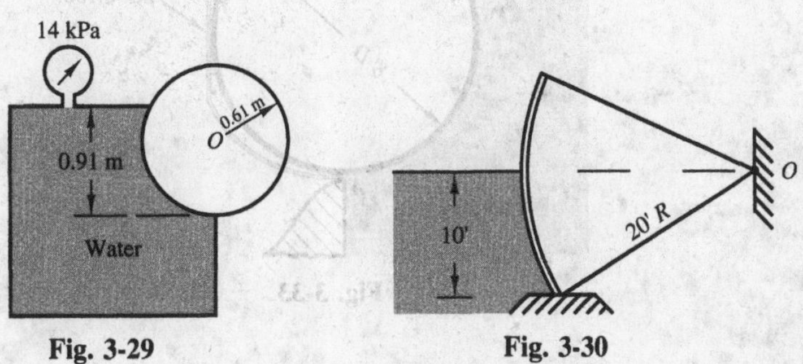

Fig. 3-29 Fig. 3-30

3.38. Find the vertical force acting on the semicylindrical dome shown in Fig. 3-31 when gage *A* reads 58.3 kPa. The dome is 1.83 m long. *Ans.* 113 kPa

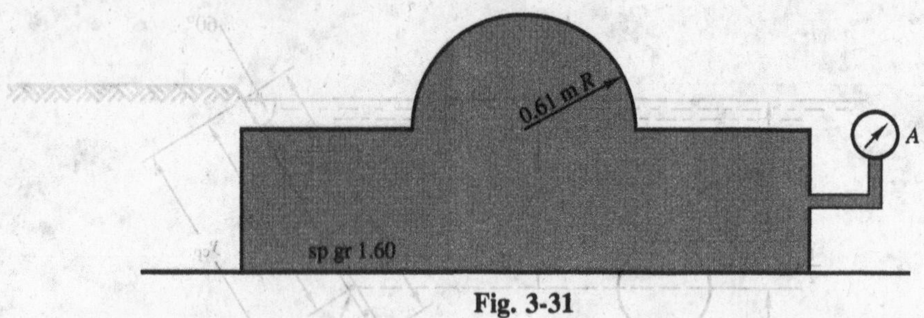

Fig. 3-31

3.39. If the dome in Problem 3.38 is changed to a hemispherical dome of the same diameter, what is the vertical force acting? *Ans.* 60 kPa

3.40. Referring to Fig. 3-32, determine (*a*) the force exerted by water on bottom plate AB of the 1-m-diameter riser pipe and (*b*) the total force on plane C *Ans.* 38.45 kN, 269 kN

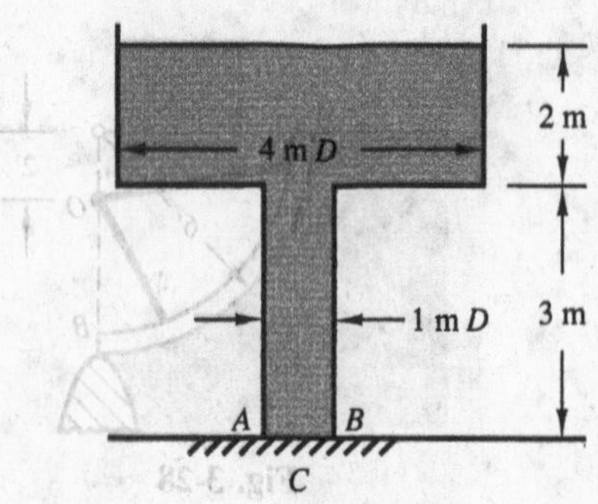

Fig. 3-32

3.41. The cylinder shown in Fig. 3-33 is 10 ft long. Assuming a watertight condition at A and no rotation of the cylinder, what weight of cylinder is required to impede motion upward? *Ans.* 12,700 lb

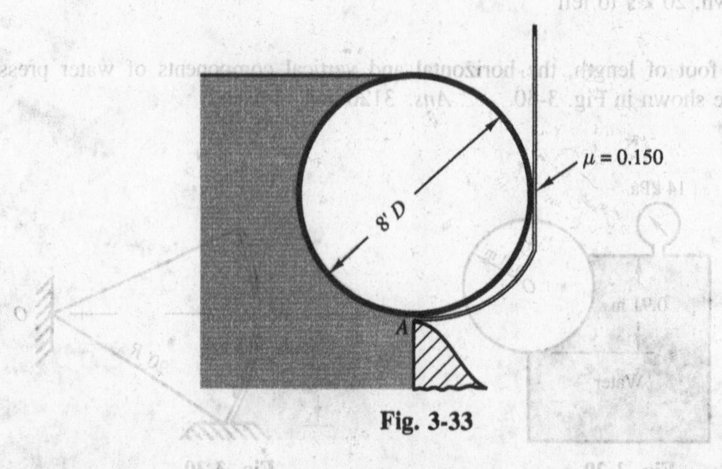

Fig. 3-33

3.42. A wood stave pipe, 48 in inside diameter, is bound by flat steel bands 4 in wide and $\frac{3}{4}$ in thick. For an allowable stress of 16,000 psi in the steel and an internal pressure of 160 psi, determine the spacing of the bands. *Ans.* 12.5 in

3.43. For the parabolic seawall shown in Fig. 3-34, what moment about A per ft of wall is created by the 10-ft depth of water? ($\gamma = 64.0\ \text{lb/ft}^3$) *Ans.* 25,200 ft-lb counterclockwise

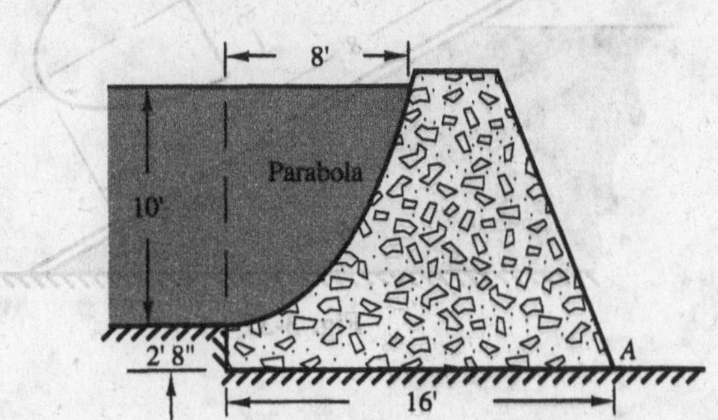

Fig. 3-34

3.44. The tank shown in Fig. 3-35 is 10 ft long, and sloping bottom BC is 8′ wide. What depth of mercury will cause the resultant moment about C due to the liquids to be 101,300 ft-lb clockwise? *Ans.* 2 ft

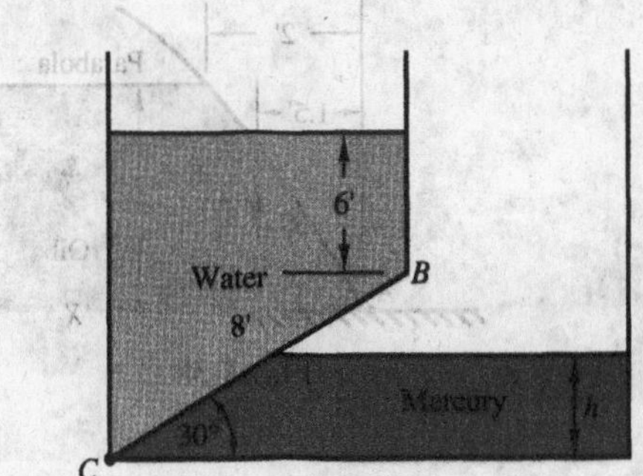

Fig. 3-35

3.45. The gate shown in Fig. 3-36 is 6.10 m long. What are the reactions at hinge O due to the water? Check to see that the torque about O is zero. *Ans.* 136 kN, 272 kN

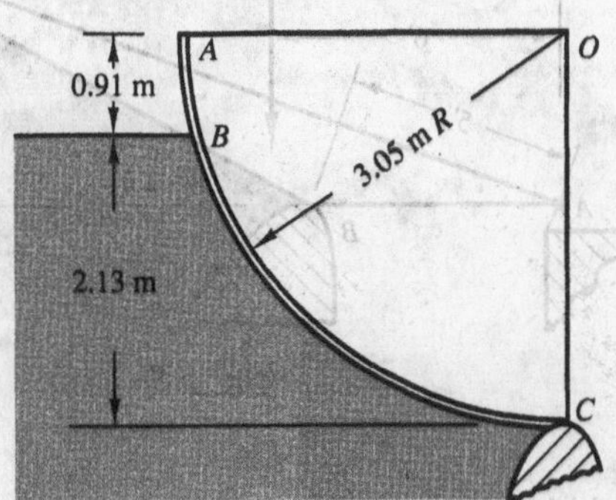

Fig. 3-36

3.46. Refer to Fig. 3-37. A flat plate hinged at C has a configuration satisfying the equation $x^2 + 1.5y = 9$. What is the force of the oil on the plate, and what is the torque about hinge C due to the oil? *Ans.* 6240 lb; 16,400 ft-lb

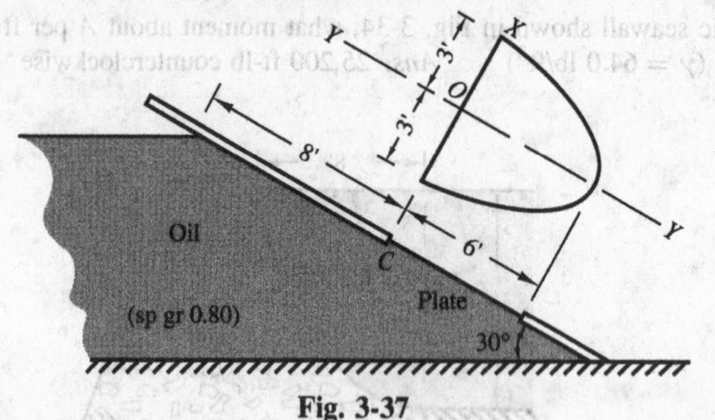

Fig. 3-37

3.47. In Fig. 3-38, parabolic gate ABC is hinged at A and is acted upon by oil weighing 50 lb/ft^3. If the gate's center of gravity is at B, what must the gate weigh per ft of length (perpendicular to the paper) in order for equilibrium to exist? Vertex of parabola is at A. *Ans.* 408 lb/ft

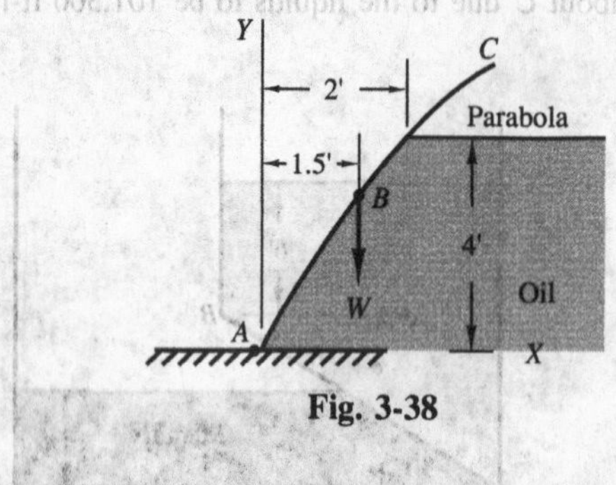

Fig. 3-38

3.48. In Fig. 3-39, automatic gate ABC weighs 1.50 tons/ft of length and its center of gravity is 6 ft to the right of hinge A. Will the gate turn open due to the depth of water shown? *Ans.* Yes

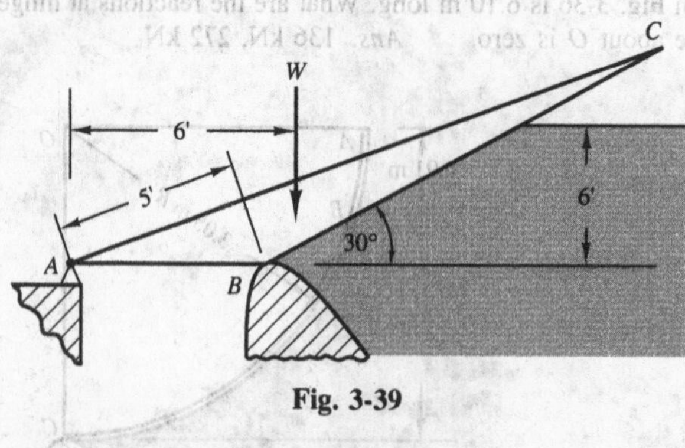

Fig. 3-39

3.49. Referring to Fig. 3-40, calculate the width of concrete wall that is necessary to prevent the wall from sliding. The unit weight of the concrete is 23.6 kN/m^3, and the coefficient of friction between the base of the wall and the foundation soil is 0.42. Use 1.5 as the factor of safety against sliding. Will it also be safe against overturning? *Ans.* 3.09 m, yes

3.50. Solve Problem 3.20 assuming there is hydrostatic uplift that varies uniformly from full hydrostatic head at the heel of the dam to zero at the toe. *Ans.* (*a*) 1.02; (*b*) 1.81; (*c*) $p_A = 173.5$ kPa, $p_B = 14.5$ kPa

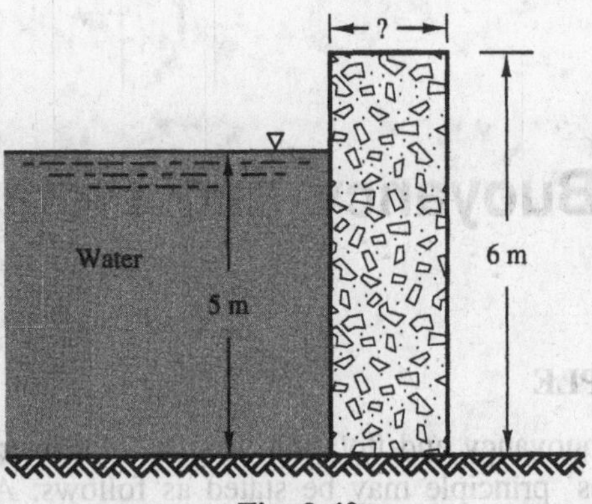

Fig. 3-40

3.51. For the dam retaining water as shown in Fig. 3-41, find (*a*) the factor of safety against sliding, (*b*) the factor of safety against overturning, and (*c*) the pressure intensity on the base of the dam. The foundation soil is permeable; assume hydrostatic uplift varies from full hydrostatic head at the heel of the dam to zero at the toe. The unit weight of the concrete is 23.5 kN/m^3.
Ans. (*a*) 1.36; (*b*) 2.20; (*c*) $p_A = 85.1$ kPa, $p_B = 300.3$ kPa

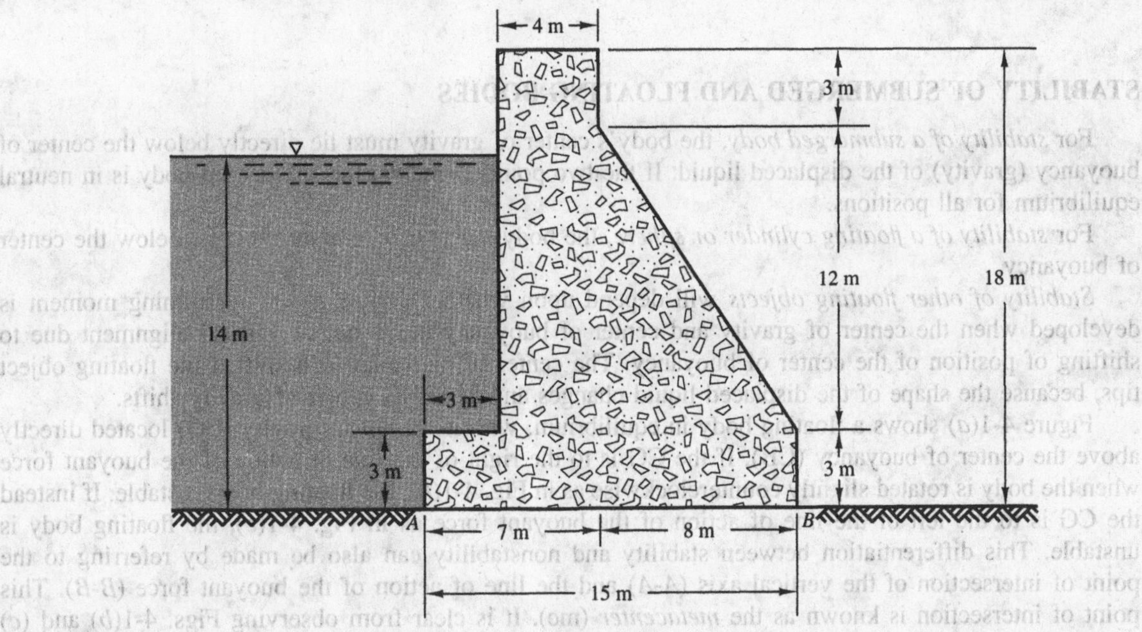

Fig. 3-41

Chapter 4

Buoyancy and Flotation

ARCHIMEDES' PRINCIPLE

The basic principle of buoyancy and flotation was first discovered and stated by Archimedes over 2200 years ago. Archimedes' principle may be stated as follows: A body floating or submerged in a fluid is buoyed (lifted) upward by a force equal to the weight of the fluid that would be in the volume displaced by the fluid. This force is known as the *buoyant force*. It follows, then, that a floating body displaces its own weight of the fluid in which it floats. Stated another way, a floating body displaces a sufficient volume of fluid to just balance its own weight. The point through which the buoyant force acts is called the *center of buoyancy*; it is located at the center of gravity of the displaced fluid.

By applying Archimedes' principle, volumes of irregular solids can be found by determining the apparent loss of weight when a body is wholly immersed in a liquid of known specific gravity. Specific gravities of liquids can be determined by observing the depth of flotation of a hydrometer. Further applications include problems of general flotation and of naval architectural design.

STABILITY OF SUBMERGED AND FLOATING BODIES

For *stability of a submerged body*, the body's center of gravity must lie directly below the center of buoyancy (gravity) of the displaced liquid. If the two points coincide, the submerged body is in neutral equilibrium for all positions.

For *stability of a floating cylinder or sphere*, the body's center of gravity must lie below the center of buoyancy.

Stability of other floating objects will depend upon whether a righting or overturning moment is developed when the center of gravity and center of buoyancy move out of vertical alignment due to shifting of position of the center of buoyancy. The center of buoyancy will shift if the floating object tips, because the shape of the displaced liquid changes and hence its center of gravity shifts.

Figure 4-1(*a*) shows a floating body in equilibrium, with its center of gravity (CG) located directly above the center of buoyancy (CB). If the CG is to the right of the line of action of the buoyant force when the body is rotated slightly counterclockwise as in Fig. 4-1(*b*), the floating body is stable. If instead the CG is to the left of the line of action of the buoyant force as in Fig. 4-1(*c*), the floating body is unstable. This differentiation between stability and nonstability can also be made by referring to the point of intersection of the vertical axis (*A-A*) and the line of action of the buoyant force (*B-B*). This point of intersection is known as the *metacenter* (mc). It is clear from observing Figs. 4-1(*b*) and (*c*) that a floating body is stable if its CG is below the mc and unstable if its CG is above the mc.

The determination as to whether the CG is below or above the mc (and therefore stable or unstable, respectively) can be made more quantitatively by using the following equation to determine the distance from the CB to the mc:

$$\overline{MB} = I/V_d \qquad (1)$$

where $\overline{MB}$ = distance from the CB to the mc [see Fig. 4-1(*d*)]

I = moment of inertia of a horizontal section of the body taken at the surface of the fluid when the floating body is on an even keel

V_d = volume of fluid displaced

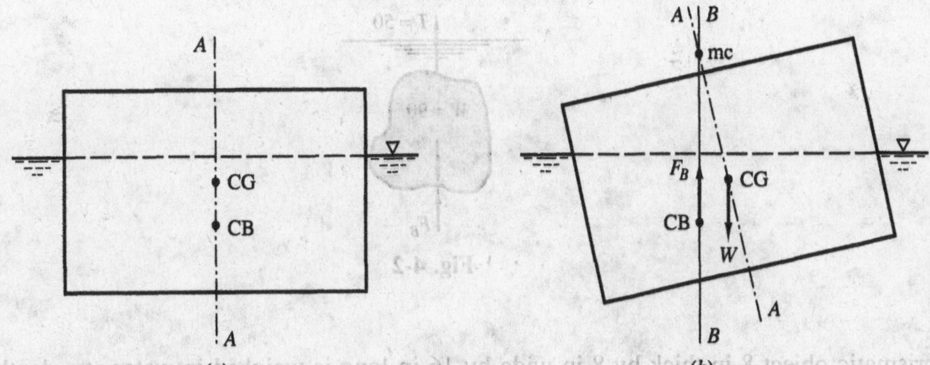

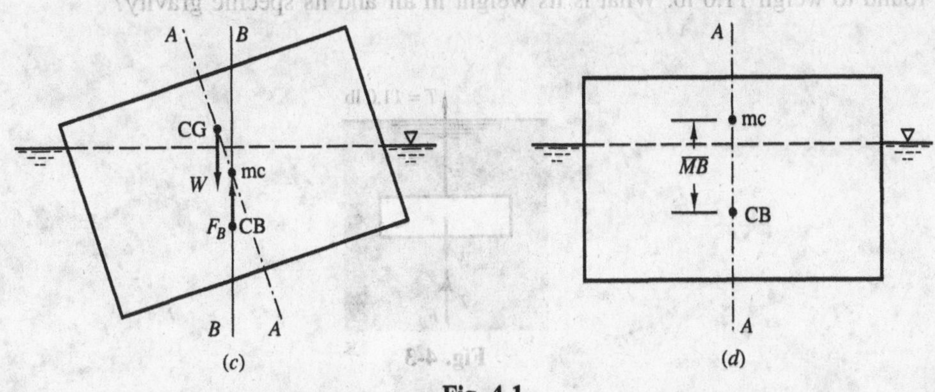

Fig. 4-1

Once distance $\overline{MB}$ is determined, the body can be judged to be stable if the mc is above the body's CG or unstable if it is below the CG.

Solved Problems

4.1. A stone weighs 90 N in air, and when immersed in water it weighs 50 N. Compute the volume of the stone and its specific gravity.

Solution:

Many problems in engineering work can best be analyzed using free-body diagrams. Reference to Fig. 4-2 indicates the total weight of 90 N acting downward, tension in the cord attached to the scales of 50 N upward, and net buoyant force F_B acting upward. From

$$\Sigma Y = 0$$

we have $$90 - 50 - F_B = 0, \qquad \text{and } F_B = 40 \text{ N}$$

Since

$$\text{buoyant force} = \text{the weight of the displaced fluid}$$

$$40 \text{ N} = 9790 \text{ N/m}^3 \times v \qquad \text{and} \qquad v = 0.00409 \text{ m}^3$$

$$\text{specific gravity} = \frac{\text{weight of the stone}}{\text{weight of an equal volume of water}} = \frac{90 \text{ N}}{40 \text{ N}} = 2.25$$

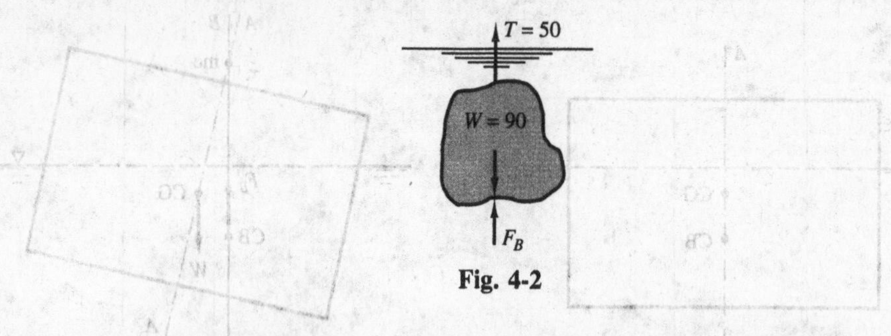

Fig. 4-2

4.2. A prismatic object 8 in thick by 8 in wide by 16 in long is weighed in water at a depth of 20 in and found to weigh 11.0 lb. What is its weight in air and its specific gravity?

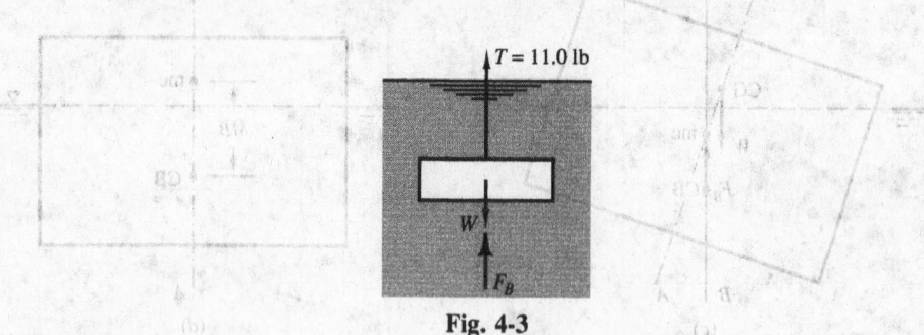

Fig. 4-3

Solution:

Referring to the free-body diagram in Fig. 4-3, $\Sigma Y = 0$; then

$$W - F_B - 11.0 = 0 \qquad \text{or} \qquad (A) \quad W = 11.0 + F_B$$

and

$$\text{buoyant force } F_B = \text{weight of displaced liquid}$$

$$= (62.4)(8 \times 8 \times 16)/1728 = 37.0 \text{ lb}$$

Therefore, from (A), $W = 11 + 37 = 48$ lb and sp gr $= 48/37 = 1.30$.

4.3. A hydrometer weighs 0.0216 N and has a stem at the upper end that is cylindrical and 2.8 mm in diameter. How much deeper will it float in oil of sp gr 0.780 than in alcohol of sp gr 0.821?

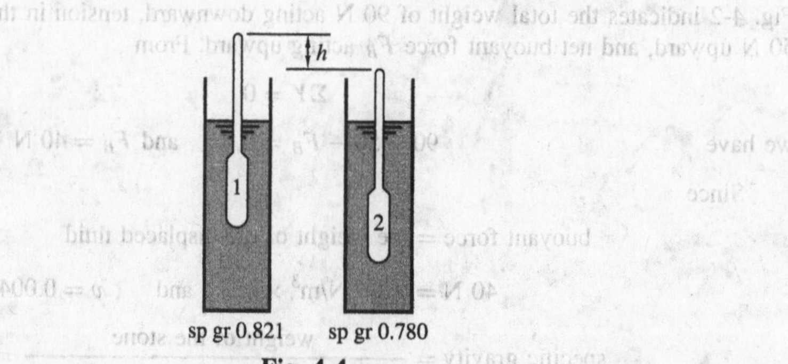

Fig. 4-4

Solution:

For position 1 in Fig. 4-4 in the alcohol,

$$\text{weight of hydrometer} = \text{weight of displaced liquid}$$

$$0.0216 = 0.821 \times 9790 \times v_1$$

from which $v_1 = 2.69 \times 10^{-6}\ \text{m}^3$ (in alcohol).

For position 2,

$$0.0216 = 0.780 \times (9790)\,(v_1 + Ah)$$

$$= 0.780 \times 9790\left[\left(2.69 \times 10^{-6}\right) + \left(\tfrac{1}{4}\pi\right)(2.8/1000)^2 h\right]$$

from which $h = 0.0225\ \text{m} = 22.5\ \text{mm}$.

4.4. A piece of wood of sp gr 0.651 is 80 mm square and 1.5 m long. How many newtons of lead weighing 110 kN/m³ must be fastened at one end of the stick so that it will float upright with 0.3 m out of water?

Solution:

$$\text{total weight of wood and lead} = \text{weight of displaced water}$$

$$\left[0.651 \times 9.79 \times (1.5)(80/1000)^2\right] + 110v = (9.79)\left[(80/1000)^2 \times 1.2 + v\right]$$

from which $v = 0.000140\ \text{m}^3$ and weight of lead $= 110\,v = 110 \times 0.000140 = 0.0154\ \text{kN} = 15.4\ \text{N}$.

4.5. What fraction of the volume of a solid piece of metal of sp gr 7.25 floats above the surface of a container of mercury of sp gr 13.57?

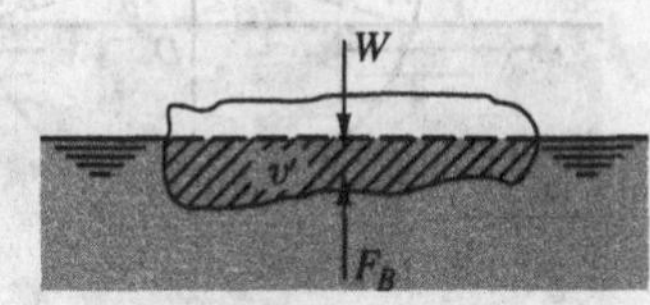

Fig. 4-5

Solution:

The free-body diagram in Fig. 4-5 indicates that, from $\Sigma Y = 0,\ W - F_B = 0$ or

$$\text{weight of body} = \text{buoyant force (weight of displaced mercury)}$$

$$7.25 \times 62.4\,v = 13.57 \times 62.4 v'$$

and the ratio of the volumes is thus $v'/v = 7.25/13.57 = 0.534$.

Hence the fraction of the volume above the mercury $= 1 - 0.534 = 0.466$.

4.6. A rectangular open box, 7.6 m by 3 m in plan and 3.7 m deep, weighs 350 kN and is launched in fresh water. (*a*) How deep will it sink? (*b*) If the water is 3.7 m deep, what weight of stone placed in the box will cause it to rest on the bottom?

Solution:

(*a*) weight of box = weight of displaced water

$$350 = (9.79)(7.6 \times 3 \times Y) \qquad Y = 1.57 \text{ m submerged}$$

(*b*) weight of box plus stone = weight of displaced water

$$350 + W_S = (9.79)(7.6 \times 3 \times 3.7) \qquad W_S = 476 \text{ kN stone}$$

4.7. A block of wood floats in water with 50 mm projecting above the water surface. When placed in glycerin of sp gr 1.35, the block projects 76 mm above the surface of that liquid. Determine the sp gr of the wood.

Solution:

Total weight of block is (*a*) $W = \text{sp gr} \times (9.79)(A \times h)$, and weights of displaced water and glycerin, respectively, are (*b*) $W_W = (9.79A)(h - 50)/1000$ and (*c*) $W_G = 1.35 \times (9.79A)(h - 76)/1000$. Since the weight of each displaced liquid equals the total weight of the block, (*b*) = (*c*), or

$$(9.79A)(h - 50)/1000 = 1.35 \times (9.79A)(h - 76)/1000 \qquad h = 150 \text{ mm}$$

Since (*a*) = (*b*), $\quad \text{sp gr} \times 9.79A \times (150/1000) = 9.79 \times A(150 - 50)/1000 \qquad \text{sp gr} = 0.667$

4.8. To what depth will an 8-ft-diameter log 15 ft long and of sp gr 0.425 sink in fresh water?

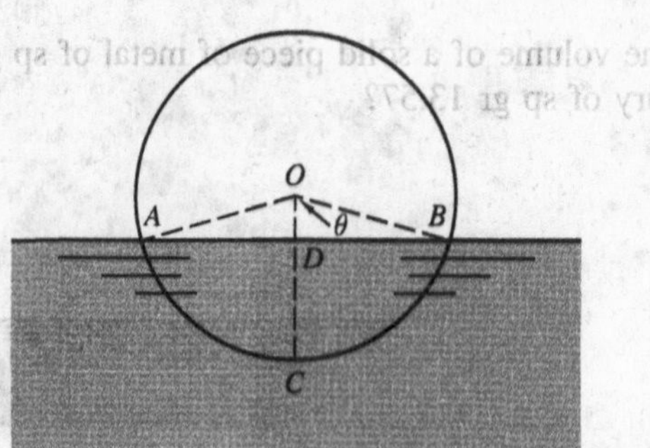

Fig. 4-6

Solution:

Figure 4-6 is drawn with center O of the log above the water surface because its specific gravity is less than 0.500. Had the sp gr been 0.500, then the log would be half-submerged.

total weight of log = weight of displaced liquid sector − 2 triangles

$$0.425 \times 62.4 \times \pi 4^2 \times 15 = 62.4 \times 15\left(\frac{2\theta}{360}16\pi - 2 \times \frac{1}{2} \times 4 \sin\theta \times 4\cos\theta\right)$$

Simplifying and substituting $\frac{1}{2}\sin 2\theta$ for $\sin\theta\cos\theta$,

$$0.425\pi = \theta\pi/180 - \frac{1}{2}\sin 2\theta$$

Solving by successive trials:

Try $\theta = 85°$:

$$1.335 \stackrel{?}{=} 85\pi/180 - \frac{1}{2}(0.1736)$$

$$1.335 \neq 1.397$$

Try $\theta = 83°$:

$$1.335 \stackrel{?}{=} 1.449 - \frac{1}{2}(0.242)$$

$$1.335 \neq 1.328$$

The trial values have straddled the answer.

Try $\theta = 83°10'$:

$$1.335 \stackrel{?}{=} 1.451 - \frac{1}{2}(0.236) = 1.333 \text{ (close check)}$$

The depth of flotation $DC = r - OD = 4.00 - 4.00 \cos 83°10'$

$$= (4.00)(1 - 0.119) = 3.52 \text{ ft.}$$

4.9. (*a*) Neglecting the thickness of the tank walls in Fig. 4-7(*a*), if the tank floats in the position shown, what is its weight?

(*b*) If the tank is held so that the top is 10 ft below the surface of the water, what is the force on the inside top of the tank?

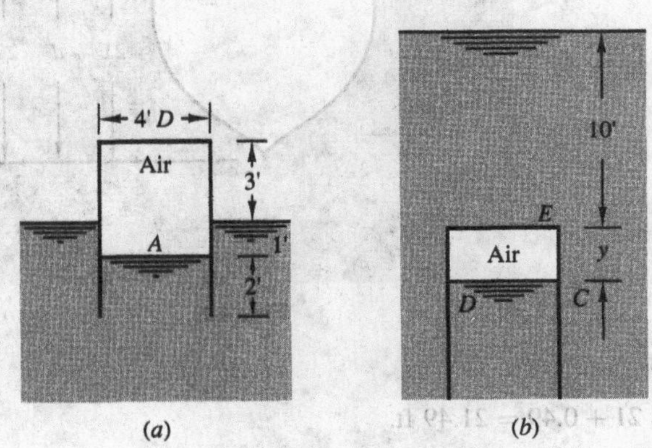

Fig. 4-7

Solution:

(*a*) Weight of tank = weight of displaced liquid = $62.4\pi 2^2(1) = 784$ lb.

(*b*) The space occupied by the air will be less at the new depth shown in Fig. 4-7(*b*). Assuming that the temperature of the air is constant, then for positions (*a*) and (*b*),

$$p_A v_A = p_D v_D \text{ (absolute pressure units must be used)}$$

$$\gamma(34 + 1)(4 \times \text{area}) = \gamma(34 + 10 + y)(y \times \text{area})$$

which yields $y^2 + 44y - 140 = 0$, whose required positive root is $y = 2.98$ ft.

The pressure at $D = 12.98$ ft of water gage = pressure at E. Hence the force on the inside top of the cylinder is $\gamma hA = 62.4(12.98)(\pi 2^2) = 10{,}200$ lb.

4.10. A ship with vertical sides near the waterline weighs 4000 tons and draws 22 ft in salt water ($\gamma = 64.0$ lb/ft^3). Discharge of 200 tons of water ballast decreases the draft to 21 ft. What would be the draft d of the ship in fresh water?

Solution:

Because the shape of the underwater section of the ship is not known, it is best to solve the problem on the basis of volumes displaced.

A 1-ft decrease in draft was caused by a reduction in weight of 200 tons, or

$$200 \times 2000 = \gamma v = 64.0(A \times 1)$$

where v represents the volume between drafts of 22 ft and 21 ft, and $(A \times 1)$ represents the waterline area $\times$ 1 ft, or the same volume v. Then

$$v = A \times 1 = (200)(2000)/64.0 = 6250 \text{ ft}^3/\text{ft depth}$$

Buoyant force $F_B = \gamma \times$ volume of displaced liquid. Then $F_B/\gamma =$ volume of displaced liquid.

From Fig. 4-8, the vertically hatched volume is the difference in displaced fresh water and salt water. This difference can be expressed as $\left(\dfrac{3800 \times 2000}{62.4} - \dfrac{3800 \times 2000}{64.0}\right)$, and this volume is also equal to $6250y$. Equating these values, $y = 0.49$ ft.

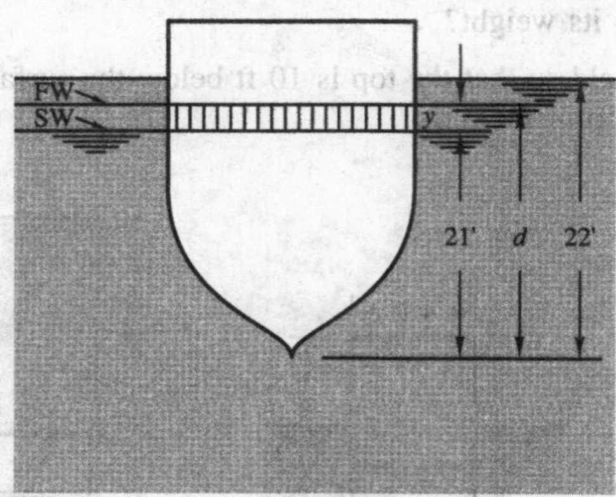

Fig. 4-8

The draft $d = 21 + 0.49 = 21.49$ ft.

4.11. A barrel containing water weighs 1.260 kN. What will be the reading on the scales if a 50 mm by 50 mm piece of wood is held vertically in the water to a depth of 0.60 m?

Solution:

For every acting force there must be an equal and opposite reacting force. The buoyant force exerted by the water upward against the bottom of the piece of wood is opposed by the 50 mm by 50 mm area of wood acting downward on the water with equal magnitude. This force will measure the increase in scale reading.

$F_B = 9.79 \times (50/1000) \times (50/1000) \times 0.60 = 0.015$ kN. The scale reading $= 1.260 + 0.015 =$ 1.275 kN.

4.12. A block of wood 6 ft by 8 ft by 10 ft floats on oil of sp gr 0.751. A clockwise couple holds the block in the position shown in Fig. 4-9. Determine (*a*) the buoyant force acting on the block and its position, (*b*) the magnitude of the couple acting on the block, and (*c*) the location of the metacenter for the tilted position.

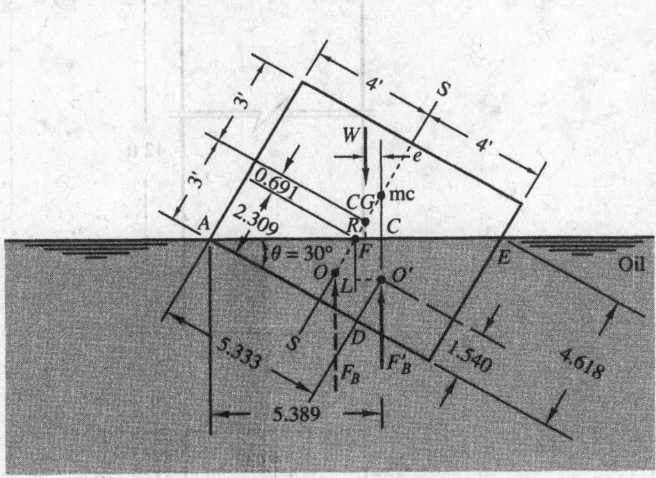

Fig. 4-9

Solution:

(*a*) weight of block = weight of triangular prism of oil (or the buoyant force)

$$W = F'_B = (0.751 \times 62.4)\left(\frac{1}{2} \times 8 \times 4.618 \times 10\right) = 8656 \text{ lb}$$

Then $F'_B = 8656$ lb acting upward through the center of gravity O' of the displaced oil. The center of gravity lies 5.333 ft from A and 1.540 ft from D, as shown in Fig. 4-9.

$$AC = AR + RC = AR + LO' = 5.333 \cos 30° + 1.540 \sin 30° = 5.389 \text{ ft}$$

The buoyant force of 8656 lb acts upward through the center of gravity of the displaced oil, which is 5.39 ft to the right of A.

(*b*) One method of obtaining the magnitude of the righting couple (which must equal the magnitude of the external couple for equilibrium) is to find the eccentricity e. This dimension is the distance between the two parallel, equal forces W and F'_B that form the righting couple.

$$e = FC = AC - AF = 5.389 - AF = 5.389 - 4.963 = 0.426 \text{ ft}$$

since $$AF = AR + RF = 4.618 + (0.691)(\sin 30°) = 4.963 \text{ ft}$$

The couple We or $F'_B e = 8656 \times 0.426 = 3687$ ft-lb. Thus the moment or couple to hold the block in the position shown is 3687 ft-lb clockwise.

(*c*) The point of intersection of the buoyant force and the axis of symmetry SS is called the metacenter (point mc in the figure). If the metacenter is located above the center of gravity of a floating object, the weight of the object and the buoyant force form a righting moment in tilted positions.

$$\text{The metacentric distance} = \frac{RC}{\sin 30°} - 0.691 = \frac{0.770}{0.5000} - 0.691 = 0.849 \text{ ft.}$$

In naval architecture, an extreme angle of some 10° is taken as the limit of heel for which the metacentric distance can be considered constant.

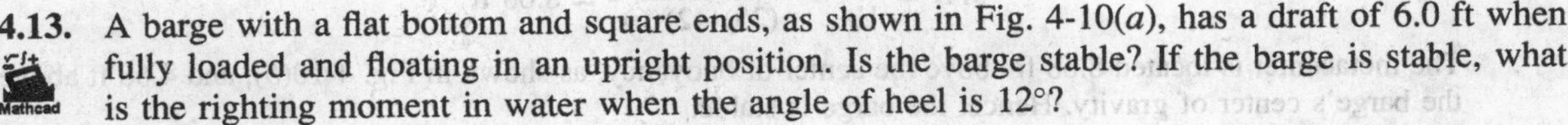

4.13. A barge with a flat bottom and square ends, as shown in Fig. 4-10(*a*), has a draft of 6.0 ft when fully loaded and floating in an upright position. Is the barge stable? If the barge is stable, what is the righting moment in water when the angle of heel is 12°?

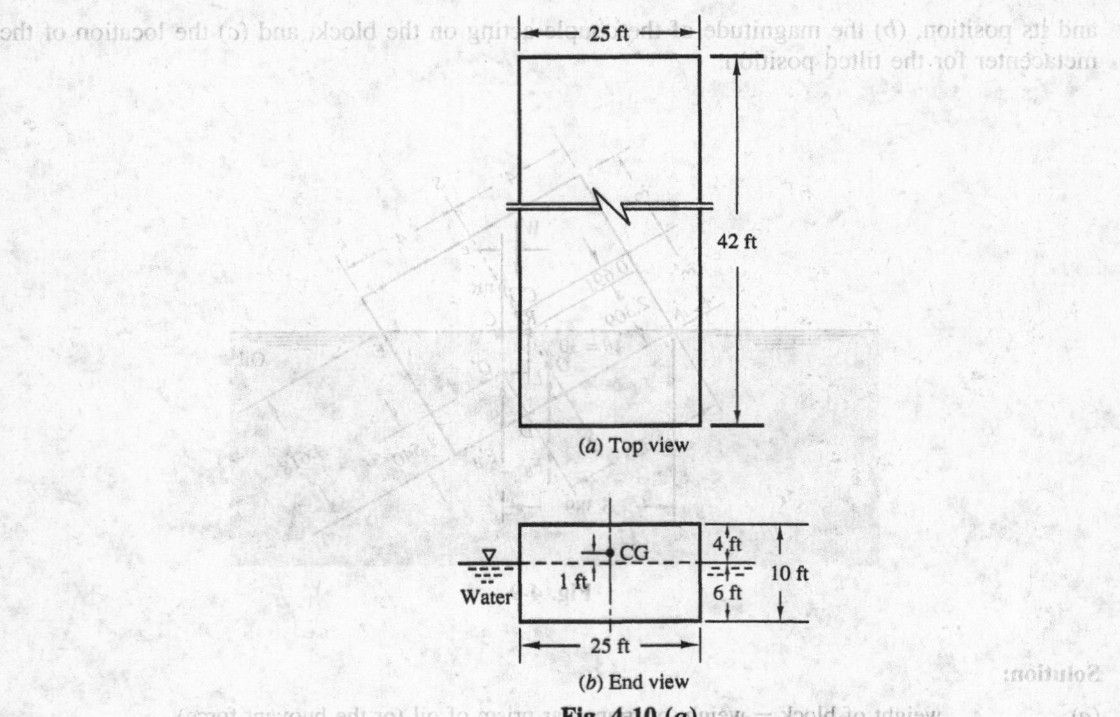

Fig. 4-10 (*a*)

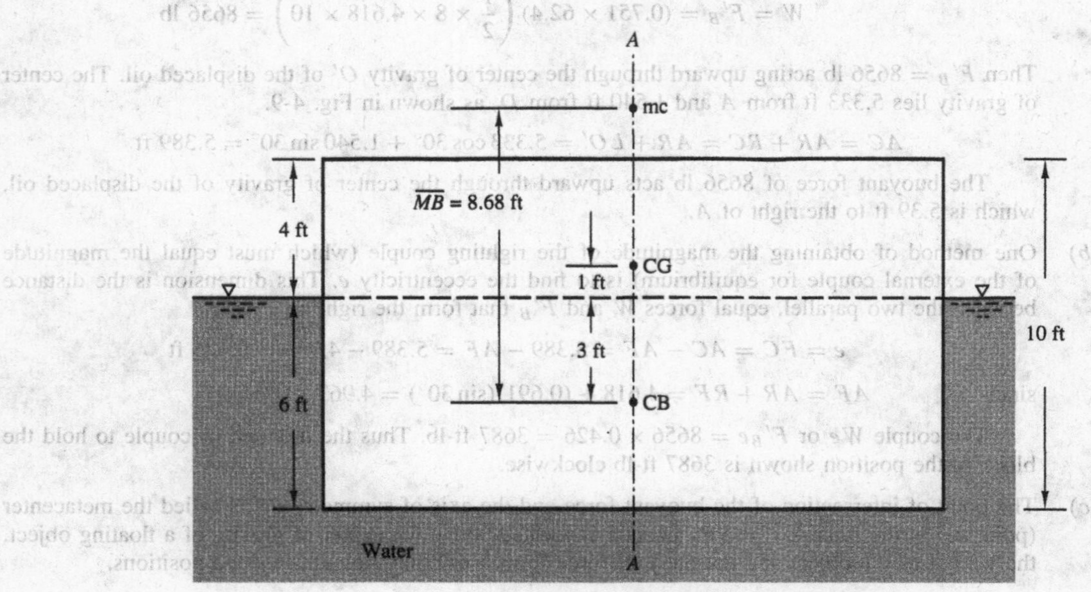

Fig. 4-10 (*b*)

Solution:

$$\overline{MB} = \frac{I}{V_d} = \frac{(42)(25)^3/12}{(25)(42)(6)} = 8.68 \text{ ft}$$

The metacenter is located 8.68 ft above the center of buoyancy, as shown in Fig. 4-10(*b*), and 4.68 ft above the barge's center of gravity. Hence, the barge is stable.

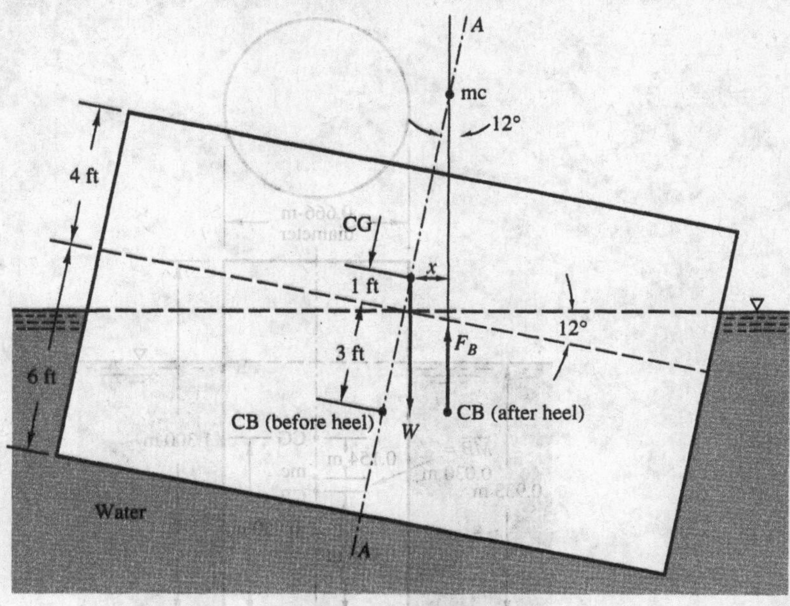

Fig. 4-10 (c)

The end view of the barge when the angle of heel is 12° is shown in Fig. 4-10(c).
Righting moment $= F_B x = [(62.4)(25 \times 42 \times 6)](4.68 \sin 12°) = 383{,}000$ lb-ft.

4.14. Would the solid wood cylinder in Fig. 4-11(a) be stable if placed vertically in oil, as shown in the figure? The specific gravity of the wood is 0.61.

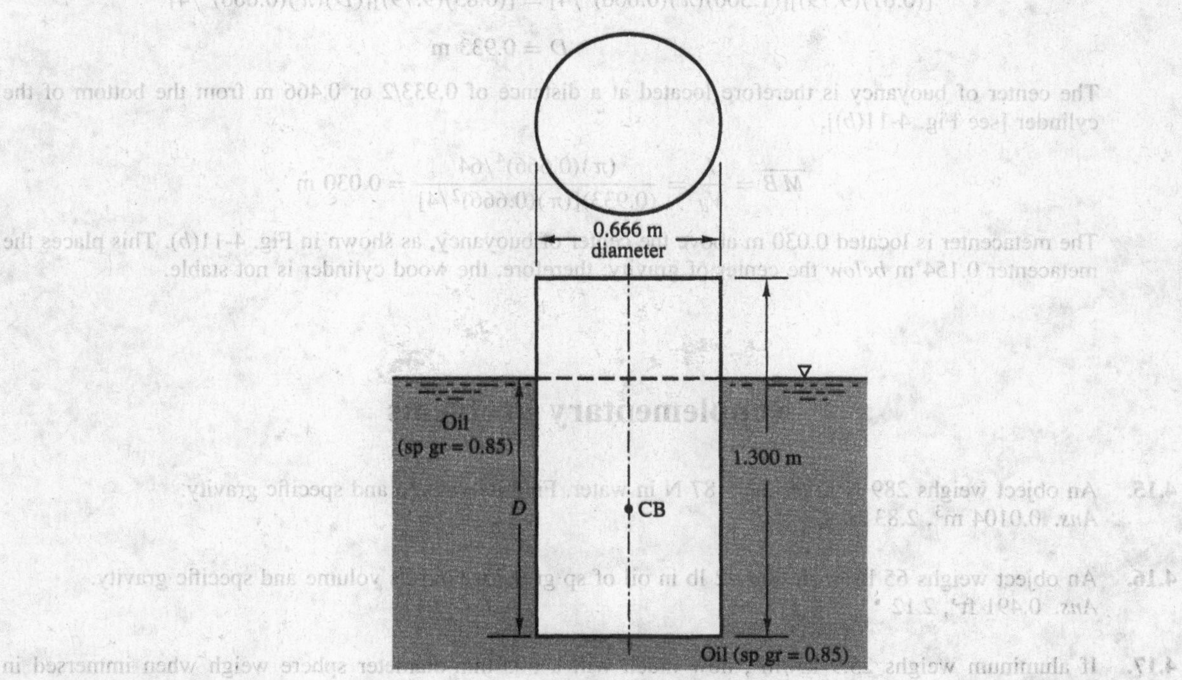

Fig. 4-11 (a)

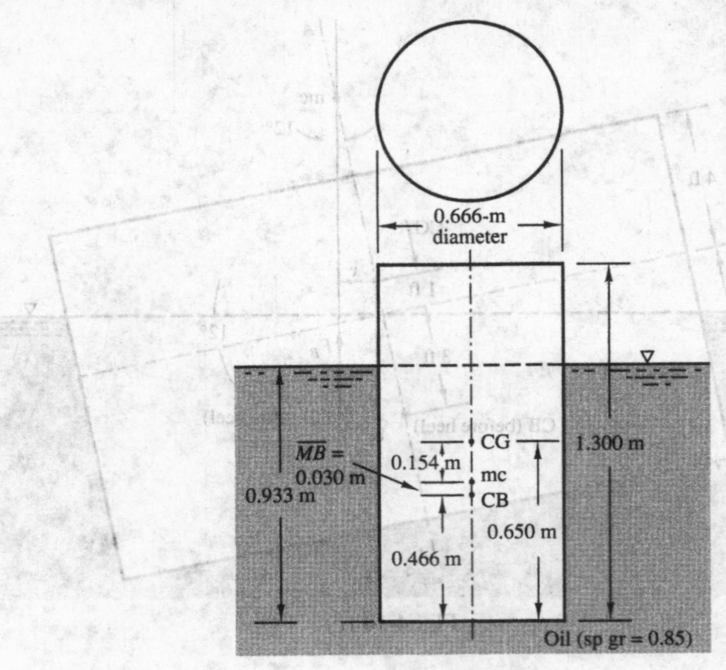

Fig. 4-11 (*b*)

Solution:

First, find the submerged depth of the cylinder [D in Fig. 4-11(*a*)] when placed in the oil.

$$\text{weight of cylinder in air} = \text{buoyant force}$$

$$[(0.61)(9.79)][(1.300)(\pi)(0.666)^2/4] = [(0.85)(9.79)][(D)(\pi)(0.666)^2/4]$$

$$D = 0.933 \text{ m}$$

The center of buoyancy is therefore located at a distance of 0.933/2 or 0.466 m from the bottom of the cylinder [see Fig. 4-11(*b*)].

$$\overline{MB} = \frac{I}{V_d} = \frac{(\pi)(0.666)^4/64}{(0.933)[(\pi)(0.666)^2/4]} = 0.030 \text{ m}$$

The metacenter is located 0.030 m above the center of buoyancy, as shown in Fig. 4-11(*b*). This places the metacenter 0.154 m *below* the center of gravity; therefore, the wood cylinder is not stable.

Supplementary Problems

4.15. An object weighs 289 N in air and 187 N in water. Find its volume and specific gravity.
Ans. 0.0104 m^3, 2.83

4.16. An object weighs 65 lb in air and 42 lb in oil of sp gr 0.75. Find its volume and specific gravity.
Ans. 0.491 ft^3, 2.12

4.17. If aluminum weighs 25.9 kN/m^3, how much will a 305-mm-diameter sphere weigh when immersed in water? When immersed in oil of sp gr 0.75? *Ans.* 238 N, 276 N

4.18. A 6-in cube of aluminum weighs 12.2 lb when immersed in water. What will be its apparent weight when immersed in a liquid of sp gr 1.25? *Ans.* 10.25 lb

4.19. A stone weighs 600 N, and when it was lowered into a square tank 0.610 m on a side, the weight of the stone in water was 323 N. How much did the water rise in the tank? *Ans.* 76 mm

4.20. A hollow cylinder 3 ft in diameter and 5 ft long weighs 860 lb. (*a*) How many pounds of lead weighing 700 lb/ft^3 must be fastened to the outside bottom to make the cylinder float vertically with 3 ft submerged? (*b*) How many pounds if placed inside the cylinder? *Ans.* 510 lb, 465 lb

4.21. A hydrometer weighs 0.0250 lb, and its stem is 0.0250 in^2 in cross-sectional area. What is the difference in depth of flotation for liquids of sp gr 1.25 and 0.90? *Ans.* 8.62 in

4.22. What length of 76.2 mm by 304.8 mm timber, sp gr 0.50, will support a 445-N boy in salt water if he stands on the timber? *Ans.* 3.72 m

4.23. An object that has a volume of 6 ft^3 requires a force of 60 lb to keep it immersed in water. If a force of 36 lb is required to keep it immersed in another liquid, what is the sp gr of that liquid? *Ans.* 0.937

4.24. A cube of steel 0.30 m on each side floats in mercury. Using specific gravities of steel and mercury of 7.8 and 13.6, respectively, find the submerged depth of the cube. *Ans.* 0.172 m

4.25. A barge 10 ft deep has a trapezoidal cross section of 30 ft top width and 20 ft bottom width. The barge is 50 ft long, and its ends are vertical. Determine (*a*) its weight if it draws 6 ft of water and (*b*) the draft if 84.5 tons of stone is placed in the barge. *Ans.* 431,000 lb; 8 ft

4.26. A 1.22-m-diameter sphere floats half-submerged in salt water ($\gamma = 10.05$ kN/m^3). What minimum weight of concrete ($\gamma = 23.56$ kN/m^3) used as an anchor will submerge the sphere completely? *Ans.* 8.34 kN

4.27. An iceberg weighing 57 lb/ft^3 floats in the ocean (64 lb/ft^3) with a volume of 21,000 ft^3 above the surface. What is the total volume of the iceberg? *Ans.* 192,000 ft^3

4.28. A hollow cube 1.0 m on each side weighs 2.4 kN. The cube is tied to a solid concrete block weighing 10.0 kN. Will these two objects tied together float or sink in water? (Show all necessary calculations and explain.) The specific gravity of the concrete is 2.40. *Ans.* float

4.29. An empty balloon and its equipment weigh 100 lb. When inflated with gas weighing 0.0345 lb/ft^3 the balloon is spherical and 20 ft in diameter. What is the maximum weight of cargo that the balloon can lift, assuming that air weighs 0.0765 lb/ft^3? *Ans.* 76 lb

4.30. A cubical float, 1.22 m on a side, weighs 1.78 kN and is anchored by means of a concrete block that weighs 6.67 kN in air. If 229 mm of the float is submerged when the chain connected to the concrete is taut, what rise in water level will lift the concrete off the bottom? Concrete weighs 23.56 kN/m^3. *Ans.* 161 mm

4.31. A rectangular barge with outside dimensions of 20 ft width, 60 ft length, and 10 ft height weighs 350,000 lb. It floats in salt water ($\gamma = 64.0$ lb/ft^3), and the center of gravity of the loaded barge is 4.50 ft from the top. Locate the center of buoyancy (*a*) when floating on an even keel and (*b*) when the barge lists at 10°, and (*c*) locate the metacenter for the 10° list.
Ans. (*a*) 2.28 ft from bottom on centerline, (*b*) 11.28 ft to right, (*c*) 4.17 ft above CG

4.32. A concrete cube 0.5 m on each side is to be held in equilibrium under water by attaching a light foam buoy to it. What is the minimum volume of the foam buoy? The unit weights of the concrete and the foam are 23.58 kN/m^3 and 0.79 kN/m^3, respectively. *Ans.* 0.192 m^3

4.33. A cube 152 mm on a side is made of aluminum and suspended by a string. The cube is submerged, half of it being in oil (sp gr = 0.80) and the other half being in water. Find the tension in the string if aluminum weighs 25.9 kN/m^3. *Ans.* 60.4 N

4.34. If the cube in the preceding problem were half in air and half in oil, what would be the tension in the string? *Ans.* 77.8 N

4.35. Figure 4-12 shows the cross section of a boat, the hull of which is solid. Is the boat stable? If the boat is stable, compute the righting moment in the water when the angle of heel is 10°. *Ans.* stable; 12,480 lb-ft

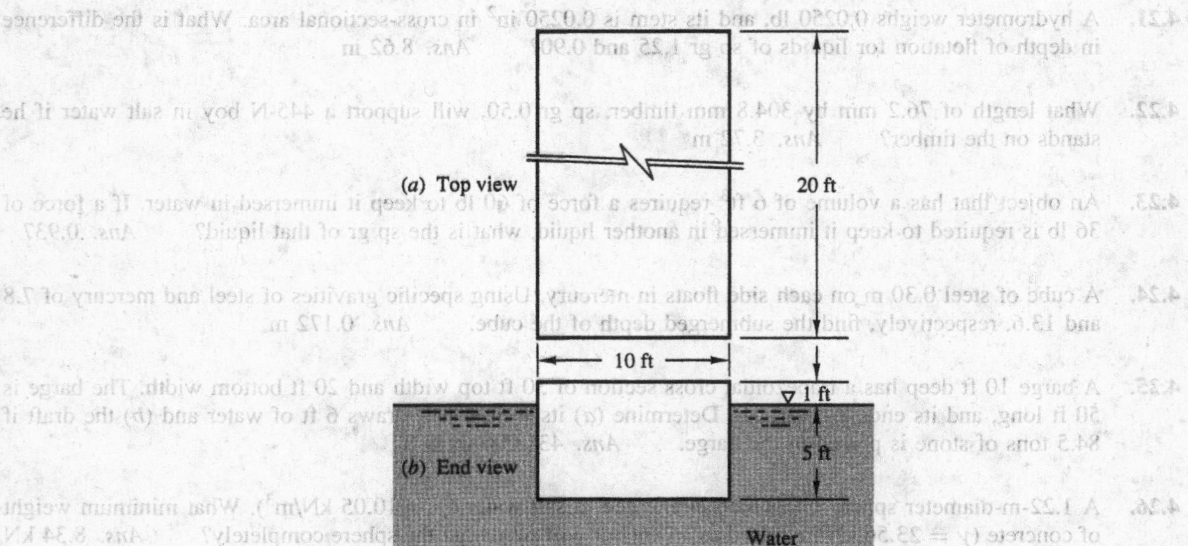

Fig. 4-12

4.36. A solid wood cylinder has a diameter of 2.0 ft and a height of 4.0 ft. The specific gravity of the wood is 0.60. If the cylinder is placed vertically in oil (sp gr = 0.85), will it be stable? *Ans.* no

Chapter 5

Translation and Rotation of Liquid Masses

INTRODUCTION

A fluid may be subjected to translation or rotation at constant accelerations without relative motion between particles. This condition is one of relative equilibrium, and the fluid is free from shear. There is generally no motion between the fluid and the containing vessel. Laws of fluid statics still apply, modified to allow for the effects of acceleration.

HORIZONTAL MOTION

For horizontal motion, the surface of the liquid will become an inclined plane. The slope of the plane will be determined by

$$\tan\theta = \frac{a\,(\text{linear acceleration of vessel, ft/sec}^2 \text{ or m/s}^2)}{g\ (\text{gravitational acceleration, ft/sec}^2 \text{ or m/s}^2)}$$

VERTICAL MOTION

For vertical motion, the pressure (psf or Pa) at any point in the liquid is given by

$$p = \gamma h\left(1 \pm \frac{a}{g}\right)$$

where the positive sign is used with a constant upward acceleration and the negative sign with a constant downward acceleration.

ROTATION OF FLUID MASSES—OPEN VESSELS

The form of the free surface of a liquid in a rotating vessel is that of a paraboloid of revolution. Any vertical plane through the axis of rotation that cuts the fluid will produce a parabola. The equation of the parabola is

$$y = \frac{\omega^2}{2g}x^2$$

where x and y are coordinates, in feet or meters, of any point in the surface measured from the vertex in the axis of revolution and ω is the constant angular velocity in rad/sec. Proof of this equation is given in Problem 5.7.

ROTATION OF FLUID MASSES—CLOSED VESSELS

The pressure in a closed vessel will be increased by rotating the vessel. The pressure increase between a point in the axis of rotation and a point x feet away from the axis is

$$p = \gamma\frac{\omega^2}{2g}x^2$$

and the increase in pressure head (ft or m) is

$$\frac{p}{\gamma} = y = \frac{\omega^2}{2g}x^2$$

The latter equation is similar to the equation for rotating open vessels. Since the linear velocity $V = x_\omega$, the term $x^2\omega^2/2g = V^2/2g$, which we will later recognize as the velocity head, in ft or m.

Solved Problems

5.1. A rectangular tank 20 ft long by 6 ft deep by 7 ft wide contains 3 ft of water. If the linear acceleration horizontally in the direction of the tank's length is 8.05 ft/sec^2, (*a*) compute the total force due to the water acting on each end of the tank and (*b*) show that the difference between these forces equals the unbalanced force necessary to accelerate the liquid mass. Refer to Fig. 5-1.

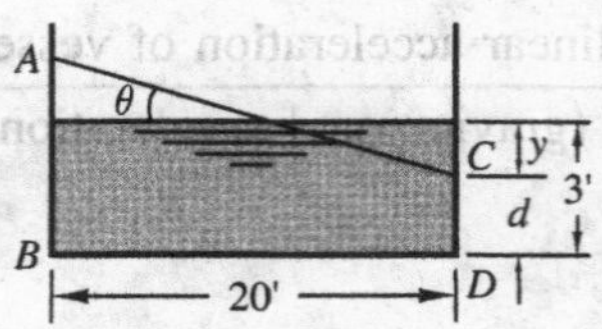

Fig. 5-1

Solution:

(*a*) $$\tan\theta = \frac{\text{linear acceleration}}{\text{gravitational acceleration}} = \frac{8.05}{32.2} = 0.250 \quad \text{and} \quad \theta = 14°02'$$

From the figure, depth d at the shallow end is $d = 3 - y = 3 - 10\tan 14°02' = 0.500$ ft, and the depth at the deep end is 5.50 ft. Then

$$F_{AB} = \gamma h_{cg} A = (62.4)(5.50/2)(5.50 \times 7) = 6607 \text{ lb}$$

$$F_{CD} = \gamma h_{cg} A = (62.4)(0.500/2)(0.500 \times 7) = 54.6 \text{ lb}$$

(*b*) Force needed = mass of water × linear acceleration $= \dfrac{20 \times 7 \times 3 \times 62.4}{32.2} \times 8.05 = 6552$ lb, and $F_{AB} - F_{CD} = 6607 - 55 = 6552$ lb.

5.2. If the tank in Problem 5.1 is filled with water and accelerated in the direction of its length at the rate of 5.00 ft/sec^2, how many gallons of water are spilled? Refer to Fig. 5-2.

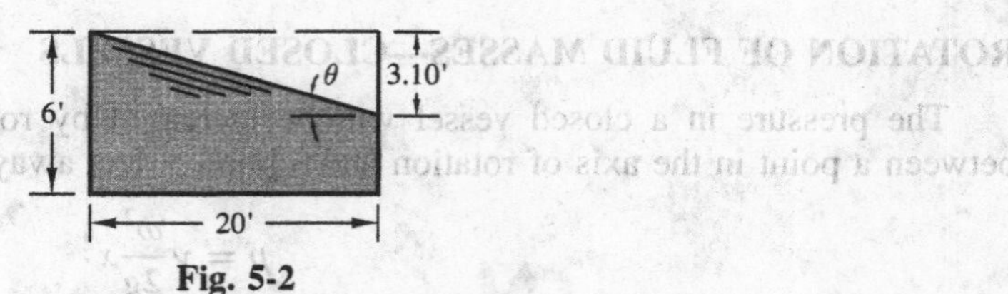

Fig. 5-2

Solution:

Slope of surface $= \tan\theta = 5.00/32.2 = 0.155$, and drop in surface $= 20\tan\theta = 3.10$ ft.

Volume spilled $= 7 \times$ triangular cross section shown in Fig. 5-2

$$= (7)\left(\frac{1}{2} \times 20.0 \times 3.10\right) = 217\ \text{ft}^3 = 217\ \text{ft}^3 \times 7.48\ \text{gal/ft}^3 = 1623\ \text{gal.}$$

5.3. A tank is 1.5 m square and contains 1.0 m of water. How high must its sides be if no water is to be spilled when the acceleration is 4.0 m/s² parallel to a pair of sides?

Solution:

$$\text{slope of surface} = \tan\theta = 4.0/9.81 = 0.408$$

$$\text{rise (or fall) in surface} = 0.75\tan\theta = (0.75)(0.408) = 0.306\ \text{m}$$

The tank must be at least $1 + 0.306 = 1.306$ m deep.

5.4. An open vessel of water accelerates up a 30° plane at 3.66 m/s². What is the angle the water surface makes with the horizontal?

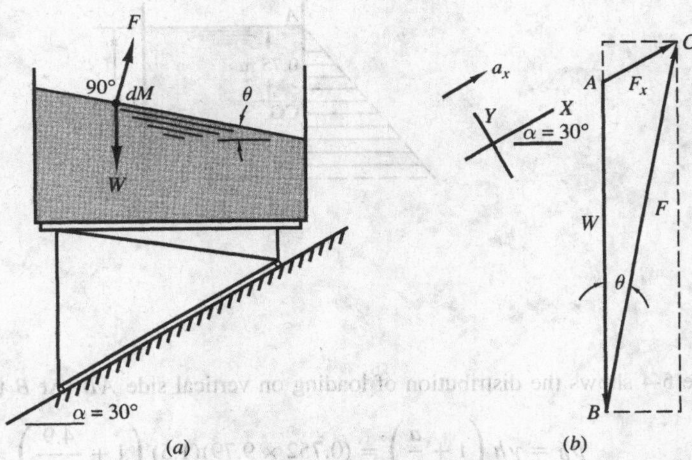

Fig. 5-3

Solution:

Referring to Fig. 5-3, the forces acting on each mass dM are weight W vertically downward and force F exerted by the surrounding particles of liquid. This force F is normal to the liquid surface because no frictional component is acting. The resultant force F_x (due to W and F) for each particle of liquid must be up the plane XX at an angle of $\alpha = 30°$ with the horizontal and must cause the common acceleration a_x. Figure 5-3(*b*) shows this vector relationship. The following equations can now be established:

$$F_x = \frac{W}{g}a_x \quad \text{or} \quad \frac{F_x}{W} = \frac{a_x}{g} \tag{1}$$

$$F_x \sin\alpha = F\cos\theta - W \tag{2}$$

$$F_x \cos\alpha = F\sin\theta \qquad \text{from the vector diagram} \tag{3}$$

Multiplying (2) by $\sin\theta$ and (3) by $\cos\theta$ and solving simultaneously,

$$F_x \sin\alpha \sin\theta + W \sin\theta - F_x \cos\alpha \cos\theta = 0 \quad \text{and} \quad \frac{F_x}{W} = \frac{\sin\theta}{\cos\alpha\cos\theta - \sin\alpha\sin\theta}$$

Substituting in (1) and simplifying,

$$\frac{a_x}{g} = \frac{1}{(\cos\alpha)(\cot\theta) - \sin\alpha} \tag{4}$$

from which, since $\alpha = 30°$,

$$\cot\theta = \tan 30° + \frac{g}{a_x \cos 30°} = 0.577 + \frac{9.81}{3.66 \times 0.866} = 3.67 \quad \text{and} \quad \theta = 15°14' \tag{A}$$

Note: For a horizontal plane, angle α becomes 0° and equation (4) becomes $a/g = \tan\theta$, the equation given for horizontally accelerated motion. For acceleration down the plane, the sign in front of tan 30° becomes minus in equation (A).

5.5. A cubic tank is filled with 1.5 m of oil, sp gr 0.752. Find the force acting on the side of the tank when the acceleration is (a) 4.9 m/s² vertically upward and (b) 4.9 m/s² vertically downward.

Mathcad

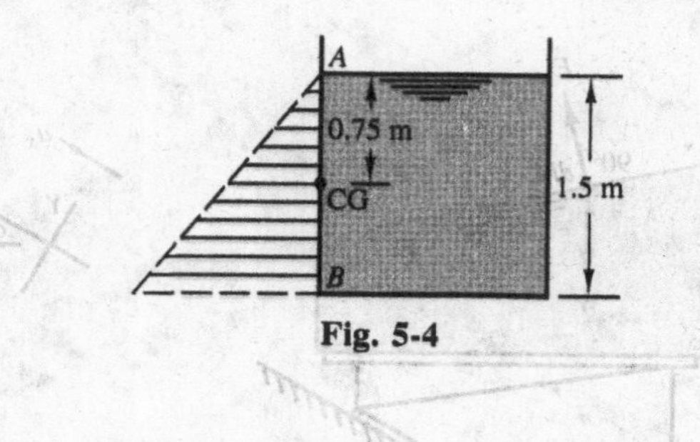

Fig. 5-4

Solution:

(a) Figure 5-4 shows the distribution of loading on vertical side AB. At B the intensity of pressure is

$$p_B = \gamma h\left(1 + \frac{a}{g}\right) = (0.752 \times 9.79)(1.5)\left(1 + \frac{4.9}{9.81}\right) = 16.56 \text{ kPa}$$

$$\text{force } F_{AB} = \text{area of loading diagram} \times 1.5 \text{ m}$$

$$= \left(\frac{1}{2} \times 16.56 \times 1.5\right)(1.5) = 18.63 \text{ kN}$$

Alternative solution:

$$F_{AB} = \gamma h_{cg} A = p_{cg} A = \left[(0.752 \times 9.79)(0.75)\left(1 + \frac{4.9}{9.81}\right)\right](1.5 \times 1.5)$$

$$= 18.63 \text{ kN}$$

(b) $$F_{AB} = \left[(0.752 \times 9.79)(0.75)\left(1 - \frac{4.9}{9.81}\right)\right](1.5 \times 1.5) = 6.22 \text{ kN.}$$

5.6. Determine the pressure at the bottom of the tank in Problem 5.5 when the acceleration is 9.81 m/s^2 vertically downward.

Solution:

$$p_B = (0.752 \times 9.79)(1.5)[1 - (9.81/9.81)] = 0 \text{ kPa}$$

Hence, for a liquid mass falling freely, the pressure within the mass at any point is zero, i.e., that of the surrounding atmosphere. This conclusion is important in considering a stream of water falling through space.

5.7. An open vessel partly filled with a liquid rotates about a vertical axis at constant angular velocity. Determine the equation of the free surface of the liquid after it has acquired the same angular velocity as the vessel.

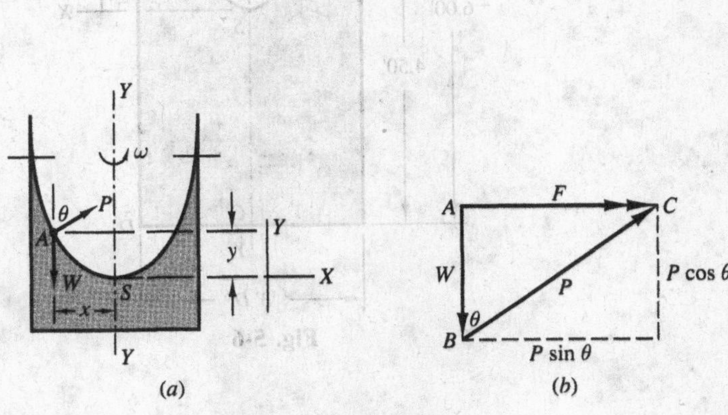

Fig. 5-5

Solution:

Figure 5-5(a) represents a section through the rotating vessel, and any particle A is at a distance x from the axis of rotation. Forces acting on mass A are the weight W vertically downward and P, which is normal to the surface of the liquid since no friction is acting. The acceleration of mass A is $x\omega^2$, directed toward the axis of rotation. The direction of the resultant of forces W and P must be in the direction of this acceleration, as shown in Fig. 5-5(b).

From Newton's second law, $F_x = Ma_x$, or $\quad P\sin\theta = \dfrac{W}{g}x\omega^2 \qquad (1)$

From $\Sigma Y = 0$, $\quad P\cos\theta = W \qquad (2)$

Dividing *(1)* by *(2)*, $\quad \tan\theta = \dfrac{x\omega^2}{g} \qquad (3)$

Now θ is also the angle between the X axis and a tangent drawn to the curve at A in Fig. 5-5(a). The slope of this tangent is $\tan\theta$ or dy/dx. Substituting in *(3)* above,

$$\frac{dy}{dx} = \frac{x\omega^2}{g} \quad \text{from which, by integration,} \quad y = \frac{\omega^2}{2g}x^2 + C_1$$

To evaluate the constant of integration, C_1: When $x = 0$, $y = 0$ and $C_1 = 0$.

5.8.  An open cylindrical tank, 6 ft high and 3 ft in diameter, contains 4.50 ft of water. If the cylinder rotates about its geometric axis, (*a*) what constant angular velocity can be attained without spilling any water? (*b*) What is the pressure at the bottom of the tank at *C* and *D* (Fig. 5-6), when $\omega = 6.00$ rad/sec?

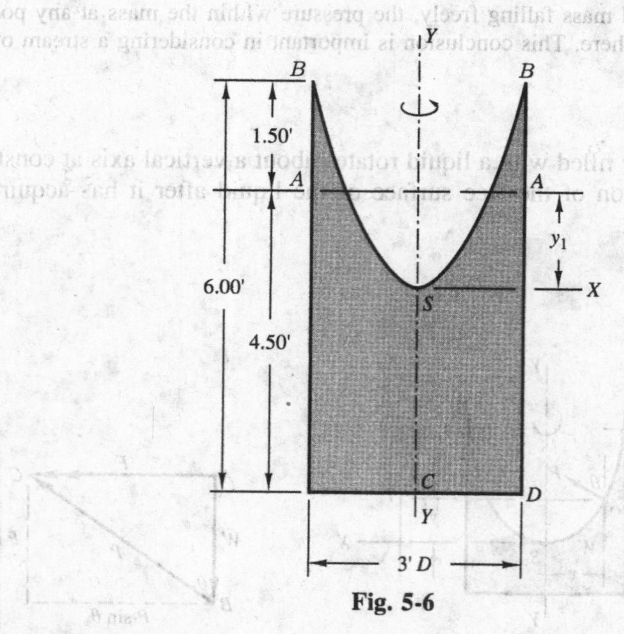

Fig. 5-6

Solution:

(*a*) Volume of paraboloid of revolution $= \frac{1}{2}$(volume circumscribed cylinder) $= \frac{1}{2}\left[\frac{1}{4}\pi 3^2(1.50 + y_1)\right]$.

If no liquid is spilled, this volume equals the volume above the original water level *AA*, or

$$\frac{1}{2}\left[\frac{1}{4}\pi 3^2(1.50 + y_1)\right] = \frac{1}{4}\pi 3^2(1.50)$$

and $y_1 = 1.50$ ft.

To generalize, the point in the axis of rotation drops by an amount equal to the rise of the liquid at the walls of the vessel.

From this information, the *x* and *y* coordinates of points *B* are respectively 1.50 and 3.00 ft from origin *S*. Then

$$y = \frac{\omega^2}{2g}x^2$$

$$3.00 = \frac{\omega^2}{2 \times 32.2}(1.50)^2$$

and $\omega = 9.27$ rad/sec.

(*b*) For $\omega = 6.00$ rad/sec,

$$y = \frac{\omega^2}{2g}x^2 = \frac{(6.00)^2}{(2)(32.2)}(1.50)^2 = 1.26 \text{ ft from } S$$

Origin S drops $\frac{1}{2}y = 0.63$ ft, and S is now $4.50 - 0.63 = 3.87$ ft from the bottom of the tank. At the walls of the tank the depth $= 3.87 + 1.26 = 5.13$ ft (or $4.50 + 0.63 = 5.13$ ft).

At C, $p_C = \gamma h = 62.4 \times 3.87 = 241$ psf

At D, $p_D = \gamma h = 62.4 \times 5.13 = 320$ psf

5.9. Consider the tank in Problem 5.8 closed with the air space subjected to a pressure of 15.5 psi. When the angular velocity is 12.0 rad/sec, what are the pressures in psi at points C and D in the Fig. 5-7?

Mathcad

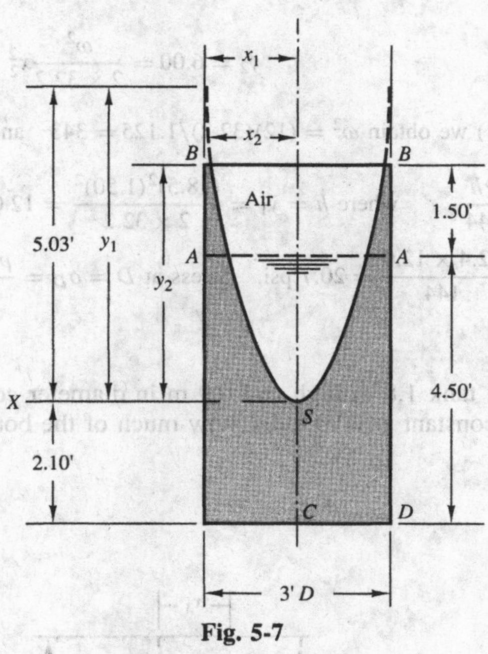

Fig. 5-7

Solution:

Since there is no change in the volume of air within the vessel,

$$\text{volume above level } AA = \text{volume of paraboloid}$$

or
$$\frac{1}{4}\pi 3^2 \times 1.50 = \frac{1}{2}\pi x_2^2 y_2 \qquad (1)$$

Also,
$$y_2 = \frac{(12.0)^2}{(2)(32.2)} x_2^2 \qquad (2)$$

Solving (*1*) and (*2*) simultaneously, $x_2^4 = 3.02$. Then $x_2 = 1.32$ ft and $y_2 = 3.90$ ft.

From the figure, S is located $6.00 - 3.90 = 2.10$ ft above C. Then

$$p_c' = 15.5 + \gamma h/144 = 15.5 + (62.4)(2.10)/144 = 16.4 \text{ psi}$$

To evaluate the pressure at D, pressure head $y_1 = \dfrac{(12.0)^2}{2 \times 32.2}(1.50)^2 = 5.03$ ft above S, and

$$p_{D'} = (62.4)(5.03 + 2.10)/144 + 15.5 = 18.6 \text{ psi}$$

5.10. (*a*) At what speed must the tank in Problem 5.9 be rotated for the center of the bottom to have zero depth of water?

(*b*) If the bottom circumferential plate is $\frac{1}{4}$ in thick, what is the stress in that plate?

Solution:

(*a*) Origin *S* will now be at point *C* in Fig. 5-7.

$$\text{volume above liquid surface} = \text{volume of paraboloid}$$

or
$$\frac{1}{4}\pi 3^2 \times 1.50 = \frac{1}{2}\pi x_2^2(6.00) \qquad (1)$$

Also
$$y_2 = 6.00 = \frac{\omega^2}{2 \times 32.2} x_2^2 \qquad (2)$$

From (*1*) and (*2*) we obtain $\omega^2 = (12)(32.2)/1.125 = 343$ and $\omega = 18.5$ rad/sec.

(*b*) $p'_D = 15.5 + \dfrac{\gamma h}{144}$, where $h = y_1 = \dfrac{(18.5)^2(1.50)^2}{2 \times 32.2} = 12.0$ ft

$= 15.5 + \dfrac{62.4 \times 12.0}{144} = 20.7$ psi. Stress at $D = \sigma_D = \dfrac{p'r}{t} = \dfrac{20.7 \times 18}{\frac{1}{4}} = 1490$ psi.

5.11. A closed cylindrical tank 1.8 m high and 0.9 m in diameter contains 1.4 m of water. When the angular velocity is constant at 20.0 rad/s, how much of the bottom of the tank is uncovered?

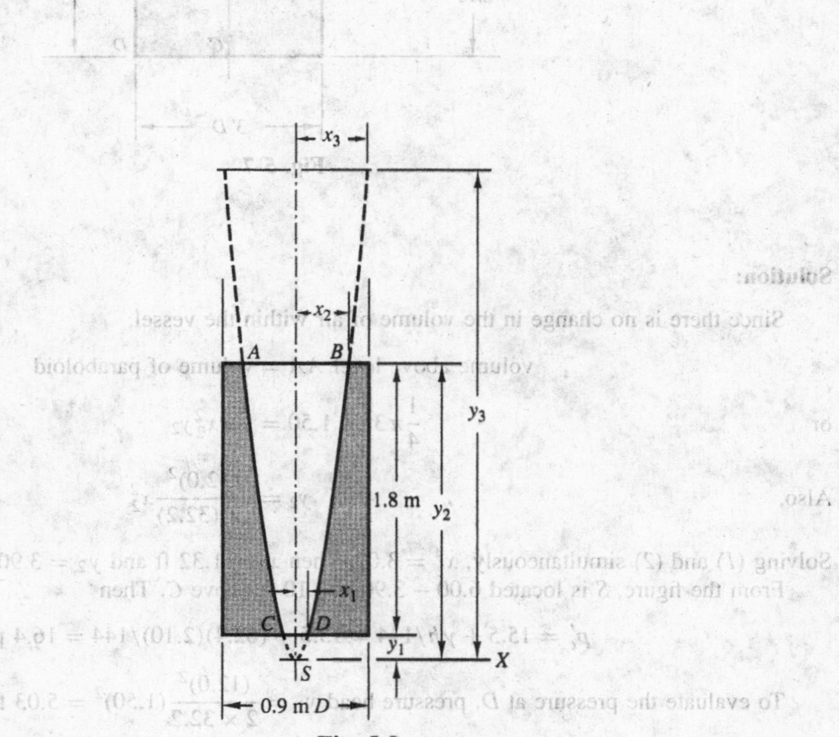

Fig. 5-8

Solution:

In order to estimate the parabolic curve to be drawn in Fig. 5-8, the value of y_3 was calculated first. Now,

$$y_3 = \frac{(20)^2}{2 \times 9.81}(0.45)^2 = 4.13 \text{ m}$$

and the curved surface of the water can now be sketched, showing S below the bottom of the tank. Next,

$$y_1 = \frac{(20)^2}{2 \times 9.81}x_1^2 \qquad (1)$$

$$y_2 = 1.8 + y_1 = \frac{(20)^2}{2 \times 9.81}x_2^2 \qquad (2)$$

and, since the volume of the air is constant,

$$\frac{1}{4}\pi 0.9^2 \times 0.45 = \text{volume (paraboloid } SAB - \text{ paraboloid } SCD)$$

$$= \frac{1}{2}\pi x_2^2 y_2 - \frac{1}{2}\pi x_1^2 y_1 \qquad (3)$$

Substituting values from (1) and (2) and solving,

$$x_1^2 = 0.00649 \quad \text{and} \quad x_1 = 0.0806 \text{ m}$$

Hence the area uncovered $= \pi(0.0806)^2 = 0.0204 \text{ m}^2$.

5.12. A cylinder 6 ft in diameter and 9 ft high is completely filled with glycerin, sp gr 1.60, under a pressure of 35.2 psi at the top. The steel plates that form the cylinder are $\frac{1}{2}$ in thick and can withstand an allowable unit stress of 12,000 psi. What maximum speed in rpm can be imposed on the cylinder?

Solution:

From the specifications for the tank and the hoop tension formula $\sigma = p'r/t$,

$$p'_A = \sigma t/r = (12,000)\left(\frac{1}{2}\right)/36 = 167 \text{ psi}$$

Also, $p'_A = \Sigma$ pressures (35.2 imposed + due to 9 ft glycerin + due to rotation)

or
$$167 = 35.2 + \frac{1.60 \times 62.4 \times 9}{144} + \left(\frac{\omega^2}{2 \times 32.2}\right)(3^2)\left(\frac{1.60 \times 62.4}{144}\right) \text{ psi}$$

Solving, $\omega = 36.0$ rad/sec or 344 rpm.

The pressure conditions are shown graphically in Fig. 5-9, not to scale. Line RST indicates a pressure head of 50.8 ft of glycerin above the top of the tank before rotation. The parabolic pressure curve with vertex at S is caused by the constant angular velocity of 36.0 rad/sec. If the vessel were full but not under pressure, vertex S would coincide with the inside top of the vessel.

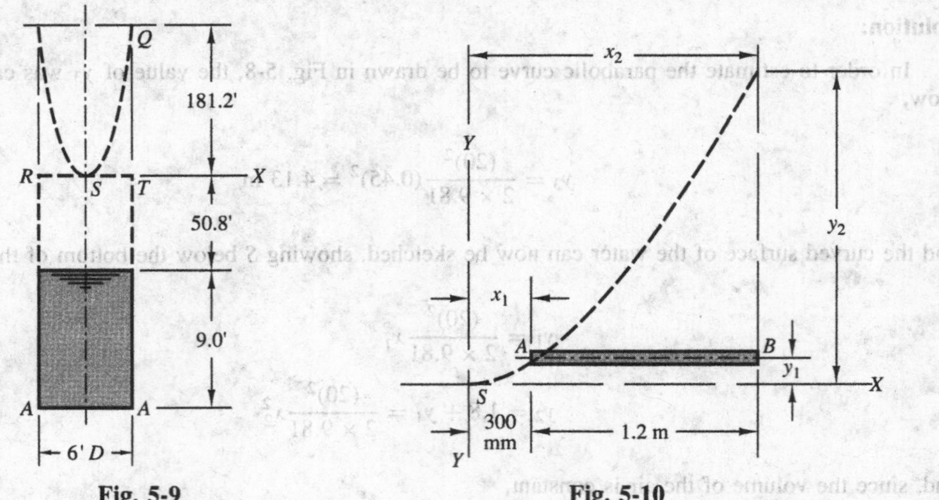

Fig. 5-9 Fig. 5-10

5.13. A 70-mm-diameter pipe, 1.2 m long, is just filled with oil of sp gr 0.822 and then capped. Placed in a horizontal position, it is rotated at 27.5 rad/s about a vertical axis 300 mm from one end. What pressure is developed at the far end of the pipe?

Solution:

As previously noted, the pressure throughout length AB in Fig. 5-10 will be increased by rotation. At some speed of rotation the increased pressure would tend to compress the element of liquid and cause the pressure at A to decrease. Since liquids are practically incompressible, rotation will not lower the pressure at A nor will it increase the pressure at A. Between A and B the pressure will increase as the square of the distance from axis YY.

To evaluate the pressure at B:

$$y_1 = \frac{(27.5)^2}{2g} \times (0.30)^2 = 3.47 \text{ m} \tag{1}$$

$$y_2 = \frac{(27.5)^2}{2g} \times (1.5)^2 = 86.73 \text{ m} \tag{2}$$

and

$$p_B = (0.822)(9.79)(86.73 - 3.47) = 670 \text{ kPa}$$

Supplementary Problems

5.14. A vessel partly filled with water is accelerated horizontally at a constant rate. The inclination of the water surface is 30°. What is the acceleration of the vessel? *Ans.* 5.67 m/s^2

5.15. An open tank is 6 ft square, weighs 770 lb, and contains 3 ft of water. It is acted upon by an unbalanced force of 2330 lb parallel to a pair of sides. What must be the height of the sides of the tank so that no water will be spilled? What is the force acting on the side where the greatest depth occurs? *Ans.* 3.93 ft, 2890 lb

5.16. An open tank 30 ft long by 4 ft wide by 4 ft deep is filled with 3.25 ft of oil, sp gr 0.822. It is accelerated uniformly from rest to 45.0 ft/sec. What is the shortest time in which the tank can be accelerated without spilling any oil? *Ans.* 28.0 sec

5.17. When an open rectangular tank, 1.52 m wide, 3.05 m long, and 1.83 m deep, containing 1.22 m of water is accelerated horizontally parallel to its length at the rate of 4.91 m/s^2, how much water is spilled? *Ans.* 0.71 m^3

5.18. At what acceleration must the tank in Problem 5.17 move for the depth at the forward edge to be zero? *Ans.* 5.88 m/s^2

5.19. An open tank of water accelerates down a 15° inclined plane at 16.1 ft/sec^2. What is the slope of the water surface? *Ans.* 29°01′

5.20. A vessel containing oil of sp gr 0.762 moves vertically upward with an acceleration of +2.5 m/s^2. What is the pressure at a depth of 2 m? *Ans.* 18.7 kPa

5.21. If the acceleration in Problem 5.20 is −2.5 m/s^2, what is the pressure at a depth of 2 m? *Ans.* 11.1 kPa

5.22. An unbalanced vertical force of 60.0 lb upward accelerates a volume of 1.55 ft^3 of water. If the water is 3 ft deep in a cylindrical tank, what is the force acting on the bottom of the tank? *Ans.* 157 lb

5.23. An open cylindrical tank 4 ft in diameter and 6 ft deep is filled with water and rotated about its axis at 60 rpm. How much liquid is spilled, and how deep is the water at the axis? *Ans.* 15.3 ft^3, 3.55 ft

5.24. At what speed should the tank in Problem 5.23 be rotated in order for the center of the bottom of the tank to have zero depth of water? *Ans.* 9.83 rad/sec

5.25. A closed vessel 1 m in diameter is completely filled with water. If the vessel is rotated at 1200 rpm, what increase in pressure will occur at the top of the tank at the circumference? *Ans.* 1970 kPa

5.26. An open vessel 18 in in diameter and filled with water is rotated about its vertical axis at such a velocity that the water surface 4 in from the axis makes an angle of 40° with the horizontal. Compute the speed of rotation. *Ans.* 9.00 rad/sec

5.27. A U-tube with right angle bends is 305 mm wide and contains mercury that rises 229 mm in each leg when the tube is at rest. At what speed must the tube be rotated about an axis 76 mm from one leg so that there will be no mercury in that leg of the tube? *Ans.* 13.9 rad/s

5.28. A 7-ft length of 2-in-diameter pipe is capped and is filled with water under 12.5 psi pressure. Placed in a horizontal position, it is rotated about a vertical axis through one end at the rate of 3 rad/sec. What will be the pressure at the outer end? *Ans.* 15.5 psi

5.29. The 2.0-m-diameter impeller of a closed centrifugal water pump is rotated at 1500 rpm. If the casing is full of water, what pressure head is developed by rotation? *Ans.* 1258 m

Chapter 6

Dimensional Analysis and Hydraulic Similitude

INTRODUCTION

Mathematical theory and experimental data have developed practical solutions to many hydraulic problems. Important hydraulic structures are now designed and built only after extensive model studies have been made. Application of dimensional analysis and hydraulic similitude enables engineers to organize and simplify experiments and analyze the results therefrom.

DIMENSIONAL ANALYSIS

Dimensional analysis is the mathematics of dimensions of quantities and is another useful tool of modern fluid mechanics. In an equation expressing a physical relationship between quantities, absolute numerical and dimensional equality must exist. In general, all such physical relationships can be reduced to the fundamental quantities of force F, length L, and time T (or mass M, length L, and time T). Applications include (1) converting one system of units to another, (2) developing equations, (3) reducing the number of variables required in an experimental program, and (4) establishing principles of model design.

The Buckingham pi theorem is outlined and illustrated in Problems 6.13 through 6.17.

HYDRAULIC MODELS

Hydraulic models, in general, may be either true models or distorted models. True models have all the significant characteristics of the prototype reproduced to scale (geometrically similar) and satisfy design restrictions (kinematic and dynamic similitude). Model-prototype comparisons have clearly shown that the correspondence of behavior is often well beyond expected limitations, as has been attested by the successful operation of many structures designed from model tests.

GEOMETRIC SIMILITUDE

Geometric similitude exists between model and prototype if the ratios of all corresponding dimensions in model and prototype are equal. Such ratios may be written

$$\frac{L_{\text{model}}}{L_{\text{prototype}}} = L_{\text{ratio}} \qquad \text{or} \qquad \frac{L_m}{L_p} = L_r \tag{1}$$

and

$$\frac{A_{\text{model}}}{A_{\text{prototype}}} = \frac{L^2_{\text{model}}}{L^2_{\text{prototype}}} = L^2_{\text{ratio}} = L^2_r \tag{2}$$

KINEMATIC SIMILITUDE

Kinematic similitude exists between model and prototype (1) if the paths of homologous moving particles are geometrically similar and (2) if the ratios of the velocities of homologous particles are equal. A few useful ratios follow.

Velocity: $$\frac{V_m}{V_p} = \frac{L_m/T_m}{L_p/T_p} = \frac{L_m}{L_p} \div \frac{T_m}{T_p} = \frac{L_r}{T_r} \tag{3}$$

Acceleration: $$\frac{a_m}{a_p} = \frac{L_m/T_m^2}{L_p/T_p^2} = \frac{L_m}{L_p} \div \frac{T_m^2}{T_p^2} = \frac{L_r}{T_r^2} \tag{4}$$

Discharge: $$\frac{Q_m}{Q_p} = \frac{L_m^3/T_m}{L_p^3/T_p} = \frac{L_m^3}{L_p^3} \div \frac{T_m}{T_p} = \frac{L_r^3}{T_r} \tag{5}$$

DYNAMIC SIMILITUDE

Dynamic similitude exists between geometrically and kinematically similar systems if the ratios of all homologous forces in model and prototype are the same.

The conditions required for complete similitude are developed from Newton's second law of motion, $\Sigma F_x = M\,a_x$. The forces acting may be any one, or a combination of several, of the following: viscous forces, pressure forces, gravity forces, surface tension forces, and elasticity forces. The following relation between forces acting on model and prototype develops:

$$\frac{\Sigma \text{ forces (viscous} \nrightarrow \text{pressure} \nrightarrow \text{gravity} \nrightarrow \text{surface tension} \nrightarrow \text{elasticity)}_m}{\Sigma \text{ forces (viscous} \nrightarrow \text{pressure} \nrightarrow \text{gravity} \nrightarrow \text{surface tension} \nrightarrow \text{elasticity)}_p} = \frac{M_m a_m}{M_p a_p}$$

THE INERTIA FORCE RATIO is developed into the following form:

$$F_r = \frac{\text{force}_{\text{model}}}{\text{force}_{\text{prototype}}} = \frac{M_m a_m}{M_p a_p} = \frac{\rho_m L_m^3}{\rho_p L_p^3} \times \frac{L_r}{T_r^2} = \rho_r L_r^2 \left(\frac{L_r}{T_r}\right)^2$$

$$F_r = \rho_r L_r^2 V_r^2 = \rho_r A_r V_r^2 \tag{6}$$

This equation expresses the general law of dynamic similarity between model and prototype and is referred to as the *Newtonian equation*.

INERTIA–PRESSURE FORCE RATIO (*Euler number*) gives the relationship (using $T = L/V$)

$$\frac{M\,a}{p\,A} = \frac{\rho L^3 \times L/T^2}{p\,L^2} = \frac{\rho L^4 (V^2/L^2)}{p\,L^2} = \frac{\rho L^2\,V^2}{p\,L^2} = \frac{\rho V^2}{p} \tag{7}$$

INERTIA–VISCOUS FORCE RATIO (*Reynolds number*) is obtained from

$$\frac{M\,a}{\tau A} = \frac{M\,a}{\mu\left(\dfrac{d\,V}{d\,y}\right)A} = \frac{\rho L^2\,V^2}{\mu\left(\dfrac{V}{L}\right)L^2} = \frac{\rho V\,L}{\mu} \tag{8}$$

INERTIA–GRAVITY FORCE RATIO is obtained from

$$\frac{M\,a}{M\,g} = \frac{\rho L^2 V^2}{\rho L^3 g} = \frac{V^2}{L\,g} \tag{9}$$

The square root of this ratio, $\dfrac{V}{\sqrt{L\,g}}$, is known as the *Froude number*.

INERTIA–ELASTICITY FORCE RATIO (*Cauchy number*) is obtained from

$$\frac{M\,a}{E\,A} = \frac{\rho L^2 V^2}{E\,L^2} = \frac{\rho V^2}{E} \tag{10}$$

The square root of this ratio, $\dfrac{V}{\sqrt{E/\rho}}$, is known as the *Mach number*.

INERTIA–SURFACE TENSION RATIO (*Weber number*) is obtained from

$$\frac{M\,a}{\sigma\,L} = \frac{\rho L^2 V^2}{\sigma L} = \frac{\rho L\,V^2}{\sigma} \tag{11}$$

In general, engineers are concerned with the effect of the dominant force. In most fluid flow problems, gravity, viscosity, and/or elasticity govern predominantly, but not necessarily simultaneously. Solutions in this book will cover cases where one predominant force influences the flow pattern with other forces causing negligible or compensating effects. If several forces jointly affect flow conditions, the problem becomes involved and is beyond the scope of this book.

TIME RATIOS

The time ratios established for flow patterns governed essentially by viscosity, gravity, surface tension, and elasticity are, respectively,

$$T_r = \frac{L_r^2}{\nu_r} \tag{12}$$

$$T_r = \sqrt{\frac{L_r}{g_r}} \tag{13}$$

$$T_r = \sqrt{L_r^3 \times \frac{\rho_r}{\sigma_r}} \tag{14}$$

$$T_r = \frac{L_r}{\sqrt{E_r/\rho_r}} \tag{15}$$

Solved Problems

6.1. Express each of the following quantities (*a*) in terms of force F, length L, and time T and (*b*) in terms of mass M, length L, and time T.

Solution:

	Quantity	Symbol	(a) F-L-T	(b) M-L-T
(a)	Area A in ft^2 or m^2	A	L^2	L^2
(b)	Volume v in ft^3 or m^3	v	L^3	L^3
(c)	Velocity V in ft/sec or m/s	V	$L\,T^{-1}$	$L\,T^{-1}$
(d)	Acceleration a or g in ft/sec^2 or m/s^2	a, g	$L\,T^{-2}$	$L\,T^{-2}$
(e)	Angular velocity ω in rad/sec	ω	T^{-1}	T^{-1}
(f)	Force F in lb or N	F	F	$M\,L\,T^{-2}$
(g)	Mass M in slugs or kg	M	$F\,T^2\,L^{-1}$	M
(h)	Specific weight γ in lb/ft^3 or N/m^3	γ	$F\,L^{-3}$	$M\,L^{-2}\,T^{-2}$
(i)	Density ρ in slugs/ft^3 or kg/m^3	ρ	$F\,T^2\,L^{-4}$	$M\,L^{-3}$
(j)	Pressure p in lb/ft^2 or Pa	p	$F\,L^{-2}$	$M\,L^{-1}T^{-2}$
(k)	Absolute viscosity μ in lb-sec/ft^2 or N·s/m^2	μ	$F\,T\,L^{-2}$	$M\,L^{-1}T^{-1}$
(l)	Kinematic viscosity ν in ft^2/sec or m^2/s	ν	$L^2\,T^{-1}$	$L^2\,T^{-1}$
(m)	Modulus of elasticity E in lb/ft^2 or Pa	E	$F\,L^{-2}$	$M\,L^{-1}T^{-2}$
(n)	Power P in ft-lb/sec or N·m/s	P	$F\,L\,T^{-1}$	$M\,L^2\,T^{-3}$
(o)	Torque T in ft-lb or N·m	T	$F\,L$	$M\,L^2\,T^{-2}$
(p)	Rate of flow Q in ft^3/sec or m^3/s	Q	$L^3\,T^{-1}$	L^3T^{-1}
(q)	Shearing stress τ in lb/ft^2 or Pa	τ	$F\,L^{-2}$	$M\,L^{-1}\,T^{-2}$
(r)	Surface tension σ in lb/ft or N/m	σ	$F\,L^{-1}$	$M\,T^{-2}$
(s)	Weight W in lb or N	W	F	$M\,L\,T^{-2}$
(t)	Weight rate of flow W in lb/sec or N/s	W	$F\,T^{-1}$	$M\,L\,T^{-3}$

6.2. **Develop an equation for the distance traveled by a freely falling body in time T, assuming the distance depends upon the weight of the body, the acceleration of gravity, and the time.**

Solution:

$$\text{distance } s = f(W, g, T)$$

or

$$s = K\,W^a\,g^b\,T^c$$

where K is a dimensionless coefficient, generally determined experimentally.

This equation must be dimensionally homogeneous. The exponents of each of the quantities must be the same on each side of the equation. We may write

$$F^0\,L^1\,T^0 = \left(F^a\right)\left(L^b\,T^{-2b}\right)\left(T^c\right)$$

Equating exponents of F, L, and T, respectively, we obtain $0 = a$, $1 = b$, and $0 = -2b + c$, from which $a = 0$, $b = 1$, and $c = 2$. Substituting,

$$s = K\,W^0\,g\,T^2 \qquad \text{or} \qquad s = K\,g\,T^2$$

Note that the exponent of the weight W is zero, signifying that the distance is independent of the weight. Factor K must be determined by physical analysis and/or experimentation.

6.3. **The *Reynolds number* (Re) is a function of density, viscosity, and velocity of a fluid and a characteristic length. Establish the Reynolds number relation by dimensional analysis.**

Solution:

$$\text{Re} = f(\rho, \mu, V, L)$$

or

$$\text{Re} = K\rho^a\,\mu^b\,V^c\,L^d$$

Then, dimensionally, $$F^0\,L^0\,T^0 = \left(F^a\,T^{2a}\,L^{-4a}\right)\left(F^b\,T^b\,L^{-2b}\right)\left(L^cT^{-c}\right)\left(L^d\right)$$

Equating exponents of F, L, and T, respectively, we obtain

$$0 = a + b, \quad 0 = -4a - 2b + c + d, \quad 0 = 2a + b - c$$

from which $a = -b$, $c = -b$, $d = -b$. Substituting, we obtain

$$\text{Re} = K\rho^{-b}\mu^{b}V^{-b}L^{-b} = K\left(\frac{VL\rho}{\mu}\right)^{-b}$$

Values of K and b must be determined by physical analysis and/or experimentation. Here $K = 1$ and $b = -1$.

6.4. For an ideal liquid, express the flow Q through an orifice in terms of the density of the liquid, the diameter of the orifice, and the pressure difference.

Solution:

$$Q = f(\rho, p, d) \qquad \text{or} \qquad Q = K\rho^{a}p^{b}d^{c}$$

Then, dimensionally, $$F^0L^3T^{-1} = \left(F^aT^{2a}L^{-4a}\right)\left(F^bL^{-2b}\right)\left(L^c\right)$$

and $$0 = a + b, \quad 3 = -4a - 2b + c, \quad -1 = 2a$$

from which $a = -\frac{1}{2}$, $b = \frac{1}{2}$, $c = 2$. Substituting,

$$Q = K\rho^{-1/2}p^{1/2}d^2 \qquad \text{or} \qquad \text{ideal } Q = Kd^2\sqrt{p/\rho}$$

Factor K must be obtained by physical analysis and/or experimentation.

For an orifice in the side of a tank under head h, $p = \gamma h$. To obtain the familiar orifice formula in Chapter 12, let $K = \sqrt{2}(\pi/4)$. Then

$$\text{ideal } Q = \sqrt{2}(\pi/4)d^2\sqrt{\gamma h/\rho}$$

But $g = \gamma/\rho$; hence $$\text{ideal } Q = \frac{1}{4}\pi d^2\sqrt{2gh}$$

6.5. Determine the dynamic pressure exerted by a flowing incompressible fluid on an immersed object, assuming the pressure is a function of the density and the velocity.

Solution:

$$p = f(\rho, V) \qquad \text{or} \qquad p = K\rho^{a}V^{b}$$

Then, dimensionally, $$F^1L^{-2}T^0 = \left(F^aT^{2a}L^{-4a}\right)\left(L^bT^{-b}\right)$$

and $1 = a$, $-2 = -4a + b$, $0 = 2a - b$, from which $a = 1$, $b = 2$. Substituting,

$$p = K\rho V^2$$

6.6. Assuming the power delivered to a pump is a function of the specific weight of the fluid, the flow in cfs, and the head delivered, establish an equation by dimensional analysis.

Solution:

$$P = f(\gamma, Q, H)$$

or $$P = K\gamma^{a}Q^{b}H^{c}$$

Then, dimensionally, $$F^1L^1T^{-1} = \left(F^aL^{-3a}\right)\left(L^{3b}T^{-b}\right)\left(L^c\right)$$

and $1 = a$, $1 = -3a + 3b + c$, $-1 = -b$ from which $a = 1$, $b = 1$, $c = 1$. Substituting,

$$P = K\gamma QH$$

6.7. A projectile is fired at an angle θ with initial velocity V. Find the range R in the horizontal plane, assuming the range is a function of V, θ, and g.

Solution:

$$R = f(V, g, \theta) = K\,V^a g^b \theta^c \tag{A}$$

Dimensionally, $$L^1 = \left(L^aT^{-a}\right)\left(L^bT^{-2b}\right) \tag{B}$$

Since θ is dimensionless, it does not appear in (B).

Solving for a and b, $a = 2$ and $b = -1$. Substituting, $R = K V^2/g$. Obviously this equation is unsatisfactory in that it is lacking in some designation of angle θ. Problem 6.8 will show how a solution can be attained.

6.8. Solve Problem 6.7, using a vector-directional notation.

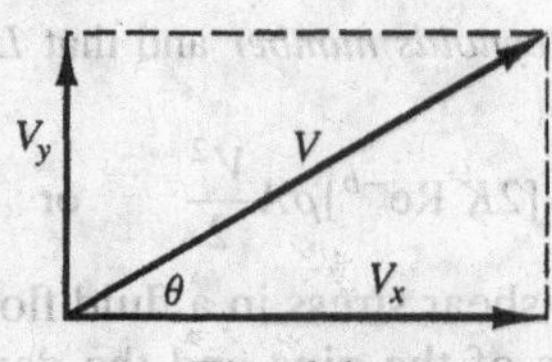

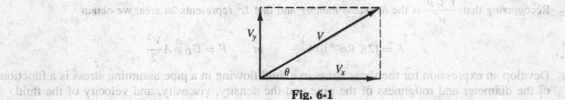

Fig. 6-1

Solution:

In cases of two-dimensional motion, X and Y components can be introduced to provide a more complete analysis. Then line (A) in Problem 6.7 may be written

$$R_x = K\, V_x^a V_y^b g_y^c \theta^d \qquad (C)$$

Dimensionally, $$L_x^1 = (L_x^a T^{-a})(L_y^b T^{-b})(L_y^c T^{-2c})$$

which gives
$$L_x:\quad 1 = a$$
$$T:\quad 0 = -a - b - 2c$$
$$L_y:\quad 0 = b + c$$

Then $a = 1$, $b = 1$, and $c = -1$. Substituting in (C),

$$R = K\left(\frac{V_x V_y}{g}\right) \qquad (D)$$

From the vector diagram (Fig. 6-1), $\cos\theta = V_x/V$, $\sin\theta = V_y/V$, and $\cos\theta\sin\theta = V_x V_y/V^2$. Substituting in ($D$),

$$R = K\frac{V^2\cos\theta\sin\theta}{g} = K\frac{V^2\sin 2\theta}{2g} \qquad (E)$$

From static mechanics, R is usually written $\dfrac{V^2\sin 2\theta}{g}$; hence $K = 2$ in equation (E).

6.9. Assuming the drag force exerted by a flowing fluid on a body is a function of the density, viscosity, and velocity of the fluid and a characteristic length of the body, develop a general equation.

Mathcad

Solution:

$$F = f(\rho, \mu, L, V)$$

or $$F = K\rho^a\mu^b L^c V^d$$

Then $$F^1L^0T^0 = (F^aT^{2a}L^{-4a})(F^bT^bL^{-2b})(L^c)(L^dT^{-d})$$

and $1 = a + b$, $0 = -4a - 2b + c + d$, $0 = 2a + b - d$.

It will be noted that there are more unknown exponents than equations. One method of attack is to express three of the unknowns in terms of the fourth unknown. Solving in terms of b yields

$$a = 1 - b, \quad d = 2 - b, \quad c = 2 - b$$

Substituting, $$F = K\rho^{1-b}\mu^b L^{2-b}V^{2-b}$$

In order to express this equation in the commonly used form, multiply by 2/2 and rearrange the terms as follows:

$$F = 2\,K\,\rho\left(\frac{V\,L\,\rho}{\mu}\right)^{-b} L^2\frac{V^2}{2}$$

Recognizing that $\dfrac{V\,L\,\rho}{\mu}$ is the *Reynolds number* and that L^2 represents an area, we obtain

$$F = [2K\,\mathrm{Re}^{-b}]\rho A\frac{V^2}{2} \qquad \text{or} \qquad F = C_D\,\rho\,A\frac{V^2}{2}$$

6.10. Develop an expression for the shear stress in a fluid flowing in a pipe assuming stress is a function of the diameter and roughness of the pipe and the density, viscosity, and velocity of the fluid.

Solution:

$$\tau = f(V, d, \rho, \mu, K) \qquad \text{or} \qquad \tau = C V^a d^b \rho^c \mu^d K^e$$

Roughness K is usually expressed as a ratio of the size of the surface protuberances to the diameter of the pipe, ϵ/d, a dimensionless number.

Then $$F^1L^{-2}T^0 = (L^aT^{-a})(L^b)(F^cT^{2c}L^{-4c})(F^dT^dL^{-2d})(L^e/L^e)$$

and $1 = c + d$, $-2 = a + b - 4c - 2d + e - e$, $0 = -a + 2c + d$. Solving in terms of d yields

$$c = 1 - d, \qquad a = 2 - d, \qquad b = -d$$

Substituting, $$\tau = C\,V^{2-d}d^{-d}\rho^{1-d}\mu^d K^e$$

Collecting terms, $$\tau = C\left(\frac{V\,d\,\rho}{\mu}\right)^{-d} K^eV^2\rho$$

or $$\tau = (C'\mathrm{Re}^{-d})V^2\rho$$

6.11. Develop the expression for lost head in a horizontal pipe for turbulent incompressible flow.

Solution:

For any fluid, lost head is represented by the drop in the pressure gradient and is a measure of the resistance to flow through the pipe. The resistance is a function of the diameter of the pipe, the viscosity and density of the fluid, the length of the pipe, the velocity of the fluid, and the roughness K of the pipe. We may write

$$(p_1 - p_2) = f(d, \mu, \rho, L, V, K)$$

or $$p_1 - p_2 = C\,d^a\mu^b\rho^cL^dV^e(\epsilon/d)^f \qquad (1)$$

From experiment and observation, the exponent of the length L is unity. The value of K is usually expressed as a ratio of the size of the surface protuberances ϵ to the diameter d of the pipe, a dimensionless number. We may now write

$$F^1L^{-2}T^0 = (L^a)(F^bT^bL^{-2b})(F^cT^{2c}L^{-4c})(L^1)(L^eT^{-e})(L^f/L^f)$$

and $1 = b + c$, $-2 = a - 2b - 4c + 1 + e + f - f$, $0 = b + 2c - e$, from which the values of a, b, and c can be determined in terms of e, or

$$c = e - 1, \quad b = 2 - e, \quad a = e - 3$$

Substituting in (*1*),

$$p_1 - p_2 = C d^{e-3}\mu^{2-e}\rho^{e-1}L^1V^e(\epsilon/d)^f$$

Dividing the left side of the equation by γ and the right side by its equivalent $\rho\,g$,

$$\frac{p_1 - p_2}{\gamma} = \text{lost head} = \frac{C(\epsilon/d)^fL(d^{e-3}V^e\rho^{e-1}\mu^{2-e})}{\rho\,g}$$

which becomes (introducing 2 into numerator and denominator)

$$\text{lost head} = 2C\left(\frac{\epsilon}{d}\right)^f\left(\frac{L}{d}\right)\left(\frac{V^2}{2g}\right)\left(\frac{d^{e-2}V^{e-2}\rho^{e-2}}{\mu^{e-2}}\right)$$

$$= K'(\text{Re}^{e-2})\left(\frac{L}{d}\right)\left(\frac{V^2}{2g}\right) = f\left(\frac{L}{d}\right)\left(\frac{V^2}{2g}\right) \quad \text{(Darcy formula)}$$

6.12. Establish an expression for the power input to a propeller assuming power can be expressed in terms of the mass density of the air, diameter, velocity of the air stream, rotational speed, coefficient of viscosity, and speed of sound.

Solution:

$$\text{power} = K\rho^a d^b V^c \omega^d \mu^e c^f$$

and, using mass, length, and time as fundamental dimensions,

$$ML^2T^{-3} = (M^aL^{-3a})(L^b)(L^cT^{-c})(T^{-d})(M^eL^{-e}T^{-e})(L^fT^{-f})$$

$$1 = a + e \qquad a = 1 - e$$

Then $\quad 2 = -3a + b + c - e + f \quad$ from which $\quad b = 5 - 2e - c - f$

$$-3 = -c - d - e - f \qquad d = 3 - c - e - f$$

Substituting, $\quad \text{power} = K\rho^{1-e}d^{5-2e-c-f}V^c\omega^{3-c-e-f}\mu^e c^f$

Rearranging and collecting terms with like exponents, we obtain

$$\text{power} = K\left[\left(\frac{\rho d^2\omega}{\mu}\right)^{-e}\left(\frac{d\omega}{V}\right)^{-c}\left(\frac{d\omega}{c}\right)^{-f}\right]\omega^3 d^5\rho$$

Examination of terms in parentheses indicates they are all dimensionless. The first term can be written as a *Reynolds number* because linear velocity = radius × angular velocity. The second term is a propeller ratio, and the third term, velocity to celerity, is the Mach number. Combining all these terms, the equation becomes

$$\text{power} = C'\rho\omega^3 d^5$$

6.13. Outline the procedure to be followed when the Buckingham pi theorem is used.

Introduction:

Where the number of physical quantities or variables is four or more, the Buckingham pi theorem provides an excellent tool by which these quantities can be organized into the smallest number of significant dimensionless groupings, from which an equation can be evaluated. The dimensionless groupings are called pi terms. Written in mathematical form, if there are n physical quantities q (such as velocity, density, viscosity, pressure, and area) and k fundamental dimensions (such as force, length, and time or mass, length, and time), then mathematically

$$f_1(q_1, q_2, q_3, \ldots, q_n) = 0$$

This expression can be replaced by the equation

$$\phi(\pi_1, \pi_2, \pi_3, \ldots, \pi_{n-k}) = 0$$

where any one π term depends on not more than $(k+1)$ physical quantities q and all of the π terms are independent, dimensionless, monomial functions of the quantities q.

Procedure:

(1) List the n physical quantities q entering into a particular problem, noting their dimensions and the number k of fundamental dimensions. There will be $(n-k)$ π terms.

(2) Select k of these quantities, none dimensionless and no two having the same dimensions. All fundamental dimensions must be included collectively in the quantities chosen.

(3) The first π term can be expressed as the product of the chosen quantities, each to an unknown exponent, and one other quantity to a known power (usually taken as one).

(4) Retain the quantities chosen in (2) as *repeating* variables, and choose one of the remaining variables to establish the next π term. Repeat this procedure for the successive π terms.

(5) For each π term, solve for the unknown exponents by dimensional analysis.

Helpful Relationships:

(*a*) If a quantity is dimensionless, it is a π term without going through the above procedure.

(*b*) If any two physical quantities have the same dimensions, their ratio will be one of the π terms. For example, L/L is dimensionless and a π term.

(*c*) Any π term may be replaced by any power of that term, including π^{-1}. For example, π_3 may be replaced by π_3^2, or π_2 by $1/\pi_2$.

(*d*) Any π term may be replaced by multiplying it by a numerical constant. For example, π_1 may be replaced by $3\pi_1$.

(*e*) Any π term may be expressed as a function of the other π terms. For example, if there are two π terms, $\pi_1 = \phi(\pi_2)$.

6.14. Solve Problem 6.2, using the Buckingham pi theorem.

Solution:

The problem can be expressed by stating that some function of distance s, weight W, gravitational acceleration g, and time T equals zero or, written mathematically

$$f_1(s, W, g, T) = 0$$

Step 1

List the quantities and the units

$$s = \text{length } L, \quad W = \text{force } F, \quad g = \text{acceleration } L/T^2, \quad T = \text{time } T$$

There are 4 physical quantities and 3 fundamental units, hence (4 − 3) or one π term.

Step 2

Choosing s, W, and T as the physical quantities provides the three fundamental dimensions F, L, and T.

Step 3

Since physical quantities of dissimilar dimensions cannot be added or subtracted, the π term is expressed as a product, as follows:

$$\pi_1 = (s^{x_1})(W^{y_1})(T^{z_1})(g) \tag{1}$$

Using dimensional homogeneity produces

$$F^0L^0T^0 = (L^{x_1})(F^{y_1})(T^{z_1})\left(LT^{-2}\right)$$

Equating exponents of F, L, and T, respectively, we obtain $0 = y_1$, $0 = x_1 + 1$, $0 = z_1 - 2$ from which $x_1 = -1$, $y_1 = 0$, $z_1 = 2$. Substituting in (*1*),

$$\pi_1 = s^{-1}W^0T^2g = \frac{W^0T^2g}{s}$$

Solving for s and noting that $1/\pi_1 = K$, we obtain $s = K\,gT^2$.

6.15. Solve Problem 6.6, using the Buckingham pi theorem.

Solution:

The problem can be written mathematically as

$$f(P, \gamma, Q, H) = 0$$

The physical quantities with their dimensions in F, L, and T units are

power $P = F\,L\,T^{-1}$ flow $Q = L^3T^{-1}$

specific weight $\gamma = F\,L^{-3}$ head $H = L$

There are 4 physical quantities and 3 fundamental units, hence (4 – 3) or one π term.

Choosing Q, γ, and H as the quantities with the unknown exponents, we establish the π term as follows:

$$\pi_1 = (Q^{x_1})(\gamma^{y_1})(H^{z_1})P \tag{1}$$

or

$$\pi_1 = \left(L^{3x_1}T^{-x_1}\right)\left(F^{y_1}L^{-3y_1}\right)\left(L^{z_1}\right)\left(F\,L\,T^{-1}\right)$$

Equating exponents of F, L, and T, respectively, we obtain $0 = y_1 + 1, 0 = 3x_1 - 3y_1 + z_1 + 1$, $0 = -x_1 - 1$, from which $x_1 = -1$, $y_1 = -1$, $z_1 = -1$. Substituting in *(1)*,

$$\pi_1 = Q^{-1}\gamma^{-1}H^{-1}P = \frac{P}{\gamma\,Q\,H} \qquad \text{or} \qquad P = K\,\gamma\,Q\,H$$

6.16. Solve Problem 6.9, using the Buckingham pi theorem.

Solution:

The problem can be expressed as

$$\phi(F, \rho, \mu, L, V) = 0$$

The physical quantities with their dimensions in F, L, and T units are

force $F = F$ length $L = L$

Density $\rho = F\,T^2\,L^{-4}$ velocity $V = LT^{-1}$

absolute viscosity $\mu = F\,T\,L^{-2}$

There are 5 physical quantities and 3 fundamental units, hence (5 – 3) or two π terms.

Choosing length L, velocity V, and density ρ as the three repeating variables with unknown exponents, we establish the π terms as follows:

$$\pi_1 = (L^{a_1})\left(L^{b_1}T^{-b_1}\right)\left(F^{c_1}T^{2c_1}L^{-4c_1}\right)(F) \tag{1}$$

Equating exponents of F, L, and T, respectively, we obtain $0 = c_1 + 1, 0 = a_1 + b_1 - 4c_1, 0 = -b_1 + 2c_1$, from which $c_1 = -1$, $b_1 = -2$, $a_1 = -2$. Substituting in *(1)*, $\pi_1 = F/L^2V^2\rho$.

To evaluate the second π term, retain the first three physical quantities and add another quantity, in this case absolute viscosity μ. [See Problem 6.13, item (4).]

$$\pi_2 = \left(L^{a_2}\right)\left(L^{b_2}T^{-b_2}\right)\left(F^{c_2}T^{2c_2}L^{-4c_2}\right)\left(F\,T\,L^{-2}\right) \tag{2}$$

Equating exponents of F, L, and T, respectively, we obtain $0 = c_2 + 1, 0 = a_2 + b_2 - 4c_2 - 2$, $0 = -b_2 + 2c_2 + 1$, from which $c_2 = -1$, $b_2 = -1$, $a_2 = -1$. Thus $\pi_2 = \mu/L\,V\,\rho$. This expression can be written $\pi_2 = L\,V\,\rho/\mu$, which we recognize as the Reynolds number.

The new relationship, written in terms of π_1, and π_2, is

$$f_1\left(\frac{F}{L^2V^2\rho}, \frac{L\,V\,\rho}{\mu}\right) = 0$$

or $$\text{force } F = (L^2V^2\rho)\, f_2\left(\frac{LV\rho}{\mu}\right)$$

which can be written $$F = (2K\text{Re})\rho L^2 \frac{V^2}{2}$$

Recognizing L^2 as an area, the final equation can be stated as $F = C_D \rho A \frac{V^2}{2}$. (See Chapter 13.)

6.17. Solve Problem 6.11, using the Buckingham pi theorem.

Solution:

This problem can be written mathematically as

$$f(\Delta p, d, \mu, \rho, L, V, K) = 0$$

where K is the relative roughness or the ratio of the size of the surface irregularities ϵ to the diameter of the pipe d. (See Chapter 8.)

The physical quantities with their dimensions in F, L, and T units are

pressure drop $\Delta p = F L^{-2}$	length $L = L$
diameter $d = L$	velocity $V = L T^{-1}$
absolute viscosity $\mu = F T L^{-2}$	relative roughness $K = L_1/L_2$
density $\rho = F T^2 L^{-4}$	

There are 7 physical quantities and 3 fundamental units, hence (7 − 3) or four π terms. Choosing diameter, velocity, and density as the repeating variables with unknown exponents, the π terms are

$$\pi_1 = (L^{x_1})(L^{y_1}T^{-y_1})\left(F^{z_1}T^{2z_1}L^{-4z_1}\right)\left(F L^{-2}\right)$$

$$\pi_2 = (L^{x_2})(L^{y_2}T^{-y_2})\left(F^{z_2}T^{2z_2}L^{-4z_2}\right)\left(F T L^{-2}\right)$$

$$\pi_3 = (L^{x_3})(L^{y_3}T^{-y_3})\left(F^{z_3}T^{2z_3}L^{-4z_3}\right)(L)$$

$$\pi_4 = K = L_1/L_2$$

Evaluating the exponents, term by term, yields

π_1: $0 = z_1 + 1, 0 = x_1 + y_1 - 4z_1 - 2, 0 = -y_1 + 2z_1$; then $x_1 = 0$, $y_1 = -2$, $z_1 = -1$.

π_2: $0 = z_2 + 1, 0 = x_2 + y_2 - 4z_2 - 2, 0 = -y_2 + 2z_2 + 1$; then $x_2 = -1$, $y_2 = -1$, $z_2 = -1$.

π_3: $0 = z_3, 0 = x_3 + y_3 - 4z_3 + 1, 0 = -y_3 + 2z_3$; then $x_3 = -1$, $y_3 = 0$, $z_3 = 0$.

Hence the π terms are

$$\pi_1 = d^0V^{-2}\rho^{-1}\Delta p = \frac{\Delta p}{\rho V^2} \qquad \text{(Euler number)}$$

$$\pi_2 = \frac{\mu}{dV\rho} \text{ or } \frac{dV\rho}{\mu} \qquad \text{(Reynolds number)}$$

$$\pi_3 = d^{-1}V^0\rho^0 L = \frac{L}{d} \qquad \text{[as might be expected; see item (b), Problem 6.13]}$$

$$\pi_4 = \frac{L_1}{L_2} = \frac{\epsilon}{d} \qquad \text{(see Chapter 8)}$$

The new relationship can now be written

$$f_1\left(\frac{\Delta p}{\rho V^2}, \frac{dV\rho}{\mu}, \frac{L}{d}, \frac{\epsilon}{d}\right) = 0$$

Solving for Δp,

$$\Delta p = \frac{\gamma}{g} V^2 f_2\left(\text{Re}, \frac{L}{d}, \frac{\epsilon}{d}\right)$$

where $\rho = \gamma/g$. Hence the pressure head drop would be

$$\frac{\Delta p}{\gamma} = \frac{V^2}{2g}(2) \times f_2\left(\text{Re}, \frac{L}{d}, \frac{\epsilon}{d}\right)$$

If it were desirable to obtain the Darcy-type expression, experiment and analysis indicate that the pressure drop is a function of L/d to the first power; hence

$$\frac{\Delta p}{\gamma} = \left(\frac{V^2}{2g}\right)\left(\frac{L}{d}\right)(2)(f_3)\left(\text{Re}, \frac{\epsilon}{d}\right)$$

which can be expressed as

$$\frac{\Delta p}{\gamma} = (\text{factor } f)\left(\frac{L}{d}\right)\left(\frac{V^2}{2g}\right)$$

Notes :

(1) If the flow were compressible, another physical quantity, bulk modulus E, would be included and the fifth π term would yield the dimensionless ratio $\frac{E}{\rho V^2}$. This is usually rewritten in the form $\frac{V}{\sqrt{E/\rho}}$, which is the Mach number.

(2) If gravity entered into the general flow problem, the gravitational force would be another physical quantity, and a sixth π term would yield the dimensionless ratio $\frac{V^2}{gL}$. This term is recognized as the Froude number.

(3) If the surface tension σ entered into the general flow problem, another physical quantity would be added that would yield a seventh π term. This π term would take the form $\frac{V^2 L\rho}{\sigma}$, which is the Weber number.

6.18. For model and prototype, show that when gravity and inertia are the only influences the ratio of flows Q is equal to the ratio of the length dimension to the 5/2 power.

Solution:

$$\frac{Q_m}{Q_p} = \frac{L_m^3/T_m}{L_p^3/T_p} = \frac{L_r^3}{T_r}$$

The time ratio must be established for the conditions influencing the flow. Expressions can be written for the gravitation and inertia forces, as follows.

Gravity: $$\frac{F_m}{F_p} = \frac{W_m}{W_p} = \frac{\gamma_m}{\gamma_p} \times \frac{L_m^3}{L_p^3} = \gamma_r L_r^3$$

Inertia: $$\frac{F_m}{F_p} = \frac{M_m a_m}{M_p a_p} = \frac{\rho_m}{\rho_p} \times \frac{L_m^3}{L_p^3} \times \frac{L_r}{T_r^2} = \rho_r L_r^3 \times \frac{L_r}{T_r^2}$$

Equating the force ratios,

$$\gamma_r L_r^3 = \rho_r L_r^3 \times \frac{L_r}{T_r^2}$$

which, when solved for the time ratio, yields

$$T_r^2 = L_r \times \frac{\rho_r}{\gamma_r} = \frac{L_r}{g_r} \tag{1}$$

Recognizing that the value of g_r is unity, substitution in the flow ratio expression gives

$$Q_r = \frac{Q_m}{Q_p} = \frac{L_r^3}{L_r^{1/2}} = L_r^{5/2} \qquad (2)$$

6.19. For the conditions laid down in Problem 6.18, establish (*a*) the velocity ratio and (*b*) the pressure ratio and force ratio.

Solution:

(*a*) Dividing both sides of equation (*1*) of Problem 6.18 by L_r^2 gives

$$\frac{T_r^2}{L_r^2} = \frac{L_r}{L_r^2 g_r} \quad \text{or, since} \quad V = \frac{L}{T}, \qquad V_r^2 = L_r g_r$$

But the value of g_r may be considered unity. This means that, for model and prototype, $V_r^2 = L_r$, which may be called the *Froude model law* for velocity ratios.

(*b*) The force ratio for pressure forces $= \dfrac{p_m L_m^2}{P_p L_p^2} = p_r L_r^2$.

The force ratio for inertia forces $= \dfrac{\rho_r L_r^4}{T_r^2} = \gamma_r L_r^3$.

Equating these, we obtain $\qquad p_r L_r^2 = \gamma_r L_r^3$

$$p_r = \gamma_r L_r \qquad (1)$$

For model studies with a free surface, the Froude numbers of model and prototype are the same. The Euler numbers of model and prototype are also the same.

Using $V_r^2 = L_r$ we may write equation (*1*) as

$$p_r = \gamma_r V_r^2$$

and, since force $F = pA$, $\qquad F_r = p_r L_r^2 = \gamma_r L_r^3 \qquad (2)$

6.20. Develop the Reynolds model law for time and velocity ratios for incompressible liquids.

Solution:

For flow patterns subjected to inertia and viscous forces only (other effects negligible), these forces for model and prototype must be evaluated.

For inertia: $\qquad \dfrac{F_m}{F_p} = \rho_r L_r^3 \times \dfrac{L_r}{T_r^2} \qquad$ (from Problem 6.18)

For viscosity: $\qquad \dfrac{F_m}{F_p} = \dfrac{\tau_m A_m}{\tau_p A_p} = \dfrac{\mu_m (dV/dy)_m A_m}{\mu_p (dV/dy)_p A_p} = \dfrac{\mu_m (L_m/T_m \times 1/L_m) L_m^2}{\mu_p (L_p/T_p \times 1/L_p) L_p^2}$

$$= \frac{\mu_m L_m^2/T_m}{\mu_p L_p^2/T_p} = \frac{\mu_r L_r^2}{T_r}$$

Equating the two force ratios, we obtain $\rho_r \dfrac{L_r^4}{T_r^2} = \dfrac{\mu_r L_r^2}{T_r}$, from which $T_r = \dfrac{\rho_r L_r^2}{\mu_r}$

Since $\nu = \dfrac{\mu}{\rho}$, we may write $\qquad T_r = \dfrac{L_r^2}{\nu_r} \qquad (1)$

The velocity ratio is $\qquad V_r = \dfrac{L_r}{T_r} = \dfrac{L_r}{L_r^2}\nu_r = \dfrac{\nu_r}{L_r} \qquad (2)$

Writing these ratio values in terms of model and prototype, we obtain from (2)

$$\frac{V_m}{V_p} = \frac{\nu_m}{\nu_p} \times \frac{L_p}{L_m}$$

Collecting the terms for model and prototype yields $V_m L_m/\nu_m = V_p L_p/\nu_p$, which the reader will recognize as Reynolds number for model = Reynolds number for prototype.

6.21. Oil of kinematic viscosity 4.65×10^{-5} m²/s is to be used in a prototype in which both viscous and gravity forces dominate. A model scale of 1:5 is also desired. What viscosity of model liquid is necessary to make both the Froude number and the Reynolds number the same in model and prototype?

Solution:

Using the scale ratios for velocity for the Froude and Reynolds model laws (see Problems 6.19 and 6.20), we equate

$$(L_r g_r)^{1/2} = \nu_r / L_r$$

Since $g_r = 1$, $L_r^{3/2} = \nu_r$ and $\nu_r = (1/5)^{3/2} = 0.0894$.
This means that $\dfrac{\nu_m}{\nu_p} = 0.0894 = \dfrac{\nu_m}{4.65 \times 10^{-5}}$, and thus $\nu_m = 4.16 \times 10^{-6}$ m²/s.
Using the time, acceleration, and discharge scale ratios will produce the same results. For example, equating the time ratios (Problems 6.18 and 6.20) yields

$$\frac{L_r^{1/2}}{g_r^{1/2}} = \frac{\rho_r L_r^2}{\mu_r} \qquad \text{or, since } g_r = 1, \qquad \frac{\mu_r}{\rho_r} = \nu_r = L_r^{3/2}, \text{ as before.}$$

6.22. Water at 60°F flows at 12.0 ft/sec in a 6″ pipe. At what velocity must medium fuel oil at 90°F flow in a 3″ pipe for the two flows to be dynamically similar?

Solution:

Since the flow pattern in pipes is subject to viscous and inertia forces only, Reynolds number is the criterion for similarity. Other properties of the fluid flowing, such as elasticity and surface tension, as well as the gravity forces, do not affect the flow picture. Thus, for dynamic similarity,

$$\text{Reynolds number for the water} = \text{Reynolds number for the oil}$$

$$\frac{V d}{\nu} = \frac{V' d'}{\nu'}$$

Obtaining values of kinematic viscosity from Table 2 in the Appendix and substituting,

$$\frac{12.0 \times 6/12}{1.217 \times 10^{-5}} = \frac{V' \times 3/12}{3.19 \times 10^{-5}}$$

and $V' = 62.9$ ft/sec for the oil.

6.23. Air at 68°F is to flow through a 24″ pipe at an average velocity of 6.00 ft/sec. For dynamic similarity, what size pipe carrying water at 60°F at 3.65 ft/sec should be used?

Solution:

Equate the two Reynolds numbers: $\dfrac{6.00 \times 2.0}{16.0 \times 10^{-5}} = \dfrac{3.65 \times d}{1.217 \times 10^{-5}}$, $d = 0.250 \text{ ft} = 3.0 \text{ in}$

6.24. A 1:15 model of a submarine is to be tested in a towing tank containing salt water. If the submarine moves at 20 km/h, at what velocity should the model be towed for dynamic similarity?

Solution:

Equate the Reynolds numbers for prototype and model: $\dfrac{20 \times L}{\nu} = \dfrac{V \times L/15}{\nu}$, $V = 300$ km/h

6.25. A 1:80 model of an airplane is tested in 68°F air that has a velocity of 150 ft/sec. (*a*) At what speed should the model be towed when fully submerged in 80°F water? (*b*) What prototype drag in air would a model resistance of 1.25 lb in water represent?

Solution:

(*a*) Equating Reynolds numbers, $\dfrac{150 \times L}{16.0 \times 10^{-5}} = \dfrac{V \times L}{0.930 \times 10^{-5}}$ or $V = 8.72$ ft/sec in water.

(*b*) Since p varies with ρV^2, equating the Euler numbers will produce

$$\frac{\rho_m V_m^2}{p_m} = \frac{\rho_p V_p^2}{p_p} \quad \text{or} \quad \frac{p_m}{p_p} = \frac{\rho_m V_m^2}{\rho_p V_p^2}$$

But forces acting are (pressure × area), or $p L^2$; hence

$$\frac{F_m}{F_p} = \frac{p_m L_m^2}{p_p L_p^2} = \frac{\rho_m V_m^2 L_m^2}{\rho_p V_p^2 L_p^2}$$

or

$$F_r = \rho_r V_r^2 L_r^2 \qquad \text{[equation (6)]}$$

To obtain the velocity of the prototype in air, equate Reynolds numbers. We obtain

$$\frac{V_m L_m}{\nu_{\text{air}}} = \frac{V_p L_p}{\nu_{\text{air}}} \quad \text{or} \quad \frac{150 \times L_p/80}{\nu_{\text{air}}} = \frac{V_p L_p}{\nu_{\text{air}}} \quad \text{and} \quad V_p = 1.875 \text{ ft/sec}$$

Then

$$\frac{1.25}{F_p} = \left(\frac{1.94}{0.00233}\right)\left(\frac{8.72}{1.875}\right)^2\left(\frac{1}{80}\right)^2 \quad \text{and} \quad F_p = 0.444 \text{ lb}$$

6.26. A model of a torpedo is tested in a towing tank at a velocity of 24.4 m/s. The prototype is expected to attain a velocity of 6.1 m/s in 16°C water. (*a*) What model scale has been used? (*b*) What would be the model speed if tested in a wind tunnel under a pressure of 20 atm and at a constant temperature of 27°C?

Solution:

(*a*) Equating the Reynolds numbers for prototype and model, $\dfrac{6.1 \times L}{\nu} = \dfrac{24.4 \times L/x}{\nu}$ or $x = 4$. The model scale is 1:4.

(*b*) For the air, from Table 1B the absolute viscosity is $1.845 \times 10^{-5} N \cdot s/m^2$ and the density $\rho = \dfrac{\gamma}{g} = \dfrac{p}{g\,RT} = \dfrac{20 \times 101,400}{(9.81)(29.3)(273+27)} = 23.5\ \text{kg/m}^3$.

$$\frac{6.1 \times L}{1.13 \times 10^{-6}} = \frac{V \times L/4}{(1.845 \times 10^{-5})/23.5} \quad \text{and} \quad V = 17.0\ \text{m/s}$$

6.27. A centrifugal pump pumps medium lubricating oil at 60°F while rotating at 1200 rpm. A model pump, using 68°F air, is to be tested. If the model's diameter is 3 times the prototype's diameter, at what speed should the model run?

Solution:

Using the peripheral speeds (which equal radius times angular velocity in radians/sec) as the velocities in Reynolds number, we obtain

$$\frac{(d/2)\left(\omega_p\right)(d)}{188 \times 10^{-5}} = \frac{(3d/2)\left(\omega_m\right)(3d)}{16.0 \times 10^{-5}}$$

Hence $\omega_p = 106\omega_m$, and model speed $= 1200/106 = 11.3$ rpm.

6.28. An airplane wing of 3-ft chord is to move at 90 mph in air. A model of 3-in chord is to be tested in a wind tunnel with air velocity at 108 mph. For air temperature of 68°F in each case, what should be the pressure in the wind tunnel?

Solution:

Equate the Reynolds numbers, model and prototype, using velocities in identical units.

$$\frac{V_m L_m}{\nu_m} = \frac{V_p L_p}{\nu_p}, \quad \frac{108 \times 3/12}{\nu_{\text{tunnel}}} = \frac{90 \times 3}{16.0 \times 10^{-5}}, \quad \nu_{\text{tunnel}} = 1.60 \times 10^{-5}\ \text{ft}^2/\text{sec}$$

The pressure that produces this kinematic viscosity of 68°F air can be found by remembering that the absolute viscosity is not affected by pressure changes. The kinematic viscosity equals absolute viscosity divided by density. But density increases with pressure (temperature constant); then

$$\nu = \frac{\mu}{\rho} \quad \text{and} \quad \frac{\nu_m}{\nu_p} = \frac{16.0 \times 10^{-5}}{1.60 \times 10^{-5}} = 10.0$$

Thus the density of the air in the tunnel must be 10 times that of standard (68°F) air, and the resulting pressure in the tunnel must be 10 atmospheres.

6.29. A ship whose hull length is 140 m is to travel at 7.6 m/s. (*a*) Compute the Froude number Fr. (*b*) For dynamic similarity, at what velocity should a 1:30 model be towed through water?

Solution:

(*a*) $$Fr = \frac{V}{\sqrt{g\,L}} = \frac{7.6}{\sqrt{9.81 \times 140}} = 0.205$$

(*b*) When two flow patterns with geometrically similar boundaries are influenced by inertia and gravity forces, the Froude number is the significant ratio in model studies. Then

Froude number of prototype = Froude number of model

or $$\frac{V}{\sqrt{g\,L}} = \frac{V'}{\sqrt{g'L'}}$$

Since $g = g'$ in practically all cases, we may write

$$\frac{V}{\sqrt{L}} = \frac{V'}{\sqrt{L'}}, \qquad \frac{7.6}{\sqrt{140}} = \frac{V'}{\sqrt{140/30}}, \qquad V' = 1.39 \text{ m/s for the model}$$

6.30. A spillway model is to be built to a scale of 1:25 across a flume that is 2 ft wide. The prototype is 37.5 ft high, and the maximum head expected is 5.0 ft. *(a)* What height of model and what head on the model should be used? *(b)* If the flow over the model at 0.20 ft head is 0.70 cfs, what flow per ft of prototype can be expected? *(c)* If the model shows a measured hydraulic jump of 1.0 in, how high is the jump in the prototype? *(d)* If the energy dissipated in the model at the hydraulic jump is 0.15 horsepower, what would be the energy dissipation in the prototype?

Solution:

(a) Since $\dfrac{\text{lengths in model}}{\text{lengths in prototype}} = \dfrac{1}{25}$, height of model $= \dfrac{1}{25} \times 37.5 = 1.5$ ft and

$$\text{head on model} = \frac{1}{25} \times 5.0 = 0.20 \text{ ft.}$$

(b) From Problem 6.18, $Q_r = L_r^{5/2}$, since gravity forces predominate; then

$$Q_p = \frac{Q_m}{L_r^{5/2}} = (0.70)/(1/25)^{5/2} = 2188 \text{ cfs}$$

This quantity can be expected over 2 × 25 or 50 ft of length of prototype. Thus the flow per ft of prototype = 2188/50 = 43.8 cfs.

(c)
$$\frac{h_m}{h_p} = L_r \quad \text{or} \quad h_p = \frac{h_m}{L_r} = \frac{1.0}{1/25} = 25 \text{ in (height of jump)}$$

(d) Power ratio $P_r = (\text{ft-lb/sec})_r = \dfrac{F_r L_r}{T_r} = \dfrac{\gamma_r L_r^3 L_r}{\sqrt{L_r/g_r}}$. But $g_r = 1$ and $\gamma_r = 1$. Then

$$\frac{P_m}{P_p} = L_r^{7/2} = \left(\frac{1}{25}\right)^{7/2} \quad \text{and} \quad P_p = P_m(25)^{7/2} = (0.15)(25)^{7/2} = 11{,}700 \text{ hp}$$

6.31. A model of a reservoir is drained in 4 min by opening a sluice gate. The model scale is 1:225. How long should it take to empty the prototype?

Solution:

Since gravity is the dominant force, the Q ratio, from Problem 6.18, is equal to $L_r^{5/2}$.

Also, $Q_r = \dfrac{Q_m}{Q_p} = \dfrac{L_m^3}{L_p^3} \div \dfrac{T_m}{T_p}$. Then $L_r^{5/2} = L_r^3 \times \dfrac{T_p}{T_m}$ and $T_p = T_m / L_r^{1/2} = (4)(225)^{1/2} = 60$ min.

6.32. A rectangular pier in a river is 1.22 m wide by 3.66 m long, and the average depth of water is 2.74 m. A model is built to a scale of 1:16. A velocity of flow of 0.76 m/s is maintained in the model, and the force acting on the model is 4.0 N. *(a)* What are the values of velocity in and force on the prototype? *(b)* If a standing wave in the model is 0.049 m high, what height of wave should be expected at the nose of the pier? *(c)* What is the coefficient of drag resistance?

Solution:

(a) Since gravity forces predominate, from Problem 6.19 we obtain

$$\frac{V_m}{V_p} = \sqrt{L_r} \quad \text{and} \quad V_p = \frac{0.76}{(1/16)^{1/2}} = 3.04 \text{ m/s}$$

Also, $\dfrac{F_m}{F_p} = \gamma_r L_r^3$ and $F_p = \dfrac{4.0}{(1.0)(1/16)^3} = 16.4 \text{ kN}$

(*b*) Since $\dfrac{V_m}{V_p} = \dfrac{\sqrt{L_m}}{\sqrt{L_p}}$, $\sqrt{h_p} = \sqrt{0.049} \times \dfrac{3.04}{0.76}$ and $h_p = 0.784$ m wave height.

(*c*) Drag force $= C_D \rho A \dfrac{V^2}{2}$, $4.0 = C_D(1000)\left(\dfrac{1.22}{16} \times \dfrac{2.74}{16}\right)\dfrac{(0.76)^2}{2}$, and $C_D = 1.06$.

Had the prototype values been used for this calculation, we would have the following:

$$(16.4)(1000) = C_D(1000)(1.22 \times 2.74)\frac{(3.04)^2}{2} \quad \text{and} \quad C_D = 1.06, \text{ as expected}$$

6.33. The measured resistance in fresh water of an 8-ft ship model moving at 6.50 ft/sec was 9.60 lb. (*a*) What would be the velocity of the 128-ft prototype? (*b*) What force would be required to drive the prototype at this speed in salt water?

Solution:

(*a*) Since gravity forces predominate, we obtain

$$\frac{V_m}{V_p} = \sqrt{L_r} = \sqrt{8/128} \quad \text{and} \quad V_p = \frac{6.50}{(1/16)^{1/2}} = 26.0 \text{ ft/sec}$$

(*b*)
$$\frac{F_m}{F_p} = \gamma_r L_r^3 \quad \text{and} \quad F_p = \frac{9.60}{(62.4/64.0)(1/16)^3} = 40{,}330 \text{ lb}$$

This latter value can be obtained by using the drag formula: Drag force $= C_f \rho \dfrac{A}{2} V^2$.

For model, $9.60 = C_f\left(\dfrac{62.4}{2g}\right)\left(\dfrac{A}{(16)^2}\right)(6.50)^2$ and $\dfrac{C_f A}{2g} = \dfrac{(9.60)(16)^2}{(62.4)(6.50)^2}$ (*1*)

For prototype, force $= C_f \dfrac{64.0}{2g} A(26.0)^2$ and $\dfrac{C_f A}{2g} = \dfrac{\text{force}}{(64.0)(26.0)^2}$ (*2*)

Equating (*1*) and (*2*) since the value of C_f is the same for model and prototype, we obtain

$$\frac{(9.60)(16)^2}{(62.4)(6.50)^2} = \frac{\text{force}}{(64.0)(26.0)^2}, \quad \text{from which force} = 40{,}330 \text{ lb, as before}$$

6.34. (*a*) Evaluate the model scale when both viscous and gravity forces are necessary to secure similitude.

(*b*) What should be the model scale if oil of viscosity 9.29×10^{-5} m²/s is used in the model tests and the prototype liquid has a viscosity of 74.3×10^{-5} m²/s?

(*c*) What would be the velocity and flow ratios for these liquids for the 1:4 model-prototype scale?

Solution:

(*a*) For this situation, both the Reynolds number and the Froude number must be satisfied simultaneously. We shall equate the velocity ratios for each model law. Using information from Problems 6.19 and 6.20,

$$\text{Reynolds number } V_r = \text{Froude number } V_r$$

$$(\nu/L)_r = \sqrt{L_r g_r}$$

Since $g_r = 1$, we obtain $L_r = \nu_r^{2/3}$.

(*b*) Using the above length ratio, $L_r = \left(\dfrac{9.29 \times 10^{-5}}{74.3 \times 10^{-5}}\right)^{2/3} = \dfrac{1}{4}$. Model scale is 1:4.

(*c*) Using the Froude model laws (see Problems 6.18 and 6.19),

$$V_r = \sqrt{L_r g_r} = \sqrt{L_r} = \sqrt{\frac{1}{4}} = \frac{1}{2} \quad \text{and} \quad Q_r = L_r^{5/2} = \left(\frac{1}{4}\right)^{5/2} = \frac{1}{32}$$

Or, using the Reynolds model laws (see Problem 6.20),

$$V_r = \frac{\nu_r}{L_r} = \frac{9.29/74.3}{1/4} = \frac{1}{2} \quad \text{and} \quad Q_r = A_r V_r = L_r^2 \times \frac{\nu_r}{L_r} = L_r \nu_r = \left(\frac{1}{4}\right)\left(\frac{9.29}{74.3}\right) = \frac{1}{32}$$

Supplementary Problems

6.35. Check the expression $\tau = \mu(dV/dy)$ dimensionally.

6.36. Show that the kinetic energy of a body equals $K\,M\,V^2$, using methods of dimensional analysis.

6.37. Using methods of dimensional analysis, prove that centrifugal force equals $K\,M\,V^2/r$.

6.38. A body falls freely for distance s, from rest. Develop an equation for velocity. *Ans.* $V = K\sqrt{s\,g}$

6.39. A body falls freely from rest for time T. Develop an equation for velocity. *Ans.* $V = K\,g\,T$

6.40. Develop an expression for the frequency of a simple pendulum, assuming it is a function of the length and mass of the pendulum and gravitational acceleration. *Ans.* frequency $= K\sqrt{g/L}$

6.41. Assuming that flow Q over a rectangular weir varies directly with length L and is a function of head H and acceleration of gravity g, establish a weir formula. *Ans.* $Q = K\,L\,H^{3/2} g^{1/2}$

6.42. Develop the relationship for distance s traveled by a freely falling body, assuming distance depends upon initial velocity V, time T, and acceleration of gravity g. *Ans.* $s = K\,V\,T\,(g\,T/V)^b$

6.43. Establish the expression for the Froude number if it is a function of velocity V, gravitational acceleration g, and length L. *Ans.* $\mathrm{Fr} = K(V^2/L\,g)^{-c}$

6.44. Establish the expression for the Weber number if it is a function of velocity V, density ρ, length L, and surface tension σ. *Ans.* $\mathrm{We} = K(\rho L V^2/\sigma)^{-d}$

6.45. Establish a dimensionless number if it is a function of gravitational acceleration g, surface tension σ, absolute viscosity μ, and density ρ. *Ans.* number $= K\,(\sigma^3\rho/g\mu^4)^d$

6.46. Assuming a ship's drag force is a function of absolute viscosity μ and density ρ of the fluid, velocity V, gravitational acceleration g, and size (length factor L) of the ship, establish a drag force formula. *Ans.* force $= K(\mathrm{Re}^{-a}\mathrm{Fr}^{-d}\rho V^2 L^2)$

6.47. Solve Problem 6.9 including the compressibility effect by adding the variable *celerity* c, the speed of sound. *Ans.* force $= K'\mathrm{Re}^{-b} Ma^{-e} \rho A\,V^2/2$ (Ma = Mach number; see page 84.)

6.48. Show that for geometrically similar orifices the velocity ratio is essentially the square root of the head ratio.

6.49. Show that the time and velocity ratios when surface tension is the dominant force are

$$T_r = \sqrt{L_r^3 \times \frac{\rho_r}{\sigma_r}} \quad \text{and} \quad V_r = \sqrt{\frac{\sigma_r}{L_r \rho_r}}, \text{ respectively}$$

6.50. Show that the time and velocity ratios when elasticity is the dominant force are

$$T_r = \frac{L_r}{\sqrt{E_r/\rho_r}} \quad \text{and} \quad V_r = \sqrt{\frac{E_r}{\rho_r}}, \text{ respectively}$$

6.51. The model of a spillway is built to a scale of 1:36. If the model velocity and discharge are 0.381 m/s and 0.0708 m^3/s, respectively, what are the corresponding values for the prototype? *Ans.* 2.29 m/s, 550 m^3/s

6.52. At what velocity should tests be run in a wind tunnel on a model of an airplane wing of 152-mm chord in order that the Reynolds number shall be the same as that of the prototype of 0.914-m chord moving at 145 km/h? Air is under atmospheric pressure in the wind tunnel. *Ans.* 869 km/h

6.53. Oil ($\nu = 6.09 \times 10^{-5}$ ft^2/sec) is to flow at 12.0 ft/sec in a 6-in pipe. At what velocity should 60°F water flow in a 12-in pipe in order for the Reynolds numbers to be the same? *Ans.* 1.20 ft/sec

6.54. Gasoline at 16°C flows in a 100-mm pipe at 3.05 m/s. What size pipe should be used to carry 16°C water at 1.52 m/s in order for the Reynolds numbers to be the same? *Ans.* 338 mm

6.55. Water at 60°F flows at 12.0 ft/sec in a 6-in pipe. For dynamic similarity, (*a*) what should be the velocity of medium fuel oil at 90°F in a 12-in pipe? (*b*) What size pipe should be used if the oil has a velocity of 63.0 ft/sec? *Ans.* 15.75 ft/sec, $d = 3.0$ in

6.56. A model is tested in standard (68°F) air at a speed of 90.0 ft/sec. At what speed should it be tested when fully submerged in water at 60°F in a towing tank to obtain similar dynamic conditions? *Ans.* 6.84 ft/sec

6.57. A surface vessel 156 m long is to move at the rate of 6.83 m/s. At what velocity should a geometrically similar model 2.44 m long be tested? *Ans.* 0.853 m/s

6.58. What force would be exerted against a seawall if a 1:36 model 0.914 m long experienced a wave force of 120 N? *Ans.* 171 kN/m

6.59. A submerged object is anchored in fresh water (60°F) that flows at the rate of 8.0 ft./sec. The resistance of a 1:5 model in a wind tunnel at standard conditions is 4.5 lb. What force acts on the prototype under dynamically similar conditions? *Ans.* 21.7 lb

6.60. For flow with viscous and pressure forces dominant, evaluate an expression for velocity ratio and lost head ratio for model and prototype. *Ans.* $V_r = p_r L_r/\mu_r$; $LH_r = V_r \mu_r/\gamma_r L_r$

6.61. Develop the relation for friction factor f if this coefficient depends upon pipe diameter d, average velocity V, fluid density ρ, fluid viscosity μ, and absolute pipe roughness ϵ. Use the Buckingham pi theorem. *Ans.* $f = \phi(\text{Re}, \epsilon/d)$

Chapter 7

Fundamentals of Fluid Flow

INTRODUCTION

Chapters 1 through 5 considered fluids at rest in which fluid weight was the only property of significance. This chapter will outline additional concepts required for the study of fluids in motion. Fluid flow is complex and not always subject to exact mathematical analysis. Unlike solids, the elements of a flowing fluid may move at different velocities and may be subject to different accelerations. Three significant concepts in fluid flow are:

(*a*) the principle of conservation of mass, from which the equation of continuity is developed,

(*b*) the principle of kinetic energy, from which certain flow equations are derived, and

(*c*) the principle of momentum, from which equations evaluating dynamic forces exerted by flowing fluids may be established (see Chapters 13 and 14).

FLUID FLOW

Fluid flow may be steady or unsteady; uniform or nonuniform; laminar or turbulent (Chapter 8); one-dimensional, two-dimensional, or three-dimensional; and rotational or irrotational.

True one-dimensional flow of an incompressible fluid occurs when the direction and magnitude of the velocity at all points are identical. However, one-dimensional flow analysis is acceptable when the single dimension is taken along the central streamline of the flow and when velocities and accelerations normal to the streamline are negligible. In such cases, average values of velocity, pressure, and elevation are considered to represent the flow as a whole, and minor variations can be neglected. For example, flow in curved pipelines is analyzed by means of one-dimensional flow principles despite the fact that the structure has three dimensions and the velocity varies across any cross section normal to the flow.

Two-dimensional flow occurs when fluid particles move in planes or parallel planes and the streamline patterns are identical in each plane.

For an ideal fluid in which no shear stresses occur and hence no torques exist, fluid particles cannot experience rotational motion about their own mass centers. Such ideal flow, which can be represented by a flow net, is called *irrotational flow*.

In Chapter 5, liquid in the rotating tanks illustrates rotational flow where the velocity of each particle varies directly as the distance from the center of rotation.

STEADY FLOW

Steady flow occurs if, at any point, the velocity of successive fluid particles is the same at successive periods of time. Thus, the velocity is constant with respect to time, or $\partial V/\partial t = 0$, but it may vary at different points or with respect to distance. This statement implies that other fluid variables will not vary with time, or $\partial p/\partial t = 0$, $\partial \rho/\partial t = 0$, $\partial Q/\partial t = 0$, etc. Most practical engineering flow problems involve steady flow conditions. For example, pipelines carrying liquids under constant head conditions or orifices flowing under constant heads illustrate steady flow. These flows may be uniform or nonuniform.

The complexities of unsteady flow are beyond the scope of a book on introductory fluid mechanics. Flow is unsteady when conditions at any point in a fluid change with time, or $\partial V/\partial t \neq 0$. Problem 7.7 will develop a general equation for unsteady flow, and Chapter 12 will present a few simple problems in which head and flow vary with time.

UNIFORM FLOW

Uniform flow occurs when the magnitude and direction of the velocity do not change from point to point in the fluid, or $\partial V/\partial s = 0$. This statement implies that other fluid variables do not change with distance, or $\partial y/\partial s = 0$, $\partial \rho/\partial s = 0$, $\partial p/\partial s = 0$, etc. Flow of liquids under pressure through long pipelines of constant diameter is uniform flow whether the flow is steady or unsteady.

Nonuniform flow occurs when velocity, depth, pressure, etc., change from point to point in the fluid flow, or $\partial V/\partial s \neq 0$, etc. (See Chapter 10.)

STREAMLINES

Streamlines are imaginary curves drawn through a fluid to indicate the direction of motion in various sections of the flow of the fluid system. A tangent at any point on the curve represents the instantaneous direction of the velocity of the fluid particles at that point. The average direction of velocity may likewise be represented by tangents to streamlines. Since the velocity vector has a zero component normal to the streamline, it should be apparent that there can be no flow across a streamline at any point.

STREAMTUBES

A streamtube represents elementary portions of a flowing fluid bounded by a group of streamlines that confine the flow. If the streamtube's cross-sectional area is sufficiently small, the velocity of the midpoint of any cross section may be taken as the average velocity for the section as a whole. The streamtube will be used to derive the equation of continuity for steady one-dimensional incompressible flow (Problem 7.1).

EQUATION OF CONTINUITY

The equation of continuity results from the principle of conservation of mass. For steady flow, the mass of fluid passing all sections in a stream of fluid per unit of time is the same. This can be evaluated as

$$\rho_1 A_1 V_1 = \rho_2 A_2 V_2 = \text{constant} \tag{1}$$

or

$$\gamma_1 A_1 V_1 = \gamma_2 A_2 V_2 \quad \text{(in lb/sec or N/s)} \tag{2}$$

For *incompressible* fluids and where $\gamma_1 = \gamma_2$ for all practical purposes, the equation becomes

$$Q = A_1 V_1 = A_2 V_2 = \text{constant} \qquad (\text{in ft}^3\text{/sec or m}^3\text{/s}) \tag{3}$$

where A_1 and V_1 are respectively the cross-sectional area and average velocity of the stream at section 1, with similar terms for section 2. (See Problem 7.1.) Units of flow commonly used are cubic feet per second (cfs), although gallons per minute (gpm) and million gallons per day (mgd) are used in water supply work. Cubic meters per second (m^3/s) or liters per minute (L/min) are also commonly used.

The equation of continuity for steady two-dimensional incompressible flow is

$$A_{n_1} V_1 = A_{n_2} V_2 = A_{n_3} V_3 = \text{constant} \tag{4}$$

where A_n terms represent areas normal to the respective velocity vectors (see Problems 7.10 and 7.11).

The equation of continuity for three-dimensional flow is derived in Problem 7.7 for steady and unsteady flow. The general equation is reduced to steady flow conditions for two- and one-dimensional flows also.

FLOW NETS

Flow nets are drawn to indicate flow patterns in cases of two-dimensional flow, or even three-dimensional flow. The flow net consists of (*a*) a system of streamlines so spaced that rate of flow q is the same between each successive pair of lines, and (*b*) another system of lines normal to the streamlines and so spaced that the distance between normal lines equals the distance between adjacent streamlines. An infinite number of streamlines are required to describe completely the flow under given boundary conditions. However, it is usual practice to use a small number of such streamlines, as long as acceptable accuracy is obtained.

Although the technique of drawing flow nets may be beyond the scope of introductory fluid mechanics, the significance of flow nets is important (see Problems 7.13 and 7.14). After a flow net for a particular boundary configuration has been obtained, it may be used for all other irrotational flows as long as the boundaries are geometrically similar.

ENERGY AND HEAD

Energy is defined as the ability to do work. Work is the result of the application of a force through a distance and is generally defined mathematically as the product of a force and the distance traversed in the direction of application. Both energy and work can therefore be expressed in units of ft-lb or N·m. One N·m is a joule (J).

Moving fluids possess energy. In analyzing fluid flow problems, three forms of energy must be considered: potential, kinetic, and pressure energy. These will be considered separately.

Consider the fluid element within the conduit shown in Fig. 7-1. The element is located a distance z above a reference datum and has a velocity V and pressure p. *Potential energy* refers to the energy possessed by the element of fluid due to its elevation above a reference datum. Potential energy (PE) is determined quantitatively by multiplying the weight (W) of the element by the distance the element is located above the reference datum (z). Therefore,

$$PE = \mathrm{Wz} \tag{5}$$

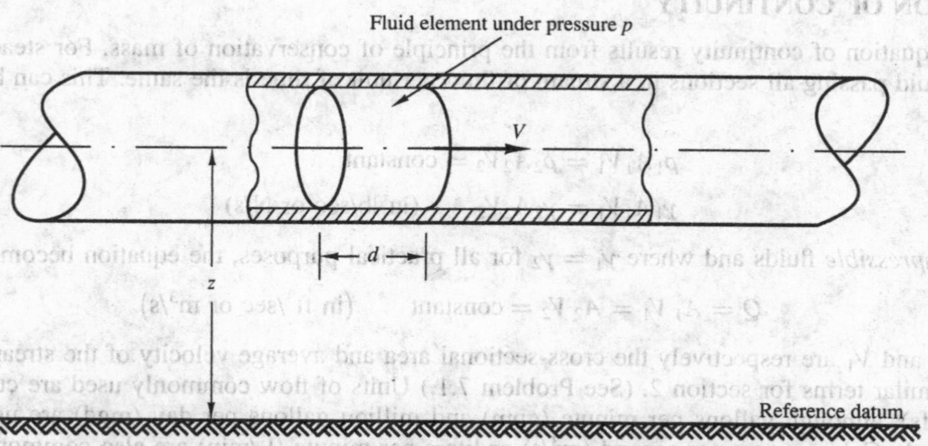

Fig. 7-1

Kinetic energy refers to the energy possessed by the element of fluid due to its velocity. Kinetic energy (KE) is determined quantitatively by multiplying the mass (m) of the element by the square of the velocity (V) and taking half the product. Therefore,

$$\mathrm{KE} = \frac{1}{2}mV^2 \tag{6}$$

The mass term (m) may be replaced by W/g (where W is weight and g is the acceleration of gravity), giving

$$\text{KE} = \left(\frac{1}{2}\right)\left(\frac{WV^2}{g}\right) \tag{7}$$

Pressure energy, sometimes called *flow energy*, is the amount of work required to force the element of fluid across a certain distance against the pressure. The pressure energy (FE) can be evaluated by determining the work done in moving the fluid element a distance equal to the segment's length (d). The force causing work is the product of pressure (p) and cross-sectional area (A) of the element. Hence,

$$\text{FE} = pAd \tag{8}$$

Term Ad is, in fact, the volume of the element, which can itself be replaced by W/γ, where γ is the specific weight of the fluid. Hence,

$$\text{FE} = pW/\gamma \tag{9}$$

Total energy (E) is the sum of PE, KE, and FE, or

$$E = Wz + \left(\frac{1}{2}\right)\left(\frac{WV^2}{g}\right) + \frac{pW}{\gamma} \tag{10}$$

Careful examination of equation (*10*) reveals that each term (and hence total energy) can be expressed in ft-lb of N·m. In fluid mechanics and hydraulics problems, it is convenient to work with energy expressed as a "head"—i.e., the amount of energy per unit weight of fluid. Technically, units for head would be ft-lb/lb of fluid and N·m/N of fluid. Mathematically, these units are ft and m, respectively.

Equation (*10*) can be modified to express total energy as a "head" (H) by dividing each term on the right-hand side of the equation by W, the weight of the fluid. This gives

$$H = z + V^2/2g + p/\gamma \tag{11}$$

The term z is known as the *elevation head*; $V^2/2g$ is known as the *velocity head*; and p/γ is known as the *pressure head*. As indicated previously, each term in equation (*11*) is expressed in length units such as feet or meters.

ENERGY EQUATION

The energy equation results from application of the principle of conservation of energy to fluid flow. The energy possessed by a flowing fluid consists of internal energy and energies due to pressure, velocity, and position. In the direction of flow, the energy principle is summarized by the general equation

$$\begin{matrix}\text{energy at}\\ \text{section 1}\end{matrix} + \begin{matrix}\text{energy}\\ \text{added}\end{matrix} - \begin{matrix}\text{energy}\\ \text{lost}\end{matrix} - \begin{matrix}\text{energy}\\ \text{extracted}\end{matrix} = \begin{matrix}\text{energy at}\\ \text{section 2}\end{matrix}$$

This equation, for steady flow of incompressible fluids in which the change in internal energy is negligible, simplifies to

$$\left(\frac{p_1}{\gamma} + \frac{V_1^2}{2g} + z_1\right) + H_A - H_L - H_E = \left(\frac{p_2}{\gamma} + \frac{V_2^2}{2g} + z_2\right) \tag{12}$$

This equation is known as the *Bernoulli theorem*. Proof of equation (*12*) and its modifications for compressible fluids will be found in Problem 7.21.

The units used are feet or meters of the fluid. Many problems dealing with flow of liquids use this equation as the basis of solution. Flow of gases, in many instances, involves principles of thermodynamics and heat transfer that are beyond the scope of this book.

VELOCITY HEAD

Velocity head represents the kinetic energy per unit weight that exists at a particular point. If the velocity at a cross section were uniform, then the velocity head calculated with this uniform or average velocity would be the true kinetic energy per unit weight of fluid. But, in general, the velocity distribution is not uniform. The true kinetic energy is found by integrating the differential kinetic energies from streamline to streamline (see Problem 7.17). The kinetic energy correction factor α to be applied to the $V_{av}^2/2g$ term is given by the expression

$$\alpha = \frac{1}{A}\int_A \left(\frac{v}{V}\right)^3 dA \tag{13}$$

where V = average velocity in the cross section
v = velocity at any point in the cross section
A = area of the cross section

Studies indicate that $\alpha = 1.0$ for uniform distribution of velocity, $\alpha = 1.02$ to 1.15 for turbulent flows, and $\alpha = 2.00$ for laminar flow. In most fluid mechanics computations, α is taken as 1.0, without serious error being introduced into the result, since the velocity head is generally a small percentage of the total head (energy).

APPLICATION OF THE BERNOULLI THEOREM

Application of the Bernoulli theorem should be rational and systematic. Suggested procedure is as follows.

(1) Draw a sketch of the system, choosing and labeling all cross sections of the stream under consideration.

(2) Apply the Bernoulli equation in the direction of flow. Select a datum plane for each equation written. The low point is logical in that minus signs are avoided and mistakes reduced in number.

(3) Evaluate the energy upstream at section 1. The energy is in units of ft-lb/lb (or N·m/N), which reduce to feet (or meters) of fluid. For liquids the pressure head may be expressed in gage or absolute units, but the same basis must be used for the pressure head at section 2. Gage units are simpler for liquids and will be used throughout this book. Absolute pressure head units must be used where specific weight is not constant. As in the equation of continuity, V_1 is taken as the average velocity at the section, without loss of acceptable accuracy.

(4) Add, in feet (or meters) of the fluid, any energy contributed by mechanical devices, such as pumps.

(5) Subtract, in feet (or meters) of the fluid, any energy lost during flow.

(6) Subtract, in feet (or meters) of the fluid, any energy extracted by mechanical devices, such as turbines.

(7) Equate this summation of energy to the sum of the pressure head, velocity head, and elevation head at section 2.

(8) If the two velocity heads are unknown, relate them to each other by means of the equation of continuity.

ENERGY LINE

The energy line is a graphical representation of the energy at each section. With respect to a chosen datum, the total energy (as a linear value in feet or meters of fluid) can be plotted at each representative section, and the line so obtained is a valuable tool in many flow problems. The energy line will slope (drop) in the direction of flow except where energy is added by mechanical devices.

HYDRAULIC GRADE LINE

The hydraulic grade line lies below the energy line by an amount equal to the velocity head at the section. The two lines are parallel for all sections of equal cross-sectional area. The ordinate between the center of the stream and the hydraulic grade line is the pressure head at the section.

POWER

Power is calculated by multiplying the number of pounds (or newtons) of fluid flowing per second (γQ) by energy H in ft-lb/lb (or m·N/N). There result the equations

$$\text{power } P = \gamma Q H = \text{lb/ft}^3 \times \text{ft}^3\text{/sec} \times \text{ft-lb/lb} = \text{ft-lb/sec}$$

$$\text{horsepower} = \gamma Q H/550$$

or

$$P = \text{N/m}^3 \times \text{m}^3\text{/s} \times \text{N}\cdot\text{m/N} = \text{N}\cdot\text{m/s, or watts (W)}$$

Solved Problems

7.1. Develop the equation of continuity for steady flow of (*a*) a compressible fluid and (*b*) incompressible fluid.

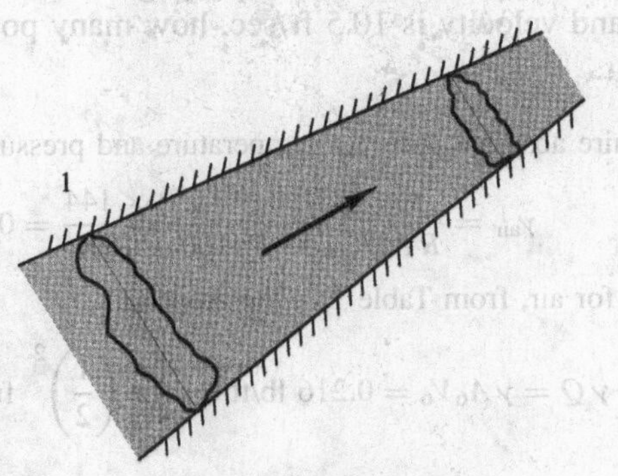

Fig. 7-2

Solution:

(*a*) Consider flow through the streamtube of Fig. 7-2 in which sections 1 and 2 are normal to the streamlines composing the tube. For values of mass density ρ_1 and normal velocity V_1, the mass per unit of time passing section 1 is $\rho_1 V_1 dA_1$, since $V_1 dA_1$ is the volume per unit of time. Similarly, the mass passing section 2 is $\rho_2 V_2 dA_2$. Since for steady flow the mass cannot change with respect to time, and since no flow can pass through the boundaries of a streamtube, the mass flowing through the streamtube is constant. Therefore,

$$\rho_1 V_1 dA_1 = \rho_2 V_2 dA_2 \tag{A}$$

Mass densities ρ_1 and ρ_2 are constant over any cross section dA, and velocities V_1 and V_2 represent the velocities of the streamtube at sections 1 and 2, respectively. Then

$$\rho_1 V_1 \int_{A_1} dA_1 = \rho_2 V_2 \int_{A_2} dA_2$$

Integrating, $\rho_1 V_1 A_1 = \rho_2 V_2 A_2$ or $\gamma_1 V_1 A_1 = \gamma_2 V_2 A_2$ (*B*)

(*b*) For incompressible fluids (and for some cases of compressible flow), mass density is constant, or $\rho_1 = \rho_2$. Therefore,

$$Q = A_1 V_1 = A_2 V_2 = \text{constant} \qquad (C)$$

Thus the discharge is constant along a collection of streamtubes. In many cases of fluid flow, average velocity at a cross section can be used in the equations of continuity (*B*) and (*C*).

7.2. When 0.03 m³/s flows through a 300-mm pipe that later reduces to a 150-mm pipe, calculate the average velocities in the two pipes.

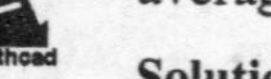

Solution:

$$V_{300} = \frac{Q}{A} = \frac{0.03}{\frac{1}{4}\pi(0.300)^2} = 0.42 \text{ m/s} \quad \text{and} \quad V_{150} = \frac{0.03}{\frac{1}{4}\pi(0.150)^2} = 1.70 \text{ m/s}$$

7.3. If the velocity in a 300-mm pipe is 0.50 m/s, what is the velocity in a 75-mm-diameter jet issuing from a nozzle attached to the pipe?

Solution:

$Q = A_{300}V_{300} = A_{75}V_{75}$, or, since area varies as diameter squared, $(300)^2 V_{300} = (75)^2 V_{75}$. Then $V_{75} = (300/75)^2 V_{300} = 16 \times 0.50 = 8.00$ m/s.

7.4. Air flows in a 6″ pipe at a pressure of 30.0 psi gage and a temperature of 100°F. If barometric pressure is 14.7 psi and velocity is 10.5 ft/sec, how many pounds of air per second are flowing?

Solution:

The gas laws require absolute units for temperature and pressure (psf). Thus

$$\gamma_{\text{air}} = \frac{p}{RT} = \frac{(30.0 + 14.7) \times 144}{(53.3)(100 + 460)} = 0.216 \text{ lb/ft}^3$$

where $R = 53.3$ ft/°R for air, from Table 1 in the Appendix.

$$W \text{ in lb/sec} = \gamma Q = \gamma A_6 V_6 = 0.216 \text{ lb/ft}^3 \times \frac{1}{4}\pi\left(\frac{1}{2}\right)^2 \text{ ft}^2 \times 10.5 \text{ ft/sec} = 0.445 \text{ lb/sec}$$

7.5. Carbon dioxide passes point *A* in a 3″ pipe at a velocity of 15 ft/sec. The pressure at *A* is 30 psi and the temperature is 70°F. At point *B* downstream the pressure is 20 psi and the temperature 90°F. For a barometric reading of 14.7 psi, calculate the velocity at *B* and compare the flows at *A* and *B*. The value of *R* for carbon dioxide is 34.9 ft/°R, from Table 1 in the Appendix.

Solution:

$$\gamma_A = \frac{p_A}{RT} = \frac{44.7 \times 144}{34.9 \times 530} = 0.348 \text{ lb/ft}^3, \qquad \gamma_B = \frac{34.7 \times 144}{34.9 \times 550} = 0.260 \text{ lb/ft}^3$$

(*a*) W in lb/sec $= \gamma_A A_A V_A = \gamma_B A_B V_B$. But since $A_A = A_B$,

$$\gamma_A V_A = \gamma_B V_B = 0.348 \times 15.0 = 0.260 V_B \quad \text{and} \quad V_B = 20.1 \text{ ft/sec}$$

(*b*) The number of pounds per second flowing is constant, but the flows in cfs will differ because the specific weight is not constant.

$$Q_A = A_A V_A = \frac{1}{4}\pi\left(\frac{1}{4}\right)^2 \times 15.0 = 0.736 \text{ cfs}, \quad Q_B = A_B V_B = \frac{1}{4}\pi\left(\frac{1}{4}\right)^2 \times 20.1 = 0.987 \text{ cfs}$$

7.6. What minimum diameter of pipe is necessary to carry 2.22 N/s of air with a maximum velocity of 5.64 m/s? The air is at 30°C and under an absolute pressure of 230 kPa.

Solution:

$$\gamma_{\text{air}} = \frac{p}{RT} = \frac{(230)(1000)}{(29.3)(273+30)} = 25.9 \text{ N/m}^3$$

$$W = 2.22 \text{ N/s} = \gamma Q \quad \text{or} \quad Q = \frac{W}{\gamma} = \frac{2.22}{25.9} = 0.0857 \text{m}^3/\text{s}$$

$$\text{Minimum area } A \text{ needed} = \frac{\text{flow } Q}{\text{average velocity } v} = \frac{0.0857}{5.64} = 0.0152 \text{ m}^2.$$

Hence, minimum diameter = 0.139 m (or 139 mm).

7.7. Develop the general equation of continuity for three-dimensional flow of a compressible fluid for (*a*) unsteady flow, and (*b*) steady flow.

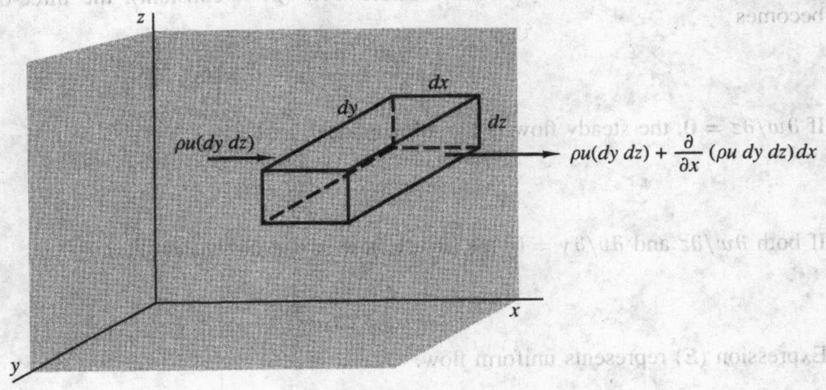

Fig. 7-3

Solution:

(*a*) Let the components of velocity in the x, y, and z directions be u, v, and w, respectively. Consider the flow through the parallelepiped whose dimensions are dx, dy, and dz. The mass of fluid flowing into any face of this volume in unit time is the density of the fluid times the cross-sectional area of the face times the velocity normal to the face, or, in the x direction, $\rho u(dy\,dz)$. In the x direction the approximate flows are (see Fig. 7-3).

$$\text{inflow } \rho u(dy\,dz) \qquad \text{and} \qquad \text{outflow } \rho u(dy\,dz) + \frac{\partial}{\partial x}(\rho u\,dy\,dz)dx$$

or the approximate net inflow is $-\dfrac{\partial}{\partial x}(\rho u\,dy\,dz)dx$ or $-\dfrac{\partial}{\partial x}(\rho u\,dx\,dy\,dz)$.

If we write similar expressions for the net inflow terms in the y and z directions and sum these net inflow values, we obtain

$$-\left[\frac{\partial}{\partial x}\rho u+\frac{\partial}{\partial y}\rho v+\frac{\partial}{\partial z}\rho w\right]dx\,dy\,dz$$

These quantities become more accurate as dx, dy, and dz approach zero.

The positive *rate of change* of mass within the parallelepiped is

$$\frac{\partial}{\partial t}(\rho\,dx\,dy\,dz)\quad\text{or}\quad\frac{\partial\rho}{\partial t}(dx\,dy\,dz)$$

where $\partial\rho/\partial t$ is the rate of change of density within the volume with respect to time. Since the net inflow is identical with the rate of change of mass, we obtain

$$-\left[\frac{\partial}{\partial x}\rho u+\frac{\partial}{\partial y}\rho v+\frac{\partial}{\partial z}\rho w\right]dx\,dy\,dz=\frac{\partial\rho}{\partial t}(dx\,dy\,dz)$$

Thus the continuity equation for three-dimensional unsteady flow of a compressible fluid becomes

$$-\left[\frac{\partial}{\partial x}\rho u+\frac{\partial}{\partial y}\rho v+\frac{\partial}{\partial z}\rho w\right]=\frac{\partial\rho}{\partial t} \tag{A}$$

(*b*) For steady flow, the fluid properties do not change with respect to time, or $\partial\rho/\partial t=0$. The continuity equation for steady compressible flow is

$$\left[\frac{\partial}{\partial x}\rho u+\frac{\partial}{\partial y}\rho v+\frac{\partial}{\partial z}\rho w\right]=0 \tag{B}$$

Furthermore, for steady incompressible flow (ρ = constant), the three-dimensional equation becomes

$$\frac{\partial u}{\partial x}+\frac{\partial v}{\partial y}+\frac{\partial w}{\partial z}=0 \tag{C}$$

If $\partial w/\partial z=0$, the steady flow is two-dimensional, and

$$\frac{\partial u}{\partial x}+\frac{\partial v}{\partial y}=0 \tag{D}$$

If both $\partial w/\partial z$ and $\partial v/\partial y=0$, the steady flow is one-dimensional, and

$$\frac{\partial u}{\partial x}=0 \tag{E}$$

Expression (*E*) represents uniform flow.

7.8. Is the continuity equation for steady incompressible flow satisfied if the following velocity components are involved?

$$u=2x^2-xy+z^2,\qquad v=x^2-4xy+y^2,\qquad w=-2xy-yz+y^2$$

Solution:

Differentiating each component with respect to the appropriate dimension,

$$\partial u/\partial x=4x-y,\quad \partial v/\partial y=-4x+2y,\quad \partial w/\partial z=-y$$

Substituting in equation (*C*) of Problem 7.7, $(4x-y)+(-4x+2y)+(-y)=0$. Satisfied.

7.9. The velocity components for steady incompressible flow are $u=(2x-3y)t$, $v=(x-2y)t$, and $w=0$. Is the equation of continuity satisfied?

Solution:

Differentiating each component with respect to the appropriate dimension,

$$\partial u/\partial x = 2t, \quad \partial v/\partial y = -2t, \quad \partial w/\partial z = 0$$

Substituting in equation (*C*) of Problem 7.7 gives 0. Satisfied.

7.10. For steady incompressible flow, are the following values of u and v possible?

$$(a)\ u = 4xy + y^2,\ v = 6xy + 3x \qquad (b)\ u = 2x^2 + y^2,\ v = -4xy$$

Solution:

For the two-dimensional flows indicated, equation (*D*) of Problem 7.7 must be satisfied.

(*a*) $\partial u/\partial x = 4y,\ \partial v/\partial y = 6x,\ 4y + 6x \neq 0$ Flow is not possible.

(*b*) $\partial u/\partial x = 4x,\ \partial v/\partial y = -4x,\ 4x - 4x = 0$ Flow is possible.

7.11. A fluid flows between two converging plates that are 450 mm wide, and the velocity varies according to the expression

$$\frac{v}{v_{\max}} = 2\frac{n}{n_0}\left(1 - \frac{n}{n_0}\right)$$

For values of $n_0 = 50$ mm and $v_{\max} = 0.3$ m/s, determine (*a*) the total flow in m³/s, (*b*) the mean velocity for the section, and (*c*) the mean velocity for the section where $n = 20$ mm. See Fig. 7-4.

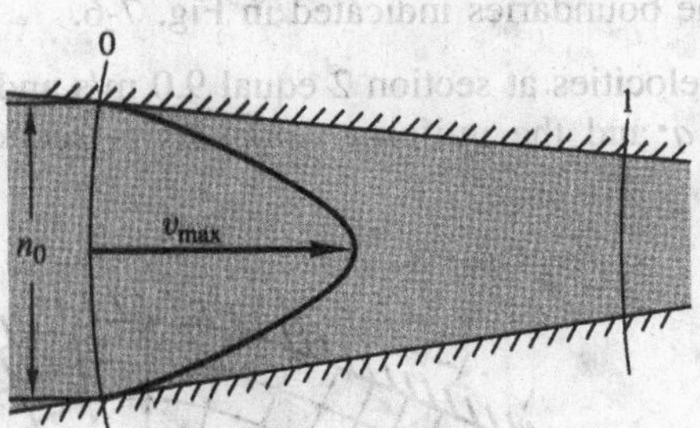

Fig. 7-4

Solution:

(*a*) The flow per unit of width perpendicular to the paper is

$$q = \int_0^{n_0} v\,dn = \frac{2v_{\max}}{n_0}\int_0^{n_0}\left(\frac{n - n^2}{n_0}\right)dn = \frac{1}{3}v_{\max}n_0 = 0.00500 \text{ m}^3\text{/s per m width}$$

and the total flow $Q = (0.00500)(450/1000) = 0.00225$ m³/s.

(*b*) The mean velocity $V_0 = q/n_0 = 0.100$ m/s, where $n_0 = \dfrac{50}{1000}$ m.

(*c*) Using equation (*4*), $V_0 A_{n_0} = V_1 A_{n_1}$, $(0.100)(50/1000)(450/1000) = V_1(20/1000)(450/1000)$, and $V_1 = 0.250$ m/s.

7.12. If magnitudes and directions of velocities are measured in a vertical plane YY at distances Δy apart, show that the flow q per unit of width can be expressed as $\Sigma v_x\,\Delta y$.

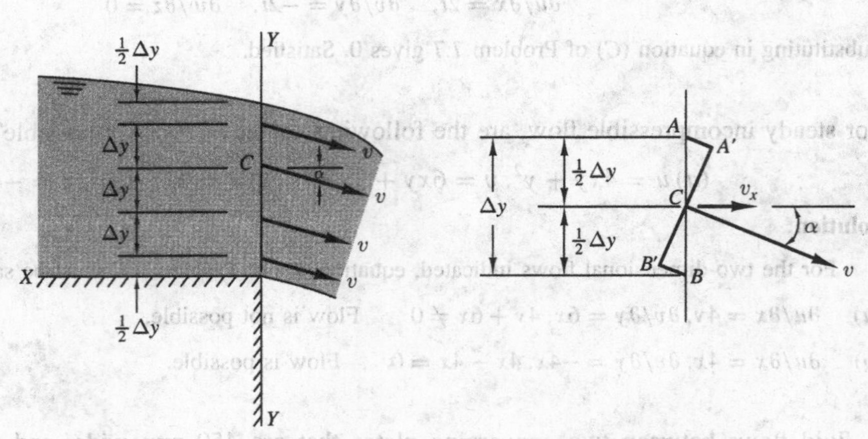

Fig. 7-5

Solution:

Flow per unit width $= q = \Sigma\,\Delta q$, where each Δq can be expressed as $v(\Delta A_n)$.
From Fig. 7-5(*b*), $A'B' = \Delta A_n = \Delta y \cos\alpha$. Then $q = \Sigma\, v(\Delta y \cos\alpha) = \Sigma\, v_x \Delta y$ per unit width.

7.13. (*a*) Outline the procedure for drawing the flow net for two-dimensional steady flow of an ideal fluid between the boundaries indicated in Fig. 7-6.

(*b*) If the uniform velocities at section 2 equal 9.0 m/s and the values of Δn_2 are each 0.03 m, determine flow q and the uniform velocities at section 1, where the values of Δn_1 are 0.09 m.

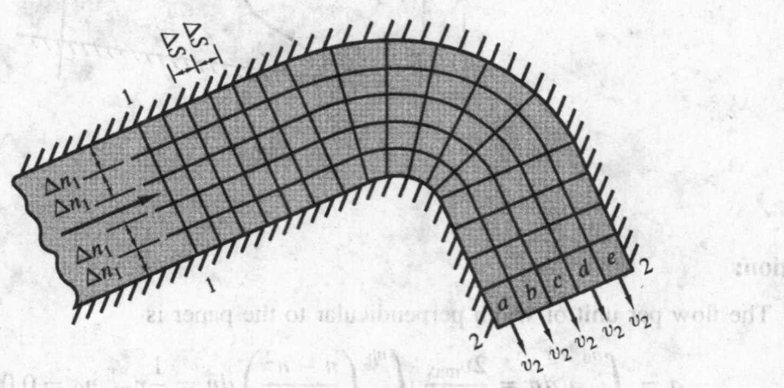

Fig. 7-6

Solution:

(*a*) The procedure of preparing the flow net for this case can be applied to more complex cases. For an *ideal* fluid we proceed as follows:

(1) Divide the flow at a cross section between parallel boundaries into a number of strips of equal width Δn (using a unit thickness perpendicular to the paper). Each strip represents a streamtube bounded by streamlines or by a streamline and a boundary. The flow between the boundaries is equally divided among the tubes, and $\Delta q \cong v(\Delta n) \cong$ constant, where Δn is measured normal to the local velocity. Since $\Delta q \cong v_1 \Delta n_1 \cong v_2 \Delta n_2, v_1/v_2 \cong \Delta n_2/\Delta n_1 \cong \Delta S_2/\Delta S_1$. The smaller the values of Δn and ΔS, the closer the approximation is to an exact relationship. A sufficient number of streamlines should be chosen that the accuracy will be acceptable without unnecessary refinement and detail.

(2) To estimate the *direction* of the streamlines, normal or equipotential lines are drawn. These lines are spaced such that $\Delta S = \Delta n$. The equipotential lines must be perpendicular to the streamlines at each intersection, and also to the boundaries because boundaries are streamlines. The diagram will resemble a group of (approximate) squares throughout the flow net.

(3) At or near changes in boundary shape it is impossible to draw good squares. Changes in initial sketching will be necessary, and a useful check is to draw diagonal lines through all the "squares." These diagonal lines, drawn in both directions, should also form approximate squares.

(4) The boundaries themselves usually represent actual streamlines. If this is not so, the flow net will not represent the true flow pattern. For example, where the flow "separates" from the boundary, the boundary itself in this region cannot be used as a streamline. In general, where divergent flows occur, separation zones may develop.

Mathematical solutions for irrotational flows are based upon the definition of the *stream function*; this definition includes the continuity principle and the properties of a streamline. Flow rate ψ of a streamline is a constant (since no flow can cross the streamline), and if ψ can be established as a function of x and y, the streamline can be plotted. Similarly, the equipotential lines can be defined by $\phi(x, y) =$ a constant. From these expressions we may deduce that

$$u = \partial\psi/\partial y \quad \text{and} \quad v = -\partial\psi/\partial x \qquad \text{for the streamlines}$$

and

$$u = -\partial\phi/\partial x \quad \text{and} \quad v = -\partial\phi/\partial y \qquad \text{for the equipotential lines}$$

These equations must satisfy the Laplace equation, i.e.,

$$\frac{\partial^2\psi}{\partial x^2} + \frac{\partial^2\psi}{\partial y^2} = 0 \quad \text{or} \quad \frac{\partial^2\phi}{\partial x^2} + \frac{\partial^2\phi}{\partial y^2} = 0$$

and the continuity equation

$$\frac{\partial u}{\partial x} + \frac{\partial v}{\partial y} = 0$$

In general, equipotential functions are evaluated and plotted. Then the orthogonal streamlines are sketched to indicate the flow.

These exact solutions are presented in texts on advanced fluid mechanics, hydrodynamics, and complex variables.

(*b*) Flow per unit width $= q = \Sigma\,\Delta q = q_a + q_b + q_c + q_d + q_e = 5v_2 A_{n_2}$.

For 1 unit width, $A_{n_2} = (1)(\Delta n_2)$ and $q = (5)(9.0)(1 \times 0.03) = 1.35\ \text{m}^3/\text{s}$ per unit width.

Then for $\Delta n_1 = 0.09$ m, $5v_1(0.09 \times 1) = 1.35$ or $v_1 = 3.00$ m/s

v_1 may also be found from $v_1/v_2 \cong \Delta n_2/\Delta n_1 : v_1/9.0 \cong 0.03/0.09,\ v_1 = 3.00$ m/s.

7.14. Sketch the streamlines and equipotential lines for the boundary conditions shown in Fig. 7-7. (The area left unmarked is reserved for the reader to use.)

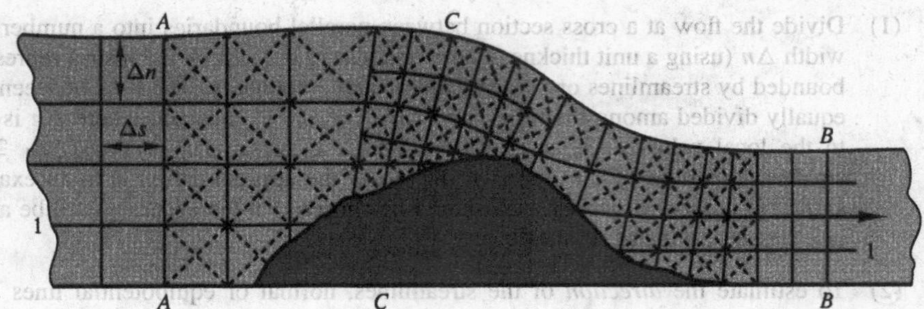

Fig. 7-7

Solution:

(1) Where the flow occurs between parallel boundaries, divide each width into four equal parts or streamtubes (at *AA* and *BB*). Try to picture the path of a particle along one of these streamlines, sketching line 1-1, for example (see Problem 7.13). Proceed in the same way with the other two streamlines.

(2) The equipotential lines must be perpendicular to the streamlines at all points and to the boundaries also. They should be so located as to form approximate squares. Starting at the center section, sketch these orthogonal lines in each direction. Frequent use of the eraser will be necessary before a satisfactory flow net is obtained.

(3) Draw diagonals (dotted) to check on the reasonableness of the flow net. These diagonals should produce squares.

(4) In Fig. 7-7, area *C* is divided into eight streamtubes. It will be seen that the smaller quadrangles are more nearly squares than the larger quadrangles. The more numerous the streamtubes, the more satisfactory will be the "squares" obtained in the flow net.

7.15. Figure 7-8 depicts a streamline for two-dimensional flow and its associated equipotential lines 1 to 10, each normal to the streamline. The spacing between the equipotential lines is given in the second column of the table on page 115. If the average velocity between 1 and 2 is 0.500 ft/sec, calculate (*a*) the average velocities of the streamline between each pair of equipotential lines and (*b*) the time it will take for a particle to move from 1 to 10 along the streamline.

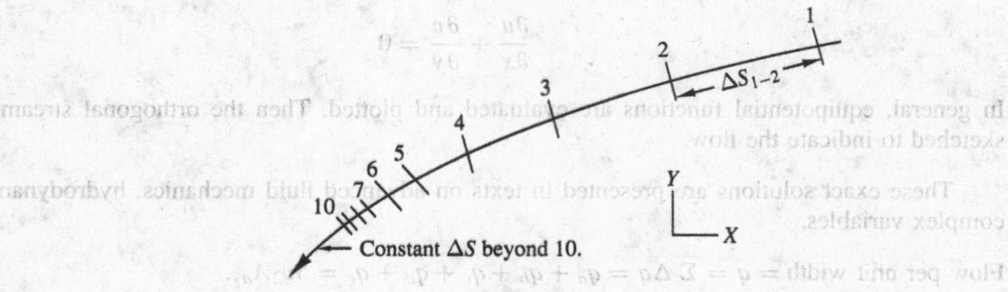

Fig. 7-8

Solution:

(*a*) Referring to the relation between velocity and Δn in Problem 7.13,

$$V_{1-2}\Delta n_{1-2} = V_{2-3}\Delta n_{2-3} = V_{3-4}\Delta n_{3-4} = \cdots$$

Also $$\Delta S_{1-2} \cong \Delta n_{1-2}, \qquad \Delta S_{2-3} \cong \Delta n_{2-3}, \ldots$$

Thus $V_{2-3} \cong V_{1-2}(\Delta S_{1-2}/\Delta S_{2-3}) = (0.500)(0.500/0.400) = 0.625$ ft/sec. Similarly, $V_{3-4} = (0.500)(0.500/0.300) = 0.833$ ft/sec, etc. The values of the average velocities are tabulated below.

Position	ΔS (ft)	$\Delta S_{1-2}/\Delta S$	$V = (0.500)(0.500/\Delta S)$ (ft/sec)	$t = (\Delta S)/V$ (sec)
1-2	0.500	1.000	0.500	1.000
2-3	0.400	1.250	0.625	0.640
3-4	0.300	1.667	0.833	0.360
4-5	0.200	2.500	1.250	0.160
5-6	0.100	5.000	2.500	0.040
6-7	0.0700	7.143	3.571	0.020
7-8	0.0450	11.11	5.56	0.008
8-9	0.0300	16.67	8.33	0.004
9-10	0.0208	24.04	12.02	0.002
				$\Sigma = 2.234$ sec

(*b*) The time to travel from 1 to 2 equals distance 1-2 divided by the average velocity from 1 to 2, or $t_{1-2} = 0.500/0.500 = 1.000$ sec. Similarly, $t_{2-3} = 0.400/0.625 = 0.640$ sec. The total time to travel from 1 to 10 is the sum of the last column, 2.234 sec.

7.16. A gas flows through a square conduit. At one point along the conduit, the conduit sides are 0.100 m, the velocity is 7.55 m/s, and the gas's mass density is (for its particular pressure and temperature) 1.09 kg/m³. At a second point, the conduit sides are 0.250 m and the velocity is 2.02 m/s. Find the mass flow rate of the gas and the gas's mass density at the second point.

Solution:

$$M = \rho_1 A_1 V_1 = (1.09)[(0.100)(0.100)](7.55) = 0.0823 \text{ kg/s}$$

$$\rho_1 A_1 V_1 = \rho_2 A_2 V_2$$

$$0.0823 = (\rho_2)[(0.250)(0.250)](2.02)$$

$$\rho_2 = 0.652 \text{ kg/m}^3$$

7.17. Derive the expression for the kinetic energy correction factor α for steady incompressible flow.

Solution:

The true kinetic energy of a particle is $\frac{1}{2}(dM)v^2$, and so the total kinetic energy of a flowing fluid is

$$\frac{1}{2}\int_A (dM)v^2 = \frac{1}{2}\int_A \frac{\gamma}{g}(dQ)v^2 = \frac{\gamma}{2g}\int_A (v\,dA)v^2$$

To evaluate this expression it must be integrated over area A.

The kinetic energy calculated by means of the average velocity at a cross section is $\frac{1}{2}(\gamma Q/g)V_{av}^2 = \frac{1}{2}(\gamma A/g)V_{av}^3$. Applying a correction factor α to this expression and equating it to the true kinetic energy, we obtain

$$\alpha\left(\frac{\gamma A}{2g}\right)(V_{av}^3) = \frac{\gamma}{2g}\int_A (v\,dA)v^2 \quad \text{or} \quad \alpha = \frac{1}{A}\int_A \left(\frac{v}{V_{av}}\right)^3 dA$$

7.18. A liquid flows through a circular pipe (see Fig. 7-9). For a velocity profile satisfying the equation $v = v_{max}(r_o^2 - r^2)/r_o^2$, evaluate the kinetic energy correction factor α.

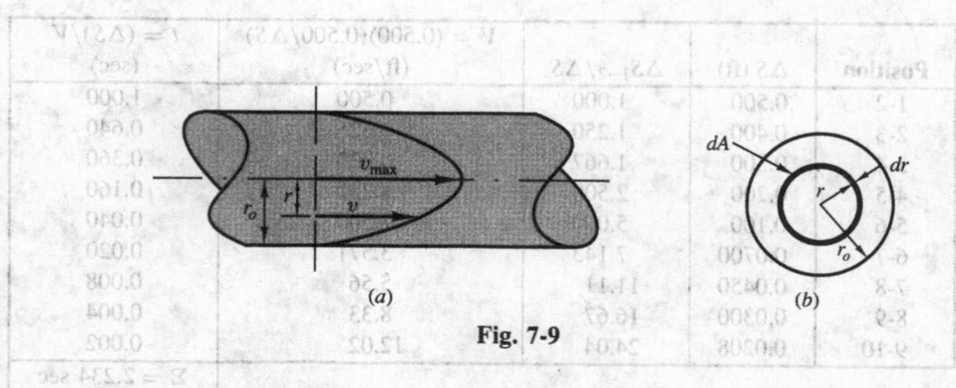

Fig. 7-9

Solution:

Evaluate the average velocity so that the equation in Problem 7.17 can be used. From the equation of continuity,

$$V_{av} = \frac{Q}{A} = \frac{\int v\,dA}{\pi r_o^2} = \frac{\int (v_{max}/r_o^2)(r_o^2 - r^2)(2\pi r\,dr)}{\pi r_o^2} = \frac{2v_{max}}{r_o^4}\int_o^{r_o}(r_o^2 r - r^3)\,dr = \frac{v_{max}}{2}$$

This value can also be obtained by considering that the given equation represents a parabola and that the volume under the generated paraboloid is half the volume of the circumscribed cylinder. Thus,

$$V_{av} = \frac{\text{volume/sec}}{\text{area of base}} = \frac{\frac{1}{2}(\pi r_o^2)v_{max}}{\pi r_o^2} = \frac{v_{max}}{2}$$

Using the value of the average velocity in the equation for α,

$$\alpha = \frac{1}{A}\int_A\left(\frac{v}{V_{av}}\right)^3 dA = \frac{1}{\pi r_o^2}\int_0^{r_0}\left(\frac{v_{max}(r_o^2 - r^2)/r_o^2}{\frac{1}{2}v_{max}}\right)^3 2\pi r\,dr = 2.00$$

(See Laminar Flow in Chapter 8.)

7.19. Oil of sp gr 0.750 is flowing through a 150-mm pipe under a pressure of 103 kPa. If the total energy relative to a datum plane 2.40 m below the center of the pipe is 17.9 m·kN/kN, determine the flow of oil.

Solution:

$$\text{energy per kN of oil} = \frac{\text{pressure}}{\text{energy}} + \frac{\text{kinetic (velocity head)}}{\text{energy}} + \frac{\text{potential}}{\text{energy}}$$

$$17.9 = \frac{103}{0.750 \times 9.79} + \frac{V^2}{2g} + 2.40$$

from which $V = 5.37$ m/s. Thus $Q = AV = \frac{1}{4}\pi(150/1000)^2 \times 5.37 = 0.095\ \text{m}^3/\text{s}$.

7.20. A turbine is rated at 450 kW when the flow of water through it is 0.609 m^3/s. Assuming an efficiency of 87%, what head is acting on the turbine?

Solution:

$$\text{rated power} = \text{extracted power} \times \text{efficiency} = (\gamma Q H_T) \times \text{efficiency}$$

$$450 = (9.79 \times 0.609 \times H_T)(0.87) \qquad \text{and} \qquad H_T = 86.8 \text{ m}$$

7.21. Derive the equations of motion for steady flow of any fluid.

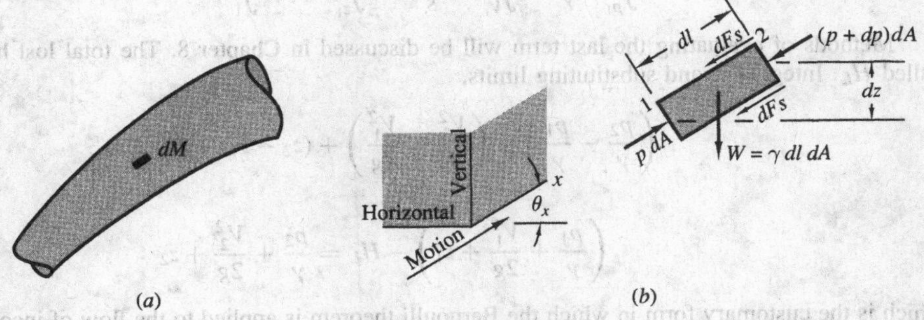

(a) (b)

Fig. 7-10

Solution:

Consider as a free body the elementary mass of fluid dM shown in Fig. 7-10(*a*) and (*b*). Motion is in the plane of the paper, and the x axis is chosen in the direction of motion. The forces normal to the direction of motion are not shown acting on the free body dM. Forces acting in the x direction are due to (*1*) the pressures acting on the end areas, (*2*) the component of the weight, and (*3*) the shearing forces (dF_s in pounds) exerted by adjacent fluid particles.

From the equation of motion $\Sigma F_x = Ma_x$, we obtain

$$[+p\,dA - (p + dp)dA - \gamma\, dA\, dl\, \sin\theta_x - dF_s] = \left(\frac{\gamma\, dA\, dl}{g}\right)\left(\frac{dV}{dt}\right) \qquad (1)$$

Dividing (*1*) by $\gamma\, dA$ and replacing dl/dt by velocity V,

$$\left[\frac{p}{\gamma} - \frac{p}{\gamma} - \frac{dp}{\gamma} - dl\, \sin\theta_x - \frac{dF_s}{\gamma\, dA}\right] = \frac{V\, dV}{g} \qquad (2)$$

The term $\dfrac{dF_s}{\gamma\, dA}$ represents the resistance to flow in length dl. The shearing forces dF_s can be replaced by the intensity of the shear τ times the area over which it acts (perimeter × length), or $dF_s = \tau dP\, dl$. Then $\dfrac{dF_s}{\gamma dA} = \dfrac{\tau dP\, dl}{\gamma dA} = \dfrac{\tau dl}{\gamma R}$, where R is called the *hydraulic radius*, which is defined as the cross-sectional area divided by the wetted perimeter or, in this case dA/dP. The sum of all the shearing forces is the measure of energy lost due to the flow, and, in ft-lb/lb, is

$$\text{lost head } dh_L = \frac{\tau dl}{\gamma R} = \frac{\text{lb/ft}^2 \times \text{ft}}{\text{lb/ft}^3 \times \text{ft}^2/\text{ft}} = \text{ft}$$

For future reference,

$$\tau = \gamma R\left(\frac{dh_L}{dl}\right) \qquad (3)$$

Returning to expression (2), since $dl \sin \theta_x = dz$, it is written in final form as

$$\frac{dp}{\gamma} + \frac{V\,dV}{g} + dz + dh_L = 0 \tag{4}$$

This expression is known as *Euler's equation* when applied to an ideal fluid (lost head = 0). When integrated for fluids of constant density, it is known as the *Bernoulli equation*. This differential equation (4) for steady flow is a fundamental fluid flow equation.

CASE 1: Flow of Incompressible Fluids

For *incompressible* fluids, the integration is simple, as follows:

$$\int_{p_1}^{p_2} \frac{dp}{\gamma} + \int_{V_1}^{V_2} \frac{V\,dV}{g} + \int_{z_1}^{z_2} dz + \int_1^2 dh_L = 0 \tag{A}$$

Methods of evaluating the last term will be discussed in Chapter 8. The total lost head term will be called H_L. Integrating and substituting limits,

$$\left(\frac{p_2}{\gamma} - \frac{p_1}{\gamma}\right) + \left(\frac{V_2^2}{2g} - \frac{V_1^2}{2g}\right) + (z_2 - z_1) + H_L = 0$$

$$\left(\frac{p_1}{\gamma} + \frac{V_1^2}{2g} + z_1\right) - H_L = \frac{p_2}{\gamma} + \frac{V_2^2}{2g} + z_2$$

which is the customary form in which the Bernoulli theorem is applied to the flow of incompressible fluids (no external energy added).

CASE 2: Flow of Compressible Fluids

For compressible fluids, the term $\int_{p_1}^{p_2} \frac{dp}{\gamma}$ cannot be integrated until γ is expressed in terms of variable p. The relationship between γ and p will depend upon the thermodynamic conditions involved.

(*a*) For *isothermal* (constant temperature) conditions, the general gas law can be expressed as

$$p_1/\gamma_1 = p/\gamma = \text{constant} \quad \text{or} \quad \gamma = (\gamma_1/p_1)p$$

where γ_1/p_1 is a constant and p must be in lb/ft^2 or *Pa absolute*. Substituting in (*A*) above.

$$\int_{p_1}^{p_2} \frac{dp}{(\gamma_1/p_1)p} + \int_{V_1}^{V_2} \frac{V\,dV}{g} + \int_{z_1}^{z_2} dz + \int_1^2 dh_L = 0$$

Integrating and substituting limits, $\frac{p_1}{\gamma_1} \ln \frac{p_2}{p_1} + \left(\frac{V_2^2}{2g} - \frac{V_1^2}{2g}\right) + (z_2 - z_1) + H_L = 0$ or, converting to the customary form,

$$\frac{p_1}{\gamma_1} \ln p_1 + \frac{V_1^2}{2g} + z_1 - H_L = \frac{p_1}{\gamma_1} \ln p_2 + \frac{V_2^2}{2g} + z_2 \tag{B}$$

Combining this equation with the equation of continuity and the gas law for isothermal conditions yields an expression with only one unknown velocity. Thus, for steady flow,

$$\gamma_1 A_1 V_1 = \gamma_2 A_2 V_2 \text{ and } \frac{p_1}{\gamma_1} = \frac{p_2}{\gamma_2} = RT, \text{ from which } V_1 = \frac{\gamma_2 A_2 V_2}{(\gamma_2/p_2) p_1 A_1} = \left(\frac{A_2}{A_1}\right)\left(\frac{p_2}{p_1}\right) V_2$$

Substituting in the Bernoulli form (*B*) above,

$$\left[\frac{p_1}{\gamma_1} \ln p_1 + \left(\frac{A_2}{A_1}\right)^2 \left(\frac{p_2}{p_1}\right)^2 \left(\frac{V_2^2}{2g}\right) + z_1\right] - H_L = \left[\frac{p_1}{\gamma_1} \ln p_2 + \frac{V_2^2}{2g} + z_2\right] \tag{C}$$

(*b*) For *adiabatic* (no heat gained or lost) conditions, the general gas law reduces to

$$\left(\frac{\gamma}{\gamma_1}\right)^k = \frac{p}{p_1} \quad \text{or} \quad \frac{p_1^{1/k}}{\gamma_1} = \frac{p^{1/k}}{\gamma} = \text{constant}, \quad \text{and thus } \gamma = \gamma_1\left(\frac{p}{p_1}\right)^{1/k}$$

where k is the adiabatic exponent.

Establishing and integrating the term dp/γ separately, we obtain

$$\int_{p_1}^{p_2} \frac{dp}{\gamma_1(p/p_1)^{1/k}} = \frac{p_1^{1/k}}{\gamma_1}\int_{p_1}^{p_2}\frac{dp}{p^{1/k}} = \left(\frac{k}{k-1}\right)\left(\frac{p_1}{\gamma_1}\right)\left[\left(\frac{p_2}{p_1}\right)^{(k-1)/k} - 1\right]$$

and the Bernoulli equation in customary form becomes

$$\left[\left(\frac{k}{k-1}\right)\left(\frac{p_1}{\gamma_1}\right) + \frac{V_1^2}{2g} + z_1\right] - H_L = \left[\left(\frac{k}{k-1}\right)\left(\frac{p_1}{\gamma_1}\right)\left(\frac{p_2}{p_1}\right)^{(k-1)/k} + \frac{V_2^2}{2g} + z_2\right] \qquad (D)$$

Combining this equation with the equation of continuity and the gas law for adiabatic conditions yields an expression with only one unknown velocity.

Using $\gamma_1 A_1 V_1 = \gamma_2 A_2 V_2$ and $\dfrac{p_1^{1/k}}{\gamma_1} = \dfrac{p_2^{1/k}}{\gamma_2} = \text{constant}$, $V_1 = \dfrac{\gamma_2 A_2 V_2}{\gamma_1 A_1} = \left(\dfrac{p_2}{p_1}\right)^{1/k}\left(\dfrac{A_2}{A_1}\right)V_2$, and the Bernoulli equation becomes

$$\left[\left(\frac{k}{k-1}\right)\left(\frac{p_1}{\gamma_1}\right) + \left(\frac{p_2}{p_1}\right)^{2/k}\left(\frac{A_2}{A_1}\right)^2\left(\frac{V_2^2}{2g}\right) + z_1\right] - H_L$$

$$= \left[\left(\frac{k}{k-1}\right)\left(\frac{p_1}{\gamma_1}\right)\left(\frac{p_2}{p_1}\right)^{(k-1)/k} + \frac{V_2^2}{2g} + z_2\right] \qquad (E)$$

7.22. In Fig. 7-11, water flows from A to B at the rate of 13.2 cfs and the pressure at A is 22.1 ft. Considering no loss of energy from A to B, find the pressure head at B. Draw the energy line.

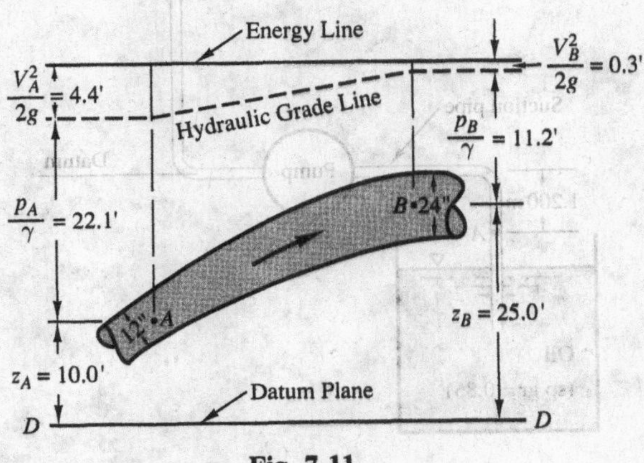

Fig. 7-11

Solution:

Apply the Bernoulli equation, A to B, datum A.

$$\text{energy at } A + \text{energy added} - \text{energy lost} = \text{energy at } B$$

$$\left(\frac{p_A}{\gamma} + \frac{V_{12}^2}{2g} + z_A\right) + 0 - 0 = \left(\frac{p_B}{\gamma} + \frac{V_{24}^2}{2g} + z_B\right)$$

where $V_{12} = Q/A_{12} = 13.2/\left(\frac{1}{4}\pi 1^2\right) = 16.8$ and $V_{24} = \left(\frac{1}{2}\right)^2(16.8) = 4.20$ ft/sec. Substituting,

$$\left(22.1 + \frac{(16.8)^2}{2g} + 0\right) - 0 = \left(\frac{p_B}{\gamma} + \frac{(4.20)^2}{2g} + 15.0\right) \quad \text{and} \quad \frac{p_B}{\gamma} = 11.2 \text{ ft water}$$

Total energy at any section can be plotted above a chosen datum plane. Using DD in this case,

$$\text{energy at } A = p_A/\gamma + V_{12}^2/2g + z_A = 22.1 + 4.4 + 10.0 = 36.5 \text{ ft}$$

$$\text{energy at } B = p_B/\gamma + V_{24}^2/2g + z_B = 11.2 + 0.3 + 25.0 = 36.5 \text{ ft}$$

Note: Transformation from one form of energy to another occurs during flow. In this case a portion of both the pressure energy and the kinetic energy at A is transformed to potential energy at B.

7.23. For the 100-mm-diameter suction pipe leading to a pump shown in Fig. 7-12, the pressure at point A in the suction pipe is a vacuum of 180 mm of mercury. If the discharge is 0.0300 m^3/s of oil (sp gr = 0.85), find the total energy head at point A with respect to a datum at the pump.

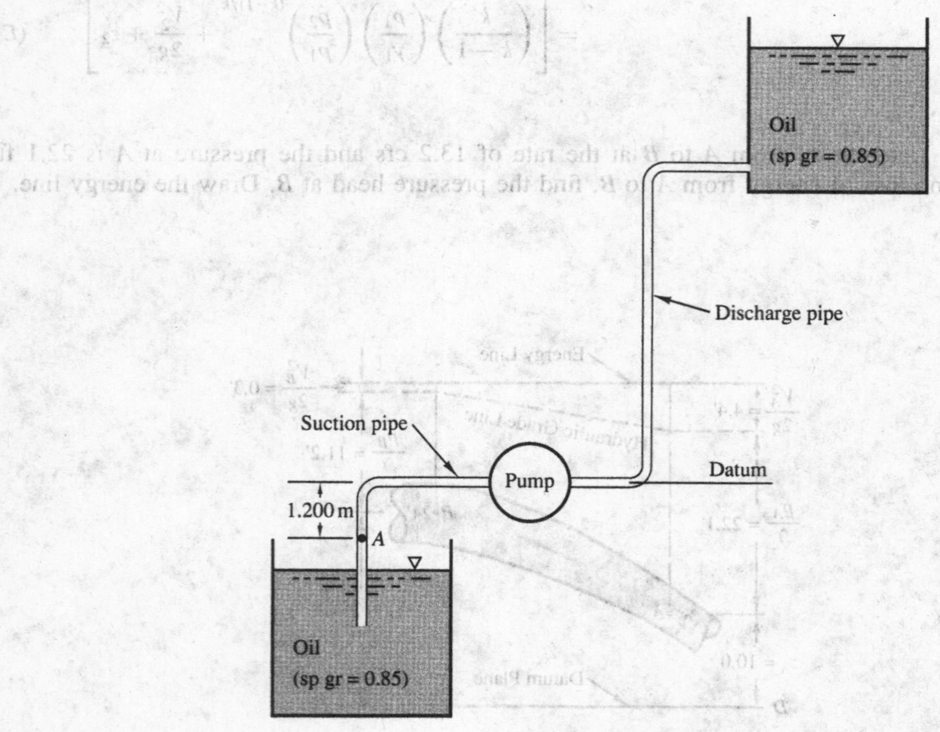

Fig. 7-12

Solution:

$$\text{energy at } A = p_A/\gamma + V_A^2/2g + z_A$$

$$V_A = Q/A = 0.0300/\left[(\pi)(0.100)^2/4\right] = 3.820 \text{ m/s}$$

$$p_A = \gamma h = [(13.6)(9.79)](-0.180) = -23.97 \text{ kPa}$$

$$\text{energy at } A = \frac{-23.97}{(0.85)(9.79)} + \frac{(3.820)^2}{(2)(9.81)} + (-1.200) = -3.337 \text{ m}$$

7.24. For the Venturi meter shown in Fig. 7-13, the deflection of mercury in the differential gage is 14.3 in. Determine the flow of water through the meter if no energy is lost between A and B.

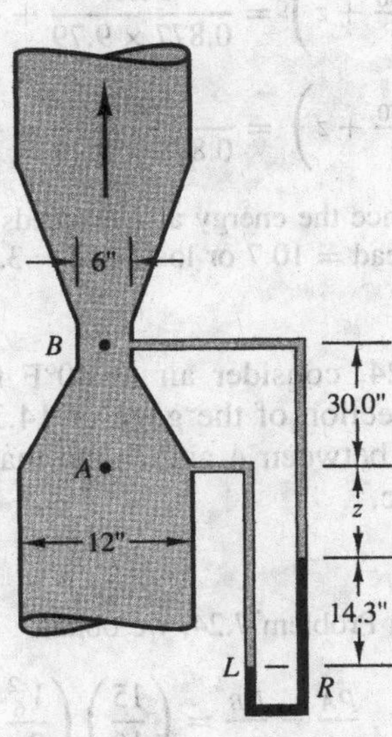

Fig. 7-13

Solution:

Apply the Bernoulli equation, A to B, datum A.

$$\left(\frac{p_A}{\gamma} + \frac{V_{12}^2}{2g} + 0\right) - 0 = \left(\frac{p_B}{\gamma} + \frac{V_6^2}{2g} + 2.50\right)$$

and
$$\left(\frac{p_A}{\gamma} - \frac{p_B}{\gamma}\right) = \left(\frac{V_6^2}{2g} - \frac{V_{12}^2}{2g} + 2.50\right) \qquad (1)$$

The equation of continuity yields $A_{12}V_{12} = A_6V_6$, or $V_{12} = \left(\frac{6}{12}\right)^2 V_6 = \frac{1}{4}V_6$, and $V_{12}^2 = \frac{1}{16}V_6^2$. For the gage,

$$\text{pressure head at } L = \text{pressure head at } R \text{ (ft of water)}$$

$$p_A/\gamma + z + 14.3/12 = p_B/\gamma + 2.50 + z + (14.3/12)(13.6)$$

from which $(p_A/\gamma - p_B/\gamma) = 17.5$ ft water. Substituting in (*1*), we obtain $V_6 = 32.1$ ft/sec and $Q = \frac{1}{4}\pi\left(\frac{1}{2}\right)^2 \times 32.1 = 6.30$ cfs.

7.25. A pipe carrying oil of sp gr 0.877 changes in size from 150 mm at section E to 450 mm at section R. Section E is 3.66 m lower than R, and the pressures are 91.0 kPa and 60.3 kPa, respectively. If the discharge is 0.146 m³/s, determine the lost head and the direction of flow.

Solution:

The average velocity at each cross section $= Q/A$. Then

$$V_{150} = \frac{0.146}{\frac{1}{4}\pi(150/1000)^2} = 8.26 \text{ m/s} \quad \text{and} \quad V_{450} = \frac{0.146}{\frac{1}{4}\pi(450/1000)^2} = 0.92 \text{ m/s}$$

Using the lower section E as the datum plane, the energy at each section is

$$\text{At } E, \left(\frac{p}{\gamma} + \frac{V_{150}^2}{2g} + z\right) = \frac{91.0}{0.877 \times 9.79} + \frac{(8.26)^2}{2g} + 0 \quad = 14.1 \text{ m} \cdot \text{kN/kN}$$

$$\text{At } R, \left(\frac{p}{\gamma} + \frac{V_{450}^2}{2g} + z\right) = \frac{60.3}{0.877 \times 9.79} + \frac{(0.92)^2}{2g} + 3.66 = 10.7 \text{ m} \cdot \text{kN/kN}$$

Flow occurs from E to R since the energy at E exceeds that at R. The lost head can be found, using E to R, datum E : 14.1 − lost head = 10.7 or lost head = 3.4 m, E to R.

7.26. For the meter in Problem 7.24, consider air at 80°F flowing with the pressure at A equal to 37.5 psi gage. Consider a deflection of the gage of 14.3 in of water. Assuming that the specific weight of air does not change between A and B and that energy loss is negligible, determine the amount of air flowing in lb/sec.

Solution:

Using A to B, datum A, as in Problem 7.24, we obtain

$$\frac{p_A}{\gamma} - \frac{p_B}{\gamma} = \left(\frac{15}{16}\right)\left(\frac{V_6^2}{2g}\right) + 2.50 \qquad (1)$$

To obtain the pressure head of fluid flowing, the specific weight of the air must be calculated.

$$\gamma = \frac{p}{RT} = \frac{(37.5 + 14.7)(144)}{(53.3)(80 + 460)} = 0.261 \text{ lb/ft}^3$$

For the differential gage, $\qquad p_L = p_R$ (in psf gage)

or $$p_A + (0.261)(z + 14.3/12) = p_B + (0.261)(2.50 + z) + (62.4)(14.3/12)$$

and $(p_A - p_B) = 74.7$ psf. Substituting in (*1*), we obtain $V_6 = 139.6$ ft/sec and

$$W = \gamma Q = (0.261)\left[\frac{1}{4}\pi\left(\frac{1}{2}\right)^2 \times 139.6\right] = 7.15 \text{ lb/sec of air}$$

7.27. A horizontal air duct reduces in cross-sectional area from 0.070 m² to 0.020 m². Assuming no losses, what pressure change will occur when 6.67 N/s of air flows? (Use $\gamma = 31.4$ N/m³ for pressure and temperature conditions involved.)

Solution:

$$Q = \frac{6.67\ \text{N/s}}{31.4\ \text{N/m}^3} = 0.212\ \text{m}^3/\text{s}, \qquad V_1 = \frac{Q}{A_1} = \frac{0.212}{0.070} = 3.03\ \text{m/s},$$

$$V_2 = \frac{Q}{A_2} = \frac{0.212}{0.020} = 10.6\ \text{m/s}.$$

Applying the Bernoulli theorem, section 1 to section 2, gives

$$\left(\frac{p_1}{\gamma} + \frac{(3.03)^2}{2g} + 0\right) - 0 = \left(\frac{p_2}{\gamma} + \frac{(10.6)^2}{2g} + 0\right) \quad \text{or} \quad \left(\frac{p_1}{\gamma} - \frac{p_2}{\gamma}\right) = 5.26\ \text{m air}$$

and $p_1 - p_2 = 5.26 \times 31.4 = 165$ Pa change. This small pressure change justifies the assumption of constant density of the fluid.

7.28. A 6″ pipe 600 ft long carries water from A at elevation 80.0 ft to B at elevation 120.0 ft. The frictional stress between the liquid and the pipe walls is 0.62 lb/ft². Determine the pressure change in the pipe and the lost head.

Mathcad

Solution:

(*a*) The forces acting on the mass of water are the same as those shown in Fig. 7-10(*b*) of Problem 7.21.

Using $F_1 = p_1 A_6$, $F_2 = p_2 A_6$ we obtain, from $\Sigma F_x = 0$,

$$p_1 A_6 - p_2 A_6 - W \sin\theta_x - \tau(\pi d)L = 0$$

Now $W = \gamma(\text{volume}) = 62.4\left[\frac{1}{4}\pi(1/2)^2 \times 600\right]$ and $\sin\theta_x = (120 - 80)/600$. Then

$$p_1\left[\frac{1}{4}\pi\left(\frac{1}{2}\right)^2\right] - p_2\left[\frac{1}{4}\pi\left(\frac{1}{2}\right)^2\right] - 62.4\left[\frac{1}{4}\pi\left(\frac{1}{2}\right)^2 \times 600\right] \times 40/600 - 0.62(\pi \times \frac{1}{2} \times 600) = 0$$

from which $p_1 - p_2 = 5472$ psf $= 38.0$ psi.

(*b*) Using the energy equation, datum at A,

$$\text{energy at } A - \text{lost head} = \text{energy at } B$$

$$\left(\frac{p_A}{\gamma} + \frac{V_A^2}{2g} + 0\right) - \text{lost head} = \left(\frac{p_B}{\gamma} + \frac{V_B^2}{2g} + 40\right)$$

or

$$\text{lost head} = (p_A/\gamma - p_B/\gamma) - 40 = 5472/62.4 - 40 = 47.7\ \text{ft}$$

Another method:

Using (*3*) of Problem 7.21, lost head $= \dfrac{\tau L}{\gamma R} = \dfrac{(0.62)(600)}{(62.4)(0.50/4)} = 47.7$ ft.

7.29. Water at 90°F is to be lifted from a sump at a velocity of 6.50 ft/sec through the suction pipe of a pump. Calculate the theoretical maximum height of the pump setting under the following conditions: atmospheric pressure = 14.25 psia, vapor pressure = 0.70 psia (see Table 1C), and lost head in the suction pipe = 3 velocity heads.

Solution:

The specific weight of water at 90°F from Table 1C is 62.1 lb/ft³. The minimum pressure at entrance to the pump cannot exceed the vapor pressure of the liquid. The energy equation will be applied from the surface of the water outside the suction pipe to the entrance to the pump, using absolute pressure heads.

$$\text{energy at water surface} - \text{lost head} = \text{energy at entrance to pump}$$

$$\left(\frac{(14.25)(144)}{62.1}+0+0\right)-\frac{(3)(6.50)^2}{2g}=\left(\frac{(0.70)(144)}{62.1}+\frac{(6.50)^2}{2g}+z\right)$$

from which $z = 28.8$ ft above water surface.

Under these conditions serious damage due to cavitation will probably occur. See Chapter 14.

7.30. A 150-mm-diameter jet of water is discharged from a nozzle into the air. The velocity of the jet is 36.0 m/s. Find the power in the jet.

Solution:

$$Q = AV = \left[(\pi)(0.150)^2/4\right](36.0) = 0.6362\ \text{m}^3/\text{s}$$

$$H = z + V^2/2g + p/\gamma = 0 + 36.0^2/[(2)(9.81)] + 0 = 66.06\ \text{m}$$

$$P = Q\gamma H = (0.6362)(9.79)(66.06) = 411\ \text{kN}\cdot\text{m/s, or } 411\ \text{kW}$$

7.31. For the system shown in Fig. 7-14, pump *BC* must deliver 5.62 cfs of oil, sp gr = 0.762, to reservoir *D*. Assuming that the energy lost from *A* to *B* is 8.25 ft-lb/lb and from *C* to *D* is 21.75 ft-lb/lb, (*a*) how many horsepower units must the pump supply to the system? (*b*) Plot the energy line.

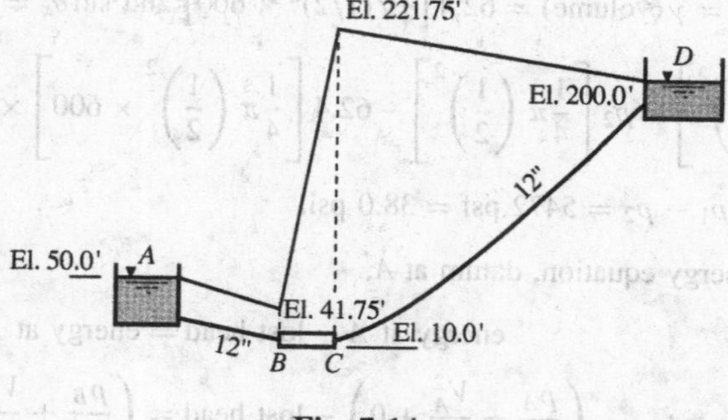

Fig. 7-14

Solution:

(*a*) The velocity of particles at *A* and *D* will be very small, and hence the velocity head terms will be negligible.

A to *D*, datum *BC* (*A* would be satisfactory also),

$$\left(\frac{p_A}{\gamma}+\frac{V_A^2}{2g}+z_A\right)+H_{\text{pump}}-H_{\text{lost}}=\left(\frac{p_D}{\gamma}+\frac{V_D^2}{2g}+z_D\right)$$

$$(0+\text{negl}+40.0)+H_{\text{pump}}-(8.25+21.75)=(0+\text{negl}+190.0)$$

and $H_{\text{pump}} = 180.0$ ft (or ft-lb/lb).

Horsepower $= \gamma Q H_{\text{pump}}/550 = (0.762\times 62.4)(5.62)(180.0)/550 = 87.5$ delivered to system.

Note that the pump has supplied a head sufficient to raise the oil 150.0 ft and also has overcome 30.0 ft of losses in the piping. Thus 180 ft is delivered to the system.

(*b*) The energy line at A is at elevation 50.0 above datum zero. From A to B the energy loss is 8.25 ft and the energy line drops by this amount, the elevation at B being 41.75. The pump adds 180.0 ft of energy, and the elevation at C is thus 221.75. Finally, the loss of energy between C and D being 21.75 ft, the elevation at $D = 221.75 - 21.75 = 200.0$. These data are shown graphically in Fig. 7-14.

7.32. Water flows through the turbine in Fig. 7-15 at the rate of 0.214 m^3/s and the pressures at A and B, respectively, are 147.5 kPa and −34.5 kPa. Determine the power delivered to the turbine by the water.

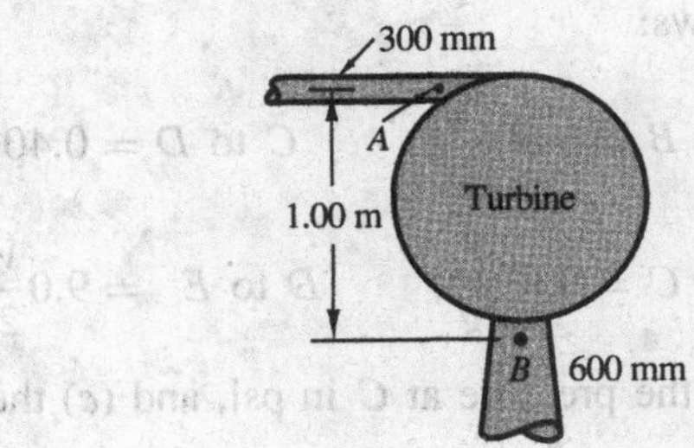

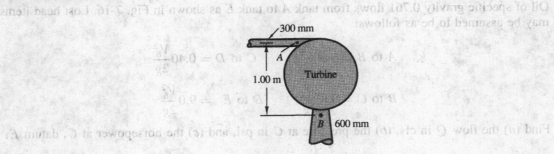

Fig. 7-15

Solution:

Using A to B (datum B), with

$$V_{300} = 0.214/A_{300} = 3.03 \text{ m/s} \qquad \text{and} \qquad V_{600} = 3.03/4 = 0.758 \text{ m/s}$$

$$\left(\frac{p_A}{\gamma} + \frac{V_{300}^2}{2g} + z_A\right) + 0 - H_{\text{turbine}} = \left(\frac{p_B}{\gamma} + \frac{V_{600}^2}{2g} + z_B\right)$$

$$\left(\frac{147.5}{9.79} + \frac{3.03^2}{2g} + 1.00\right) - H_T = \left(\frac{-34.5}{9.79} + \frac{0.758^2}{2g} + 0\right) \quad \text{and } H_T = 20.0 \text{ m}$$

Power $= \gamma Q H_r = (9.79)(0.214)(20.0) = 41.9$ kW.

7.33. For the turbine of Problem 7.32, if 48.8 kW are extracted while the pressure gages at A and B are reading 141.3 kPa and −33.1 kPa respectively, how much water is flowing?

Solution:

Using A to B (datum B), $\left(\frac{141.3}{9.79} + \frac{V_{300}^2}{2g} + 1.00\right) - H_T = \left(\frac{-33.1}{9.79} + \frac{V_{600}^2}{2g} + 0\right)$ and

$$H_T = \left(\frac{174.4}{9.79} + 1.00 + \frac{V_{300}^2}{2g} - \frac{V_{600}^2}{2g}\right) \tag{A}$$

$$A_{300}V_{300} = A_{600}V_{600} \quad \text{or} \quad \frac{V_{600}^2}{2g} = \left(\frac{1}{2}\right)^4 \frac{V_{300}^2}{2g} = \frac{1}{16}\frac{V_{300}^2}{2g} \tag{B}$$

$$48.8 = \gamma Q H_T = 9.79 \times \frac{1}{4}\pi(0.300)^2 V_{300} \times H_T \quad \text{or} \quad H_T = \frac{70.5}{V_{300}} \tag{C}$$

Equating (*A*) and (*C*) (with velocity head substitution), $70.5/V_{300} = 18.81 + \frac{15}{16}(V_{300}^2/2g)$ or

$$18.81\,V_{300} + 0.0478\,V_{300}^3 = 70.5$$

Solving this equation by successive trials:

Try $V_{300} = 3.000$ m/s, $\quad 56.4 + 1.3 \neq 70.5$ (must increase V)

Try $V_{300} = 3.70$ m/s, $\quad 69.6 + 2.4 \neq 70.5$ (answer straddled)

Try $V_{300} = 3.63$ m/s, $\quad 68.3 + 2.3 \cong 70.6$ (good result)

The flow $Q = A_{300}V_{300} = \frac{1}{4}\pi(0.300)^2 \times 3.63 = 0.257\ \text{m}^3/\text{s}$.

7.34. Oil of specific gravity 0.761 flows from tank A to tank E as shown in Fig. 7-16. Lost head items may be assumed to be as follows:

$$A \text{ to } B = 0.60\frac{V_{12}^2}{2g}, \qquad C \text{ to } D = 0.40\frac{V_6^2}{2g}$$

$$B \text{ to } C = 9.0\frac{V_{12}^2}{2g}, \qquad D \text{ to } E = 9.0\frac{V_6^2}{2g}$$

Find (a) the flow Q in cfs, (b) the pressure at C in psi, and (c) the horsepower at C, datum E.

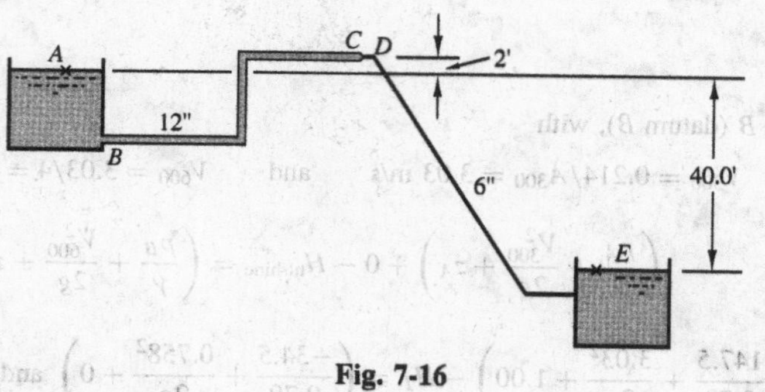

Fig. 7-16

Solution:

(a) Using A to E, datum E,

At A $\quad$ A to B $\quad$ B to C $\quad$ C to D $\quad$ D to E $\quad$ at E

$$(0 + \text{negl} + 40.0) - \left[\left(0.60\frac{V_{12}^2}{2g} + 9.0\frac{V_{12}^2}{2g}\right) + \left(0.40\frac{V_6^2}{2g} + 9.0\frac{V_6^2}{2}\right)\right] = (0 + \text{negl} + 0)$$

or $40.0 = 9.6(V_{12}^2/2g) + 9.4(V_6^2/2g)$. Also, $V_{12}^2 = (\frac{1}{2})^4 V_6^2 = \frac{1}{16}V_6^2$.

Substituting and solving,

$$V_6^2/2g = 4 \text{ ft}, \qquad V_6 = 16.0 \text{ ft/sec}, \qquad \text{and} \qquad Q = \frac{1}{4}\pi\left(\frac{1}{2}\right)^2 \times 16.0 = 3.14 \text{ cfs}$$

(b) Using A to C, datum A,

$$(0 + \text{negl} + 0) - (0.60 + 9.0)\frac{V_{12}^2}{2g} = \left(\frac{p_C}{\gamma} + \frac{V_{12}^2}{2g} + 2\right) \text{ and } \frac{V_{12}^2}{2g} = \frac{1}{16}\frac{V_6^2}{2g} = \frac{1}{16}(4) = \frac{1}{4} \text{ ft}$$

Then $p_C/\gamma = -4.65$ ft of oil gage and $p_C = (0.761 \times 62.4)(-4.65)/144 = -1.53$ psi gage.

The Bernoulli equation could have been applied from C to E with equally satisfactory results. The two equations obtained by the two choices would *not* be independent, simultaneous equations.

(*c*) Horsepower at $C = \dfrac{\gamma Q H_C}{550} = \dfrac{(0.761 \times 62.4)(3.14)(-4.65 + 0.25 + 42.0)}{550} = 10.19$, datum E.

7.35. For the 50-mm-diameter siphon drawing oil (sp gr = 0.82) from the oil reservoir as shown in Fig. 7-17, the head loss from point 1 to point 2 is 1.50 m and from point 2 to point 3 is 2.40 m. Find the discharge of oil from the siphon and the oil pressure at point 2.

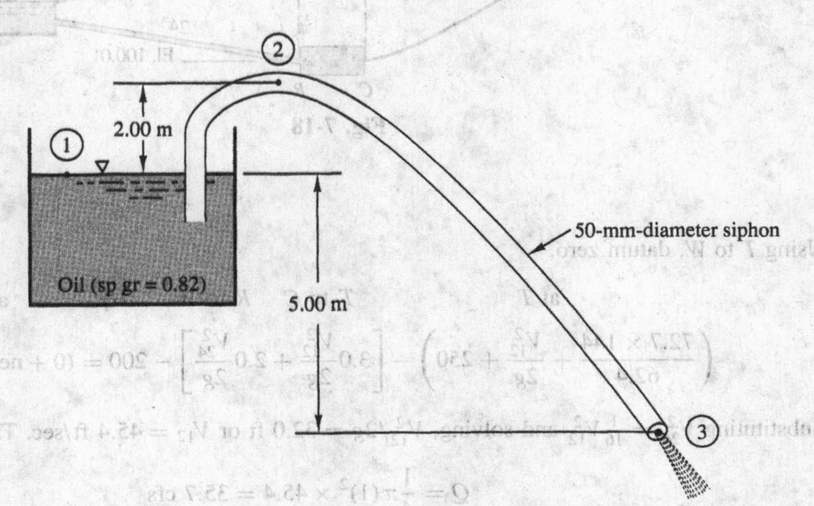

Fig. 7-17

Solution:

$$p_1/\gamma + V_1^2/2g + z_1 = p_3/\gamma + V_3^2/2g + z_3 + H_L$$

$$0 + 0 + 5.00 = 0 + V_3^2/[(2)(9.81)] + 0 + 3.90$$

$$V_3 = 4.646 \text{ m/s}$$

$$Q = AV = \left[(\pi)(0.050)^2/4\right](4.646) = 0.00912 \text{ m}^3\text{/s}$$

$$p_1/\gamma + V_1^2/2g + z_1 = p_2/\gamma + V_2^2/2g + z_2 + H_L$$

$$0 + 0 + 5.00 = p_2/\gamma + (4.646)^2/[(2)(9.81)] + 7.00 + 1.50$$

$$p_2/\gamma = -4.600 \text{ m}$$

$$p_2 = [(9.79)(0.82)](-4.600) = -36.9 \text{ kPa}$$

7.36. The head extracted by turbine CR in Fig. 7-18 is 200 ft, and the pressure at T is 72.7 psi. For losses of $2.0(V_{24}^2/2g)$ between W and R and $3.0(V_{12}^2/2g)$ between C and T, determine (*a*) how much water is flowing and (*b*) the pressure head at R. Draw the energy line.

Solution:

Because the energy line at T is at elevation $\left(250.0 + \dfrac{72.7 \times 144}{62.4} + \dfrac{V_{12}^2}{2g}\right)$ and well above the elevation at W, the water flows into reservoir W.

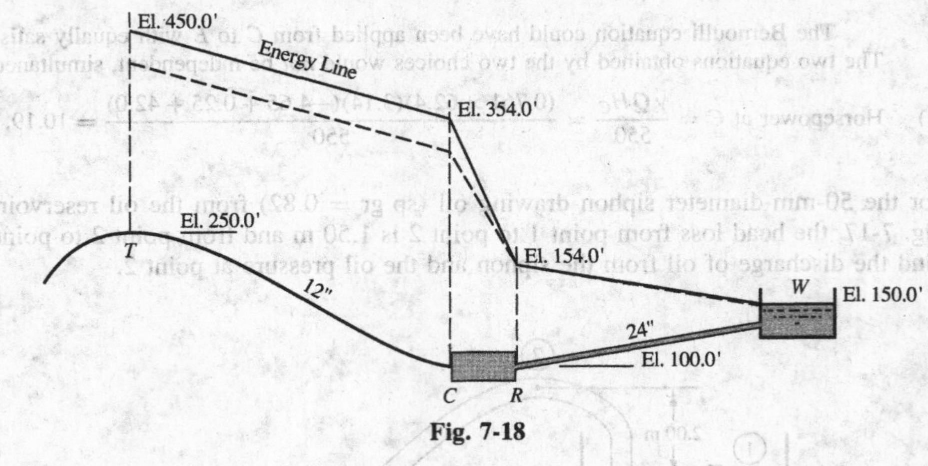

Fig. 7-18

(*a*) Using T to W, datum zero,

$$\overset{\text{at } T}{\left(\frac{72.7\times 144}{62.4}+\frac{V_{12}^2}{2g}+250\right)} - \overset{T \text{ to } C \quad R \text{ to } W}{\left[3.0\frac{V_{12}^2}{2g}+2.0\frac{V_{24}^2}{2g}\right]} - \overset{H_T}{200} = \overset{\text{at } W}{(0+\text{negl}+150)}$$

Substituting $V_{24}^2 = \frac{1}{16}V_{12}^2$ and solving, $V_{12}^2/2g = 32.0$ ft or $V_{12} = 45.4$ ft/sec. Then

$$Q = \frac{1}{4}\pi(1)^2 \times 45.4 = 35.7 \text{ cfs}$$

(*b*) Using R to W, datum R, $(p_R/\gamma + \frac{1}{16}\times 32.0 + 0) - 2\left(\frac{1}{16}\times 32.0\right) = (0+\text{negl}+50)$ and $p_R/\gamma = 52.0$ ft. The reader may check this pressure head by applying the Bernoulli equation between T and R.

To plot the energy line in the figure, evaluate the energy at the four sections indicated.

$$\text{elevation of energy line at } T = 168.0 + 32.0 + 250.0 = 450.0$$
$$\text{at } C = 450.0 - 3\times 32.0 = 354.0$$
$$\text{at } R = 354.0 - 200.0 = 154.0$$
$$\text{at } W = 154.0 - 2\times\frac{1}{16}\times 32 = 150.0$$

It will be shown in the following chapter that the energy line is a straight line for steady flow in a pipe of constant diameter. The hydraulic grade line will be parallel to the energy line and $V^2/2g$ below it (shown dotted).

7.37. (*a*) What is the pressure on the nose of a torpedo moving in salt water at 100 ft/sec at a depth of 30.0 ft?

(*b*) If the pressure at a point C on the side of the torpedo at the same elevation as the nose is 10.0 psi gage, what is the relative velocity at that point?

Solution:

(*a*) In this case, greater clarity in the application of the Bernoulli equation can be attained by considering the relative motion of a stream of water past the stationary torpedo. The velocity at the nose of the torpedo will then be zero. Assuming no lost head in the streamtube from a point A in the undisturbed water just ahead of the torpedo to a point B on the nose of the torpedo, the Bernoulli equation becomes

$$\left(\frac{p_A}{\gamma}+\frac{V_A^2}{2g}+z_A\right)-0=\left(\frac{p_B}{\gamma}+\frac{V_B^2}{2g}+z_B\right) \quad \text{or} \quad 30.0+\frac{(100)^2}{2g}+0=\frac{p_B}{\gamma}+0+0$$

Then $p_B/\gamma = 185$ ft of salt water, and $p_B = \gamma h/144 = (64.0)(185)/144 = 82.2$ psi gage.

This pressure is called the *stagnation pressure* and can be expressed as $p_s = p_0 + \frac{1}{2}\rho V_0^2$ in psf units. For further discussion, see Chapters 12 and 13.

(*b*) The Bernoulli equation may be applied either from point A to point C or from point B to point C. Considering A and C,

$$\left(\frac{p_A}{\gamma}+\frac{V_A^2}{2g}+z_A\right)-0=\left(\frac{p_C}{\gamma}+\frac{V_C^2}{2g}+z_C\right) \quad \text{or} \quad \left(30.0+\frac{(100)^2}{2g}+0\right)=\left(\frac{10.0\times 144}{64.0}+\frac{V_C^2}{2g}+0\right)$$

from which $V_C = 102.4$ ft/sec.

7.38. A sphere is placed in an air stream that is at atmospheric pressure and moving at 30 m/s. Using the density of air constant at 1.23 kg/m^3, calculate (*a*) the stagnation pressure and (*b*) the pressure on the surface of the sphere at a point B, 24°C from the stagnation point, if the velocity there is 67 m/s.

Solution:

(*a*) Applying the expression given in the preceding problem, we obtain

$$p_S = p_0 + \frac{1}{2}\rho V_0^2 = 101{,}400 + \left(\frac{1}{2}\right)(1.23)(30)^2 = 102{,}000 \text{ Pa} = 102.0 \text{ kPa}$$

(*b*) The specific weight of air $= \rho g = (1.23)(9.81) = 12.1$ N/m^3.

Applying the Bernoulli equation, stagnation point to point B, produces

$$\left(\frac{p_S}{\gamma}+\frac{V_S^2}{2g}+0\right)-0=\left(\frac{p_B}{\gamma}+\frac{V_B^2}{2g}+0\right) \quad \text{or} \quad \left(\frac{102{,}000}{12.1}+0+0\right)=\left(\frac{p_B}{\gamma}+\frac{(67)^2}{2g}+0\right)$$

from which $p_B/\gamma = 8200$ m of air, and $p_B = \gamma h = (12.1)(8200) = 99{,}200$ Pa $= 99.2$ kPa.

7.39. A large closed tank is filled with ammonia under a pressure of 5.30 psi gage and at 65°F. The ammonia discharges into the atmosphere through a small opening in the side of the tank. Neglecting friction losses, calculate the velocity of the ammonia leaving the tank assuming (*a*) constant density and (*b*) adiabatic flow conditions.

Mathcad

Solution:

(*a*) Apply the Bernoulli equation, tank to atmosphere.

$$\left(\frac{5.30\times 144}{\gamma_1}+0+0\right)=\left(0+\frac{V^2}{2g}+0\right)$$

$$\text{where } \gamma_1 = \frac{p_1}{RT} = \frac{(5.30+14.7)(144)}{(89.5)(460+65)} = 0.0613 \text{ lb/ft}^3$$

Substituting and solving, $V = 895$ ft/sec.

For a constant specific weight γ, either gage or absolute pressure heads may be used. Absolute pressure head *must* be used for cases where γ is not constant.

(*b*) For $V_1 = 0$ and $z_1 = z_2$, the adiabatic expression (*D*) in Problem 7.21 may be written

$$\left(\frac{k}{k-1}\right)\left(\frac{p_1}{\gamma_1}\right)\left[1-\left(\frac{p_2}{p_1}\right)^{(k-1)/k}\right] = \frac{V_2^2}{2g}$$

For ammonia, $k = 1.32$ from Table 1 in the Appendix, and

$$\frac{1.32}{0.32} \times \frac{20.0 \times 144}{0.0613}\left[1-\left(\frac{14.7 \times 144}{20.0 \times 144}\right)^{0.242}\right] = \frac{V_2^2}{2g} \quad \text{or} \quad V_2 = 947 \text{ ft/sec}$$

The error in using the velocity based upon the constant density assumption is about 5.3%.

The specific weight of the ammonia in the jet is calculated by using the expression

$$\frac{p_1}{p_2} = \left(\frac{\gamma_1}{\gamma_2}\right)^k \quad \text{or} \quad \frac{20.0}{14.7} = \left(\frac{0.0613}{\gamma_2}\right)^{1.32} \quad \text{and } \gamma_2 = 0.0485 \text{ lb/ft}^3$$

Despite this 20.7% change in density, the error in the velocity was only 5.3%.

7.40. Compare the velocities in (*a*) and (*b*) of Problem 7.39 for a pressure of 15.3 psi gage in the tank.

Solution:

(*a*) $\gamma_1 = \dfrac{p_1}{RT} = \dfrac{30.0 \times 144}{89.5 \times 525} = 0.0919 \text{ lb/ft}^3$, and from Problem 7.39,

$$\frac{15.3 \times 144}{0.0919} = \frac{V^2}{2g} \quad \text{and} \quad V = 1243 \text{ ft/sec}$$

(*b*) Using the adiabatic expression given in Problem 7.39,

$$\frac{V^2}{2g} = \frac{1.32}{0.32} \times \frac{30.0 \times 144}{0.0919}\left[1-\left(\frac{14.7 \times 144}{30.0 \times 144}\right)^{0.242}\right] \quad \text{and} \quad V = 1407 \text{ ft/sec}$$

The error in using the velocity based upon the constant density assumption is about 12%. The change in density in this case is about 42%.

Limitations on the magnitude of velocity will be discussed in Chapter 13. It should be pointed out that the limiting velocity for the temperature involved is 1412 ft/sec.

7.41. Nitrogen flows from a 2″ pipe in which the temperature and pressure are 40°F and 40 psi gage, respectively, into a 1″ pipe in which the pressure is 21.3 psi gage. Calculate the velocity in each pipe, assuming isothermal conditions apply and no losses.

Solution:

Referring to Problem 7.21, equation (*C*) for isothermal conditions can be solved for V_2, noting that $z_1 = z_2$,

$$\frac{V_2^2}{2g}\left[1-\left(\frac{A_2 p_2}{A_1 p_1}\right)^2\right] = \frac{p_1}{\gamma_1}\ln\left(\frac{p_1}{p_2}\right) = RT\ln\left(\frac{p_1}{p_2}\right) \quad \text{or} \quad V_2 = \sqrt{2g \times \frac{RT\ln(p_1/p_2)}{1-(A_2p_2/A_1p_1)^2}}$$

Substituting herein, using $R = 55.1$ ft/°R for nitrogen from Table 1 in the Appendix,

$$V_2 = \sqrt{2g \times \frac{55.1 \times 500 \ln[(54.7 \times 144)/(36.0 \times 144)]}{1-\left(\frac{1}{2}\right)^4[(36.0 \times 144)/(54.7 \times 144)]^2}} = 873 \text{ ft/sec}$$

Also $V_1 = (A_2/A_1)(p_2/p_1)V_2 = \left(\frac{1}{2}\right)^2(36.0/54.7)(873) = 144$ ft/sec.

7.42. In Problem 7.41, for a pressure, velocity, and temperature in the 2″ pipe of 38.0 psi gage, 143 ft/sec, and 32°F, respectively, calculate the velocity and pressure in the 1″ pipe. Assume no lost head and isothermal conditions.

Solution:

From Problem 7.21 for isothermal conditions, using equation (*C*) in terms of V_1 instead of V_2,

$$(a)\qquad \frac{(143)^2}{2g}\left[1-\left(\frac{4}{1}\right)^2\left(\frac{52.7\times 144}{p_2\times 144}\right)^2\right]=55.1\times 492\ \ln\frac{p_2\times 144}{52.7\times 144}$$

Only one unknown appears, yet a direct solution is difficult. The method of successive trials seems indicated, assuming a value for p_2 in the denominator of the bracket.

(1) Assume $p_2 = 52.7$ psi absolute, and solve for p_2 in the right-hand side of the equation.

$$(318)\left[1-(16)(1)^2\right]=27{,}100\ \ln(p_2/52.7)$$

from which $p_2 = 44.2$ psi absolute.

(2) The effect of using $p_2 = 44.2$ psi in (*a*) would result in an inequality. Anticipating the result, assume the new value of p_2 at 35.0 psi and solve.

$$(318)\left[1-(16)(52.7/35.0)^2\right]=27{,}100\ \ln(p_2/52.7)$$

from which $p_2 = 34.8$ psi absolute. For velocity,

$$V_2=\frac{\gamma_1 A_1}{\gamma_2 A_2}V_1\quad\text{or}\quad V_2=\frac{p_1}{p_2}\left(\frac{A_1}{A_2}\right)V_1=\frac{52.7\times 144}{34.8\times 144}\left(\frac{2}{1}\right)^2\times 143=866\text{ ft/sec}$$

Supplementary Problems

7.43. What average velocity in a 6″ pipe will produce a flow of 1.0 mgd of water? *Ans.* 7.87 ft/sec

7.44. What size pipe can carry 2.36 m^3/s at an average velocity of 3.0 m/s? *Ans.* 1 m

7.45. A 12″ pipe carrying 3.93 cfs connects to a 6″ pipe. Find the velocity head in the 6″ pipe. *Ans.* 6.21 ft

7.46. A 150-mm pipe carries 0.0813 m^3/s of water. The pipe branches into two pipes, one 50 mm in diameter and the other 100 mm in diameter. If the velocity in the 50-mm pipe is 12.2 m/s, what is the velocity in the 100-mm pipe? *Ans.* 7.32 m/s

7.47. Determine if the following expressions for velocity components satisfy the conditions for steady, incompressible flow. (*a*) $u = 3xy^2+2x+y^2$; $v = x^2-2y-y^3$. (*b*) $u = 2x^2+3y^2$; $v = -3xy$. *Ans.* (*a*) Yes (*b*) No

7.48. A 12″ diameter pipe carries oil with a velocity distribution of $v = 9(r_o^2 - r^2)$. Determine the average velocity and the value of the kinetic energy correction factor. *Ans.* $\alpha = 2.00$, $V_{av} = 1.13$ ft/sec

7.49. Show that the equation of continuity can be written in the form $1 = \dfrac{1}{A}\displaystyle\int_A\left(\frac{v}{V_{av}}\right)dA$.

7.50. A 300-mm pipe carries oil of sp gr 0.812 at a rate of 0.111 m^3/s, and the pressure at a point *A* is 18.4 kPa gage. If point *A* is 1.89 m above the datum plane, calculate the energy at *A*. *Ans.* 4.3 m · kN/kN

7.51. How many lb/sec of carbon dioxide are flowing through a 6″ pipe when the pressure is 25.0 psi gage, the temperature 80°F, and the average velocity 8.00 ft/sec? *Ans.* 0.476 lb/sec

7.52. Water flows through a 200-mm-diameter pipe at a velocity of 2.00 m/s. Determine the volume flow rate, weight flow rate, and mass flow rate. *Ans.* 0.0628 m^3/s, 0.615 kN/s, 62.8 kg/s

7.53. An 8″ pipe carries air at 80.0 ft/sec, 21.5 psi absolute, and 80°F. How many pounds of air are flowing? The 8″ pipe reduces to a 4″ pipe, and the pressure and temperature in the 4″ pipe are 19.0 psi absolute and 52°F, respectively. Find the velocity in the 4″ pipe and compare the flows in cfs in the two pipes. *Ans.* 3.00 lb/sec, 343 ft/sec, 27.9 cfs, 29.9 cfs

7.54. Air flows with a velocity of 4.88 m/s in a 100-mm pipe. A pressure gage measures 207 kPa, and the temperature is 16°C. At another point downstream a gage measures 138 kPa, and the temperature is 27°C. For a standard barometric reading, calculate the velocity downstream and compare the rate of flow at each section. *Ans.* 6.5 m/s, 0.040 m^3/s, 0.053 m^3/s

7.55. Sulfur dioxide flows through a 12″ discharge duct that reduces to 6″ in diameter at discharge into a stack. The pressures in the duct and discharge stream are, respectively, 20.0 psi absolute and atmospheric (14.7 psi). The velocity in the duct is 50.0 ft/sec, and the temperature is 80°F. Calculate the velocity in the discharge stream if the temperature of the gas is 23°F. *Ans.* 244 ft/sec

7.56. Gas flows through a conduit as shown in Fig. 7-19. For the data indicated on the figure, determine the mass flow rate of the gas and its mass density at section 2. *Ans.* 0.399 kg/s, 1.76 kg/m^3

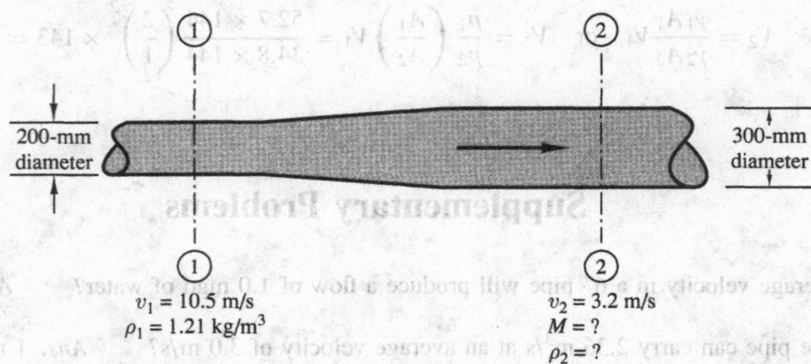

Fig. 7-19

7.57. Water flows through a horizontal 150-mm pipe under a pressure of 414 kPa. Assuming no losses, what is the flow if the pressure at a 75-mm diameter reduction is 138 kPa? *Ans.* 0.11 m^3/s

7.58. If oil of specific gravity 0.752 flows in Problem 7.57, find flow Q. *Ans.* 0.13 m^3/s

7.59. If carbon tetrachloride (sp gr 1.594) flows in Problem 7.57, find flow Q. *Ans.* 0.087 m^3/s

7.60. Water flows upward in a vertical 300-mm pipe at the rate of 0.222 m^3/s. At point A in the pipe the pressure is 210 kPa. At B, 4.57 m above A, the diameter is 600 mm, and the lost head A to B equals 1.83 m. Determine the pressure at B. *Ans.* 152 kPa

7.61. A 12″ pipe contains a short section in which the diameter is gradually reduced to 6″ and then enlarged again to 12″. The 6″ section is 2 ft below section A in the 12″ section, where the pressure is 75 psi. If a differential gage containing mercury is attached to the 12″ and 6″ sections, what is the deflection of the gage when the flow of water is 4.25 cfs downward? Assume no lost head. *Ans.* 6.46 in

7.62. A fluid is flowing in a 150-mm-diameter pipe with a velocity of 2.50 m/s. The fluid pressure is 35 kPa. The elevation of the pipe's center above a given datum is 5.0 m. Determine the total energy head if the fluid is (*a*) water, (*b*) ammonia with a specific gravity of 0.83, (*c*) gas with a specific weight of 12.5 N/m^3. *Ans.* 8.89 m, 9.63 m, 2805.32 m

7.63. A 12″ pipe line carries oil of specific gravity 0.811 at a velocity of 80.0 ft/sec. At points *A* and *B*, measurements of pressure and elevation were 52.6 psi and 100.0 ft and 42.0 psi and 110.0 ft, respectively. For steady flow, find the lost head between *A* and *B*. *Ans.* 20.2 ft

7.64. A stream of water 75 mm in diameter discharges into the atmosphere at a velocity 24.4 m/s. Find the power in the jet using the datum plane through the center of the jet. *Ans.* 33 kW

7.65. A reservoir supplies water to a horizontal 6″ pipe 800 ft long. The pipe flows full and discharges into the atmosphere at the rate of 2.23 cfs. What is the pressure in psi midway in the pipe, assuming the only lost head is 6.20 ft in each 100 ft of length? *Ans.* 10.7 psi

7.66. A 100-mm-diameter jet of water is discharged (horizontally) from a nozzle into the air. The flow rate of the water jet is 0.22 m^3/s. Determine the power in the jet. Assume the jet of water is at the datum. *Ans.* 86.2 kW

7.67. Oil of specific gravity 0.750 is pumped from a tank over a hill through a 24″ pipe with the pressure at the top of the hill maintained at 25.5 psi. The summit is 250 ft above the surface of the oil in the tank, and oil is pumped at the rate of 22.0 cfs. If the lost head from tank to summit is 15.7 ft, what horsepower must the pump supply to the liquid? *Ans.* 645 hp

7.68. A pump draws water from a sump through a vertical 6″ pipe. The pump has a horizontal discharge pipe 4″ in diameter that is 10.6 ft above the water level in the sump. While pumping 1.25 cfs, gages near the pump at entrance and discharge read −4.6 psi and +25.6 psi, respectively. The discharge gage is 3.0 ft above the suction gage. Compute the horsepower output of the pump and the head lost in the 6″ suction pipe. *Ans.* 10.7 hp, 2.4 ft

7.69. Compute the lost head in a 150-mm pipe if it is necessary to maintain a pressure of 231 kPa at a point upstream and 1.83 m below where the pipe discharges water into the atmosphere at the rate of 0.0556 m^3/s. *Ans.* 21.7 m

7.70. A large tank is partly filled with water, the air space above being under pressure. A 2″ hose connected to the tank discharges on the roof of a building 50 ft above the level in the tank. The friction loss is 18 ft. What air pressure must be maintained in the tank to deliver 0.436 cfs on the roof? *Ans.* 32.1 psi

7.71. Water flows from section 1 to section 2 in the pipe shown in Fig. 7-20. For the data given in the figure, determine the velocity of flow and the fluid pressure at section 2. Assume that the total head loss from section 1 to section 2 is 3.00 m. *Ans.* 8.00 m/s, 260 kPa

7.72. Water is pumped from reservoir *A* at elevation 750.0′ to reservoir *E* at elevation 800.0′ through a 12″ pipeline. The pressure in the 12″ pipe at point *D*, at elevation 650.0′, is 80.0 psi. The lost heads are: *A* to pump suction $B = 2.0$ ft, pump discharge *C* to $D = 38V^2/2g$, and *D* to $E = 40V^2/2g$. Find discharge *Q* and horsepower supplied by pump *BC*. *Ans.* 5.95 cfs, 82 hp

7.73. A horizontal Venturi meter has diameters at inlet and throat of 24″ and 18″, respectively. A differential gage connected to inlet and throat contains water that is deflected 4″ when air flows through the meter. Considering the specific weight of air to be constant at 0.0800 lb/ft^3 and neglecting friction, determine the flow in cfs. *Ans.* 276 cfs

7.74. Water is to be siphoned from a tank at the rate of 0.0892 m^3/s. The flowing end of the siphon pipe must be 4.27 m below the water surface. The lost head terms are $1.50V^2/2g$ from tank to summit of siphon and $1.00V^2/2g$ from summit to end of siphon. The summit is 1.52 m above the water surface. Find the size pipe needed and the pressure at the summit. *Ans.* 150 mm, −45 kPa

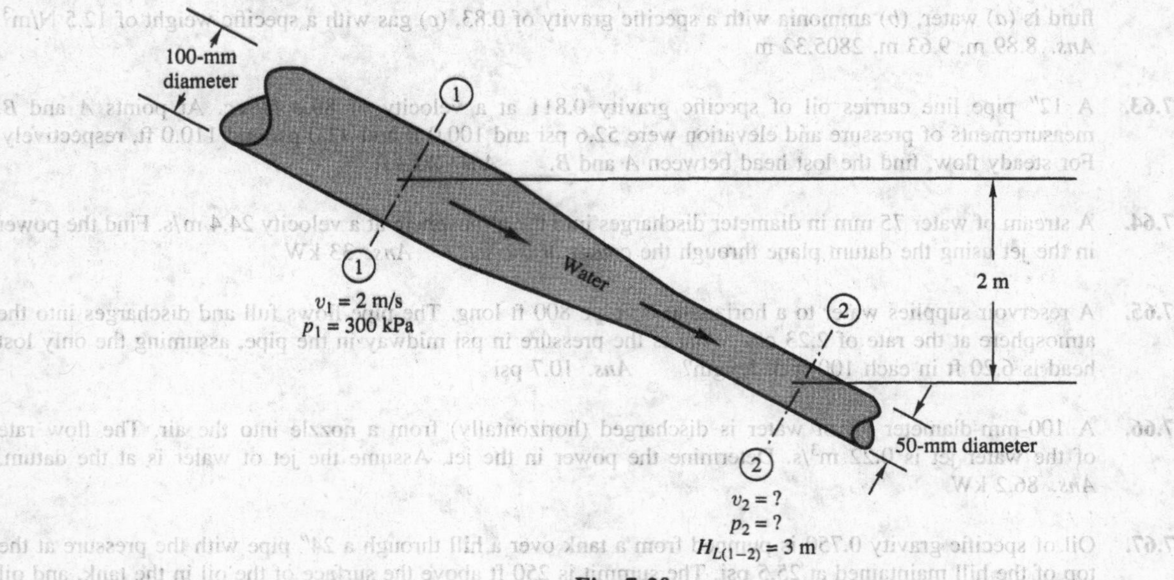

Fig. 7-20

7.75. A horizontal 24″ pipeline carries oil of sp gr 0.825 flowing at a rate of 15.7 cfs. Each of four pumps required along the line is the same, i.e., the pressure on the suction side and on the discharge side will be −8.0 psi and +350 psi, respectively. If the lost head at the discharge stated is 6.0 ft for each 1000 ft of pipe, how far apart may the pumps be placed? *Ans.* 167,000 ft

7.76. Oil with a specific gravity of 0.87 is flowing in the pipe shown in Fig. 7-21. The pressure at point 1 is 500 kPa. If the head loss from point 1 to point 2 is 5.00 m of oil and the discharge of the oil is 0.050 m^3/s, determine the pressure at point 2. *Ans.* 721 kPa

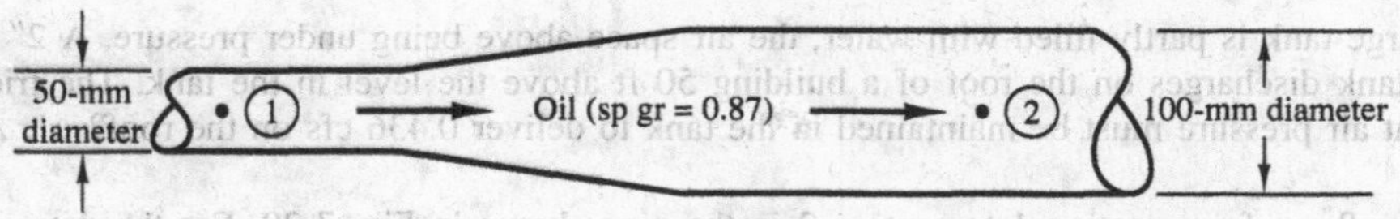

Fig. 7-21

7.77. A large closed tank is filled with air under a pressure of 5.3 psi gage and at a temperature of 65°F. The air discharges into the atmosphere (14.7 psi) through a small opening in the side of the tank. Neglecting friction losses, compute the velocity of the air leaving the tank assuming (*a*) constant density of the air and (*b*) adiabatic flow conditions. *Ans.* 692 ft/sec, 728 ft/sec

7.78. For Problem 7.77, if the pressure were 10.0 psi, what would be the velocity for (*a*) and for (*b*)? *Ans.* 855 ft/sec, 934 ft/sec

7.79. Carbon dioxide flows from a 25-mm pipe where the pressure is 414 kPa gage and the temperature is 4°C into a 12.5-mm pipe at the rate of 0.267 N/s. Neglecting friction and assuming isothermal conditions, determine the pressure in the 12.5-mm pipe. *Ans.* 19.2 kPa absolute

7.80. An air blower is to deliver 40,000 ft^3/min. Two U-tube gages measure suction and discharge pressures. The suction gage reads a negative 2″ of water. The discharge gage, located 3 ft above the point to which the suction gage is attached, reads +3″ of water. The discharge and suction ducts are of equal diameter. What size motor should be used to drive the blower if the overall efficiency is 68% ($\gamma = 0.0750$ lb/ft^3 for air)? *Ans.* 46.7 hp

7.81. A nozzle is attached to a pipe as shown in Fig. 7-22. Determine the water jet's velocity for the conditions shown in the figure. Assume head loss in the jet is negligible. *Ans.* 32.7 m/s

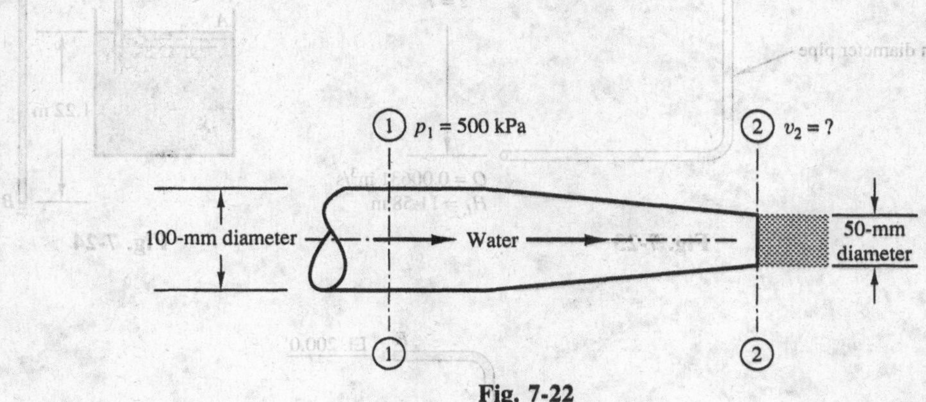

Fig. 7-22

7.82. A 12″ pipe is being tested to evaluate the lost head. When the flow of water is 6.31 cfs, the pressure at point A in the pipe is 40 psi. A differential gage is attached to the pipe at A and at a point B downstream that is 10 ft higher than A. The deflection of mercury in the gage is 3.33 ft, indicating a greater pressure at A. What is the lost head A to B? *Ans.* 42.0 ft

7.83. Prandtl has suggested that the velocity distribution for turbulent flow in conduits can be approximated by using $v = v_{\max}(y/r_o)^{1/7}$, where r_o is the pipe radius and y is the distance from the pipe wall. Determine the expression for average velocity in the pipe in terms of the center velocity $v_{\max}$. *Ans.* $V = 0.817v_{\max}$

7.84. What is the kinetic energy correction factor for the velocity distribution of Problem 7.83? *Ans.* $\alpha = 1.06$

7.85. Two large plates are spaced 1″ apart. Show that $\alpha = 1.54$ if the velocity profile is represented by $v = v_{\max}(1 - 576r^2)$, where r is measured from the centerline between the plates.

7.86. Air flows isentropically through a duct whose area varies. For steady flow, show that the velocity V_2 at any section downstream from section 1 can be written

$$V_2 = V_1(p_1/p_2)^{1/k}(A_1/A_2) \text{ for any shaped duct, and } V_2 = V_1(p_1/p_2)^{1/k}(D_1/D_2)^2 \text{ for a circular duct.}$$

7.87. Water is to be delivered from a reservoir through a pipe to a lower level and discharged into the air, as shown in Fig. 7-23. For the data given in the figure, find the vertical distance between the point of water discharge and the water surface in the reservoir. *Ans.* 12.11 m

7.88. The pressure inside the pipe at S must not fall below 23.9 kPa absolute. Neglecting losses, how high above water level A may point S be located? Refer to Fig. 7-24. *Ans.* 6.7 m

7.89. Pump B delivers a head of 140.6 ft to the water flowing to E as shown in Fig. 7-25. If the pressure at C is – 2.0 psi and the lost head from D to E is $8.0(V^2/2g)$, what flow occurs? *Ans.* 8.91 cfs

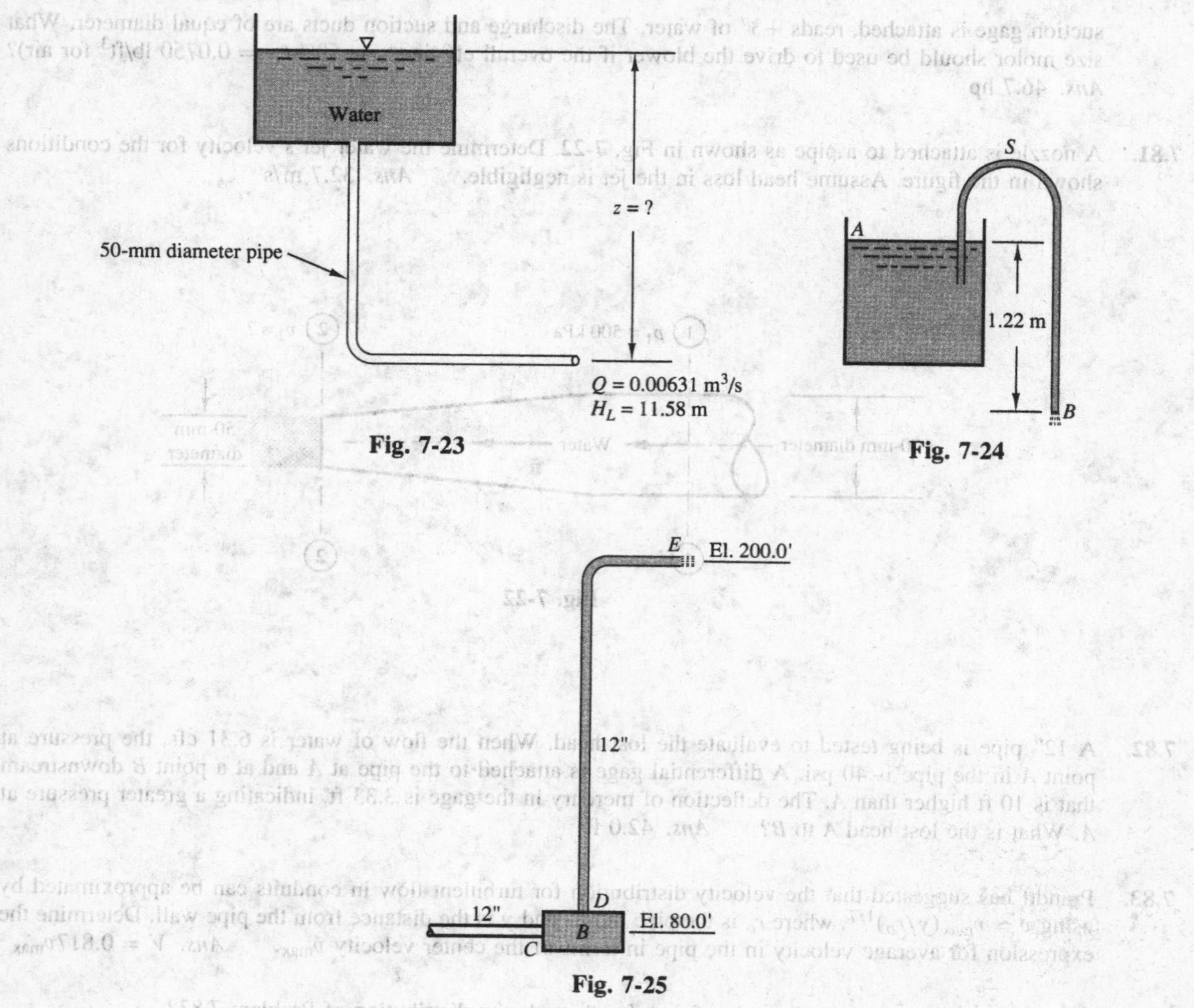

Fig. 7-23

Fig. 7-24

Fig. 7-25

7.90. Water flows radially between the two flanges at the end of a 6″ diameter pipe as shown in Fig. 7-26. Neglecting losses, if the pressure head at A is -1.0 ft, find the pressure head at B and the flow in cfs.
Ans. -0.15 ft, 3.88 cfs

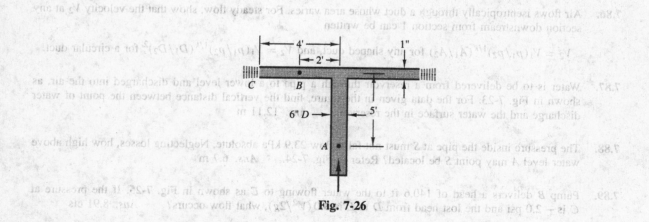

4'
2'
1"
C
B
6" D
5'
A

Fig. 7-26

7.91. The pipe shown in Fig. 7-27 has a uniform diameter of 150 mm. Assume the head loss between points 1 and 2 is 1.2 m and between points 2 and 3 is 2.0 m. Determine the discharge of water in the pipe and the pressure head at point 2. *Ans.* 0.102 m^3/s, −40.11 kPa

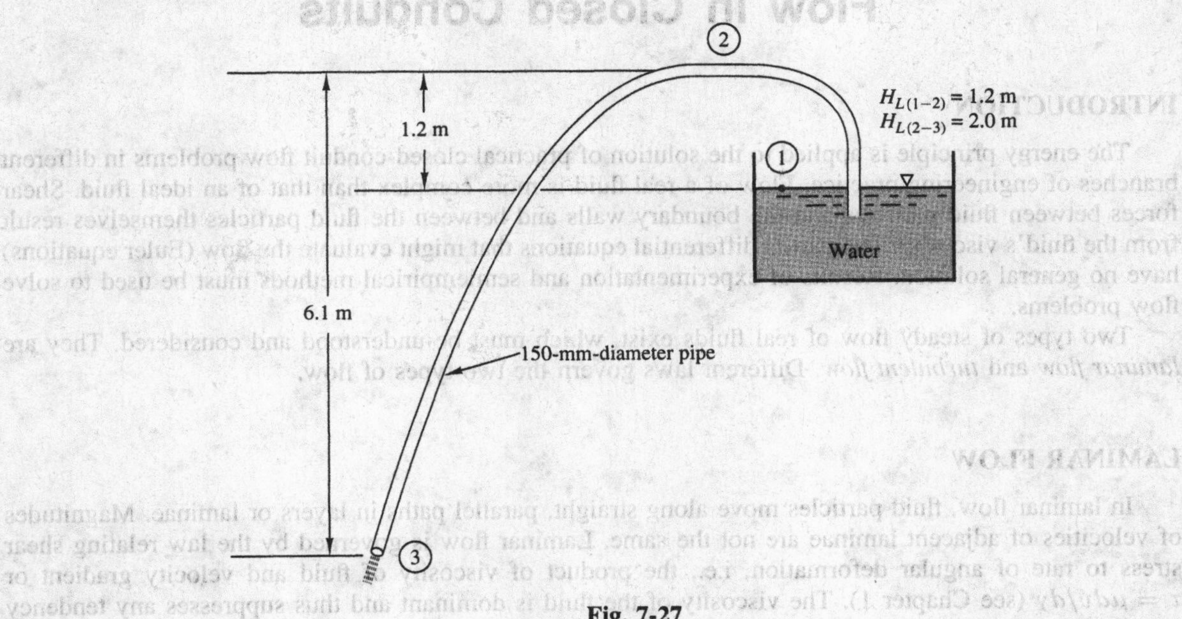

Fig. 7-27

7.92. Show that the average velocity V in a circular pipe of radius r_o equals $2v_{\max}\left[\dfrac{1}{(K+1)(K+2)}\right]$ for a velocity distribution that can be expressed as $v = v_{\max}(1 - r/r_o)^K$.

7.93. Find the kinetic energy correction factor α for Problem 7.92. *Ans.* $\alpha = \dfrac{(K+1)^3(K+2)^3}{4(3K+1)(3K+2)}$

Chapter 8

Flow in Closed Conduits

INTRODUCTION

The energy principle is applied to the solution of practical closed-conduit flow problems in different branches of engineering practice. Flow of a real fluid is more complex than that of an ideal fluid. Shear forces between fluid particles and the boundary walls and between the fluid particles themselves result from the fluid's viscosity. The partial differential equations that might evaluate the flow (Euler equations) have no general solution. Results of experimentation and semiempirical methods must be used to solve flow problems.

Two types of steady flow of real fluids exist, which must be understood and considered. They are *laminar flow* and *turbulent flow*. Different laws govern the two types of flow.

LAMINAR FLOW

In laminar flow, fluid particles move along straight, parallel paths in layers or laminae. Magnitudes of velocities of adjacent laminae are not the same. Laminar flow is governed by the law relating shear stress to rate of angular deformation, i.e., the product of viscosity of fluid and velocity gradient or $\tau = \mu dv/dy$ (see Chapter 1). The viscosity of the fluid is dominant and thus suppresses any tendency to turbulent conditions.

CRITICAL VELOCITY

The critical velocity of practical interest to engineers is the velocity below which all turbulence is damped out by the viscosity of the fluid. It is found that the upper limit of laminar flow of practical interest is represented by a Reynolds number of about 2000.

REYNOLDS NUMBER

The Reynolds number (Re), which is dimensionless, represents the ratio of inertia forces to viscous forces (see Chapter 6 on Dynamic Similitude).

For circular pipes flowing full,

$$\text{Reynolds number Re} = \frac{Vd\rho}{\mu} \quad \text{or} \quad \frac{Vd}{\nu} = \frac{V(2r_o)}{\nu} \tag{1a}$$

where V = mean velocity in ft/sec or m/s
d = diameter of pipe and r_o = radius of pipe in ft or m
ν = kinematic viscosity of the fluid in ft^2/sec or m^2/s
ρ = mass density of fluid in slugs/ft^3 or lb-sec^2/ft^4 or kg/m^3 or N·s^2/m^4
μ = absolute viscosity in lb-sec/ft^2 or N·s/m^2

For noncircular cross sections, the ratio of cross-sectional area to wetted perimeter, called the *hydraulic radius R* (in feet or meters), is used in the Reynolds number. The expression becomes

$$\text{Re} = \frac{V(4R)}{\nu} \tag{1b}$$

TURBULENT FLOW

In turbulent flow, fluid particles move in a haphazard fashion in all directions. It is virtually impossible to trace the motion of an individual particle.

The shear stress for turbulent flow can be expressed as

$$\tau = (\mu + \eta)\frac{dv}{dy} \tag{2a}$$

where η (eta) is a factor depending upon the fluid's density and fluid motion. The first factor (μ) represents effects of viscous action, and the second (η) accounts for effects of turbulent action.

Results of experimentation provide means by which solutions for shear stress in turbulent flow can be accomplished. Prandtl suggested that

$$\tau = \rho l^2 \left(\frac{dv}{dy}\right)^2 \tag{2b}$$

was a valid equation for shear stress in turbulent flow. This expression has the disadvantage that mixing length l is a function of y. The greater the distance y from the pipe wall, the greater the value of l. Later, von Karman suggested that

$$\tau = \tau_o\left(1 - \frac{y}{r_o}\right) = \rho k^2 \frac{(dv/dy)^4}{(d^2v/dy^2)^2} \tag{2c}$$

While k is not precisely constant, this dimensionless number is approximately 0.40. Integration of this expression leads to formulas of the type shown as (*7b*) below.

SHEARING STRESS AT A PIPE WALL

Shearing stress at a pipe wall, as developed in Problem 8.5, is

$$\tau_o = f\rho V^2/8 \text{ in psf or Pa} \tag{3}$$

where f is a dimensionless frictional factor (described in a subsequent paragraph).

It will be shown in Problem 8.4 that shear variation at a cross section is linear and that

$$\tau = \left(\frac{p_1 - p_2}{2L}\right) r \quad \text{or} \quad \tau = \left(\frac{\gamma h_L}{2L}\right) r \tag{4}$$

The term $\sqrt{\tau_o/\rho}$ is called *shear velocity* or *friction velocity* and is designated by the symbol v_*. From expression (*3*) we obtain

$$v_* = \sqrt{\tau_o/\rho} = V\sqrt{f/8} \tag{5}$$

VELOCITY DISTRIBUTION

Velocity distribution at a cross section will follow a parabolic law of variation for *laminar* flow. Maximum velocity is at the center and is twice the average velocity. The equation of the velocity profile for laminar flow (see Problem 8.6) can be expressed as

$$v = v_c - \left(\frac{\gamma h_L}{4\mu L}\right) r^2 \tag{6}$$

For *turbulent* flows, more uniform velocity distribution results. From experiments of Nikuradse and others, equations of velocity profiles in terms of center velocity v_c or shear velocity v_* follow.

(*a*) An empirical formula is

$$v = v_c(y/r_o)^n \tag{7a}$$

where $n = \frac{1}{7}$ for smooth tubes, up to Re $= 100{,}000$
$n = \frac{1}{8}$ for smooth tubes for Re from 100,000 to 400,000

(*b*) For *smooth* pipes,

$$v = v_*[5.5 + 5.75 \log(yv_*/\nu)] \tag{7b}$$

For the yv_*/ν term, see Problem 8.8, part (*e*).

(*c*) For *smooth* pipes (for 5000 < Re < 3,000,000) and for pipes in the wholly rough zone,

$$(v_c - v) = -2.5\sqrt{v_o/\rho}\,\ln(y/r_o) = -2.5 v_* \ln(y/r_o) \tag{7c}$$

In terms of average velocity V, Vennard suggests that V/v_c may be written

$$\frac{V}{v_c} = \frac{1}{1 + 4.07\sqrt{f/8}} \tag{8}$$

(*d*) For *rough* pipes,

$$v = v_*[8.5 + 5.75 \log(y/\epsilon)] \tag{9a}$$

where ϵ is the absolute roughness of the boundary.

(*e*) For *rough or smooth* boundaries,

$$\frac{v - V}{V\sqrt{f}} = 2 \log\frac{y}{r_o} + 1.32 \tag{9b}$$

Also

$$v_c/V = 1.43\sqrt{f} + 1 \tag{9c}$$

LOSS OF HEAD FOR LAMINAR FLOW

Loss of head for laminar flow is expressed by the Hagen-Poiseuille equation. The expression is

$$\text{lost head} = \frac{32(\text{viscosity } \mu)(\text{length } L)(\text{average velocity } V)}{(\text{specific weight } \gamma)(\text{diameter } d)^2}$$

$$= \frac{32\mu LV}{\gamma d^2} \tag{10a}$$

In terms of kinematic viscosity, we obtain, since $\mu/\gamma = \nu/g$,

$$\text{lost head} = \frac{32\nu LV}{gd^2} \tag{10b}$$

DARCY-WEISBACH FORMULA

The Darcy-Weisbach formula, as developed in Chapter 6, Problem 6.11, is the basis for evaluating lost head for fluid flow in pipes and conduits. The equation is

$$\text{lost head} = \text{friction factor } f \times \frac{\text{length } L}{\text{diameter } d} \times \text{velocity head } \frac{V^2}{2g}$$

$$= f\left(\frac{L}{d}\right)\left(\frac{V^2}{2g}\right) \tag{11}$$

As noted in Chapter 7, the exact velocity head at a cross section is obtained by multiplying the average velocity squared $(Q/A)^2$ by a coefficient α and dividing by $2g$. For turbulent flow in pipes and conduits, α may be considered as unity without causing appreciable error in the results.

FRICTION FACTOR

Friction factor f can be derived mathematically for laminar flow, but no simple mathematical relation for the variation of f with Reynolds number is available for turbulent flow. Furthermore, Nikuradse and others found that the relative roughness of a pipe (ratio of size of surface imperfections ϵ to inside diameter of the pipe) affects the value of f also.

(*a*) For *laminar* flow, equation (*10b*) can be rearranged as follows:

$$\text{lost head} = 64\frac{\nu}{Vd}\left(\frac{L}{d}\right)\left(\frac{V^2}{2g}\right) = \frac{64}{\text{Re}}\left(\frac{L}{d}\right)\left(\frac{V^2}{2g}\right) \tag{12a}$$

Thus, *for laminar flow in all pipes for all fluids*, the value of f is

$$f = 64/\text{Re} \tag{12b}$$

Re has a practical maximum value of 2000 for laminar flow.

(*b*) For *turbulent flow*, many hydraulic engineers have endeavored to evaluate f from the results of their own experiments and from those of others.

(*1*) For turbulent flow in *smooth and rough pipes*, universal resistance laws can be derived from

$$f = 8\tau_o/\rho V^2 = 8V_*^2/V^2 \tag{13}$$

(*2*) For *smooth* pipes, Blasius suggests, for Reynolds numbers between 3000 and 100,000,

$$f = 0.316/\text{Re}^{0.25} \tag{14}$$

For values of Re up to about 3,000,000, von Karman's equation modified by Prandtl is

$$1/\sqrt{f} = 2\log\left(\text{Re}\sqrt{f}\right) - 0.8 \tag{15}$$

(*3*) For *rough* pipes,

$$1/\sqrt{f} = 2\log(r_o/\epsilon) + 1.74 \tag{16}$$

(*4*) For *all* pipes, the Hydraulic Institute and many engineers consider the Colebrook equation reliable when evaluating f. This equation is

$$\frac{1}{\sqrt{f}} = -2\log\left[\frac{\epsilon}{3.7d} + \frac{2.51}{\text{Re}\sqrt{f}}\right] \tag{17}$$

Inasmuch as equation (*17*) is awkward to solve, diagrams are available to give the relation between friction factor *f*, Reynolds number Re, and relative roughness ϵ/d. Two such diagrams are included in the Appendix. Diagram A-1 (the Moody Diagram, published through the courtesy of the American Society of Mechanical Engineers) is usually used when flow Q is known, and Diagram A-2 is used when the flow is to be evaluated. The latter form was first suggested by S. P. Johnson and by Hunter Rouse.

It should be observed that for smooth pipes where the value of ϵ/d is very small, the first term in brackets of equation (*17*) can be neglected; then equation (*17*) and equation (*15*) are similar. Likewise, should the Reynolds number Re be very large, the second term in brackets in equation (*17*) can be neglected; in such cases the effect of viscosity is negligible, and f depends upon the pipe's relative roughness. This statement is indicated graphically in Diagram A-1 by the fact that the curves become horizontal at high Reynolds numbers.

Before formulas or diagrams can be used, engineers must estimate a conduit's relative roughness (ϵ/d). Diagrams A-1 and A-2 in the Appendix give some values of surface imperfections (ϵ) for various new surfaces.

MINOR HEAD LOSSES

Other head losses are generally categorized as "minor" head losses. They result when there is a significant change in flow pattern. Hence, they occur in conduit contractions and enlargements (both sudden and gradual), valves, fittings, bends, etc., and entrance to or exit from a conduit. In some cases, a "minor" loss can be quite important.

Entrance losses occur when a liquid enters a conduit from a large tank or reservoir. The amount of head loss is significantly dependent on the shape of the entrance. If an entrance is well rounded, the entrance loss will be very small. *Exit losses* occur when a liquid exits a conduit and enters a large tank or reservoir. *Sudden contraction losses* occur when there is an abrupt decrease in conduit size, and *sudden expansion losses* happen when there is an abrupt increase in conduit size. Similarly, *gradual expansion losses* occur when there is a gradual increase in conduit size, and *gradual contraction losses* happen when there is a gradual decrease in conduit size.

Theoretical considerations of the various minor losses are quite complicated; therefore, minor losses are usually evaluated by empirical methods. They are commonly expressed in terms of the applicable velocity head. In equation form,

$$\text{lost head (ft or m)} = K\left(\frac{V^2}{2g}\right) \tag{18}$$

Tables 4 and 5 of the Appendix give empirically determined values of the minor loss coefficient [K in equation (18)] for use in solving problems where minor losses are encountered.

EMPIRICAL EQUATIONS FOR WATER FLOW

There are available several empirical formulas that can be used to solve, approximately, water flow problems in closed conduits. Two such formulas will be considered here—the Hazen-Williams formula and the Manning formula.

The *Hazen-Williams formula* is given by

$$V = 1.318CR^{0.63}S^{0.54} \text{ (fps units)} \tag{19a}$$

where V = velocity in ft/sec
R = hydraulic radius in ft
C = Hazen-Williams roughness coefficient
S = slope of the energy grade line (head loss per unit length of conduit)

(The hydraulic radius was defined in Chapter 7 as the cross-sectional area divided by the wetted perimeter.) For velocity in m/s with an input hydraulic radius in m, the coefficient 1.318 in equation (*19a*) is replaced by 0.8492; that is,

$$V = 0.8492CR^{0.63}S^{0.54} \text{ (SI units)} \tag{19b}$$

Some typical values of the Hazen-Williams roughness coefficient are given in Table 6 of the Appendix.

The *Manning formula* is given by

$$V = \frac{1.486}{n}R^{2/3}S^{1/2} \text{ (fps units)} \tag{20a}$$

where V = velocity in ft/sec
R = hydraulic radius in ft
n = Manning roughness coefficient
S = slope of the energy grade line (head loss per unit length of conduit)

For velocity in m/s with an input hydraulic radius in m, the coefficient 1.486 in equation (20*a*) is replaced by 1.0; that is,

$$V = \frac{1.0}{n} R^{2/3} S^{1/2} \text{ (SI units)} \qquad (20b)$$

Some typical values of the Manning roughness coefficient are given in Table 9 of the Appendix.

Both the Hazen-Williams and Manning formulas can be used to analyze closed-conduit flow. The former has been used for designing water supply systems in the United States. The Manning formula is generally used less often in closed-conduit flow and more often in open-channel flow (Chapter 10). Both formulas have some important limitations and disadvantages. They can be used only for flow of water at normal temperatures (since fluid viscosity is not considered). They are applicable only to fairly highly turbulent flow (i.e., high Reynolds numbers).

PIPE DIAGRAMS

Prior to the advent of modern computers and hand-held calculators, computations by both the Hazen-Williams and Manning formulas were cumbersome because of the exponents involved. Accordingly, many charts, graphs, tables, diagrams, etc., were developed to facilitate pipe analyses. Nowadays, such computations are simple because of the availability of modern computing capability; nevertheless, charts, graphs, etc., can be helpful when many quick solutions are needed [e.g., when determining "equivalent pipes" (Chapter 9)].

Diagrams B-1 through B-5 in the Appendix give five charts that can be used to solve problems using the Hazen-Williams and Manning formulas. Diagram B-1 is for the Hazen-Williams formula with $C = 100$ and flow rate in millions of gallons per day (mgd). As illustrated at the bottom of the figure, the diagram can also be used to solve problems having other values of C. Diagrams B-2 and B-3 are also for the Hazen-Williams formula but with $C = 120$. The former is used with fps units; the latter, when values are in SI units. Diagrams B-4 and B-5 are for the Manning formula with $n = 0.013$. Again, the former is used with fps units and the latter when values are in SI units.

It should be emphasized again that, because these pipe diagrams are based on the Hazen-Williams and Manning formulas, their use is limited to water flow problems at normal temperatures and high Reynolds numbers. Their use is also restricted to circular conduits flowing full.

Solved Problems

8.1. Determine the critical velocity for (*a*) gasoline at 20°C flowing through a 20-mm pipe and (*b*) water at 20°C flowing in the 20-mm pipe.

Mathcad

Solution:

(*a*) For laminar flow, the maximum value of Reynolds number is 2000. From Table 2 in the Appendix, the kinematic viscosity at 20°C is 6.48×10^{-7} m²/s.

$$2000 = \text{Re} = V_c d/\nu = V_c \left(\frac{20}{1000}\right) / (6.48 \times 10^{-7}), \qquad V_c = 0.0648 \text{ m/s}$$

(*b*) From Table 2, $\nu = 1.02 \times 10^{-6}$ m²/s for 20°C water.

$$2000 = V_c\left(\frac{20}{1000}\right)/(1.02 \times 10^{-6}), \quad V_c = 0.102 \text{ m/s}$$

8.2. **Determine the type of flow occurring in a 12″ pipe when (*a*) water at 60°F flows at a velocity of 3.50 ft/sec and (*b*) heavy fuel oil at 60°F flows at the same velocity.**

Solution:

(*a*) $\text{Re} = Vd/\nu = (3.50)(1)/(1.217 \times 10^{-5}) = 288{,}000 > 2000$. The flow is turbulent.

(*b*) From Table 2 in the Appendix, $\nu = 221 \times 10^{-5}$ ft²/sec.

$\text{Re} = Vd/\nu = (3.50)(1)/(221 \times 10^{-5}) = 1580 < 2000$. The flow is laminar.

8.3. **For laminar flow conditions, what size pipe will deliver 0.0057 m³/s of medium fuel oil at 4°C? ($\nu = 6.09 \times 10^{-6}$ m²/s)**

Solution:

$$V = Q/A = Q/\tfrac{1}{4}\pi d^2 = 4Q/\pi d^2 = 0.0228/\pi d^2$$

$$\text{Re} = \frac{Vd}{\nu}, \quad 2000 = \frac{0.0228}{\pi d^2}\left(\frac{d}{6.09 \times 10^{-6}}\right), \quad d = 0.596 \text{ m}$$

8.4. **Determine the nature of the distribution of shear stress at a cross section in a horizontal circular pipe under steady flow conditions.**

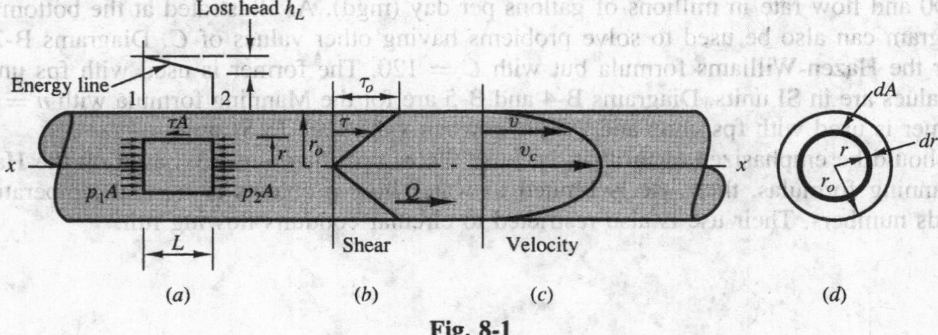

Fig. 8-1

Solution:

(*a*) For the free body in Fig. 8-1(*a*), since the flow is steady, each particle moves to the right without acceleration. Hence the summation of the forces in the x direction must equal zero.

$$p_1(\pi r^2) - p_2(\pi r^2) - \tau(2\pi r L) = 0 \quad \text{or} \quad \tau = \frac{(p_1 - p_2)r}{2L} \qquad (A)$$

When $r = 0$, the shear stress τ *is* zero; and when $r = r_o$, the stress τ_o at the wall is a maximum. The variation is linear and is so indicated in Fig. 8-1(*b*). Equation (*A*) holds for laminar and turbulent flows as no limitations concerning flow were imposed in the derivation.

Since $(p_1 - p_2)/\gamma$ represents the drop in the energy line, or the lost head h_L, multiplying equation (*A*) by γ/γ yields

$$\tau = \frac{\gamma r}{2L}\left(\frac{p_1 - p_2}{\gamma}\right) \qquad \text{or} \qquad \tau = \frac{\gamma h_L}{2L}r \tag{B}$$

8.5. Develop the expression for shear stress at a pipe wall.

Solution:

From Problem 8.4,

$$h_L = \frac{2\tau_o L}{\gamma r_o} = \frac{4\tau_o L}{\gamma d}$$

The Darcy-Weisbach formula is

$$h_L = f\left(\frac{L}{d}\right)\left(\frac{V^2}{2g}\right).$$

Equating these expressions, $\dfrac{4\tau_o L}{\gamma d} = f\left(\dfrac{L}{d}\right)\left(\dfrac{V^2}{2g}\right)$ and $\tau_o = f\left(\dfrac{\gamma}{g}\right)\left(\dfrac{V^2}{8}\right) = f\rho V^2/8.$

8.6. For steady laminar flow, (*a*) what is the relation between the velocity at a point in the cross section and the velocity at the center of the pipe, and (*b*) what is the equation for velocity distribution?

Mathcad

Solution:

(*a*) For laminar flow, the shear stress (see Chap. 1) is $\tau = -\mu(dv/dr)$. Equating this with the value of τ in equation (*A*) of Problem 8.4, we obtain

$$-\mu\frac{dv}{dr} = \frac{(p_1 - p_2)r}{2L}$$

Since $(p_1 - p_2)/L$ is not a function of r,

$$-\int_{v_c}^{v} dv = \frac{p_1 - p_2}{2\mu L}\int_o^r r\,dr \qquad \text{and} \qquad -(v - v_c) = \frac{(p_1 - p_2)r^2}{4\mu L}$$

or

$$v = v_c - \frac{(p_1 - p_2)r^2}{4\mu L} \tag{A}$$

But the lost head in L feet is $h_L = (p_1 - p_2)/\gamma$; hence

$$v = v_c - \frac{\gamma h_L r^2}{4\mu L} \tag{B}$$

(*b*) Since the velocity at the boundary is zero, when $r = r_o$, $v = 0$ in (*A*), and we have

$$v_c = \frac{(p_1 - p_2)r_o^2}{4\mu L} \quad \text{(at centerline)} \tag{C}$$

Thus, in general,

$$v = \frac{p_1 - p_2}{4\mu L}(r_o^2 - r^2) \tag{D}$$

8.7. Develop the expression for the loss of head in a pipe for steady laminar flow of an incompressible fluid. Refer to Fig. 8-1(*d*).

Mathcad

Solution:

$$V_{av} = \frac{Q}{A} = \frac{\int v\,dA}{\int dA} = \frac{\int_0^{r_o} v(2\pi r\,dr)}{\pi r_o^2} = \frac{2\pi(p_1 - p_2)}{\pi r_o^2(4\mu L)} \int_0^{r_o} (r_o^2 - r^2) r\,dr$$

from which

$$V_{av} = \frac{(p_1 - p_2) r_o^2}{8\mu L} \qquad (A)$$

Thus for laminar flow the average velocity is half the maximum velocity v_c in equation (C) of Problem 8.6. Rearranging (A), we obtain

$$\frac{p_1 - p_2}{\gamma} = \text{lost head} = \frac{8\mu L V_{av}}{\gamma r_o^2} = \frac{32\mu L V_{av}}{\gamma d^2} \qquad (B)$$

These expressions apply for *laminar flow of all fluids in all pipes and conduits.*

As stated at the beginning of this chapter, the lost head expression for laminar flow in the Darcy form is

$$\text{lost head} = \frac{64}{\text{Re}}\left(\frac{L}{d}\right)\left(\frac{V^2}{2g}\right) = f\left(\frac{L}{d}\right)\left(\frac{V^2}{2g}\right)$$

8.8. Determine (a) the shear stress at the walls of a 12″ diameter pipe when water flowing causes a measured lost head of 15 ft in 300 ft of pipe length, (b) the shear stress 2″ from the centerline of the pipe, (c) the shear velocity, (d) the average velocity for an f value of 0.050, (e) the ratio v/v_*.

Solution:

(a) Using equation (B) of Problem 8.4, when $r = r_o$ the shear stress at the wall is

$$\tau_o = \gamma h_L r_o / 2L = (62.4)(15)(\tfrac{1}{2})/600 = 0.780 \text{ psf } = 0.00542 \text{ psi}$$

(b) Since τ varies linearly from centerline to wall, $\tau = \left(\frac{2}{6}\right)(0.00542) = 0.00181$ psi

(c) By equation (5), $v_* = \sqrt{\tau_o/\rho} = \sqrt{0.780/1.94} = 0.634$ ft/sec.

(d) Using $h_L = f\left(\frac{L}{d}\right)\left(\frac{V^2}{2g}\right)$, we have $15 = 0.050\left(\frac{300}{1}\right)\left(\frac{V^2}{2g}\right)$ and $V = 8.02$ ft/sec.

Otherwise: From equation (3), $\tau_o = f\rho V^2/8$, $0.780 = 0.050(1.94)V^2/8$ and $V = 8.02$ ft/sec.

(e) From $\tau_o = \mu(v/y)$ and $\nu = \mu/\rho$ we obtain $\tau_o = \rho\nu(v/y)$ or $\tau_o/\rho = \nu(v/y)$.

Since $\tau_o/\rho = v_*^2$, we have $v_*^2 = \nu(v/y)$, $v/v_*^2 = y/\nu$, and $v/v_* = v_* y/\nu$.

8.9. If in Problem 8.8 the water is flowing through a 3 ft by 4 ft rectangular conduit of the same length with the same lost head, what is the shear stress between the water and the pipe wall?

Solution:

For noncircular conduits, the hydraulic radius is the appropriate hydraulic dimension. For a circular pipe,

$$\text{hydraulic radius } R = \frac{\text{cross-sectional area}}{\text{wetted perimeter}} = \frac{\pi d^2/4}{\pi d} = \frac{d}{4} = \frac{r_o}{2}$$

Substituting $r = 2R$ in equation (B) of Problem 8.4,

$$\tau = \frac{\gamma h_L}{L} R = \frac{(62.4)(15)}{300} \times \frac{(3 \times 4)}{(2)(3+4)} = 2.67 \text{ psf} = 0.0186 \text{ psi}$$

8.10. Medium lubricating oil, sp gr 0.860, is pumped through 300 m of horizontal 50-mm pipe at the rate of 0.00114 m^3/s. If the drop in pressure is 200 kPa, what is the absolute viscosity of the oil?

Mathcad

Solution:

Assuming laminar flow and referring to expression (*B*) in Problem 8.7, we obtain

$$p_1 - p_2 = \frac{32\mu L V_{av}}{d^2}, \text{ where } V_{av} = \frac{Q}{A} = \frac{0.00114}{\frac{1}{4}\pi(50/1000)^2} = 0.581 \text{ m/s}$$

Then

$$200,000 = 32\mu(300)(0.581)/(50/1000)^2 \quad \text{and} \quad \mu = 0.0896 \text{ N} \cdot \text{s/m}^2$$

Checking the original assumption of laminar flow means evaluating the Reynolds number for the conditions of flow. Thus

$$\text{Re} = \frac{Vd}{\nu} = \frac{Vd\gamma}{\mu g} = \frac{(0.581)(50/1000)(0.860 \times 9.79 \times 1000)}{(0.0896)(9.81)} = 278$$

Since Reynolds number < 2000, laminar flow exists and the value of μ is correct.

8.11. Oil of absolute viscosity 0.101 N·s/m^2 and sp gr 0.850 flows through 3000 m of 300-mm cast iron pipe at the rate of 0.0444 m^3/s. What is the lost head in the pipe?

Solution:

$$V = \frac{Q}{A} = \frac{0.0444}{\frac{1}{4}\pi(300/1000)^2} = 0.628 \text{ m/s}$$

$$\text{and Re} = \frac{Vd\gamma}{\mu g} = \frac{(0.628)(300/1000)(0.850 \times 9.79 \times 1000)}{(0.101)(9.81)} = 1582$$

which means laminar flow exists. Hence

$$f = \frac{64}{\text{Re}} = 0.0405 \quad \text{and} \quad \text{lost head} = f\left(\frac{L}{d}\right)\left(\frac{V^2}{2g}\right) = 0.0405 \times \frac{3000}{0.300} \times \frac{(0.628)^2}{(2)(9.81)} = 8.14 \text{ m}$$

8.12. Heavy fuel oil flows from *A* to *B* through 3000 ft of horizontal 6″ steel pipe. The pressure at *A* is 155 psi and at *B* is 5.0 psi. The kinematic viscosity is 0.00444 ft^2/sec, and the specific gravity is 0.918. What is the flow in cfs?

Solution:

The Bernoulli Equation, *A* to *B*, datum *A*, gives

$$\left(\frac{155 \times 144}{0.918 \times 62.4} + \frac{V_6^2}{2g} + 0\right) - f\left(\frac{3000}{\frac{1}{2}}\right)\left(\frac{V_6^2}{2g}\right) = \left(\frac{5 \times 144}{0.918 \times 62.4} + \frac{V_6^2}{2g} + 0\right)$$

or

$$377 = f(6000)(V_6^2/2g)$$

Both *V* and *f* are unknown and are functions of each other. If laminar flow exists, then from equation (*B*) of Problem 8.7,

$$V_{av} = \frac{(p_1 - p_2)d^2}{32\mu L} = \frac{(155 - 5)(144) \times \left(\frac{1}{2}\right)^2}{(32)(0.00444 \times 0.918 \times 62.4/32.2)(3000)} = 7.12 \text{ ft/sec}$$

and $Re = (7.12)\left(\frac{1}{2}\right)/0.00444 = 802$, hence laminar flow. Thus $Q = A_6V_6 = \frac{1}{4}\pi\left(\frac{1}{2}\right)^2 \times 7.12 = 1.40$ cfs.

Had the flow been turbulent, equation (*B*) of Problem 8.7 would not apply. Another approach will be used in Problem 8.16. Furthermore, had there been a difference in elevation between points *A* and *B*, the term $(p_1 - p_2)$ in equation (*B*) would be replaced by the drop in the hydraulic grade line, in lb/ft^2.

8.13. What size pipe should be installed to carry $0.0222\ \text{m}^3/\text{s}$ of heavy fuel oil at 16°C if the available lost head in the 300-m length of horizontal pipe is 6.7 m?

Solution:

For the oil, $\nu = 0.000205\ \text{m}^2/\text{s}$ and sp gr $= 0.912$. For such a large value of kinematic viscosity, assume laminar flow. Then

$$\text{lost head} = \frac{V_{\text{av}} \times 32\mu L}{\gamma d^2} \quad \text{and} \quad V_{\text{av}} = \frac{Q}{A} = \frac{0.0222}{\frac{1}{4}\pi d^2} = \frac{0.0283}{d^2}$$

$$\text{Substituting, } 6.7 = \frac{(0.0283/d^2)(32)(0.000205 \times 0.912 \times 9.79/9.81)(300)}{(0.912 \times 9.79)d^2}, \qquad d = 0.170\ \text{m}$$

Check assumption of laminar flow, using $d = 0.170$ m.

$$\text{Re} = \frac{Vd}{\nu} = \frac{(0.0283/d^2)d}{\nu} = \frac{0.0283}{0.170 \times 0.000205} = 812, \text{ hence laminar flow.}$$

Use a 170-mm pipe.

8.14. Gasoline is being discharged from a pipe at point 2 at elevation 66.66 m. Point 1, located 965.5 m along the pipe from point 2, is at elevation 82.65 m, and the pressure there is 2.50 kPa. If the pipe roughness is 0.500 mm, what pipe diameter is needed to discharge gasoline ($\gamma = 7.05\ \text{kN/m}^3$, $\mu = 2.92 \times 10^{-4}\ \text{N·s/m}^2$, $\rho = 719\ \text{kg/m}^3$) at a rate of $0.10\ \text{m}^3/\text{s}$?

Mathcad

Solution:

$$\text{lost head} = f(L/d)(V^2/2g) = f(965.5/d)\left[V^2/(2 \times 9.81)\right] = 49.21 f V^2/d$$

$$Q = AV = 0.10 = \left[(\pi)(d)^2/4\right]V; \quad V = 0.1273/d^2$$

$$\text{lost head} = (49.21 f)(0.1273/d^2)^2/d = 0.7975 f/d^5$$

Write the Bernoulli equation, 1 to 2, datum 2.

$$2.50/7.05 + V_1^2/2g + 82.65 = 0 + V_2^2/2g + 66.66 + 0.7975 f/d^5$$

Assume a value of $f = 0.0200$ and substitute it into the Bernoulli equation. Note that since $V_1 = V_2$ the velocity head terms cancel.

$$0.355 + 82.65 = 66.66 + (0.7975)(0.0200)/d^5$$

$$d = 0.250\ \text{m}$$

Now check to see if the assumed value $f = 0.0200$ is correct.

$$V = \frac{0.1273}{(0.250)^2} = 2.037\ \text{m/s}$$

$$\text{Re} = \frac{\rho d V}{\mu} = \frac{(719)(0.250)(2.037)}{2.92 \times 10^{-4}} = 1.25 \times 10^6$$

$$\epsilon/d = \frac{0.00050}{0.250} = 0.0020$$

From Diagram A-1, $f = 0.0235$. Evidently, the assumed value of f of 0.0200 was not the correct one. Assume a value of f of 0.0235 and repeat the computations.

$$0.355 + 82.65 = 66.66 + (0.7975)(0.0235)/d^5, \qquad d = 0.258 \text{ m}$$

Again, check to see if the assumed value of f is correct.

$$V = 0.1273/(0.258)^2 = 1.912 \text{ m/s}$$

$$\text{Re} = \frac{\rho d V}{\mu} = \frac{(719)(0.258)(1.912)}{2.92 \times 10^{-4}} = 1.21 \times 10^6 \qquad \frac{\epsilon}{d} = \frac{0.00050}{0.258} = 0.00194$$

From Diagram A-1, $f = 0.0235$. This value of f agrees with the last assumed value of f; hence the correct value of d, the required pipe diameter, is taken to be 0.258 m, or 258 mm.

8.15. Determine the lost head in 1000 ft of new, uncoated 12" inside diameter cast iron pipe when (*a*) water at 60°F flows at 5.00 ft/sec, and (*b*) medium fuel oil at 60°F flows at the same velocity.

Solution:

(*a*) When Diagram A-1 is to be used, the relative roughness must be evaluated and then the Reynolds number calculated. From the tabulation on Diagram A-1, the value of ϵ for uncoated cast iron pipe ranges from 0.0004 ft to 0.0020 ft. For an inside diameter of 1 ft and the design value of $\epsilon = 0.0008$ ft, the relative roughness $\epsilon/d = 0.0008/1 = 0.0008$.

Using the kinematic viscosity of water from Table 2 in the Appendix.

$$\text{Re} = Vd/\nu = (5.00)(1.00)/(1.217 \times 10^{-5}) = 411{,}000 \qquad \text{(turbulent flow)}$$

From Diagram A-1, for $\epsilon/d = 0.0008$ and Re = 411,000, $f = 0.0194$ and

$$\text{lost head} = (0.0194)(1000/1)(25/2g) = 7.5 \text{ ft}$$

Or, using Table 3 in the Appendix (for water only): $f = 0.0200$ and

$$\text{lost head} = f(L/d)(V^2/2g) = (0.0200)(1000/1)(25/2g) = 7.8 \text{ ft}$$

(*b*) For the oil, using Table 2, $\text{Re} = (5)(1)/(4.75 \times 10^{-5}) = 105{,}000$. For turbulent flow, from Diagram A-1, $f = 0.0213$ and lost head $= (0.0213)(1000/1)(25/2g) = 8.3$ ft.

In general, the degree of roughness of pipes *in service* cannot be estimated with great accuracy and so, in such cases, a precise value of f is not to be anticipated. For this reason, when using Diagrams A-1 and A-2 and Table 3 for other than new surfaces, it is suggested that the third significant figure in f be read or interpolated as either *zero or five*, no greater accuracy being warranted in most practical cases.

For laminar flow, for any pipe and any fluid, use $f = 64/\text{Re}$.

8.16. Points A and B are 4000 ft apart along a new 6" I.D. steel pipe. Point B is 50.5 ft higher than A, and the pressures at A and B are 123 psi and 48.6 psi, respectively. How much medium fuel oil at 70°F will flow from A to B? (From Diagram A-1, $\epsilon = 0.0002$ ft.)

Mathcad

Solution:

Reynolds number cannot be calculated immediately. Write the Bernoulli Equation, A to B, datum A.

$$\left(\frac{123 \times 144}{0.854 \times 62.4} + \frac{V_6^2}{2g} + 0\right) - f\left(\frac{4000}{\frac{1}{2}}\right)\left(\frac{V_6^2}{2g}\right) = \frac{48.6 \times 144}{0.854 \times 62.4} + \frac{V_6^2}{2g} + 50.5 \text{ and } \frac{V_6^2}{2g} = \frac{150.5}{8000f}$$

Also Re = Vd/ν. Substituting for V from above,

$$\text{Re} = \frac{d}{\nu}\sqrt{\frac{2g(150.5)}{8000f}} \quad or \quad \text{Re}\sqrt{f} = \frac{d}{\nu}\sqrt{\frac{2g(150.5)}{8000}} \qquad (A)$$

Since the term 150.5 is h_L or the drop in the hydraulic grade line, and the 8000 represents L/d, the general expression to be used *when Q is to be found* is

$$\text{Re}\sqrt{f} = \frac{d}{\nu}\sqrt{\frac{2g(d)(h_L)}{L}} \text{ (see Diagram A-2 also)} \qquad (B)$$

Then

$$\text{Re}\sqrt{f} = \frac{0.500}{4.12 \times 10^{-5}}\sqrt{\frac{64.4 \times 150.5}{8000}} = 13{,}400$$

Examination of Diagram A-2 will indicate that the flow is turbulent. Then, from Diagram A-2, $f = 0.020$ for $\epsilon/d = 0.0002/\frac{1}{2} = 0.0004$. Completing the solution, from the Bernoulli equation above,

$$\frac{V_6^2}{2g} = \frac{150.5}{(8000)(0.020)} = 0.941, \; V_6 = 7.78 \text{ ft/sec, and } Q = A_6V_6 = \frac{1}{4}\pi\left(\frac{1}{2}\right)^2 \times 7.78 = 1.53 \text{ cfs oil}$$

The reader can check the solution by calculating the Reynolds number and finding the value of f from Diagram A-1.

When the flow is laminar, methods illustrated in Problem 8.12 should be used.

8.17. How much water (60°F) would flow under the conditions of Problem 8.16? Use Table 3.

Solution:

The Bernoulli equation yields $(171.7 - 50.5) = 8000f\dfrac{V_6^2}{2g}, \; \dfrac{V_6^2}{2g} = \dfrac{121.2}{8000f}.$

The most direct solution in this case is to assume a value of f. Table 3 indicates, for new 6″ pipe, a range in f from 0.0275 to 0.0175. Try $f = 0.0225$. Then

$$V_6^2/2g = 121.2/(8000 \times 0.0225) = 0.673 \text{ ft} \qquad \text{and} \qquad V_6 = 6.59 \text{ ft/sec}$$

Check on both type of flow and f in Table 3:

$$\text{Re} = (6.59)\left(\frac{1}{2}\right)/(1.217 \times 10^{-5}) = 271{,}000, \text{ hence turbulent flow}$$

Now f by interpolation is 0.0210. Repeating the calculation,

$$V_6^2/2g = 121.2/(8000 \times 0.0210) = 0.721 \text{ ft} \qquad \text{and} \qquad V_6 = 6.82 \text{ ft/sec}$$

From Table 3 to a justifiable accuracy, $f = 0.0210$ (check). Then

$$Q = A_6V_6 = \frac{1}{4}\pi\left(\frac{1}{2}\right)^2 \times 6.82 = 1.34 \text{ cfs of water}$$

This procedure can also be applied using Diagram A-1, but the method of Problem 8.16 is preferred.

8.18. What rate of flow of air at 68°F will be carried by a new horizontal 2″ I.D. steel pipe at an absolute pressure of 3 atm and with a drop of 0.150 psi in 100 ft of pipe? Use $\epsilon = 0.00025$ ft.

Solution:

From the Appendix, for 68°F, $\gamma = 0.0752$ lb/ft^3 and $\nu = 16.0 \times 10^{-5}$ ft^2/sec at standard atmospheric pressure. At 3 atm, $\gamma = 3 \times 0.0752 = 0.2256$ lb/ft^3 and $\nu = \frac{1}{3} \times 16.0 \times 10^{-5} = 5.33 \times 10^{-5}$ ft^2/sec. This kinematic viscosity can also be obtained from

$$\mu = \frac{\gamma}{g}\nu = \frac{0.0752 \times 16.0 \times 10^{-5}}{32.2} = 3.74 \times 10^{-7}\frac{\text{lb-sec}}{\text{ft}^2} \text{ at 68°F and 14.7 psi absolute}$$

Furthermore, at 3×14.7 psi absolute, $\gamma_{air} = 0.2256\ \text{lb/ft}^3$ and

$$\nu \text{ for 3 atm} = \mu\frac{g}{\gamma} = 3.74 \times 10^{-7} \times \frac{32.2}{0.2256} = 5.34 \times 10^{-5}\ \text{ft}^2/\text{sec}$$

To find the flow, the air can be considered incompressible. Then

$$\frac{p_1 - p_2}{\gamma} = \text{lost head} = f\left(\frac{L}{d}\right)\left(\frac{V^2}{2g}\right), \quad \frac{0.150 \times 144}{0.2256} = 95.7 = f\left(\frac{100}{\frac{2}{12}}\right)\left(\frac{V^2}{2g}\right) \text{ and } \frac{V^2}{2g} = \frac{0.160}{f}$$

Also, from Problem 8.16,

$$\text{Re}\sqrt{f} = \frac{d}{\nu}\sqrt{\frac{2g(d)(h_L)}{L}} = \frac{\frac{2}{12}}{5.34 \times 10^{-5}}\sqrt{\frac{(64.4)\left(\frac{2}{12}\right)(95.7)}{100}} = 10{,}000 \text{ (turbulent).}$$

From Diagram A-2, $f = 0.025$ for $\epsilon/d = 0.00025/(2/12) = 0.0015$. Then

$$V^2/2g = 0.160/f = 6.40 \text{ ft}, \; V_2 = 20.3 \text{ ft/sec, and } Q = A_2V_2 = \frac{1}{4}\pi(2/12)^2 \times 20.3 = 0.443 \text{ cfs air}$$

8.19. What size of new cast iron pipe, 8000 ft long, will deliver 37.5 cfs of water with a drop in the hydraulic grade line of 215 ft? Use Table 3 for this calculation.

Solution:

The Bernoulli theorem gives $\left(\frac{p_A}{\gamma} + \frac{V_A^2}{2g} + z_A\right) - f\left(\frac{8000}{d}\right)\left(\frac{V^2}{2g}\right) = \frac{p_B}{\gamma} + \frac{V_B^2}{2g} + z_B$

or $$\left[\left(\frac{p_A}{\gamma} + z_A\right) - \left(\frac{p_B}{\gamma} + z_B\right)\right] = f\left(\frac{8000}{d}\right)\left(\frac{V^2}{2g}\right)$$

The left-hand term in the brackets represents the drop in the hydraulic grade line. Expressing V as Q/A and assuming turbulent flow,

$$215 = f\left(\frac{8000}{d}\right)\frac{\left(37.5/\frac{1}{4}\pi d^2\right)^2}{2g}, \text{ which simplifies to } d^5 = f\frac{(8000)(37.5)^2}{(39.7)(215)} = 1318f$$

Assume $f = 0.020$ (since both d and V are unknown, an assumption is necessary). Then

$$d^5 = f(1318) = (0.020)(1318) = 26.36, \; d = 1.92 \text{ ft}$$

From Table 3, for $V = \dfrac{37.5}{\pi(1.92)^2/4} = 13.0$ ft/sec, $f = 0.0165$.

For this magnitude of velocity in most pipes, turbulent flow of water exists. Recalculating,

$$d^5 = (0.0165)(1318) = 21.75, \; d = 1.85 \text{ ft}$$

Checking f, $V = 14.0$ ft/sec, and Table 3 gives $f = 0.0165$ (check).
Use nearest standard size: 2-ft or 24″ pipe. (Check Re using ν for 70°F water.)

8.20. Points C and D, at the same elevation, are 150 m apart in a 200-mm pipe and are connected to a differential gage by means of small tubing. When the flow of water is $0.18\ \text{m}^3/\text{s}$, the deflection of the mercury in the gage is 1.96 m. Determine the friction factor f.

Solution:

$$\left(\frac{p_C}{\gamma} + \frac{V_{200}^2}{2g} + 0\right) - f\left(\frac{150}{0.200}\right)\left(\frac{V_{200}^2}{2g}\right) = \frac{p_D}{\gamma} + \frac{V_{200}^2}{2g} + 0 \text{ or } \left(\frac{p_C}{\gamma} - \frac{p_D}{\gamma}\right) = f(750)\frac{V_{200}^2}{2g} \quad (1)$$

From the differential gage (see Chapter 2), $p_L = p_R$ or

$$p_C/\gamma + 1.96 = p_D/\gamma + (13.57)(1.96) \quad \text{and} \quad (p_C/\gamma) - (p_D/\gamma) = 24.64 \text{ m} \qquad (2)$$

Equating (1) and (2), $24.64 = f(750)(5.73)^2/2g$, from which $f = 0.0196$.

8.21. Medium fuel oil at 50°F is pumped to tank C (see Fig. 8-2) through 6000 ft of new, riveted steel pipe, 16″ inside diameter. The pressure at A is +2.00 psi when the flow is 7.00 cfs. (*a*) What power must pump AB supply to the oil, and (*b*) what pressure must be maintained at B? Draw the hydraulic grade line.

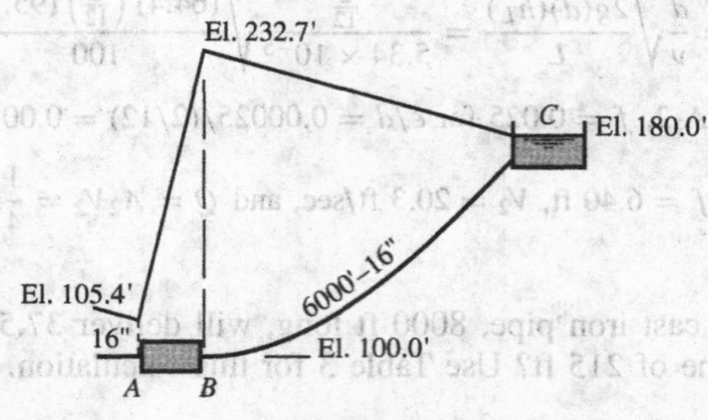

Fig. 8-2

Solution:

$$V_{16} = \frac{Q}{A} = \frac{7.00}{\pi(16/12)^2/4} = 5.01 \text{ ft/sec} \quad \text{and} \quad \text{Re} = \frac{5.01 \times (16/12)}{5.55} \times 10^5 = 120,000$$

From Diagram A-1, $f = 0.030$ for $\epsilon/d = 0.0060/(16/12) = 0.0045$.

(*a*) The Bernoulli equation, A to C, datum A, gives

$$\left(\frac{2.00 \times 144}{0.861 \times 62.4} + \frac{(5.01)^2}{2g} + 0\right) + H_p - 0.030\left(\frac{6000}{16} \times 12\right)\left(\frac{(5.01)^2}{2g}\right) - \frac{(5.01)^2}{2g} = 0 + 0 + 80$$

Solving, $H_p = 127.3$ ft and $\text{hp} = \dfrac{\gamma Q H_p}{550} = \dfrac{0.861 \times 62.4 \times 7.00 \times 127.3}{550} = 87.0.$

The last term on the left-hand side of the energy equation is the lost head from pipe to tank (see Table 4 in Appendix). In general, when the length-to-diameter ratio (L/d) is in excess of 2000 : 1, velocity heads and minor losses can be neglected in the Bernoulli equation (here they cancel).

(*b*) The pressure head at B can be calculated by using sections A and B or sections B and C. The former is less work; then

$$\left(5.4 + \frac{V_{16}^2}{2g} + 0\right) + 127.3 = \frac{p_B}{\gamma} + \frac{V_{16}^2}{2g} + 0$$

Thus $p_B/\gamma = 132.7$ ft and $p_B = (0.861 \times 62.4)(132.7)/144 = 49.5$ psi.

The hydraulic grade line elevations are shown in Fig. 8-2.

At A: elevation = (100.0 + 5.4) ft = 105.4 ft

At B: elevation = (100.0 + 132.7) ft (or 105.4 + 127.3) = 232.7 ft

At C: elevation = 180.0 ft

8.22. At a point A in a horizontal 12″ pipe ($f = 0.020$) the pressure head is 200 ft. At a distance of 200 ft from A the 12″ pipe reduces suddenly to a 6″ pipe. At a distance of 100 ft from this sudden reduction the 6″ pipe ($f = 0.015$) suddenly enlarges to a 12″ pipe, and point F is 100 ft beyond this change in size. For a velocity of 8.025 ft/sec in the 12″ pipes, draw the energy and hydraulic grade lines. Refer to Fig. 8-3.

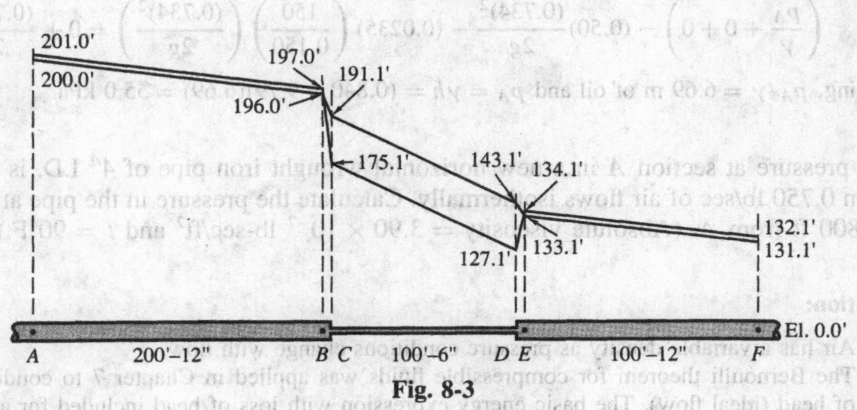

Fig. 8-3

Solution:

The velocity heads are $V_{12}^2/2g = (8.025)^2/2g = 1.00$ ft and $V_6^2/2g = 16.0$ ft.

The energy line drops in the direction of flow by the amount of lost head. The hydraulic grade line is below the energy line by the amount of the velocity head at any cross section. Note (Fig. 8-3) that the *hydraulic grade line* can rise where a change (enlargement) in size occurs.

Tabulating results to the nearest 0.1 ft,

At	Lost Head in Feet: From	Lost Head in Feet: Calculated	Elevation Energy Line	$\frac{V^2}{2g}$	Elevation of Hyd. Grade Line
A	(Elevation 0.0)		201.0	1.0	200.0
B	A to B	$0.020 \times (200/1) \times 1 = 4.0$	197.0	1.0	196.0
C	B to C	$K_C{}^* \times 16 = 0.37 \times 16 = 5.9$	191.1	16.0	175.1
D	C to D	$0.015 \times (100/\frac{1}{2}) \times 16 = 48.0$	143.1	16.0	127.1
E	D to E	$\frac{(V_6 - V_{12})^2}{2g} = \frac{(32.1 - 8.0)^2}{64.4} = 9.0$	134.1	1.0	133.1
F	E to F	$0.020 \times 100/1 \times 1 = 2.0$	132.1	1.0	131.1

* [K_C is from Table 5, Sudden Enlargement term (D to E) from Table 4]

8.23. Oil flows from tank A through 150 m of 150-mm new asphalt-dipped cast iron pipe to point B in Fig. 8-4. What pressure will be needed at A to cause 0.013 m³/s of oil to flow? (Sp gr = 0.840 and $\nu = 2.11 \times 10^{-6}$ m²/s.) Use $\epsilon = 0.00012$ m.

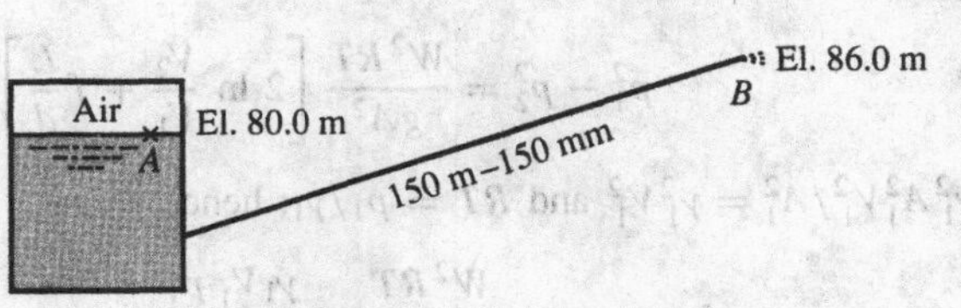

Fig. 8-4

Solution:

$$V_{150} = \frac{Q}{A} = \frac{0.013}{0.0177} = 0.734 \text{ m/s} \quad \text{and} \quad \text{Re} = \frac{Vd}{\nu} = \frac{0.734 \times 0.150}{2.11} \times 10^6 = 52{,}200$$

From Diagram A-1, $f = 0.0235$ and the Bernoulli Equation, A to B, datum A, gives

$$\left(\frac{p_A}{\gamma} + 0 + 0\right) - (0.50)\frac{(0.734)^2}{2g} - (0.0235)\left(\frac{150}{0.150}\right)\left(\frac{(0.734)^2}{2g}\right) = 0 + \frac{(0.734)^2}{2g} + 6.0$$

Solving, $p_A/\gamma = 6.69$ m of oil and $p_A = \gamma h = (0.840 \times 9.79)(6.69) = 55.0$ kPa.

8.24. The pressure at section A in a new horizontal wrought iron pipe of 4″ I.D. is 49.5 psi absolute when 0.750 lb/sec of air flows isothermally. Calculate the pressure in the pipe at section B, which is 1800 ft from A. (Absolute viscosity $= 3.90 \times 10^{-7}$ lb-sec/ft^2 and $t = 90°$F.) Use $\epsilon = 0.0003$ ft.

Solution:

Air has a variable density as pressure conditions change with flow.

The Bernoulli theorem for compressible fluids was applied in Chapter 7 to conditions involving no loss of head (ideal flow). The basic energy expression with loss of head included for a length of pipe dL and where $z_1 = z_2$ would be

$$\frac{dp}{\gamma} + \frac{V\,dV}{g} + f\left(\frac{dL}{d}\right)\left(\frac{V^2}{2g}\right) = 0$$

Dividing by $\dfrac{V^2}{2g}$,

$$\frac{2g}{V^2}\left(\frac{dp}{\gamma}\right) + \frac{2\,dV}{V} + \frac{f}{d}dL = 0$$

For steady flow, the number of lb/sec flowing is constant; then $W = \gamma Q = \gamma AV$, and $W/\gamma A$ can be substituted for V in the pressure head term, giving

$$\frac{2g\gamma^2 A^2}{W^2\gamma}dp + \frac{2dV}{V} + \frac{f}{d}dL = 0$$

For isothermal conditions, $p_1/\gamma_1 = p_2/\gamma_2 = RT$ or $\gamma = p/RT$. Substituting for γ,

$$\frac{2gA^2}{W^2RT}\int_{p_1}^{p_2} p\,dp + 2\int_{V_1}^{V_2}\frac{dV}{V} + \frac{f}{d}\int_0^L dL = 0$$

f being considered a constant, as explained below. Integrating and substituting limits,

$$\frac{gA^2}{W^2RT}(p_2^2 - p_1^2) + 2(\ln V_2 - \ln V_1) + f(L/d) = 0 \qquad (A)$$

For comparison with the familiar form of the equation ($z_1 = z_2$), we obtain

$$(Kp_1^2 + 2\ln V_1) - f(L/d) = (Kp_2^2 + 2\ln V_2) \qquad (B)$$

where $K = \dfrac{gA^2}{W^2RT}$. Rearranging (A) for ready solution,

$$p_1^2 - p_2^2 = \frac{W^2RT}{gA^2}\left[2\ln\frac{V_2}{V_1} + f\frac{L}{d}\right] \qquad (C)$$

Now $W^2/A^2 = \gamma_1^2A_1^2V_1^2/A_1^2 = \gamma_1^2V_1^2$ and $RT = p_1/\gamma_1$; hence

$$\frac{W^2RT}{gA^2} = \frac{\gamma_1V_1^2p_1}{g} \qquad (D)$$

Then (*C*) becomes $(p_1 - p_2)(p_1 + p_2) = \dfrac{\gamma_1 p_1 V_1^2}{g}\left[2 \ln \dfrac{V_2}{V_1} + f\dfrac{L}{d}\right]$ and

$$\frac{p_1 - p_2}{\gamma_1} = \frac{2\left[2 \ln \dfrac{V_2}{V_1} + f\dfrac{L}{d}\right]\dfrac{V_1^2}{2g}}{1 + p_2/p_1} = \text{lost head} \qquad (E)$$

Limiting pressures and velocities will be discussed in Chapter 13.

Before solving this expression, investigation of friction factor f is important, as velocity V is not constant for gases where changes in density may occur.

$$\text{Re} = \frac{Vd}{\mu/\rho} = \frac{Vd\rho}{\mu} = \frac{Wd\rho}{\gamma A \mu}$$

$$\text{Since } g = \frac{\gamma}{\rho},\ \text{Re} = \frac{Wd}{Ag\mu} \qquad (F)$$

It should be observed that the Reynolds number is constant for steady flow because μ does not vary if there is no temperature change. Hence friction factor f is constant for the problem even though the velocity will increase as the pressure decreases. Solving equation (*F*), using the absolute viscosity given,

$$\text{Re} = \frac{0.750 \times 4/12 \times 10^9}{(\pi/4)(4/12)^2 \times 32.2 \times 390} = 228{,}000.\ \text{From Diagram A-1, } f = 0.0205 \text{ for } \epsilon/d = 0.0009.$$

Using equation (*C*) and neglecting $2 \ln V_2/V_1$, which is very small compared to the $f(L/d)$ term,

$$(49.5 \times 144)^2 - p_2^2 = \frac{(0.750)^2 \times (53.3)(90 + 460)}{(32.2)[(\pi/4)(4/12)^2]^2}\left[\text{negl.} + (0.0205)\frac{1800}{4/12}\right]$$

from which $p_2 = 6585$ psf $= 45.7$ psi absolute.

At B:

$$\gamma_2 = \frac{6585}{(53.3)(90 + 460)} = 0.225 \text{ lb/ft}^3, \qquad V_2 = \frac{W}{\gamma_2 A} = \frac{0.750}{0.225 \times 0.0873} = 38.2 \text{ ft/sec}$$

At A:

$$\gamma_1 = \frac{49.5 \times 144}{(53.3)(90 + 460)} = 0.243 \text{ lb/ft}^3, \qquad V_1 = \frac{0.750}{0.243 \times 0.0873} = 35.4 \text{ ft/sec}$$

Hence $2 \ln V_2/V_1 = 2 \ln (38.2/35.4) = 0.152$, which is negligible in terms of the $f(L/d)$ term of 111. Therefore the pressure at section *B* is $p_2 = 45.7$ psi.

Had the air been treated as incompressible, then

$$\frac{p_1 - p_2}{\gamma_1} = f\frac{L}{d}\left(\frac{V^2}{2g}\right) = 0.0205 \times \frac{1800}{4/12} \times \frac{(35.4)^2}{2g} = 2154 \text{ ft}$$

$$\Delta p = \gamma_1 h = 0.243 \times 2154 = 523 \text{ psf} = 3.63 \text{ psi}$$

and $p_2' = 49.5 - 3.6 = 45.9$ psi, an unusually close agreement.

8.25. A horizontal wrought iron pipe of 150 mm I.D. and somewhat corroded, is transporting 20 N of air per second from *A* to *B*. At *A* the pressure is 483 kPa absolute, and at *B* the pressure must be 448 kPa absolute. Flow is isothermal at 20°C. What is the length of pipe from *A* to *B*? Use $\epsilon = 0.00040$ m.

Solution:

Calculating basic values (see Appendix for 68°F at 14.7 psi),

$$\gamma_1 = (11.8)(483/101) = 56.4\ \text{N/m}^3, \qquad \gamma_2 = (11.8)(448/101) = 52.3\ \text{N/m}^3$$

$$V_1 = \frac{W}{\gamma_1 A} = \frac{20}{56.4 \times \frac{1}{4}\pi(150/1000)^2} = 20.1\ \text{m/s}, \qquad V_2 = \frac{20}{52.3 \times \frac{1}{4}\pi(150/1000)^2} = 21.6\ \text{m/s}$$

$$\text{Re} = \frac{(20.1)(150/1000)}{(101/483)(1.49 \times 10^{-5})} = 968{,}000. \text{ From Diagram A-1, } f = 0.025 \text{ for } \epsilon/d = 0.0027.$$

Using equation (E) of Problem 8.24,

$$\frac{(483 - 448) \times 10^3}{56.4} = \frac{(2)[2\ \ln\ (21.6/20.1) + 0.025(L/0.150)](20.1)^2/2g}{1 + 448/483} \quad \text{and } L = 173\ \text{m}$$

Note: For the flow of gases in pipelines, if p_2 is not more than 10% smaller than p_1, less than 5% error in pressure drop will result from assuming the fluid to be incompressible and using the Bernoulli equation in its usual form.

8.26. The elevations of the energy line and hydraulic grade line at point G are 44.0 and 42.0 ft, respectively. For the system shown in Fig. 8-5, calculate (a) the power extracted between G and H if the energy line at H is at elevation 4.0′ and (b) the pressure heads at E and F, which are at elevation 20.0′ (c) Draw, to the nearest 0.1 ft, the energy and hydraulic grade lines, assuming K for valve CD is 0.40 and $f = 0.010$ for the 6″ pipes.

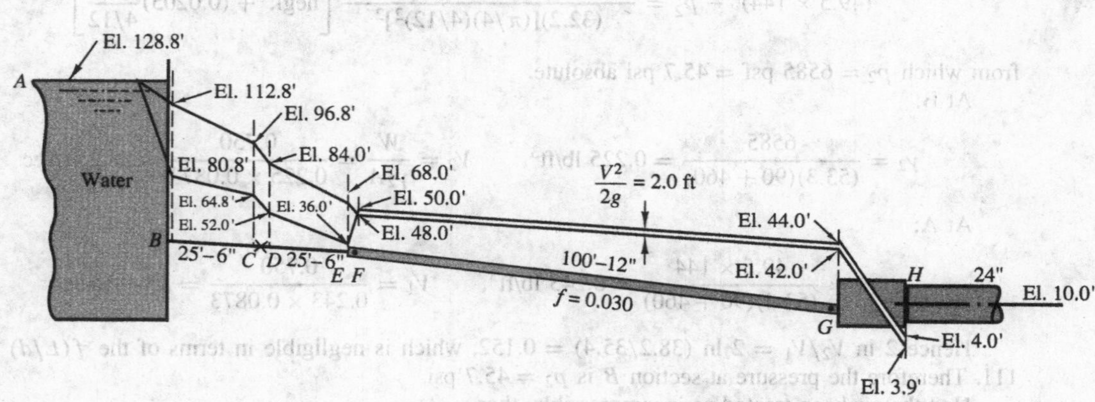

Fig. 8-5

Solution:

The flow must be from the reservoir because the energy line at G is below the reservoir level. GH is a turbine. Before the power extracted can be calculated, flow Q and the head extracted must be obtained.

(a) At G, $V_{12}^2/2g = 2.0$ (the difference between the elevations of the energy and hydraulic grade lines). Also $V_6^2/2g = 16 \times 2.0 = 32.0$ and $V_{24}^2/2g = \frac{1}{16}(2.0) = 0.13$ ft. To obtain Q,

$$V_{12} = 11.35\ \text{ft/sec} \quad \text{and} \quad Q = \frac{1}{4}\pi(1)^2 \times 11.35 = 8.91\ \text{cfs}$$

$$\text{Horsepower} = \gamma Q H_T/550 = (62.4)(8.91)(44.0 - 4.0)/550 = 40.4\ \text{hp extracted}$$

(*b*) F to G, datum zero:

$$(\text{energy at } F) - (0.030)(100/1)(2.0) = \text{energy at } G = 44.0$$

$$\text{energy at } F = 50.0 \text{ ft}$$

E to F, datum zero:

$$(\text{energy at } E) - (45.4 - 11.3)^2/2g = \text{energy at } F = 50.0$$

$$\text{energy at } E = 68.1 \text{ ft}$$

Pressure head at $E = 68.1 - (20 + 32) = 16.1$ ft water.

Pressure head at $F = 50.0 - (20 + 2) = 28.0$ ft water.

(*c*) Working back from E.

$$\begin{aligned} \text{Drop in energy, line } D\text{-}E &= (0.010)(25/\tfrac{1}{2})(32.0) = 16.0 \text{ ft} \\ \text{line } C\text{-}D &= (0.40)(32.0) = 12.8 \text{ ft} \\ \text{line } B\text{-}C &= \text{same as } D\text{-}E = 16.0 \text{ ft} \\ \text{line } A\text{-}B &= (0.50)(32.0) = 16.0 \text{ ft} \end{aligned}$$

$$\begin{aligned} &\text{elevation at } D - 16.0 = \text{elevation at } E \text{ of } 68.0, && \text{El. } D = 84.0 \\ &\text{elevation at } C - 12.8 = \text{elevation at } D \text{ of } 84.0, && \text{El. } C = 96.8 \\ &\text{elevation at } B - 16.0 = \text{elevation at } C \text{ of } 96.8, && \text{El. } B = 112.8 \\ &\text{elevation at } A - 16.0 = \text{elevation at } B \text{ of } 112.8, && \text{El. } A = 128.8 \end{aligned}$$

The hydraulic grade line is $V^2/2g$ below the energy line: 32.0 ft in the 6″, 2.0 ft in the 12″, and 0.13 ft in the 24″. The values are shown in Fig. 8.5.

8.27. A 300 mm by 460 mm rectangular duct carries air at 105 kPa absolute and 20°C through 460 m with an average velocity of 2.97 m/s. Determine the loss of head and the pressure drop, assuming the duct is horizontal and the size of the surface imperfections is 0.00055 m.

Mathcad

Solution:

The lost head term must be revised slightly to apply to noncircular cross sections. The resulting equation will apply to turbulent flow with reasonable accuracy. Substitute for the diameter the *hydraulic radius*, which is defined as the cross-sectional area divided by the wetted perimeter, or $R = A/p$.

For a circular pipe, $R = \frac{1}{4}\pi d^2/\pi d = d/4$ and the Darcy formula can be rewritten as

$$\text{lost head} = \frac{f}{4}\left(\frac{L}{R}\right)\left(\frac{V^2}{2g}\right)$$

For f and its relation to the roughness of the conduit and the Reynolds number, we use

$$\text{Re} = Vd/\nu = V(4R)/\nu$$

$$\text{For the 300 mm} \times 460 \text{ mm duct, } R = \frac{A}{p} = \frac{(0.300)(0.460)}{(2)(0.300 + 0.460)} = 0.0908 \text{ m}$$

$$\text{Re} = \frac{4VR}{\nu} = \frac{4 \times 2.97 \times 0.0908}{(101/105) \times 1.49} \times 10^5 = 75{,}300$$

From Diagram A-1, $f = 0.024$ for $\epsilon/d = \epsilon/4R = (0.00055)/(4 \times 0.0908) = 0.0015$. Then

$$\text{lost head} = \frac{0.024}{4} \times \frac{460}{0.0908} \times \frac{(2.97)^2}{2g} = 13.7 \text{ m of air}$$

and pressure drop $= \gamma h = (105/101)(11.8)(13.7) = 168$ Pa.

It can be observed that the assumption of constant density of the air is satisfactory.

8.28. A 1-m-diameter new cast iron pipe ($C = 130$) is 845 m long and has a head loss of 1.11 m. Find the discharge capacity of the pipe according to the Hazen-Williams formula.

Solution:

$$V = (0.8492)(130)(1/4)^{0.63}(1.11/845)^{0.54} = 1.281 \text{ m/s}$$

$$Q = AV = [(\pi)(1)^2/4](1.281) = 1.01 \text{ m}^3/\text{s}$$

8.29. Solve Problem 8.28 using the Manning formula.

Solution:

$$V = \frac{1.0}{0.012}\left(\frac{1}{4}\right)^{2/3}\left(\frac{1.11}{845}\right)^{1/2} = 1.199 \text{ m/s}$$

$$Q = AV = [(\pi)(1)^2/4](1.199) = 0.942 \text{ m}^3/\text{s}$$

8.30. Solve Problem 8.28 using the Hazen-Williams pipe diagram.

Solution:

From Diagram B-3 with $h_1 = 1.11/845 = 0.001314$ and $d = 1000$ mm, $Q = 0.91 \text{ m}^3/\text{s}$. This discharge is for $C = 120$.

For $C = 130$, $0.91/(Q)_{C=130} = 120/130$, $\quad (Q)_{C=130} = 0.99 \text{ m}^3/\text{s}$

8.31. Solve Problem 8.29 using the Manning pipe diagram.

Solution:

From Diagram B-5 with $h_1 = 1.11/845 = 0.001314$ and $d = 1000$ mm, $Q = 0.88 \text{ m}^3/\text{s}$. This discharge is for $n = 0.013$.

For $n = 0.012$, $0.88/(Q)_{n=0.012} = 0.012/0.013$, $\quad (Q)_{n=0.012} = 0.95 \text{ m}^3/\text{s}$

8.32. A 36-in-diameter concrete pipe ($C = 120$) is 4000 ft long and has a head loss of 12.7 ft. Find the discharge capacity of the pipe according to the Hazen-Williams formula.

Solution:

$$V = (1.318)(120)(3/4)^{0.63}(12.7/4000)^{0.54} = 5.906 \text{ ft/sec}$$

$$Q = AV = [(\pi)(3)^2/4](5.906) = 41.7 \text{ ft}^3/\text{sec}$$

8.33. Solve Problem 8.32 using the Manning formula.

Solution:

$$V = \frac{1.486}{0.013}\left(\frac{3}{4}\right)^{2/3}\left(\frac{12.7}{4000}\right)^{1/2} = 5.317 \text{ ft/sec}, \qquad Q = AV = \frac{(\pi)(3)^2}{4}(5.317) = 37.6 \text{ ft}^3/\text{sec}$$

8.34. **Solve Problem 8.32 using the Hazen-Williams pipe diagram.**

Solution:

From Diagram B-2 with $h_1 = 12.7/4000 = 0.003175$ and $d = 36$ in, $Q = 41.5\ \text{ft}^3/\text{sec}$.

8.35. **Solve Problem 8.33 using the Manning pipe diagram.**

Solution:

From Diagram B-4 with $h_1 = 12.7/4000 = 0.003175$ and $d = 36$ in, $Q = 37\ \text{ft}^3/\text{sec}$.

8.36. **What size square concrete conduit is needed to transport 4.0 m³/s of water a distance of 45 m with a head loss of 1.80 m? Use the Hazen-Williams formula.**

Solution:

$$V = 4.0/a^2 = (0.8492)(120)(a^2/4a)^{0.63}(1.80/45)^{0.54} \qquad a = 0.788\ \text{m}$$

(In practice, an 0.80 m by 0.80 m conduit would probably be specified.)

8.37. **Water is flowing in a 500-mm-diameter new cast iron pipe ($C = 130$) at a velocity of 2.0 m/s. Find the pipe friction loss per 100 m of pipe, using the Hazen-Williams pipe diagram.**

Solution:

$$2.0/(V)_{C=120} = 130/120, \qquad (V)_{C=120} = 1.85\ \text{m/s}$$

From Diagram B-3 with $d = 500$ mm and $V = 1.85$ m/s, $h_1 = 0.0067$ m per meter of pipe length. The head loss per 100 m of pipe is therefore 0.67 m. This loss can be expressed in kilopascals by

$$p = \gamma h = (9.79)(0.67) = 6.6\ \text{kPa}$$

Supplementary Problems

8.38. If the shear stress at the wall of a 12″ diameter pipe is 1.00 psf and $f = 0.040$, what is the average velocity (*a*) if water at 70°F is flowing, (*b*) if a fluid with a specific gravity of 0.70 is flowing?
Ans. 10.1 ft/sec, 12.1 ft/sec

8.39. What are the shear velocities in the preceding problem? *Ans.* 0.717 ft/sec, 0.857 ft/sec

8.40. Water flows through 61 m of 150-mm pipe, and the shear stress at the walls is 44 Pa. Determine the lost head. *Ans.* 7.2 m

8.41. SAE 10 oil at 20°C ($\rho = 869\ \text{kg/m}^3$, $\mu = 8.14 \times 10^{-2}\ \text{N·s/m}^2$) flows in a 200-mm-diameter pipe. Find the maximum velocity for which the flow will be laminar. *Ans.* 0.937 m/s

8.42. What size pipe will maintain a shear stress at the wall of 0.624 psf when water flows through 300 ft of pipe causing a lost head of 20.0 ft? *Ans.* $r = 0.30$ ft

8.43. Compute the critical velocity (lower) for a 100-mm pipe carrying water at 27°C. *Ans.* 0.017 m/s

8.44. Compute the critical velocity (lower) for a 4″ pipe carrying heavy fuel oil at 110°F. *Ans.* 2.88 ft/sec

8.45. Water flows at 20°C through a new cast iron pipe at a velocity of 4.2 m/s. The pipe is 400 m long and has a diameter of 150 mm. Determine the head loss due to friction. *Ans.* 54.20 m

8.46. What pressure head drop will occur in 300 ft of new horizontal cast iron pipe, 4″ in diameter, carrying medium fuel oil at 50°F when the velocity is 0.25 ft/sec? *Ans.* 0.037 ft

8.47. What pressure head drop will occur in Problem 8.46 if the velocity of the oil is 4.00 ft/sec? *Ans.* 6.7 ft

8.48. Considering pipe loss only, how much head is required to deliver 0.222 m^3/s of heavy fuel oil at 38°C through 914 m of new cast iron pipe, 300 mm inside diameter? Use $\epsilon = 0.00024$ m. *Ans.* 41 m

8.49. In Problem 8.48, what least value of kinematic viscosity of the oil will produce laminar flow? *Ans.* 0.00046 m^2/s

8.50. Considering pipe loss only, what difference in elevation between two tanks 800 ft apart will deliver 1.10 cfs of medium weight lubricating oil at 50°F through a 6″ diameter pipe? *Ans.* 49.9 ft

8.51. Water at 20°C flows in a 100-mm-diameter new cast iron pipe with a velocity of 5.0 m/s. Determine the pressure drop in kPa per 100 m of pipe and the power lost (in kilowatts) to friction. *Ans.* 314 kPa per 100 m, 12.36 kW per 100 m

8.52. Oil of specific gravity 0.802 and kinematic viscosity 0.000186 m^2/s flows from tank A to tank B through 305 m of new pipe at the rate of 0.089 m^3/s. The available head is 0.161 m. What size pipe should be used? *Ans.* 0.600 m or 600 mm

8.53. A pump delivers heavy fuel oil at 60°F through 1000 ft of 2″ brass pipe to a tank 10 ft higher than the supply tank. Neglecting minor losses, for a flow of 0.131 cfs, determine the size of the pump (horsepower) if its efficiency is 80%. *Ans.* 8.2 hp

8.54. Water at 38°C flows from A to B through 244 m of average 300-mm-I.D. cast iron pipe ($\epsilon = 0.00061$ m). Point B is 9.1 m above A, and the pressure at B must be maintained at 138 kPa. If 0.222 m^3/s is to flow through the pipe, what must be the pressure at A? *Ans.* 314 kPa

8.55. Determine the discharge capacity of a 150-mm-diameter new wrought iron pipe to carry water at 20°C if the pressure loss due to friction may not exceed 35 kPa per 100 m of level pipe. *Ans.* 0.0445 m^3/s

8.56. An old commercial pipe, horizontal, 36″ inside diameter and 8000 ft long, carries 44.2 cfs of heavy fuel oil with a lost head of 73.5 ft. What pressure in psi must be maintained at A upstream to keep the pressure at B at 20 psi? Use $\epsilon = 0.045$ ft. *Ans.* 48.6 psi

8.57. An old pipe, 600 mm I.D. and 1219 m long, carries medium fuel oil at 30°C from A to B. The pressures at A and B are 393 kPa and 138 kPa respectively, and point B is 18.3 m above point A. Calculate the flow, using $\epsilon = 0.00049$ m. *Ans.* 0.73 m^3/s

8.58. SAE 10 oil at 20°C ($\gamma = 8.52$ kN/m^3, $\rho = 860$ kg/m^3, $\mu = 8.14 \times 10^{-2}$ N·s/m^2) is to flow through a 300-m level concrete pipe. What size pipe will carry 0.0142 m^3/s with a pressure drop due to friction of 23.94 kPa? *Ans.* 156 mm

8.59. Water flows from tank A whose level is at elevation 84.0′ to tank B whose level is kept at elevation 60.0′. The tanks are connected by 100 ft of 12″ pipe ($f = 0.020$) followed by 100 ft of 6″ pipe ($f = 0.015$). There are two 90° bends in each pipe ($K = 0.50$ each), K for the contraction is 0.75, and the 12″ pipe projects into tank A. If the elevation at the sudden reduction of the pipes is 54.0′, find the pressure heads in the 12″ and in the 6″ at this change in size. *Ans.* 22.8 ft, 22.0 ft

8.60. In Fig. 8-6, point B is 600 ft from the reservoir. When water flows at the rate of 0.500 cfs, calculate (a) the head lost due to the partial obstruction C and (b) the pressure at B in psi absolute. *Ans.* 7.7 ft, 14.2 psia

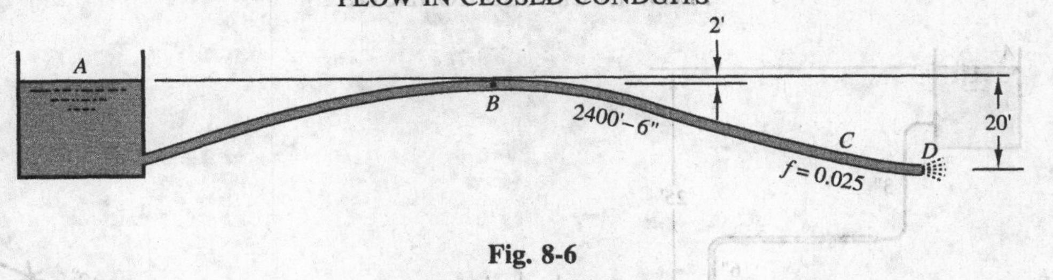

Fig. 8-6

8.61. A commercial solvent at 20°C flows from tank A to tank B through 152 m of new 150-mm asphalt-dipped cast iron pipe. The difference in elevation of the liquid levels is 7.0 m. The pipe projects into tank A, and two bends in the line cause a loss of 2 velocity heads. What flow will occur? Use $\epsilon = 0.00014$ m. *Ans.* 0.044 m^3/s

8.62. A 300-mm-diameter vitrified pipe is 100 m long. Using the Hazen-Williams formula, determine the discharge capacity of the pipe if the head loss is 2.54 m. *Ans.* 0.177 m^3/s

8.63. A steel conduit, 2″ by 4″, carries 0.64 cfs of water at 60°F average temperature and at a constant pressure, making the hydraulic grade line parallel to the slope of the conduit. How much does the conduit drop in 1000 ft, assuming the size of the surface imperfections is 0.00085 ft? (Use $\nu = 1.217 \times 10^{-5}$ ft²/sec.) *Ans.* 260 ft

8.64. When 0.042 m^3/s of medium fuel oil flows from A to B through 1067 m of new 150-mm uncoated cast iron pipe, the lost head is 44 m. Sections A and B are at elevations of 0.0 and 18.3 m respectively, and the pressure at B is 345 kPa. What pressure in psi must be maintained at A to deliver the flow stated? *Ans.* 862 kPa

8.65. (*a*) Determine the flow of water through the new cast iron pipes shown in Fig. 8-7.
(*b*) What is the pressure head at B, which is 100 ft from the reservoir? (Use Table 3.)
Ans. 3.49 cfs, 1.91 ft

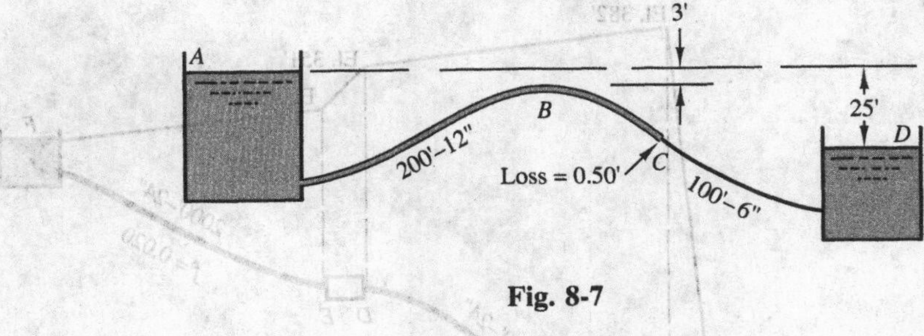

Fig. 8-7

8.66. Solve Problem 8.62, using the Manning formula. *Ans.* 0.143 m^3/s

8.67. Water at 100°F flows through the system shown in Fig. 8-8. Lengths of 3″ and 6″ new, asphalt-dipped cast iron pipe are 180 ft and 100 ft, respectively. Loss factors for fittings and valves are: 3″ bends, $K = 0.40$ each; 6″ bend, $K = 0.60$; and 6″ valve, $K = 3.0$. Determine the flow in cfs. *Ans.* 0.457 cfs

8.68. If pump BC shown in Fig. 8-9 delivers 70 hp to the system when the flow of water is 7.85 cfs, at what elevation can reservoir D be maintained? *Ans.* 76.6 ft

8.69. A pump at elevation 10.0′ delivers 7.85 cfs of water through a horizontal pipe system to a closed tank whose liquid surface is at elevation 20.0′. The pressure head at the 12″-diameter suction side of the pump is −4.0 ft and at the 6″-diameter discharge side, +193.4 ft. The 6″ pipe ($f = 0.030$) is 100 ft long, suddenly

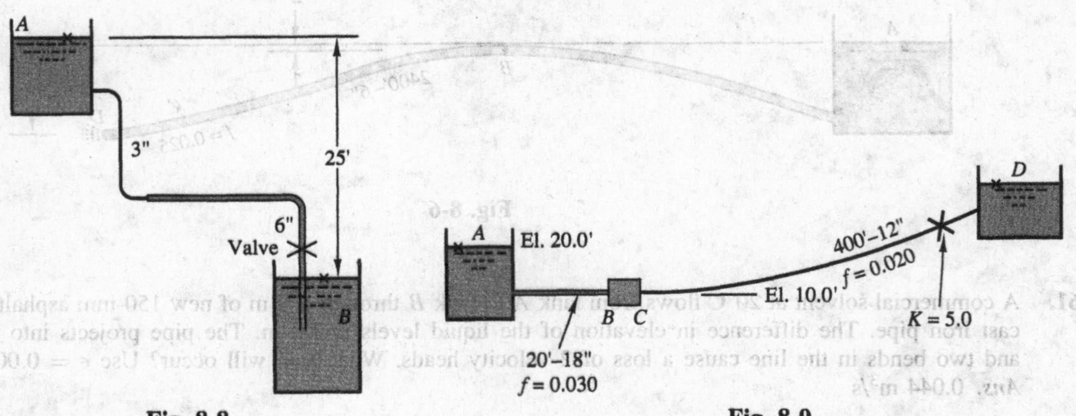

Fig. 8-8 Fig. 8-9

expanding to a 12″ pipe ($f = 0.020$) that is 600 ft long and that terminates at the tank. A 12″ valve, $K = 1.0$, is located 100 ft from the tank. Determine the pressure in the tank above the water surface. Draw the energy and hydraulic grade lines. *Ans.* 10.0 psi

8.70. A 915-m-long, 250-mm-diameter concrete pipe carries 0.142 m^3/s of water. Calculate the loss of energy due to friction, using the Hazen-Williams formula. *Ans.* 311 kPa

8.71. What size average cast iron pipe should be used to deliver 0.0283 m^3/s of water at 20°C through 1219 m with a drop in the hydraulic grade line of 21.3 m? *Ans.* $d = 0.162$ m $= 162$ mm

8.72. Pump BC delivers water to reservoir F, and the hydraulic grade line is shown in Fig. 8-10. Determine (a) the power supplied to the water by pump BC, (b) the power removed by turbine DE, and (c) the elevation maintained in reservoir F. *Ans.* 1010 hp, 71.3 hp, El. 300 ft

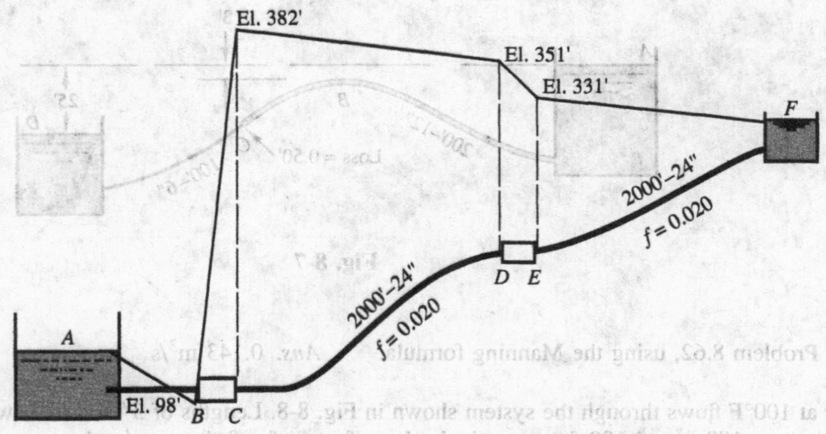

Fig. 8-10

8.73. Air is blown through a 50-mm I.D., old wrought iron pipe at a constant temperature of 20°C and at the rate of 0.667 N/s. At section A the pressure is 377 kPa absolute. What will be the pressure 152 m away from A in the horizontal pipe? Use $\epsilon = 0.00025$ m. *Ans.* 365 kPa abs

8.74. Solve Problem 8.62, using the Hazen-Williams pipe diagram. *Ans.* 0.178 m^3/s

8.75. Solve Problem 8.66, using the Manning pipe diagram. *Ans.* 0.144 m^3/s

8.76. A 12-in-diameter new cast iron pipe is 1.0 mile long. Using the Hazen-Williams formula, determine the discharge capacity of this pipe if the head loss is 24.5 ft. *Ans.* 3.09 cfs

8.77. Solve Problem 8.76 using the Manning formula. *Ans.* 2.63 cfs

8.78. Solve Problem 8.76 using the Hazen-Williams pipe diagram. *Ans.* 3.06 cfs

8.79. Solve Problem 8.76 using the Manning pipe diagram. *Ans.* 2.62 cfs

8.80. Solve Problem 8.70 using the Hazen-Williams pipe diagram. *Ans.* 305 kPa

8.81. Solve Problem 8.70 for a vitrified pipe using the Hazen-Williams pipe diagram. *Ans.* 358 kPa

8.82. Carbon dioxide at 40°C flows through a horizontal 100-mm new wrought iron pipe for 61 m. The pressure at upstream section A is 827 kPa gage, and the average velocity is 12.2 m/s. Assuming negligible density change, what is the pressure drop in the 61 m of pipe? (Absolute viscosity at 40°C = 1.58×10^{-8} $kN{\cdot}s/m^2$.) *Ans.* 12.3 kPa

8.83. Laminar flow occurs through a wide rectangular conduit 9″ high. Assuming that the velocity distribution satisfies the equation $v = 16y(3 - 4y)$, calculate (*a*) the flow per unit of width, (*b*) the kinetic energy correction factor, and (*c*) the ratio of average to maximum velocity. *Ans.* 4.50 cfs, $\alpha = 1.52$, 0.67

8.84. In the laboratory, a 25-mm-diameter plastic pipe is used to demonstrate laminar flow. If the lower critical velocity is to be 3.0 m/s, what should be the kinematic viscosity of the liquid used? *Ans.* 0.000039 m^2/s

8.85. For laminar flow in pipes, $f = 64/\text{Re}$. Using this information, develop the expression for the average velocity in terms of lost head, diameter, and other pertinent items. *Ans.* $V = gd^2h_L/32\nu L$

8.86. Determine the flow in a 12″-diameter pipe if the equation of velocity distribution is $v^2 = 400(y - y^2)$, with the origin at the pipe wall. *Ans.* 5.25 cfs

Chapter 9

Complex Pipeline Systems

INTRODUCTION

Chapter 8 covered flow in closed conduits, but the consideration was generally limited to flow through a single conduit of constant size. In many practical applications, problems are considerably more complicated, for they must involve more than one conduit or a single conduit of varying size. Chapter 9 covers some of these more complex pipeline flow problems, including equivalent pipes, pipes in series and in parallel, branching pipes, and pipe networks.

EQUIVALENT PIPES

A pipe is *equivalent* to another one or to a piping system when, for a given head loss, the same flow rate is produced in the equivalent pipe as occurred in the original. Or it may be stated that a pipe is equivalent (to another pipe or to a piping system) when, for a specific flow rate, the same head loss is produced in the equivalent pipe as occurred in the original. In actuality, there are an infinite number of equivalent pipes for a given system of pipes connected in series; hence, either the diameter of a required equivalent pipe may be specified and the necessary length determined or the length of a needed equivalent pipe may be set and the required diameter found.

Computation of equivalent pipes is rather straightforward and involves finding head losses when flow rates and conduit sizes are known or determining flow rates when head losses and conduit sizes are given. Such computations may be done using the Hazen-Williams formula [equations (*19a*) and (*19b*), Chapter 8]. Diagrams B-1, B-2, and B-3 may be used when speedy but less accurate solutions are acceptable. (*It must be strongly emphasized here that the Hazen-Williams formula is applicable only to the flow of water.*) Problems 9.1 through 9.3 illustrate computations involving equivalent pipes.

PIPES IN SERIES

Pipes are *in series* if they are connected end to end so that a fluid flows in a continuous line without any branching. The volume rate of flow through pipes in series is constant throughout.

Solutions of flow problems involving pipes in series can be accomplished by first finding an equivalent pipe (see preceding section) and then applying the methods of Chapter 8 to the equivalent pipe. Problems 9.4 through 9.6 demonstrate these kinds of solutions.

PIPES IN PARALLEL

Pipes are *in parallel* if they are connected in such a way that flow branches into two or more separate pipes and then comes together again downstream, as illustrated in Fig. 9-1. In Fig. 9-1, fluid flowing in pipe AB branches at joint B, with part of the fluid going through pipe BCE and the remainder through pipe BDE. At joint E, these two pipes combine, and the fluid flows singly through pipe EF. (It should be noted that Fig. 9-1 depicts a plan view—i.e., the pipes are in a horizontal plane.)

In solving problems involving pipes in parallel, three important principles are applicable.

(1) The total flow entering each joint must equal the total flow leaving that joint.

(2) The head loss between two joints (e.g., joints B and E in Fig. 9-1) is the same for each branch connecting these joints.

Fig. 9-1 Pipes in parallel.

(3) Within the range of velocities normally encountered, the percentage of total flow passing through each branch (i.e., pipes BCE and BDE and any additional pipes that might go from joint B to joint E in Fig. 9-1) will be constant, regardless of the head loss between the joints.

Problems involving flow in pipes in parallel can generally be solved by applying (and satisfying) the three principles just stated. Problems 9.7 through 9.11 illustrate pipe problems of this type.

BRANCHING PIPES

Branching pipes consist of one or more pipes that separate into two or more pipes (or combine to a single one) and do not come together again downstream. Figure 9-2 gives an example of a simple branching-pipe system, where three tanks under various pressures are connected by three pipes that join at J. Flow may occur from the upper left tank to the other two (one pipe separating into two pipes) or from the two upper tanks to the one at the lower left (two pipes combining to a single one). The actual direction of flow will depend on (1) the tank pressures and elevations and (2) the diameters, lengths, and kinds of pipes. (If the tanks in Fig. 9-2 are instead reservoirs, all pressures will be equal to atmospheric pressure.)

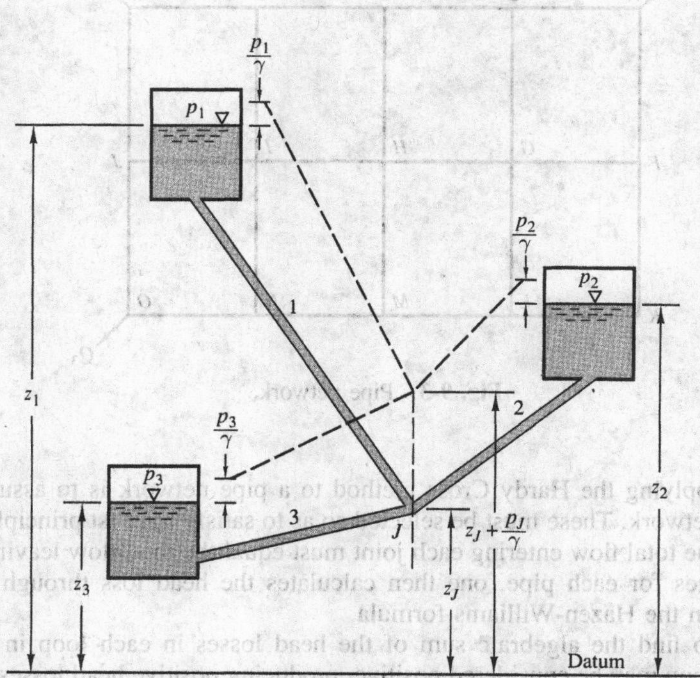

Fig. 9-2 Branching pipes.

The general problem associated with branching pipes is to find the flow rate in each pipe when other data (tank pressures and elevations, pipe data, and fluid properties) are known. This type of problem can be solved by applying the continuity equation, which states that the total flow into joint J must equal the total flow out of the joint. Thus in Fig. 9-2, Q_1 must equal $Q_2 + Q_3$, or $Q_1 + Q_2$ must equal Q_3. The flow rate in each pipe is computed by one of the pipe flow formulas, such as Darcy-Weisbach or Hazen-Williams, based on friction and minor losses and on elevation differences.

This type of problem generally requires a trial-and-error solution. The best procedure is to assume an elevation of the hydraulic grade line at joint J and then compute the flow rate in each pipe. If continuity at the joint is satisfied (total flow into the joint equals total flow out), then the computed flow rates are correct. If continuity is not satisfied, a different grade-line elevation at the joint (higher if flow into the joint is too great, lower if flow out is too great) is tried. Usually a satisfactory solution can be obtained after several trials.

Problems 9.13 through 9.15 demonstrate the solutions of branching-pipe problems.

PIPE NETWORKS

In practice, many pipe systems consist of numerous pipes connected in a complex manner with many entry and withdrawal points. For example, the pipe configuration shown in Fig. 9-3 might represent the water distribution system for a small town or subdivision. Such a pipe system is known as a *pipe network* and is actually a complex set of pipes in parallel. The analysis of pipe networks can be exceedingly complex, but solutions can be obtained using the *Hardy Cross method*, so named after the person who developed the method.

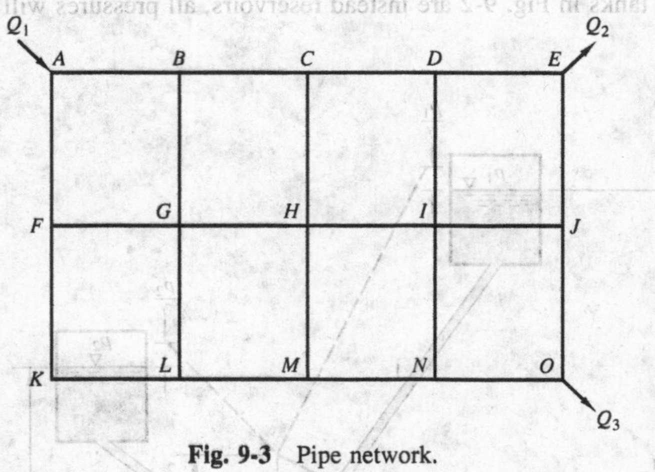

Fig. 9-3 Pipe network.

The first step in applying the Hardy Cross method to a pipe network is to assume flows for each individual pipe in the network. These must be selected so as to satisfy the first principle given previously for pipes in parallel—the total flow entering each joint must equal the total flow leaving that joint. Using these assumed flow rates for each pipe, one then calculates the head loss through each pipe; this is generally done based on the Hazen-Williams formula.

The next step is to find the algebraic sum of the head losses in each loop in the pipe network. (Clockwise flows in a loop may be considered positive, producing positive head losses; counterclockwise ones are then negative and produce negative head losses.) According to the second principle given previously for pipes in parallel—head loss between two joints is the same for each branch connecting these joints—the algebraic sum of the head losses in each loop must equal zero in order for the flow rates within pipes in the loop to be correct. Hence, if the computed head loss sum for every loop in the network is zero, the initially assumed flow rates are correct and the problem solved.

The probability of initially guessing all flow rates correctly is virtually zero, however. Hence, the next step is to compute a *flow rate correction* for each loop in the network, using the equation

$$\Delta = -\frac{\Sigma(\mathrm{LH})}{n\Sigma(\mathrm{LH}/Q_0)} \tag{1}$$

where Δ = flow rate correction for a loop, $\Sigma(\mathrm{LH})$ = algebraic sum of head losses for all pipes in the loop, n = a value that depends on which formula is being used to compute flow rates ($n = 1.85$ for Hazen-Williams formula), and $\Sigma(\mathrm{LH}/Q_0)$ = summation of head loss divided by flow rate for each pipe in the loop.

The final step is to use the flow rate corrections (one for each loop) to adjust the initially assumed flow rates for all pipes and then repeat the entire process for the adjusted flow rates. This procedure is repeated until all corrections (values of Δ) become zero or negligible.

In Problem 9.18, equation *(1)* is derived and additional details regarding the Hardy Cross method are given. Problems 9.19 through 9.21 illustrate the application of the method.

Solved Problems

9.1. For a lost head of 5.0 ft/1000 ft, and using $C = 100$ for all pipes, how many 8″ pipes are equivalent to a 16″ pipe? to a 24″ pipe?

Mathcad

Solution:

Using Diagram B-1, for $S = 5.0$ ft/1000 ft: Q for 8″ pipe = 0.55 mgd
Q for 16″ pipe = 3.40 mgd
Q for 24″ pipe = 10.0 mgd

Thus it would take 3.40/0.55 or 6.2 8″ pipes to be hydraulically equivalent to a 16″ pipe of the same relative roughness. Likewise, 10.0/0.55 or 18.2 8″ pipes are equivalent to a 24″ pipe for a lost head of 5.0 ft/1000 ft or for any other lost head condition.

9.2. A 225-m-long, 300-mm-diameter concrete pipe and a 400-m-long, 500-mm-diameter concrete pipe are connected in series. Find the diameter of a 625-m-long equivalent pipe.

Solution:

Assume a flow rate of 0.1 m³/s. For the 300-mm-diameter pipe with $Q = 0.1\ \mathrm{m^3/s}$, $h_1 = 0.0074$ m/m (from Diagram B-3). For the 500-mm-diameter pipe with $Q = 0.1\ \mathrm{m^3/s}$, $h_1 = 0.00064$ m/m (from Diagram B-3).

$$\text{total head loss} = (0.0074)(225) + (0.00064)(400) = 1.921\ \mathrm{m}$$

For a 625-m-long equivalent pipe, with this head loss,

$$h_1 = 1.921/625 = 0.00307\ \mathrm{m/m}$$

Therefore, from Diagram B-3 with $Q = 0.1\ \mathrm{m^3/s}$, $d = 360$ mm.

9.3. Convert the piping system shown in Fig. 9-4 to an equivalent length of 6″ pipe.

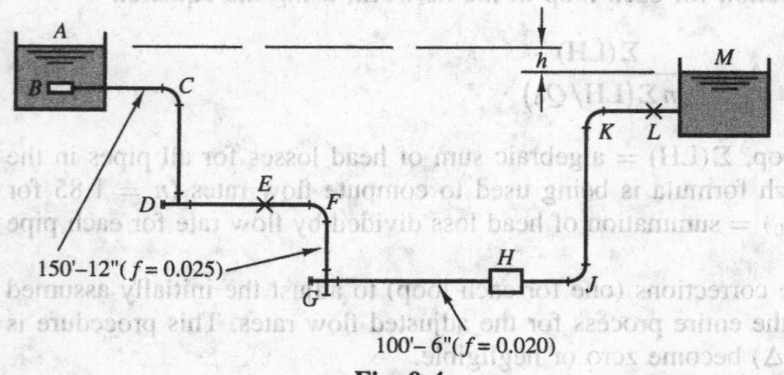

Fig. 9-4

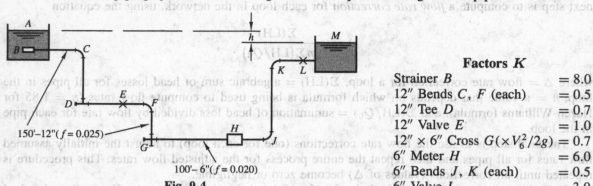

Factors K

Strainer B	$= 8.0$
12″ Bends C, F (each)	$= 0.5$
12″ Tee D	$= 0.7$
12″ Valve E	$= 1.0$
12″ × 6″ Cross $G(\times V_6^2/2g)$	$= 0.7$
6″ Meter H	$= 6.0$
6″ Bends J, K (each)	$= 0.5$
6″ Valve L	$= 3.0$

Solution:

This problem will be solved by use of the Bernoulli equation, A to M, datum M, as follows.

$$(0+0+h) - \Big(8.0 + \overset{\text{Bends}}{2\times 0.5} + 0.7 + 1.0 + 0.025 \times \frac{150}{1}\Big)\left(\frac{V_{12}^2}{2g}\right)$$

$$-\Big(0.7 + 6.0 + \overset{\text{Bends}}{2\times 0.5} + 3.0 + \overset{\text{Exit}}{1.0} + 0.020 \times \frac{100}{\frac{1}{2}}\Big)\left(\frac{V_6^2}{2g}\right) = 0+0+0$$

Then $h = 14.5\dfrac{V_{12}^2}{2g} + 15.7\dfrac{V_6^2}{2g} = (14.5 \times \dfrac{1}{16} + 15.7)\left(\dfrac{V_6^2}{2g}\right) = 16.6\dfrac{V_6^2}{2g}$.

For any available head h, the lost head is $(16.6)(V_6^2/2g)$. The lost head in L_E ft of 6″ pipe is $f(L_E/d)(V_6^2/2g)$. Equating the two values,

$$16.6\frac{V_6^2}{2g} = 0.020\left(\frac{L_E}{\frac{1}{2}}\right)\left(\frac{V_6^2}{2g}\right) \qquad \text{and} \qquad L_E = 415 \text{ ft}$$

The velocity heads in this equality canceled each other. It should be remembered that exact hydraulic equivalence depends upon f, which is not constant over wide ranges of velocity.

9.4. Three pipes in series consist of 6000 ft of 20″, 4000 ft of 16″, and 2000 ft of 12″ new cast iron pipe. Convert the system to (a) an equivalent length of 16″ pipe and (b) an equivalent size pipe 12,000 ft long.

Solution:

Use $C = 130$ for new cast iron pipe.

(a) Because the common hydraulic quantity for pipes in series is the flow, assume a value of 2.6 mgd (any convenient value will serve). In order to use Diagram B-1, change Q_{130} to Q_{100}, that is,

$$Q_{100} = (100/130)(2.6) = 2.0 \text{ mgd}$$

$$\begin{array}{llr}
S_{20} = 0.64 \text{ ft}/1000 \text{ ft} \;\text{ and} & \text{lost head} = 0.64 \times 6 = & 3.8 \text{ ft} \;\; (13.9\%) \\
S_{16} = 1.87 \text{ ft}/1000 \text{ ft} & \text{lost head} = 1.87 \times 4 = & 7.5 \text{ ft} \;\; (27.5\%) \\
S_{12} = 8.0 \;\text{ ft}/1000 \text{ ft} & \text{lost head} = 8.0 \;\times 2 = & \underline{16.0 \text{ ft} \;\; (58.6\%)} \\
\qquad \text{For } Q = 2.6 \text{ mgd.} & \text{total lost head} = & 27.3 \text{ ft} \; (100.0\%).
\end{array}$$

The equivalent 16″ pipe must carry 2.6 mgd with a lost head of 27.3 ft ($C = 130$).

$$S_{16} = 1.87 \text{ ft}/1000 \text{ ft} = \frac{\text{lost head in ft}}{\text{equivalent length in 1000 ft units}} = \frac{27.3}{L_E}$$

and $L_E = 14.6$ in 1000-ft units, or 14,600 ft.

(*b*) The 12,000 ft of pipe, $C = 130$, must carry the 2.6 mgd with a lost head of 27.3 ft.

$$S_E = \frac{\text{lost head in ft}}{\text{length in 1000-ft units}} = \frac{27.3}{12} = 2.28 \text{ ft}/1000 \text{ ft}$$

From Diagram B-1, using $Q_{100} = 2$ mgd, $D = 15.5''$ (approximately).

9.5. For the pipes in series in Problem 9.4, what flow will be produced for a total lost head of 70.0 ft, using (*a*) the equivalent pipe method and (*b*) the percentage method?

Solution:

(*a*) From Problem 9.4, 14,600 ft of 16″ pipe is equivalent to the pipes in series. For a lost head of 70.0 ft, $S_{16} = 70.0/14.6 = 4.8$ ft/1000 ft, and from Diagram B-1, $Q_{100} = 3.4$ mgd

Hence

$$Q_{130} = (130/100)(3.4) = 4.4 \text{ mgd}$$

(*b*) The percentage method requires the calculation of lost head values for an assumed flow Q. Although values are available from Problem 9.4, an additional calculation will be made to serve as a check on the solution. Assuming $Q_{130} = 3.9$ mgd, then $Q_{100} = 100/130 \times 3.9 = 3.0$ mgd, and, from Diagram B-1,

$$\begin{aligned} S_{20} &= 1.35 \text{ ft}/1000 \text{ ft} \quad \text{and} \quad \text{lost head} = 1.35 \times 6 = 8.1 \text{ ft} \quad (14.0\%) \\ S_{16} &= 4.0 \text{ ft}/1000 \text{ ft} \quad \text{and} \quad \text{lost head} = 4.0 \times 4 = 16.0 \text{ ft} \quad (27.5\%) \\ S_{12} &= 17.0 \text{ ft}/1000 \text{ ft} \quad \text{and} \quad \text{lost head} = 17.0 \times 2 = 34.0 \text{ ft} \quad (58.5\%) \\ &\text{For } Q = 3.9 \text{ mgd:} \quad \text{total lost head} = 58.1 \text{ ft} \quad (100.0\%). \end{aligned}$$

The same percentages appear here as are shown in Problem 9.4. Apply these percentages to the given total lost head of 70 ft, i.e.,

$$\begin{aligned} LH_{20} &= 70 \times 14.0\% = 9.8 \text{ ft}, \quad S = 9.8/6 = 1.63 \text{ ft}/1000 \text{ ft}, \quad Q = 130/100 \times 3.3 = 4.3 \text{ mgd} \\ LH_{16} &= 70 \times 27.5\% = 19.2 \text{ ft}, \quad S = 19.2/4 = 4.80 \text{ ft}/1000 \text{ ft}, \quad Q = 130/100 \times 3.4 = 4.4 \text{ mgd} \\ LH_{12} &= 70 \times 58.5\% = 41.0 \text{ ft}, \quad S = 41.0/2 = 20.50 \text{ ft}/1000 \text{ ft}, \quad Q = 130/100 \times 3.4 = 4.4 \text{ mgd} \end{aligned}$$

The calculation for one size is sufficient to obtain flow Q, but the check calculations give assurance that no mistakes have been made.

9.6. Water flows at a rate of 0.05 m^3/s from reservoir *A* to reservoir *B* through three concrete pipes connected in series, as shown in Fig. 9-5. Find the difference between water surface elevations in the reservoirs. Neglect all minor losses.

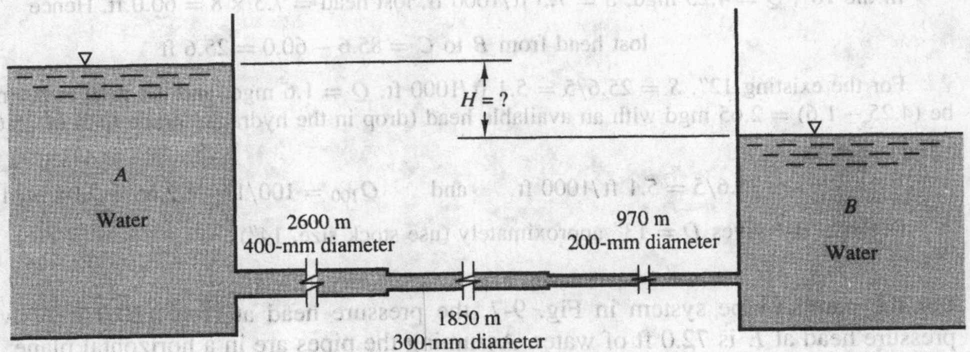

Fig. 9-5

Solution:

For the 400-mm-diameter pipe with $Q = 0.05\ \text{m}^3/\text{s}$, $h_1 = 0.00051$ m/m (from Diagram B-3). For the 300-mm-diameter pipe, $h_1 = 0.0020$ m/m; and for the 200-mm-diameter pipe, $h_1 = 0.015$ m/m.

$$\text{total head loss} = (0.00051)(2600) + (0.0020)(1850) + (0.015)(970) = 19.58\ \text{m}$$

If all minor head losses are neglected, the difference between water surface elevations in the reservoirs must be equal to the total head loss of 19.58 m.

9.7. For the system shown in Fig. 9-6, when the flow from reservoir A to main D is 3.25 mgd, the pressure at D is 20.0 psi. The flow to D must be increased to 4.25 mgd with the pressure at 40.0 psi. What size pipe, 5000 ft long, should be laid from B to C (short dashes) parallel to the existing 12″ to accomplish this result?

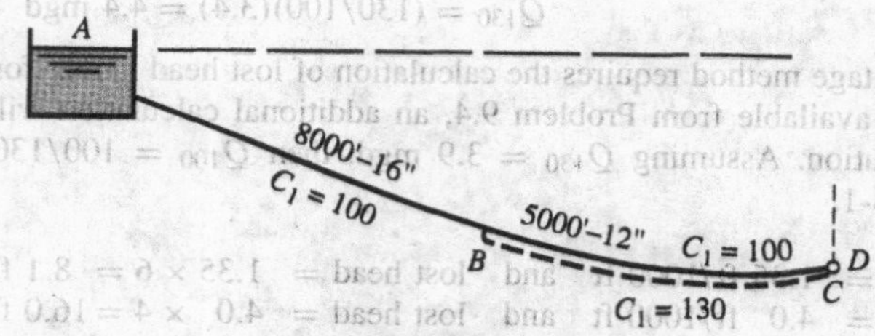

Fig. 9-6

Solution:

The elevation of reservoir A can be determined by using the data in the first sentence of the problem. From Diagram B-1,

For $Q = 3.25$ mgd,

$$S_{16} = 4.6\ \text{ft}/1000\ \text{ft}, \qquad \text{LH} = 4.6 \times 8 = 36.8\ \text{ft}$$

$$S_{12} = 19.0\ \text{ft}/1000\ \text{ft}, \qquad \text{LH} = 19.0 \times 5 = \underline{95.0\ \text{ft}}$$

$$\text{total lost head} = 131.8\ \text{ft}$$

The hydraulic grade line drops 131.8 ft to an elevation of 46.2 ft above D (equivalent to 20.0 psi). Thus reservoir A is $(131.8 + 46.2) = 178.0$ ft above point D.

For a pressure of 40.0 psi, the elevation of the hydraulic grade line at D will be 92.4 ft above D, or the available head for the flow of 4.25 mgd will be $(178.0 - 92.4) = 85.6$ ft.

In the 16″, $Q = 4.25$ mgd, $S = 7.5$ ft/1000 ft, lost head $= 7.5 \times 8 = 60.0$ ft. Hence

$$\text{lost head from } B \text{ to } C = 85.6 - 60.0 = 25.6\ \text{ft}$$

For the existing 12″, $S = 25.6/5 = 5.1$ ft/1000 ft, $Q = 1.6$ mgd, and the flow in the new pipe must be $(4.25 - 1.6) = 2.65$ mgd with an available head (drop in the hydraulic grade line) of 25.6 ft from B to C.

$$S = 25.6/5 = 5.1\ \text{ft}/1000\ \text{ft} \quad \text{and} \quad Q_{100} = 100/130 \times 2.65 = 2.04\ \text{mgd}$$

Diagram B-1 gives $D = 13''$ approximately (use stock size, 14″).

9.8. For the parallel-pipe system in Fig. 9-7, the pressure head at A is 120.0 ft of water and the pressure head at E is 72.0 ft of water. Assuming the pipes are in a horizontal plane, what are the flows in each branch of the loop?

Mathcad

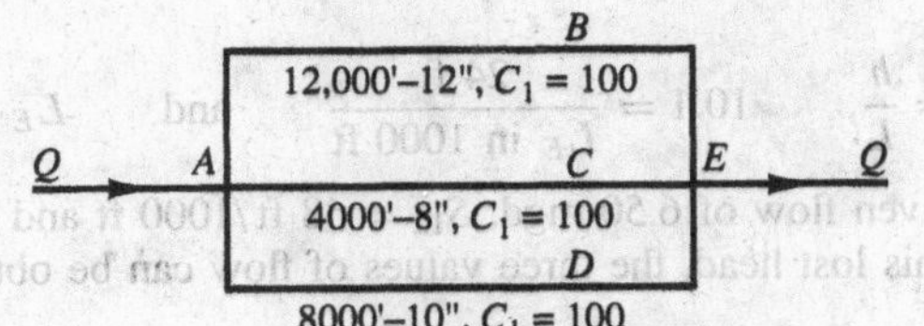

Fig. 9-7

Solution:

The drop in the hydraulic grade line A to E is $(120 - 72) = 48$ ft, neglecting the minor values of velocity head differences. The flows can be calculated inasmuch as the slopes of the grade lines are known. Thus, using Diagram B-1,

$$\begin{aligned} S_{12} &= 48/12 = 4.0 \text{ ft}/1000 \text{ ft}, & Q_{12} &= 1.4 \text{ mgd} \quad (41.1\%) \\ S_8 &= 48/4 = 12.0 \text{ ft}/1000 \text{ ft}, & Q_8 &= 0.9 \text{ mgd} \quad (26.5\%) \\ S_{10} &= 48/8 = 6.0 \text{ ft}/1000 \text{ ft}, & Q_{10} &= 1.1 \text{ mgd} \quad (32.4\%) \\ & & \text{total } Q &= 3.4 \text{ mgd} \quad (100.0\%) \end{aligned}$$

9.9. In Problem 9.8, if the total flow Q is 6.50 mgd, how much lost head occurs between A and E, and how does Q divide in the loop? Use two solutions, the percentage method and the equivalent pipe method.

Solution:

In a parallel-pipe system, the common hydraulic quantity is the lost head across the loop (A to E). The solution will proceed as if Problem 9.8 had not been solved.

Assuming a lost head, A to E, of 24 ft, the values of flow for the assumed lost head may be obtained from Diagram B-1.

$$\begin{aligned} S_{12} &= 24/12 = 2.0 \text{ ft}/1000 \text{ ft}, & Q_{12} &= 0.95 \text{ mgd} \quad (41.3\%) \\ S_8 &= 24/4 = 6.0 \text{ ft}/1000 \text{ ft}, & Q_8 &= 0.60 \text{ mgd} \quad (26.1\%) \\ S_{10} &= 24/8 = 3.0 \text{ ft}/1000 \text{ ft}, & Q_{10} &= 0.75 \text{ mgd} \quad (32.6\%) \\ & & \text{total } Q &= 2.30 \text{ mgd} \quad (100.0\%) \end{aligned}$$

(*a*) **Percentage Method**

The flow in each branch of the loop will be a constant percentage of the total flow through the loop for any reasonable range of lost heads across the loop. The percentages shown above agree favorably with the percentages tabulated in Problem 9.8 (within the accuracy of the diagram). Apply these percentages to the given flow of 6.50 mgd.

$$\begin{aligned} Q_{12} &= 41.3\% \times 6.50 = 2.68 \text{ mgd}, & S_{12} &= 13.5 \text{ ft}/1000 \text{ ft}, & \text{LH}_{A\text{-}E} &= 162 \text{ ft} \\ Q_8 &= 26.1\% \times 6.50 = 1.70 \text{ mgd}, & S_8 &= 40.0 \text{ ft}/1000 \text{ ft}, & \text{LH}_{A\text{-}E} &= 160 \text{ ft} \\ Q_{10} &= 32.6\% \times 6.50 = 2.12 \text{ mgd}, & S_{10} &= 20.0 \text{ ft}/1000 \text{ ft}, & \text{LH}_{A\text{-}E} &= 160 \text{ ft} \\ Q &= 6.50 \text{ mgd} \end{aligned}$$

This method gives a check on the calculations, as indicated above for the three lost head values. It is the preferred method.

(*b*) **Equivalent Pipe Method** (use a 12" diameter)

The calculation of flows for an assumed lost head must be made, as for the first method. Using the values above, for a lost head of 24 ft a flow of 2.30 mgd is produced in the given looping system. An equivalent pipe must produce the same flow for the lost head of 24 ft, i.e.,

$$Q = 2.30 \text{ mgd}, \quad \text{lost head} = 24 \text{ ft}, \quad \text{and} \quad S_{12} = 10.1 \text{ ft}/1000 \text{ ft from Diagram B-1}$$

From $S = \dfrac{h}{L}$, $\quad 10.1 = \dfrac{24\text{ ft}}{L_E \text{ in } 1000\text{ ft}}$ $\quad$ and $\quad L_E = 2380$ ft (of 12″ pipe, $C_1 = 100$).

For the given flow of 6.50 mgd, $S_{12} = 68$ ft/1000 ft and total lost head A to $E = 68 \times 2.38 =$ 162 ft. With this lost head, the three values of flow can be obtained.

9.10. For the system shown in Fig. 9-8, (*a*) what flow will occur when the drop in the hydraulic grade line from A to B is 200 ft? (*b*) What length of 20″ pipe ($C = 120$) is equivalent to section AB?

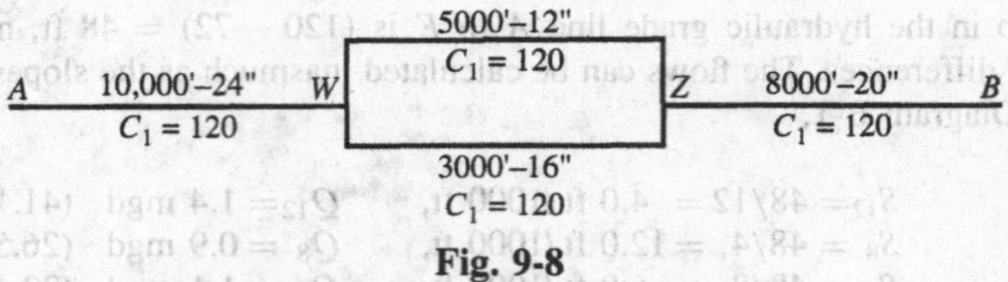

Fig. 9-8

Solution:

(*a*) The most direct solution can be obtained by assuming a drop in the hydraulic grade line (lost head) from W to Z and following that assumption to a logical conclusion.

For example, assume a lost head of 30 ft from W to Z. Then, from Diagram B-1,

$$S_{12} = 30/5 = 6.0\text{ ft}/1000\text{ ft} \quad \text{and} \quad Q_{12} = 120/100 \times 1.74 = 2.1\text{ mgd} \quad (25.9\%)$$
$$S_{16} = 30/3 = 10.0\text{ ft}/1000\text{ ft} \quad \text{and} \quad Q_{16} = 120/100 \times 5.00 = 6.0\text{ mgd} \quad (74.1\%)$$
$$\text{total } Q = 8.1\text{ mgd} \ (100.0\%)$$

The lost head from A to B can be computed for this total flow of 8.1 mgd. To employ Diagram B-1, use $Q_{100} = 100/120 \times 8.1 = 6.75$ mgd.

$$A \text{ to } W\text{: } S_{24} = 2.45\text{ ft}/1000\text{ ft, lost head} = 2.45 \times 10 = 24.5\text{ ft} \quad (23.9\%)$$
$$W \text{ to } Z\text{: (as assumed above)} = 30.0\text{ ft} \quad (29.3\%)$$
$$Z \text{ to } B\text{: } S_{20} = 6.0\text{ ft}/1000\text{ ft, lost head} = 6.0 \times 8 = 48.0\text{ ft} \quad (46.8\%)$$
$$\text{total lost head for } Q \text{ of } 8.1\text{ mgd} = 102.5\text{ ft} \ (100.0\%)$$

Applying these percentage values to the given lost head of 200 ft produces

$$\text{actual LH}_{A\text{-}W} = 200 \times 23.9\% = 47.8\text{ ft,} \quad S_{24} = 47.8/10 = 4.78\text{ ft}/1000\text{ ft}$$
$$\text{actual LH}_{W\text{-}Z} = 200 \times 29.3\% = 58.6\text{ ft,}$$
$$\text{actual LH}_{Z\text{-}B} = 200 \times 46.8\% = 93.6\text{ ft,} \quad S_{20} = 93.6/8 = 11.70\text{ ft}/1000\text{ ft}$$

From Diagram B-1, the flow in the 24″ is (120/100)(9.75) = 11.70 mgd.

As a check, in the 20″ pipe, $Q = (120/100)(9.8) = 11.76$ mgd. This flow divides in loop WZ at the percentages calculated above, namely 25.9% and 74.1%.

(*b*) Using the information above for the system from A to B, a flow of 8.1 mgd is produced with a drop in the hydraulic grade line of 102.5 ft. For 8.1 mgd in a 20″ pipe, $C = 120$,

$$S_{20} = 6.0\text{ ft}/1000\text{ ft} = 102.5/L_E \quad \text{or} \quad L_E = 17{,}100\text{ ft}$$

9.11. For the parallel-pipe system shown in Fig. 9-9, the flow rate in pipes AB and EF is 0.850 m^3/s. If all pipes are concrete, find the flow rate in pipes BCE and BDE.

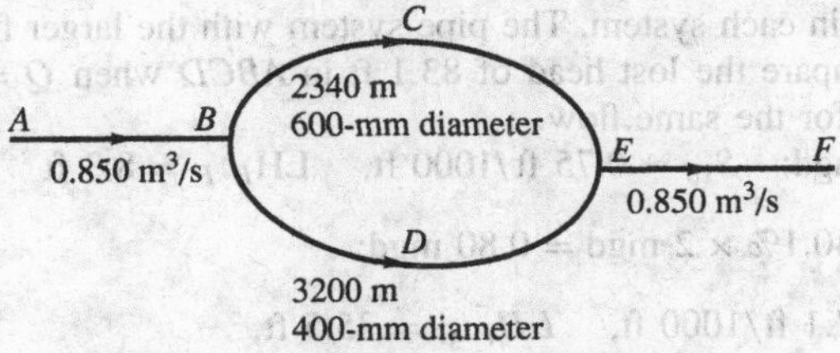

Fig. 9-9

Solution:

Assume a head loss from B to E of 1.00 m. For pipe BCE, $h_1 = 1.00/2340 = 0.00043$ m/m. From Diagram B-3, with $d = 600$ mm, $Q_{BCE} = 0.133$ m^3/s. For pipe BDE, $h_1 = 1.00/3200 = 0.00031$ m/m, and $Q_{BDE} = 0.038$ m^3/s. If the assumed head loss from B to E of 1.00 m is correct, the sum of the flow rates through pipes BCE and BDE will be equal to the flow rate through pipe AB. But

$$\left[Q_{BCE} + Q_{BDE} = 0.133 + 0.038 = 0.171\right] \neq \left[Q_{AB} = 0.850\right]$$

Since the values above are unequal, the assumed head loss of 1.00 m is incorrect; however, the actual flow rate in pipes BCE and BDE will be in the same proportion as the flow rate values determined above based on the head loss of 1.00 m. Hence,

$$Q_{BCE} = (0.133/0.171)(0.850) = 0.661 \text{ m}^3/\text{s}, \qquad Q_{BDE} = (0.038/0.171)(0.850) = 0.189 \text{ m}^3/\text{s}$$

9.12. In Fig. 9-10, which system has the greater capacity, $ABCD$ or $EFGH$? ($C = 120$ for all pipes.)

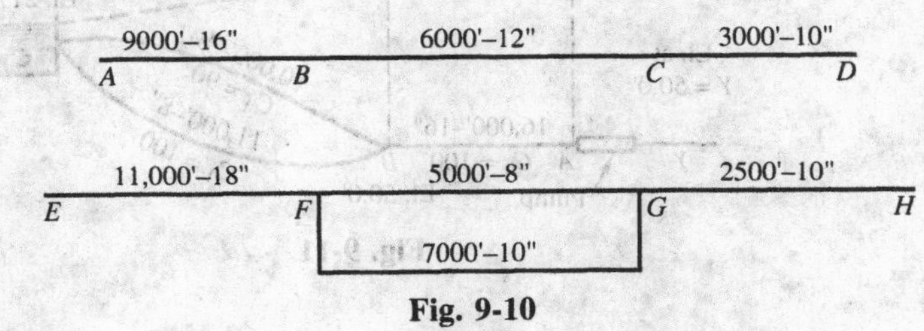

Fig. 9-10

Solution:

Assume $Q = 2$ mgd in $ABCD$. Then

$$\begin{aligned}
S_{16} &= 1.33 \text{ ft}/1000 \text{ ft}, & \text{lost head} &= 1.33 \times 9 = 12.0 \text{ ft}\\
S_{12} &= 5.35 \text{ ft}/1000 \text{ ft}, & \text{lost head} &= 5.35 \times 6 = 32.1 \text{ ft}\\
S_{10} &= 13.0 \text{ ft}/1000 \text{ ft}, & \text{lost head} &= 13.0 \times 3 = 39.0 \text{ ft}\\
&\text{For } Q = 2 \text{ mgd}, & \text{total lost head} &= 83.1 \text{ ft}
\end{aligned}$$

For loop FG in $EFGH$, find the percentage of any flow Q in each branch. Assume the lost head from F to G to be 24 ft. Then

$$\begin{aligned}
S_8 &= 24/5 = 4.80 \text{ ft}/1000 \text{ ft} \quad \text{and} \quad & Q_8 &= 0.65 \text{ mgd} \quad (40.1\%)\\
S_{10} &= 24/7 = 3.43 \text{ ft}/1000 \text{ ft} \quad \text{and} \quad & Q_{16} &= 0.97 \text{ mgd} \quad (59.9\%)\\
& & \text{total } Q &= 1.62 \text{ mgd} \quad (100.0\%)
\end{aligned}$$

To compare capacities, several alternatives present themselves. Rather than using equivalent pipes, we might calculate the lost head caused by a flow of 2 mgd through each system. The system with the smaller lost head would have the greater capacity. Or we might find flow Q caused by the identical drop in the hydraulic grade line in each system. The pipe system with the larger flow would have the greater capacity. In this problem, compare the lost head of 83.1 ft in *ABCD* when $Q = 2$ mgd with the value of lost head obtained for *EFGH* for the same flow.

(*a*) For $Q_{18} = 2$ mgd: $S_{18} = 0.75$ ft/1000 ft, $\text{LH}_{E\text{-}F} = 8.3$ ft

(*b*) For $Q_8 = 40.1\% \times 2 \text{ mgd} = 0.80$ mgd:

$$S_8 = 7.1 \text{ ft/1000 ft}, \quad LH_{F\text{-}G} = 35.5 \text{ ft},$$

or for $Q_{10} = 59.9\% \times 2$ mgd (as check) $= 1.20$ mgd:

$$S_{10} = 5.1 \text{ ft/1000 ft}, \qquad (LH = 5.1 \times 7 = 35.7 \text{ ft})$$

(*c*) For $Q_{10} = 2$ mgd: $S_{10} = 13.0$ ft/1000 ft, LH = 32.5 ft

For $Q = 2$ mgd: total lost head from E to $H = 76.3$ ft

Hence system *EFGH* has the greater capacity.

9.13. In Fig. 9-11, when pump *YA* delivers 5.00 cfs, find the pressure heads at *A* and *B*. Draw the hydraulic grade lines.

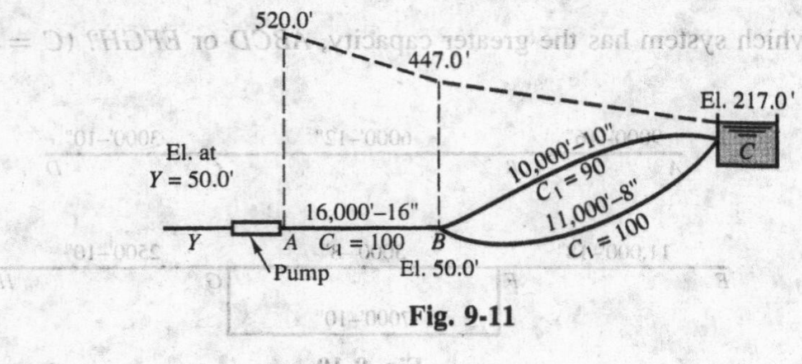

Fig. 9-11

Solution:

Reduce section (loop) *BC* to an equivalent pipe, 16″ in diameter, $C = 100$. By so doing, a single-size pipe of the same relative roughness is readily handled for all conditions of flow. Assuming a drop in the grade line of 22 ft from *B* to *C*, the following values are obtained.

$$S_{10} = 22/10 = 2.2 \text{ ft/1000 ft}, \qquad Q_{10} = 0.57 \text{ mgd}$$
$$S_8 = 22/11 = 2.0 \text{ ft/1000 ft}, \qquad Q_8 = 0.34 \text{ mgd}$$
$$\text{total } Q = 0.91 \text{ mgd}$$

For $Q = 0.91$ mgd and $D = 16''(C = 100)$, $S_{16} = 0.435 \text{ ft/1000 ft} = 22.0/L_E$ and $L_E = 50{,}600$ ft.

The flow from pump to reservoir is 5.00 cfs or 3.23 mgd. For $(50{,}600 + 16{,}000) = 66{,}600$ ft of equivalent 16″ pipe, the lost head from *A* to *C* will be

$$S_{16} = 4.55 \text{ ft/1000 ft}, \qquad \text{lost head} = 4.55 \times 66.6 = 303 \text{ ft}$$

Thus the elevation of the hydraulic grade line at *A* is $(217 + 303) = 520'$, as shown in Fig. 9-11. The drop from *A* to $B = 4.55 \times 16 = 73$ ft, and the elevation at *B* becomes $(520 - 73) = 447'$.

$$\text{pressure head at } A = 520 - 50 = 470 \text{ ft}$$

$$\text{pressure head at } B = 447 - 50 = 397 \text{ ft}$$

9.14. In Fig. 9-12, the flow from reservoir A is 10 mgd. Determine the power extracted by turbine DE if the pressure head at E is -10.0 ft. Indicate grade lines.

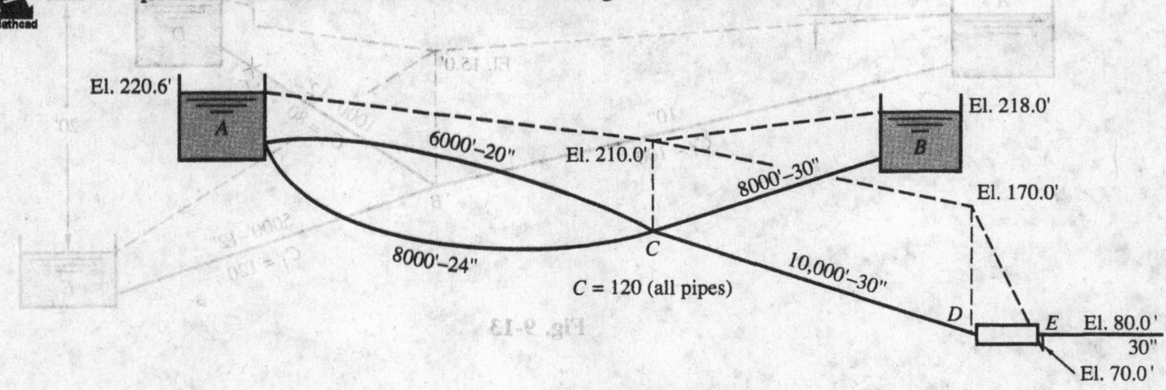

Fig. 9-12

Solution:

The analysis of a branching system of pipes should concentrate itself on the junction point C. First, the sum of the flows toward C must equal the sum of the flows away from C. Second, the elevation of the hydraulic grade line at C is often the key to the solution.

To calculate the elevation of the hydraulic grade line at C, assume the lost head, A to C, to be 24 ft. Then

$$S_{20} = 24/6 = 4.0 \text{ ft}/1000 \text{ ft}, \qquad Q_{20} = 6.5 \text{ mgd} \quad (41.9\%)$$
$$S_{24} = 24/8 = 3.0 \text{ ft}/1000 \text{ ft}, \qquad Q_{24} = 9.0 \text{ mgd} \quad (58.1\%)$$
$$\text{total } Q = 15.5 \text{ mgd} \quad (100.0\%)$$

Apply these percentages to the given 10 mgd flow from A to C.

$$Q_{20} = 4.2 \text{ mgd}, \quad S_{20} = 1.77 \text{ ft}/1000 \text{ ft}, \quad \text{lost head} = 10.6 \text{ ft}$$
$$Q_{24} = 5.8 \text{ mgd}, \quad S_{24} = 1.33 \text{ ft}/1000 \text{ ft}, \quad \text{lost head} = 10.6 \text{ ft (check)}$$

Thus the elevation of the hydraulic grade line at $C = 220.6 - 10.6 = 210.0'$. From this information, the grade line drops 8 ft from B to C, and the flow must be from B to C. Then

$$S_{30} = 8/8 = 1.0 \text{ ft}/1000 \text{ ft}, \qquad Q = 8.95 \text{ mgd}$$

Also, flow from C = flow to C

$$Q_{C\text{-}D} = 10.0 + 8.95 = 18.95 \text{ mgd}$$

Thus $S_{30} = 4.0$ ft/1000 ft, $\text{LH}_{C\text{-}D} = 40.0$ ft, and elevation of grade line at $D = 210 - 40 = 170.0'$.

$$\text{horsepower extracted} = \frac{\gamma Q H_T}{550} = \frac{(62.4)(18.95 \times 1.547)(170 - 70)}{550} = 333 \text{ hp}$$

9.15. In Fig. 9-13, valve F is partly closed, creating a lost head of 3.60 ft when the flow through the valve is 0.646 mgd. What is the length of 10″ pipe to reservoir A?

Solution:

For DB, flow $Q = 0.646$ mgd and $S_{12} = 1.40$ ft/1000 ft. The total lost head from D to $B = 1.40 \times 1 + 3.60 = 5.00$ ft, giving the grade line elevation at B of $15.0'$ (calling elevation $E = 0$).

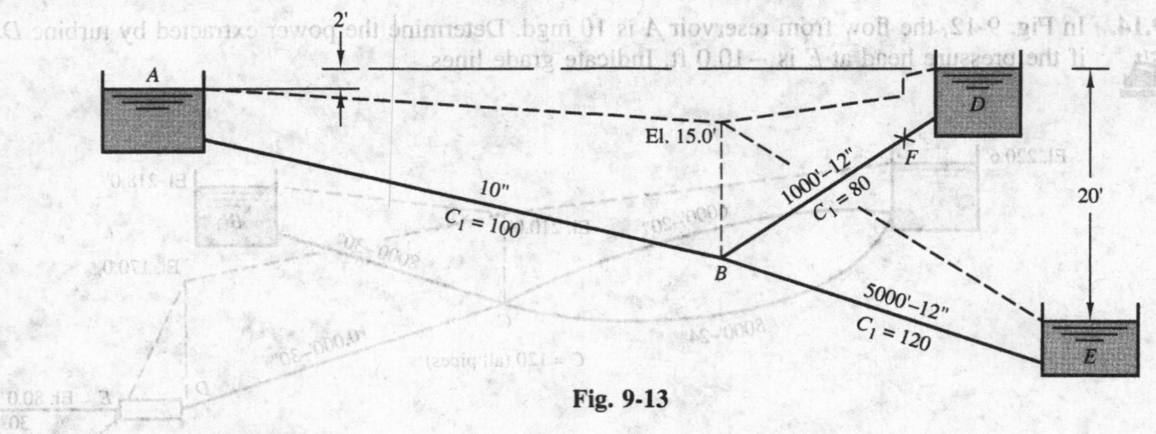

Fig. 9-13

For BE, $S_{12} = (15.0 - 0.0)/5 = 3$ ft/1000 ft and $Q = 1.46$ mgd.
For AB, flow $Q = 1.46 - 0.646 = 0.81$ mgd and $S_{10} = 3.45$ ft/1000 ft.
Then, from $S = h/L$, $L = h/S = 3.00/3.45 = 0.870$ in 1000-ft units, or $L = 870$ ft.

9.16. Water is to be pumped at the rate of 2.00 cfs through 4000 ft of new cast iron pipe to a reservoir whose surface is 120 ft above the lower water level. Assume the annual cost of pumping the 2.00 cfs is \$5.00 per ft pumped against, and the annual cost of the pipe itself is 10% of its initial cost. Also assume cast iron pipe in place costs \$140.00 per ton, with class B (200-ft head) pipe having the following weights per foot: 6″, 33.3 lb; 8″, 47.5 lb; 10″, 63.8 lb; 12″, 82.1 lb; and 16″, 125 lb. Determine the economical pipe size for this installation.

Solution:

Calculations for the 12″ pipe will be shown in detail, with a summary of all results given in the table below. The lost head in the 12″, using $C = 130$ for new cast iron pipe, is 2.06 ft/1000 ft.
Hence, total head pumped against $= 120 + 4 \times 2.06 = 128.2$ ft.

$$\text{Cost of pumping} = 128.2 \times \$5 = \$641 \text{ per year}$$

$$\text{Cost of pipe in place} = \$140 \times 4000 \times 82.1/2000 = \$23{,}000$$

$$\text{Annual cost of pipe} = 10\% \times \$23{,}000 = \$2300$$

Tabulating these values for comparison with the costs of the other sizes under consideration gives the following.

D (in)	S (ft/1000 ft)	Lost Head (ft)	Total Pumping Head = 120 + LH	Annual Costs for 2 cfs: Pumping	+ Pipe Cost	= Total
6″	59.0	236.	356. ft	\$1780	\$ 933	\$2713
8″	14.8	59.2	179.2 ft	896	1330	2226
10″	5.00	20.0	140.0 ft	700	1790	2490
12″	2.06	8.2	128.2 ft	641	2300	2941
16″	0.515	2.1	122.1 ft	611	3500	4111

The economical size is 8″.

9.17. For the constant elevations of the water surfaces shown in Fig. 9-14(*a*), what flows will occur?

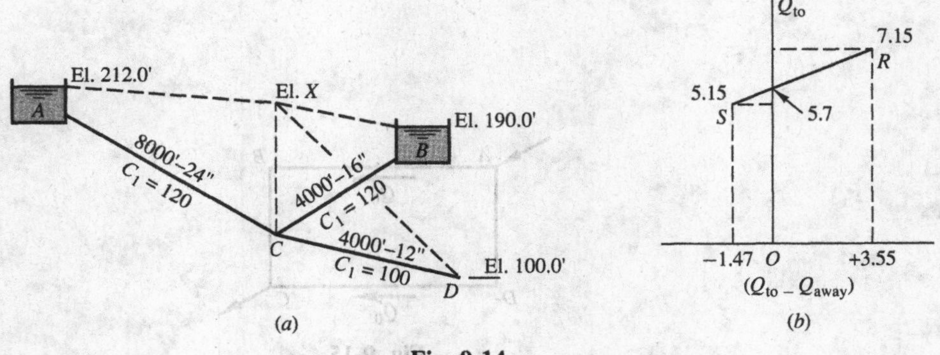

Fig. 9-14

Solution:

Because the elevation of the hydraulic grade line at C cannot be computed (all flows unknown), the problem will be solved by successive trials. A convenient assumption is to choose the elevation of the hydraulic grade line at C at 190.0′. By so assuming, flow to or from reservoir B will be zero, thus reducing the number of calculations.

For elevation of hydraulic grade line at $C = 190.0'$,

$$S_{24} = (212 - 190)/8 = 2.75 \text{ ft}/1000 \text{ ft} \quad \text{and} \quad Q = 7.15 \text{ mgd to } C$$

$$S_{12} = (190 - 100)/4 = 22.5 \text{ ft}/1000 \text{ ft} \quad \text{and} \quad Q = 3.60 \text{ mgd away from } C$$

Examination of these values of flow indicates that the grade line at C must be higher, thereby reducing the flow from A and increasing the flow to D as well as adding flow to B. In an endeavor to "straddle" the correct elevation at C, assume a value of 200.0′. Thus, for elevation at $C = 200.0'$,

$$S_{24} = (212 - 200)/8 = 1.50 \text{ ft}/1000 \text{ ft} \quad \text{and} \quad Q = 5.15 \text{ mgd to } C$$

$$S_{16} = (200 - 190)/4 = 2.50 \text{ ft}/1000 \text{ ft} \quad \text{and} \quad Q = 2.82 \text{ mgd away from } C$$

$$S_{12} = (200 - 100)/4 = 25.0 \text{ ft}/1000 \text{ ft} \quad \text{and} \quad Q = 3.80 \text{ mgd away from } C$$

The flow away from $C = 6.62$ mgd against the flow to C of 5.15 mgd. Using Fig. 9-14(*b*) to obtain a guide regarding a reasonable third assumption, connect plotted points R and S. The line so drawn intersects the $(Q_{to} - Q_{away})$ zero abscissa at $Q_{to} = 5.7$ mgd (scaled). Inasmuch as the values plotted do not vary linearly, use a flow to C slightly larger, say 6.0 mgd.

For $Q = 6.0$ mgd to C, $S_{24} = 1.98$ ft/1000 ft, $\text{LH}_{A\text{-}C} = 1.98 \times 8 = 15.8$ ft, and the hydraulic grade line at C is at elevation $(212 - 15.8) = 196.2'$. Then

$$S_{16} = 6.2/4 = 1.55 \text{ ft}/1000 \text{ ft}, \qquad Q = 2.17 \text{ mgd from } C$$

$$S_{12} = 96.2/4 = 24.05 \text{ ft}/1000 \text{ ft}, \qquad Q = \underline{3.73} \text{ mgd from } C$$

$$\text{total } Q \text{ from } C = 5.90 \text{ mgd}$$

These resulting flows agree sufficiently well as not to require further calculations. (An elevation of the hydraulic grade line at C of about 196.5′ would probably prove correct, with flows to C and from C of about 5.95 mgd.)

9.18. Develop the expression used to study flows in a pipe network.

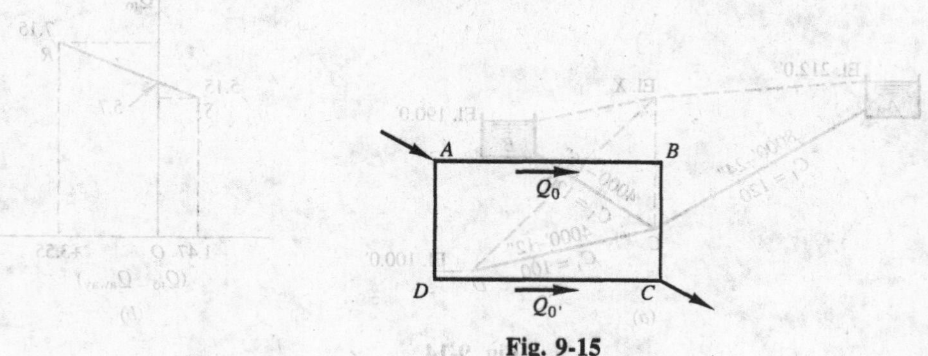

Fig. 9-15

Solution:

The method of attack developed by Professor Hardy Cross consists of assuming flows throughout the network and then balancing the calculated head losses. In the simple looping pipe system shown in Fig. 9-15, for the correct flow in each branch of the loop,

$$\mathrm{LH}_{ABC} = \mathrm{LH}_{ADC} \quad \text{or} \quad \mathrm{LH}_{ABC} - \mathrm{LH}_{ADC} = 0 \tag{1}$$

In order to use this relationship, the flow formula must be written in the form $\mathrm{LH} = kQ^n$. For the Hazen-Williams formula, this expression is $\mathrm{LH} = kQ^{1.85}$.

But, since we are assuming flows Q_0, the correct flow Q in any pipe of a network can be expressed as $Q = Q_0 + \Delta$, where Δ is the correction to be applied to Q_0. Then, using the binomial theorem,

$$kQ^{1.85} = k\left(Q_0 + \Delta\right)^{1.85} = k\left(Q_0^{1.85} + 1.85Q_0^{1.85-1}\Delta + \cdots\right)$$

Terms beyond the second can be neglected because Δ is small compared with Q_0.

For the loop above, substituting in expression (*1*) we obtain

$$k\left(Q_0^{1.85} + 1.85\,Q_0^{0.85}\Delta\right) - k\left(Q_{0'}^{1.85} + 1.85Q_{0'}^{0.85}\Delta\right) = 0$$

$$k\left(Q_0^{1.85} - Q_{0'}^{1.85}\right) + 1.85k\left(Q_0^{0.85} - Q_{0'}^{0.85}\right)\Delta = 0$$

Solving for Δ,

$$\Delta = -\frac{k\left(Q_0^{1.85} - Q_{0'}^{1.85}\right)}{1.85k\left(Q_0^{0.85} - Q_{0'}^{0.85}\right)} \tag{2}$$

In general, we may write for a more complicated loop,

$$\Delta = -\frac{\Sigma\, kQ_0^{1.85}}{1.85\Sigma\, kQ_0^{0.85}} \tag{3}$$

But $kQ_0^{1.85} = \text{LH}$ and $kQ_0^{0.85} = (\text{LH})/Q_0$. Therefore,

$$\Delta = -\frac{\Sigma\,(\text{LH})}{1.85\Sigma(\text{LH}/Q_0)} \quad \text{for each loop of a network} \tag{4}$$

In using expression (4), care must be exercised regarding the sign of the numerator. Expression (1) indicates that clockwise flows may be considered as producing clockwise losses, and counterclockwise flows, counterclockwise losses. This means that the minus sign is assigned to all counterclockwise conditions in a loop, namely flow Q and lost head LH. Hence to avoid mistakes, this sign notation must be observed in carrying out a solution. On the other hand, the denominator of (4) is always positive.

The next two problems will illustrate how equation (4) is used.

9.19. The looping system shown in Fig. 9-16 is the same system that appears in Problem 9.10. For $Q = 11.7$ mgd total, how much flow occurs in each branch of the loop, using the Hardy Cross procedure?

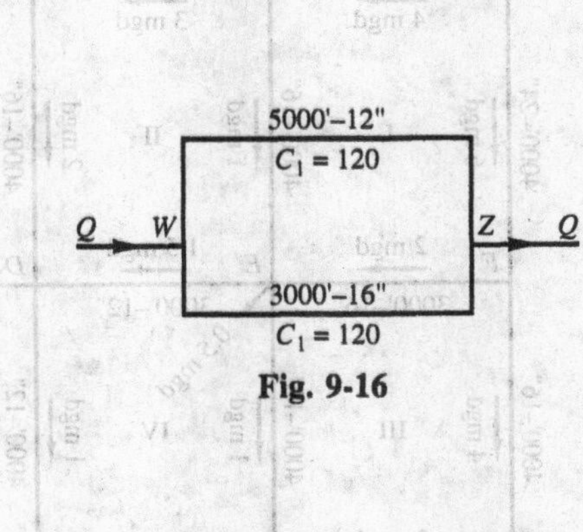

Fig. 9-16

Solution:

Values of Q_{12} and Q_{16} are assumed to be 4.0 mgd and 7.7 mgd, respectively, The tabulation below is prepared, (note the −7.70 mgd), the values of S calculated, then $\text{LH} = S \times L$, and LH/Q_0 can be calculated. Note that the large ΣLH indicates that the Q's are not well balanced. (The values were assumed deliberately to produce this large Σ LH to illustrate the procedure.)

D	L	Assumed Q_0 (mgd)	S (ft/1000′)	LH (ft)	LH/Q_0	Δ	Q_1
12″	5000′	4.00	19.5	97.5	24.4	−0.85	3.15
16″	3000′	−7.70	−16.3	−48.9	6.4	−0.85	−8.55
		Σ = 11.70		Σ = +48.6	30.8		11.70

$$\Delta = -\frac{\Sigma\text{LH}}{1.85\Sigma(\text{LH}/Q)} = -\frac{+48.6}{(1.85)(30.8)} = -0.85 \text{ mgd}$$

Then the Q_1 values becomes (4.00 − 0.85) = 3.15 mgd and (−7.70 − 0.85) = −8.55 mgd. Repeating the calculation produces

S	LH	LH/Q_1	Δ	Q_2
12.5	62.5	19.84	−0.06	3.09
−19.8	−59.4	6.95	−0.06	−8.61
	Σ = +3.1	26.79		11.70

No further calculation is necessary, since Diagram B-1 cannot be read to the accuracy of 0.06 mgd. Ideally, ΣLH should equal zero, but this goal is seldom attained.

It will be noted that in Problem 9.10 the quantity flowing in the 12″ was 26% of 11.7 mgd, or 3.04 mgd, a satisfactory check.

9.20. Water flows through the piping system shown in Fig. 9-17 with certain measured flows indicated on the sketch. At point *A*, the elevation is 200.0 ft and the pressure head is 150.0 ft. The elevation at *I* is 100.0 ft. Determine (*a*) the flows throughout the network and (*b*) the pressure head at *I*. (Use $C = 120$.)

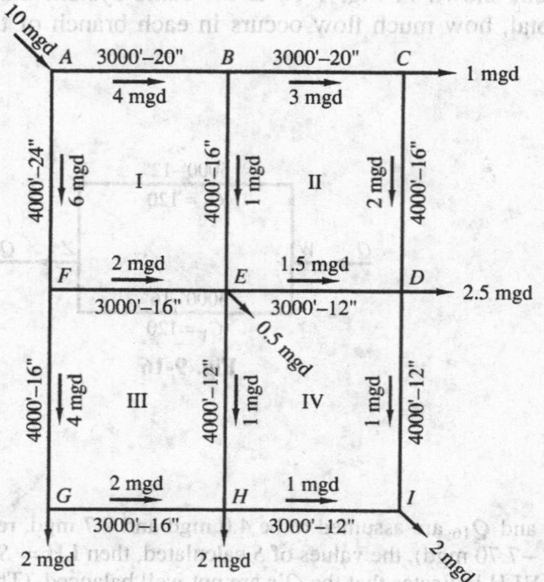

Fig. 9-17

Solution:

(*a*) The method of attack can be outlined as follows:

(1) Assume any distribution of flow, proceeding loop by loop—in this case loops I, II, III, and IV. Examine each junction point carefully so that the flow *to* the point equals the flow *away from* the point (continuity principle).

(2) Compute for each loop the head loss in each pipe of the loop (by equation or diagram).

(3) Sum the lost heads around each loop, with due regard to sign (should the sum of the lost heads for the loop be zero, flows Q_1 are correct).

(4) Sum the LH/Q_1 values, and calculate correction term Δ for each loop.

(5) Apply the Δ value to each pipeline, thereby increasing or decreasing the assumed Q's. For cases where a pipe is in two loops, the difference between the two Δ's must be applied as the proper correction to the assumed flow Q_1 (see application below).

(6) Proceed until the Δ values are negligible.

The steps listed above have been carried out in tabular form. Values of LH are obtained by multiplying S by the number of thousand feet in the pipe being considered. Values of LH divided by Q are also tabulated.

Line	D (in)	L (ft)	Assumed Q_1 (mgd)	S (ft/1000′)	LH (ft)	LH/Q_1	Δ	Q_2
AB	20	3000	4.0	1.62	4.86	1.22	+0.32	4.32
BE	16	4000	1.0	0.37	1.48	1.48	+0.32 − (0.13) = +0.19	1.19
EF	16	3000	−2.0	−1.33	−3.99	2.00	+0.32 − (0.49) = −0.17	−2.17
FA	24	4000	−6.0	−1.41	−5.64	0.94	+0.32	−5.68
					Σ = −3.29	5.64		
BC	20	3000	3.0	0.95	2.85	0.95	+0.13	3.13
CD	16	4000	2.0	1.33	5.32	2.66	+0.13	2.13
DE	12	3000	−1.5	−3.15	−9.45	6.30	+0.13 − (−0.12) = +0.25	−1.25
EB	16	4000	−1.0	−0.37	−1.48	1.48	+0.13 − (0.32) = −0.19	−1.19
					Σ = −2.76	11.39		
FE	16	3000	2.0	1.33	3.99	2.00	+0.49 − (+0.32) = +0.17	2.17
EH	12	4000	1.0	1.48	5.92	5.92	+0.49 − (−0.12) = +0.61	1.61
HG	16	3000	−2.0	−1.33	−3.99	2.00	+0.49	−1.51
GF	16	4000	−4.0	−4.85	−19.40	4.85	+0.49	−3.51
					Σ = −13.48	14.77		
ED	12	3000	1.5	3.15	9.45	6.30	−0.12 − (0.13) = −0.25	1.25
DI	12	4000	1.0	1.48	5.92	5.92	−0.12	0.88
IH	12	3000	−1.0	−1.48	−4.44	4.44	−0.12	−1.12
HE	12	4000	−1.0	−1.48	−5.92	5.92	−0.12 − (0.49) = −0.61	−1.61
					Σ = +5.01	22.58		

The Δ terms are calculated [expression (*4*), Problem 9.18], as follows:

$$\Delta_{\mathrm{I}} = \frac{-(-3.29)}{(1.85)(5.64)} = +0.32 \qquad \Delta_{\mathrm{III}} = \frac{-(-13.48)}{(1.85)(14.77)} = +0.49$$

$$\Delta_{\mathrm{II}} = \frac{-(-2.76)}{(1.85)(11.39)} = +0.13 \qquad \Delta_{\mathrm{IV}} = \frac{-(+5.01)}{(1.85)(22.58)} = -0.12$$

For line *EF* in loop I, its net Δ term is $(\Delta_{\mathrm{I}} - \Delta_{\mathrm{III}})$ or $[+0.32-(+0.49)] = -0.17$. It will be observed that the Δ for loop I is combined with that of the Δ for loop III because the line *EF* occurs in each loop. In similar fashion, for line *FE* in loop III, the net Δ term is $(\Delta_{\mathrm{III}} - \Delta_{\mathrm{I}})$ or $[+0.49-(+0.32)] = +0.17$. Note that the net Δ's have the same magnitude but *opposite signs*. This can readily be understood because flow in *EF* is counterclockwise for loop I, whereas flow in *FE* in loop III is clockwise.

In determining values of Q_2 for the second calculation,

$$Q_{AB} = (4.00 + 0.32) = 4.32 \text{ mgd}$$

whereas

$$Q_{EF} = (-2.00 - 0.17) = -2.17 \text{ mgd} \quad \text{and} \quad Q_{FA} = (-6.00 + 0.32) = -5.68 \text{ mgd}$$

The procedure is carried along until the Δ terms are insignificant with regard to the accuracy expected, keeping in mind the accuracy of the values of C and of the diagram. Reference to the last column of the table on the next page will indicate the final values of Q in the several pipelines.

Since the sums of the lost head values are small for all the loops, we may consider the flow values listed in the last column in the table on the next page as correct within the accuracy expected. The reader may practice by calculating the next values of Δ, then Q_5, etc.

(*b*) The elevation of the hydraulic grade line at A is $(200.0 + 150.0) = 350.0'$. The lost head to I can be calculated by any route from A to I, adding the losses in the usual manner, i.e., in the direction of flow. Using $ABEHI$ we obtain $\text{LH}_{A\text{-}I} = (6.06 + 2.72 + 10.72 + 3.93) = 23.43$ ft. As a check, using $ABEDI$, LH $= (6.06 + 2.72 + 8.91 + 6.72) = 24.41$ ft. Using a value of 24 ft, the elevation of the hydraulic grade line at $I = (350.0 - 24.0) = 326.0$ ft. Hence the pressure head at $I = (326.0 - 100.0) = 226.0$ ft.

Line	Q_2	S	LH	LH/Q	Δ
AB	4.32	1.86	5.58	1.29	+0.20
BE	1.19	0.51	2.04	1.71	+0.20 + negl = +0.20
EF	−2.17	−1.57	−4.71	2.17	+0.20 − (−0.06) = +0.26
FA	−5.68	−1.28	−5.12	0.90	+0.20
			$\Sigma = -2.21$	6.07	
BC	3.13	1.02	3.06	0.98	negl
CD	2.13	1.48	5.92	2.78	negl
DE	−1.25	−2.28	−6.84	5.47	negl − 0.19 = −0.19
EB	−1.19	−0.51	−2.04	1.71	negl − 0.20 = −0.20
			$\Sigma = +0.10$	10.94	
FE	2.17	1.57	4.71	2.17	−0.06 − 0.20 = −0.26
EH	1.61	3.65	14.60	9.07	−0.06 − 0.19 = −0.25
HG	−1.51	−0.79	−2.37	1.57	−0.06
GF	−3.51	−3.75	−15.00	4.27	−0.06
			$\Sigma = +1.94$	17.08	
ED	1.25	2.28	6.84	5.47	+0.19 + negl = +0.19
DI	0.88	1.18	4.72	5.36	+0.19
IH	−1.12	−1.83	−5.49	4.90	+0.19
HE	−1.61	−3.65	−14.60	9.07	+0.19 − (−0.06) = +0.25
			$\Sigma = -8.53$	24.80	

Line	Q_3	S	LH	LH/Q	Δ	Q_4
AB	4.52	2.02	6.06	1.34	−0.02	4.50
BE	1.39	0.68	2.72	1.96	−0.02 − 0.12 = −0.14	1.25
EF	−1.91	−1.25	−3.75	1.96	−0.02 − 0.12 = −0.14	−2.05
FA	−5.48	−1.20	−4.80	0.88	−0.02	−5.50
			$\Sigma = +0.23$	6.14		
BC	3.13	1.02	3.06	0.98	+0.12	3.25
CD	2.13	1.49	5.96	2.80	+0.12	2.25
DE	−1.44	−2.97	−8.91	6.19	+0.12 + 0.02 = +0.14	−1.30
EB	−1.39	−0.68	−2.72	1.96	+0.12 + 0.02 = +0.14	−1.25
			$\Sigma = -2.61$	11.93		
FE	1.91	1.25	3.75	1.96	+0.12 + 0.02 = +0.14	2.05
EH	1.36	2.68	10.72	7.88	+0.12 + 0.02 = +0.14	1.50
HG	−1.57	−0.84	−2.52	1.61	+0.12	−1.45
GF	−3.57	−3.90	−15.60	4.37	+0.12	−3.45
			$\Sigma = -3.65$	15.82		
ED	1.44	2.97	8.91	6.19	−0.02 − 0.12 = −0.14	1.30
DI	1.07	1.68	6.72	6.28	−0.02	1.05
IH	−0.93	−1.31	−3.93	4.23	−0.02	−0.95
HE	−1.36	−2.68	−10.72	7.88	−0.02 − 0.12 = −0.14	−1.50
			$\Sigma = +0.98$	24.58		

9.21. The pipe network shown in Fig. 9-18 represents a spray rinse system. Find the flow rate of water in each pipe. Assume $C = 120$ for all pipes.

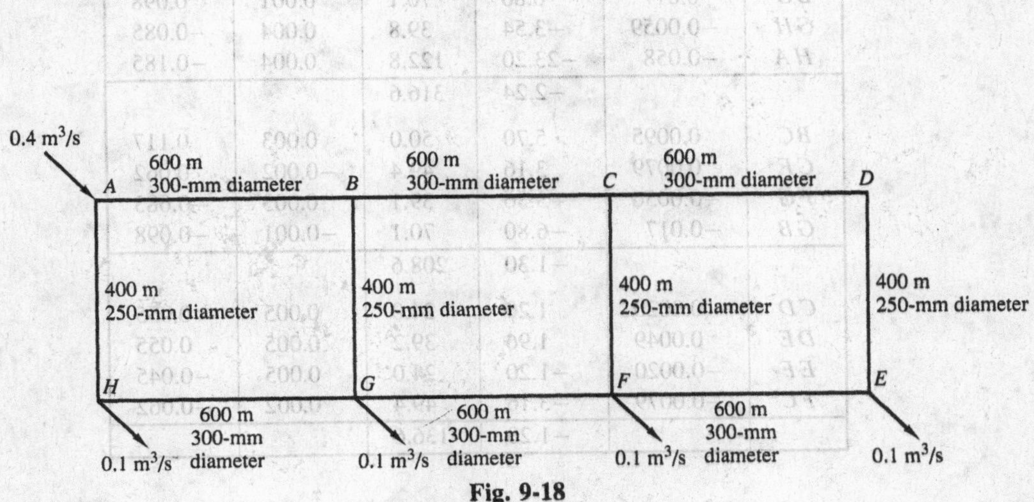

Fig. 9-18

Solution:

Pipe	D (mm)	L (m)	Q_{assumed} (m³/s)	S (m/m)	LH (m)	LH/Q	Δ	Q_{new}
AB	300	600	0.200	0.027	16.20	81.0	0.011	0.211
BG	250	400	0.100	0.0175	7.00	70.0	−0.003	0.097
GH	300	600	−0.100	−0.0074	−4.44	44.4	0.011	−0.089
HA	250	400	−0.200	−0.064	−25.60	128.0	0.011	−0.189
					−6.84	323.4		
BC	300	600	0.100	0.0074	4.44	44.4	0.014	0.114
CF	250	400	0.050	0.0049	1.96	39.2	0.014	0.064
FG	300	600	−0.100	−0.0074	−4.44	44.4	0.014	−0.086
GB	250	400	−0.100	−0.0175	−7.00	70.0	0.003	−0.097
					−5.04	198.0		
CD	300	600	0.050	0.0020	1.20	24.0	0.000	0.050
DE	250	400	0.050	0.0049	1.96	39.2	0.000	0.050
EF	300	600	−0.050	−0.0020	−1.20	24.0	0.000	−0.050
FC	250	400	−0.050	−0.0049	−1.96	39.2	−0.014	−0.064
					0.00	126.4		

$$\Delta Q_{\text{I}} = -(-6.84)/[(1.85)(323.4)] = 0.011$$

$$\Delta Q_{\text{II}} = -(-5.04)/[(1.85)(198.0)] = 0.014$$

$$\Delta Q_{\text{III}} = -0.00/[(1.85)(126.4)] = 0.00$$

Pipe	S	LH	LH/Q	Δ	Q_{new}
AB	0.0295	17.70	83.9	0.004	0.215
BG	0.017	6.80	70.1	0.001	0.098
GH	−0.0059	−3.54	39.8	0.004	−0.085
HA	−0.058	−23.20	122.8	0.004	−0.185
		−2.24	316.6		
BC	0.0095	5.70	50.0	0.003	0.117
CF	0.0079	3.16	49.4	−0.002	0.062
FG	−0.0056	−3.36	39.1	0.003	−0.083
GB	−0.017	−6.80	70.1	−0.001	−0.098
		−1.30	208.6		
CD	0.0020	1.20	24.0	0.005	0.055
DE	0.0049	1.96	39.2	0.005	0.055
EF	−0.0020	−1.20	24.0	0.005	−0.045
FC	−0.0079	−3.16	49.4	0.002	−0.062
		−1.20	136.6		

$$\Delta Q_{\mathrm{I}} = -(-2.24)/[(1.85)(316.6)] = 0.004$$

$$\Delta Q_{\mathrm{II}} = -(-1.30)/[(1.85)(208.6)] = 0.003$$

$$\Delta Q_{\mathrm{III}} = -(-1.20)/[(1.85)(136.6)] = 0.005$$

Pipe	S	LH	LH/Q	Δ	Q_{new}
AB	0.031	18.60	86.5	0.000	0.215
BG	0.0172	6.88	70.2	−0.003	0.095
GH	0.0055	−3.30	38.8	0.000	−0.085
HA	0.056	−22.40	121.1	0.000	−0.185
		−0.22	316.6		
BC	0.010	6.00	51.3	0.003	0.120
CF	0.0075	3.00	48.4	0.002	0.064
FG	0.0052	−3.12	37.6	0.003	−0.080
GB	0.0172	−6.88	70.2	0.003	−0.095
		−1.00	207.5		
CD	0.0024	1.44	26.2	0.001	0.056
DE	0.0059	2.36	42.9	0.001	0.056
EF	0.0017	−1.02	22.7	0.001	−0.044
FC	0.0075	−3.00	48.4	−0.002	−0.064
		−0.22	140.2		

$$\Delta Q_{\mathrm{I}} = -(-0.22)/[(1.85)(316.6)] = 0.000$$

$$\Delta Q_{\mathrm{II}} = -(-1.00)/[(1.85)(207.5)] = 0.003$$

$$\Delta Q_{\mathrm{III}} = -(-0.22)/[(1.85)(140.2)] = 0.001$$

Supplementary Problems

9.22. Using Diagram B-1, calculate the flow in a 16″ pipe with a drop in the hydraulic grade line of 6 ft in a mile. (Use $C = 100$.) *Ans.* 1.5 mgd

9.23. Had the pipe in Problem 9.22 been new cast iron pipe, what flow could be expected? *Ans.* 2.0 mgd

9.24. In a test of a 500-mm cast iron pipe the flow was steady at 0.175 m^3/s and the grade line dropped 1.22 m in a length of 610 m of pipe. What is the indicated value of C? *Ans.* 107

9.25. Given a 4000-m-long, 100-mm-diameter concrete pipe, determine the diameter of a 1000-m-long equivalent pipe. *Ans.* 760 mm

9.26. What size new cast iron pipe will deliver 0.552 m^3/s through 1829 m with a lost head of 9.14 m? *Ans.* 600 mm

9.27. A flow of 12.0 mgd is required through old cast iron pipe ($C = 100$) with a slope of the hydraulic grade line of 1.0 ft/1000 ft. Theoretically, how many 16″ pipes would be required? 20″ pipes? 24″ pipes? 36″ pipes? *Ans.* 8.46, 4.68, 2.90, 1

9.28. Check the ratios in Problem 9.27 by using a flow of 12 mgd with any assumed slope of the grade line.

9.29. What lost head in a 400-mm new cast iron pipe will create the same flow as occurs in a new 500-mm pipe with a drop in the hydraulic grade line of 1.0 m/1000 m? *Ans.* 2.97 m/1000 m

9.30. Water flows at a rate of 0.020 m^3/s from reservoir A to reservoir B through three concrete pipes connected in series, as shown in Fig. 9-19. Find the difference between water surface elevations in the reservoirs. Neglect minor losses. *Ans.* 16.59 m

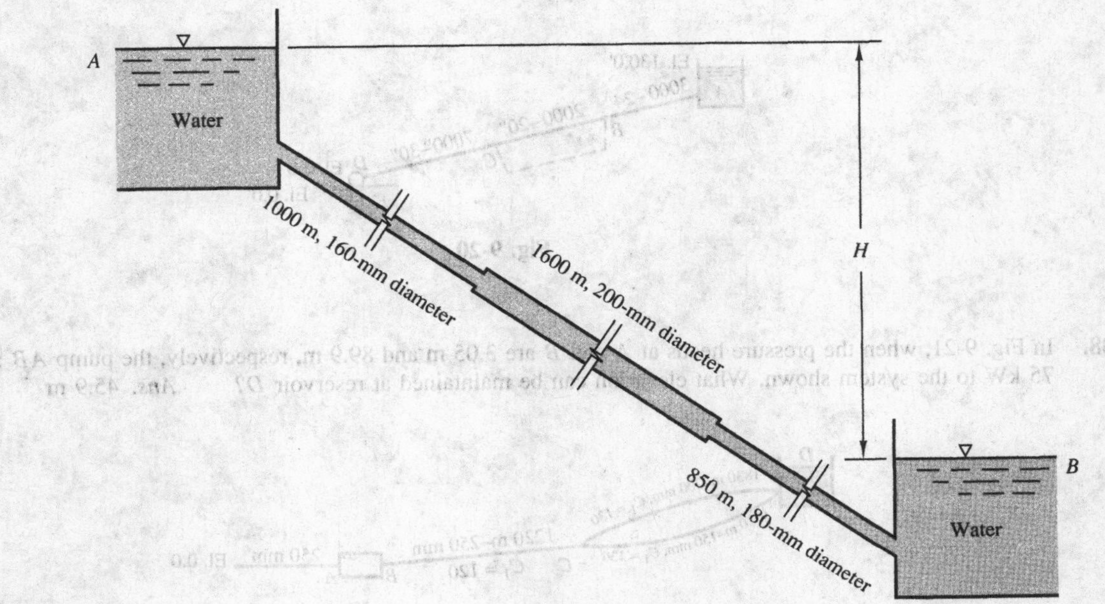

Fig. 9-19

9.31. Series pipe *ABCD* consists of 6100 m of 400-mm, 3050 m of 300-mm, and 1520 m of 200-mm ($C = 100$). (*a*) Find the flow when the lost head is 61 m from *A* to *D*. (*b*) What size pipe 1520 m long laid parallel to the existing 200 mm and joining it at *C* and *D* will make the new *CD* section equivalent to the *ABC* section (use $C = 100$)? (*c*) If 2440 m of 300-mm pipe were laid from *C* to *D*, paralleling the 200-mm pipe *CD*, what would be the total loss of head from *A* to *D* when $Q = 0.085\ \text{m}^3/\text{s}$? *Ans.* 0.061 m³/s, 168 mm, 41 m

9.32. Series pipe *ABCD* consists of 10,000 ft of 20″, 8000 ft of 16″, and *L* ft of 12″ ($C = 120$). What length *L* will make the *ABCD* pipes equivalent to a 15″ pipe, 16,500 ft long, $C = 100$? If the length of the 12″ pipe from *C* to *D* were 3000 ft, what flow would occur for a lost head of 135 ft from *A* to *D*? *Ans.* 5000 ft, 4.4 mgd

9.33. Convert 914 m of 250-mm, 457 m of 200-mm, and 152 m of 150-mm in series to an equivalent length of 200-mm pipe (all values of $C = 120$). *Ans.* 1387 m

9.34. Two concrete pipes are connected in series. The flow rate of water through the pipes is 0.14 m³/s with a total friction loss of 14.10 m for both pipes. Each pipe has a length of 300 m. If one pipe has a diameter of 300 mm, what is the diameter of the other one? Neglect minor losses. *Ans.* 250 mm

9.35. Reservoirs *A* and *D* are connected by 8000 ft of 20″ pipe (*AB*), 6000 ft of 16″ pipe (*BC*), and 2000 ft of unknown size (*CD*). The difference in levels in the reservoirs is 85 ft. (*a*) Find the size of pipe *CD* so that 4.50 mgd will flow from *A* to *D*, using $C = 120$ for all pipes. (*b*) How much flow will be produced if pipe *CD* is 14″ in diameter and if a 12″ pipe is connected at *B*, paralleling *BCD* and 9000 ft long? *Ans.* 13″, 6.45 mgd

9.36. A piping system ($C = 120$) consists of 3050 m of 750-mm (*AB*), 2440 m of 600-mm (*BC*) and, from *C* to *D*, two 400-mm pipes in parallel, each 1830 m long. (*a*) For a flow of 0.394 m³/s from *A* to *D*, what is the lost head? (*b*) If a valve in one of the 400-mm pipes were closed, what change in lost head would occur for the same flow? *Ans.* 21 m, change = 29 m

9.37. In Fig. 9-20, for a pressure head at *D* of 100 ft, (*a*) compute the power supplied the turbine *DE*. (*b*) If the dashed-line pipe were installed (3000 ft of 24″), what power would be available to the turbine if the flow were 13.0 mgd? ($C = 120$). *Ans.* 154 hp, 208 hp

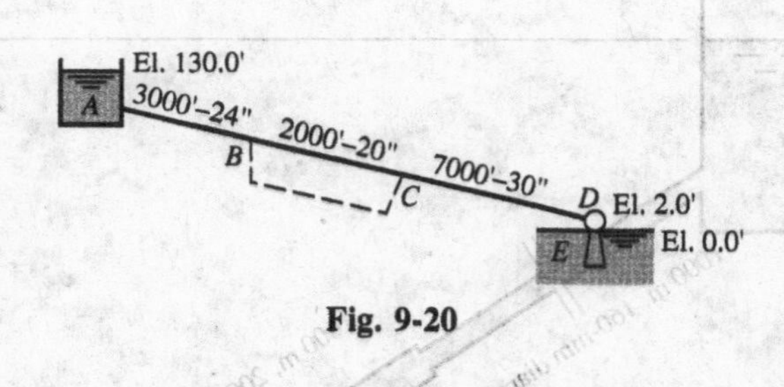

Fig. 9-20

9.38. In Fig. 9-21, when the pressure heads at *A* and *B* are 3.05 m and 89.9 m, respectively, the pump *AB* adds 75 kW to the system shown. What elevation can be maintained at reservoir *D*? *Ans.* 45.9 m

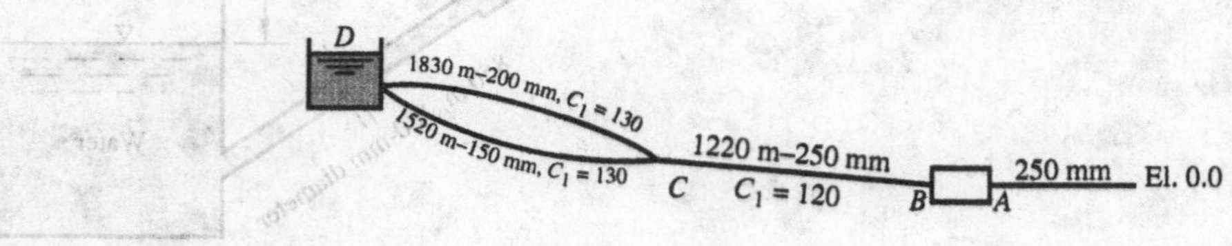

Fig. 9-21

9.39. Figure 9-22 shows a parallel-pipe system. Pressure heads at points A and E are 70.0 m and 46.0 m, respectively. Compute the flow rate of water through each branch of the loop. Assume $C = 120$ for all pipes. *Ans.* 0.105 m³/s, 0.056 m³/s, 0.070 m³/s

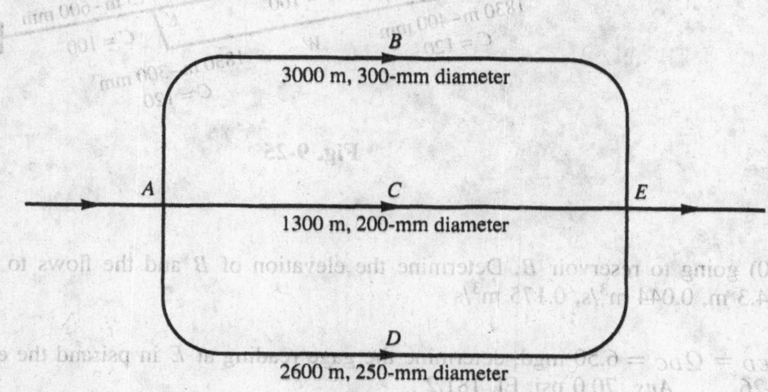

Fig. 9-22

9.40. It is necessary to deliver 21.2 cfs to point D in Fig. 9-23 at a pressure of 40 psi. Determine the pressure at A in psi. *Ans.* 45.8 psi

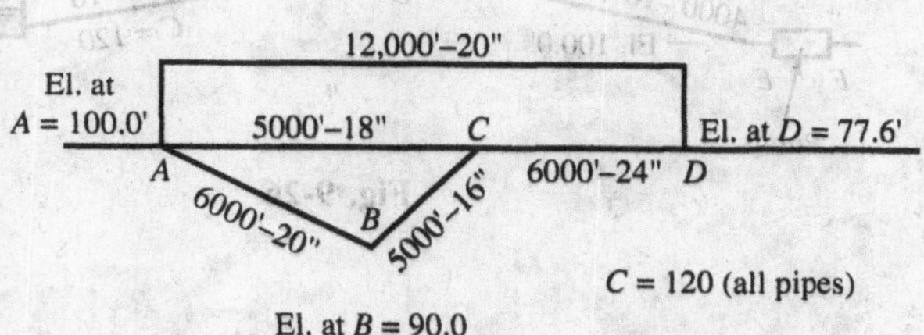

Fig. 9-23

9.41. (*a*) In Fig. 9-24, the pressure in supply main D is 207 kPa when the flow from A is 0.263 m³/s. Valves B and C are closed. Find the elevation of reservoir A. (*b*) The flow and pressure in (*a*) are unchanged, but valve C is open fully and valve B is only partly open. If the new elevation of reservoir A is 64.4 m, what is the loss through valve B? *Ans.* El. 68.3 m, 6.1 m

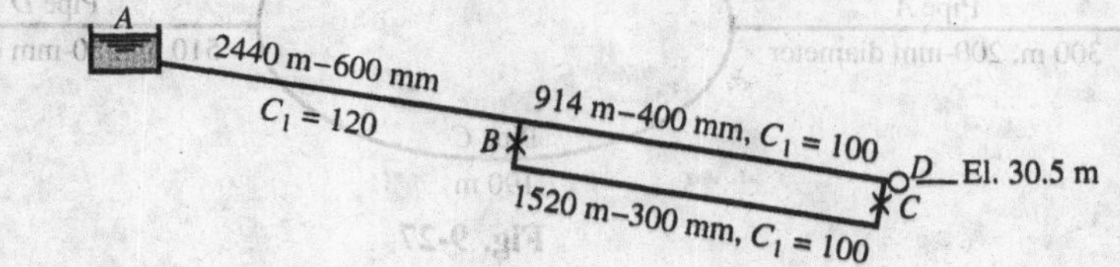

Fig. 9-24

9.42. Determine the flow through each pipe of the system shown in Fig. 9-25. *Ans.* 0.20 m³/s, 0.15 m³/s, 0.05 m³/s

9.43. Pump XY at elevation 6.1 m delivers 0.13 m³/s through 1830 m of 400-mm new cast iron pipe YW. The discharge pressure at Y is 267 kPa. At W two pipes connect to the 400-mm, 760 m of 300-mm ($C = 100$) going to reservoir A whose flow line is at elevation 30.5 m, and 610 m of 250-mm

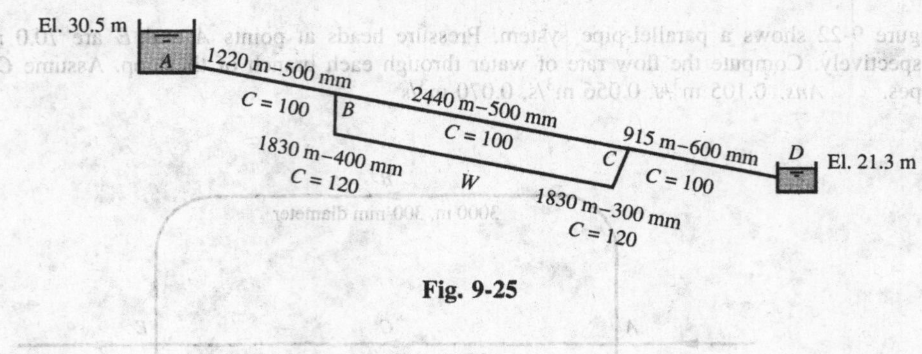

Fig. 9-25

(C = 130) going to reservoir B. Determine the elevation of B and the flows to or from the reservoirs. *Ans.* El. 4.3 m, 0.044 m³/s, 0.175 m³/s

9.44. When $Q_{ED} = Q_{DC} = 6.50$ mgd, determine the gage reading at E in psi and the elevation of reservoir B in Fig. 9-26. *Ans.* 70.0 psi, El. 181.2′

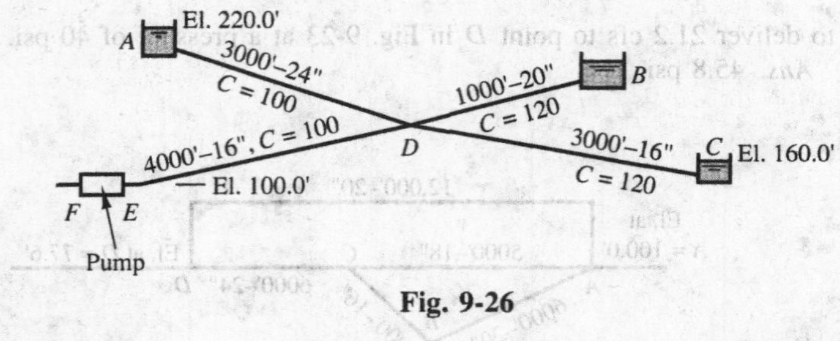

Fig. 9-26

9.45. If the flow rate of water through the system shown in Fig. 9-27 is 0.050 m³/s under a total head loss of 9.0 m, determine the diameter of pipe C. Assume C = 120 for all pipes. *Ans.* 180 mm

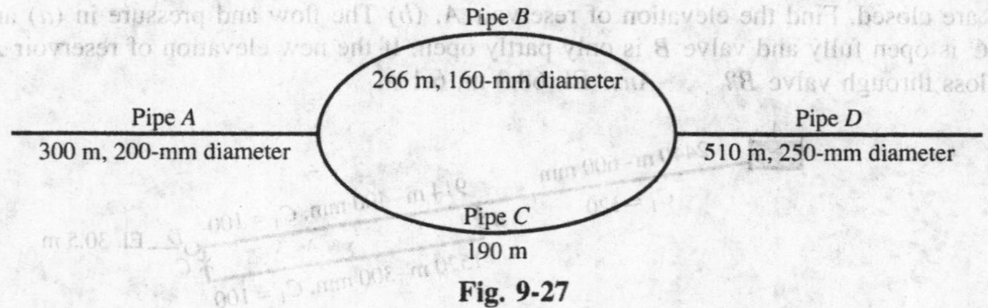

Fig. 9-27

9.46. In Fig. 9-28, water flows through the 36″ pipe at the rate of 22.3 mgd. Determine the horsepower of the pump XA (78.5% efficiency) that will produce the flows and elevations for the system if the pressure head at X is zero ft. (Draw grade lines). *Ans.* 250 hp

9.47. How much water must the pump supply when the flow through the 900-mm pipe is 1.31 m³/s, and what is the pressure head at A in Fig. 9-29? *Ans.* 1.10 m³/s, 58 m

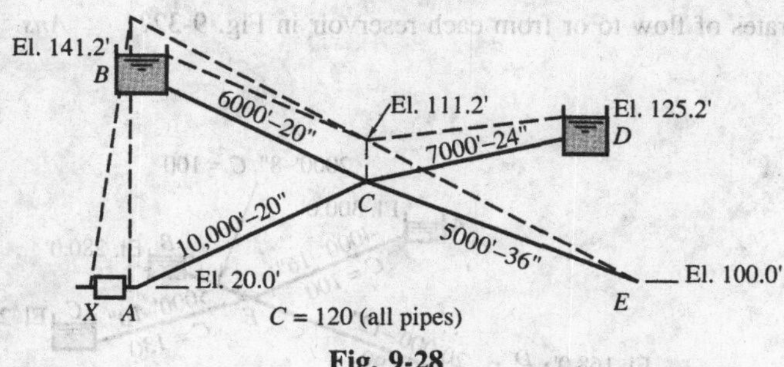

Fig. 9-28

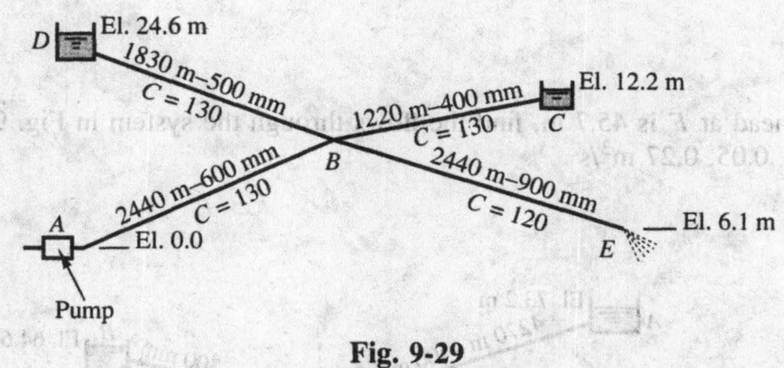

Fig. 9-29

9.48. The pressure head at A in pump AB is 120 ft when the energy change in the system (see Fig. 9-30) due to the pump is 153 hp. The lost head through valve Z is 10 ft. Find all the flows and the elevation of reservoir T. Sketch the hydraulic grade lines. *Ans.* 87.4 ft

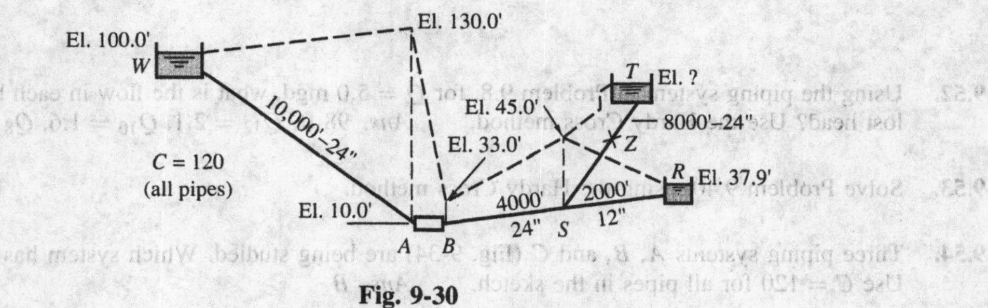

Fig. 9-30

9.49. The total flow from A in Fig. 9-31 is 0.394 m^3/s, and the flow to B is 0.300 m^3/s. Find (a) the elevation of B and (b) the length of the 600-mm pipe. *Ans.* El. 28.7 m, 2220 m

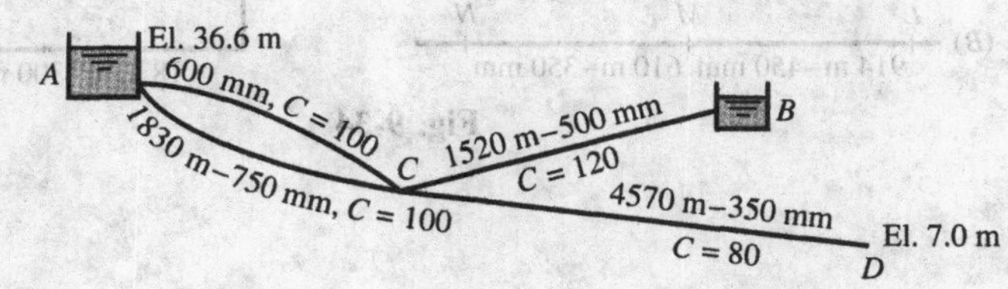

Fig. 9-31

9.50. What are the rates of flow to or from each reservoir in Fig. 9-32? *Ans.* 3.42, 0, 1.87, 1.53 mgd

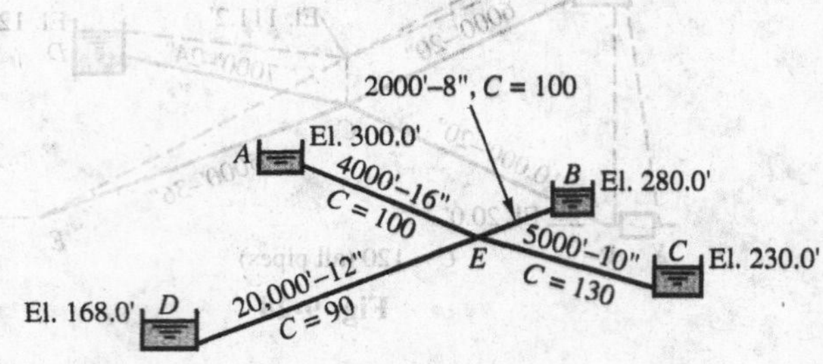

Fig. 9-32

9.51. If the pressure head at F is 45.7 m, find the flows through the system in Fig. 9-33.
Ans. 0.10, 0.11, 0.05, 0.27 m^3/s

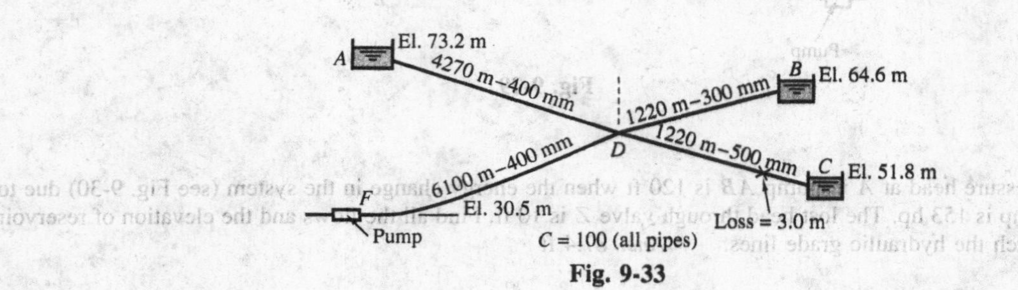

Fig. 9-33

9.52. Using the piping system in Problem 9.8, for $Q = 5.0$ mgd, what is the flow in each branch and what is the lost head? Use the Hardy Cross method. *Ans.* 98 ft, $Q_{12} = 2.1$, $Q_{10} = 1.6$, $Q_8 = 1.3$ mgd

9.53. Solve Problem 9.40, using the Hardy Cross method.

9.54. Three piping systems A, B, and C (Fig. 9-34) are being studied. Which system has the greatest capacity? Use $C = 120$ for all pipes in the sketch. *Ans.* B

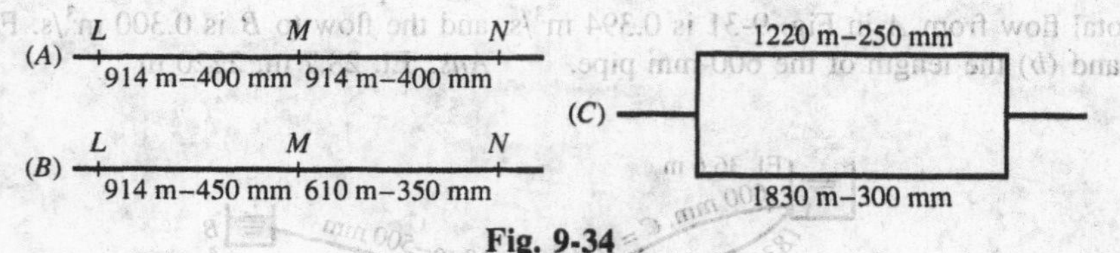

Fig. 9-34

9.55. In Problem 9.54, what size pipe, 914 m long and laid parallel to MN in system A (thus forming a loop from M to N), would make new system A have 50% more capacity than system C? *Ans.* $d = 376$ mm

9.56. Compute the flow rate of water in each pipe in the network shown in Fig. 9-35. Assume $C = 120$ for all pipes.
Ans. AB, 0.265; BE, 0.098; EF, 0.079; FA, 0.235; BC, 0.167; CD, 0.067; DE, 0.037; DI, 0.054; IH, 0.046; HE, 0.040; HG, 0.056; GF, 0.156 m³/s

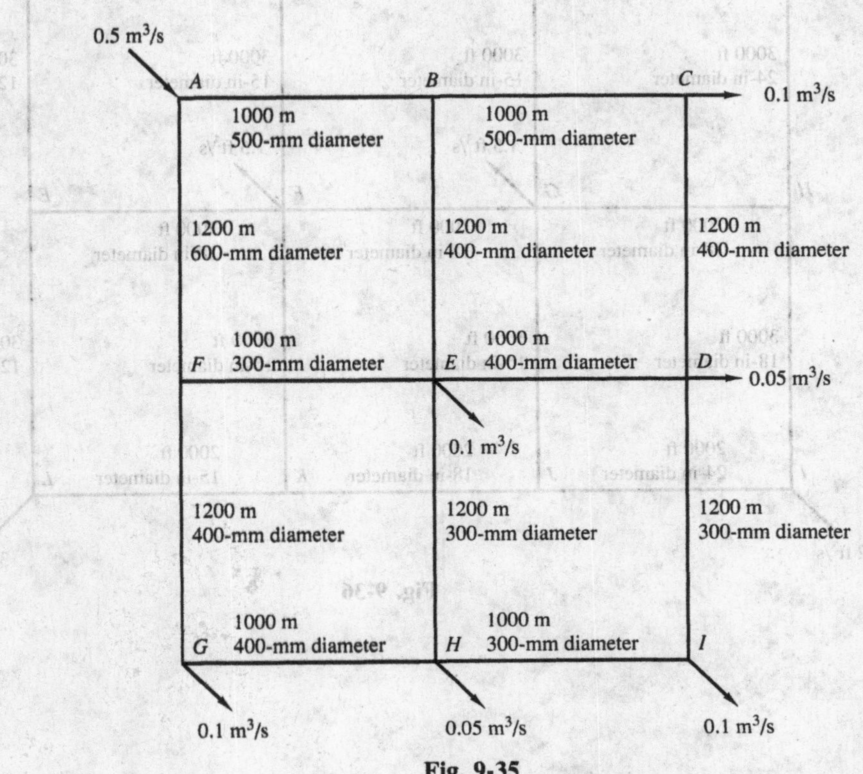

Fig. 9-35

9.57. Compute the flow rate of water in each pipe in the network shown in Fig. 9-36. Assume $C = 120$ for all pipes.
Ans. AB, 5.19; BG, 1.53; GH, 2.39; HA, 4.81; BC, 3.66; CF, 1.55; FG, 1.61; CD, 2.11; DE, 0.11; EF, 0.87; EL, 0.98, LK, 2.02; KF, 0.79; KJ, 1.23; JG, 0.81; JI, 0.42; IH, 2.42 ft³/sec

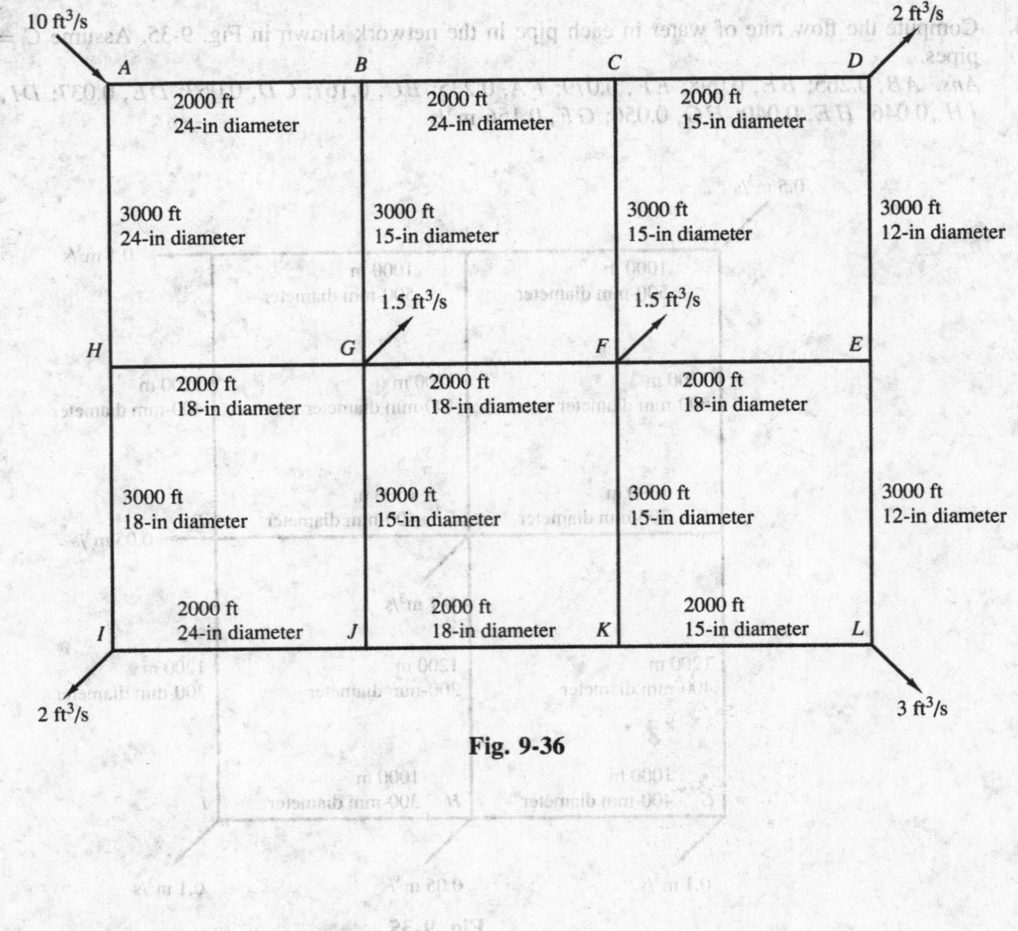
10 ft³/s
2 ft³/s
A
B
C
D
2000 ft
24-in diameter
2000 ft
24-in diameter
2000 ft
15-in diameter
3000 ft
24-in diameter
3000 ft
15-in diameter
3000 ft
15-in diameter
3000 ft
12-in diameter
1.5 ft³/s
1.5 ft³/s
H
G
F
E
2000 ft
18-in diameter
2000 ft
18-in diameter
2000 ft
18-in diameter
3000 ft
18-in diameter
3000 ft
15-in diameter
3000 ft
15-in diameter
3000 ft
12-in diameter
2000 ft
24-in diameter
2000 ft
18-in diameter
2000 ft
15-in diameter
I
J
K
L
2 ft³/s
3 ft³/s

Fig. 9-36

Chapter 10

Flow in Open Channels

INTRODUCTION

Open-channel flow occurs when a liquid flowing due to gravity is only partially enclosed by its solid boundary. In open-channel flow, the flowing liquid has a free surface, and the liquid is not under pressure other than that caused by its own weight and by atmospheric pressure. Some open-channel flow occurs naturally as in the case of creeks and rivers, which generally have irregular cross sections. Open-channel flow may also occur in artificial (i.e., of human construction) channels, such as flumes and canals. These channels often have more regularly shaped cross sections, such as rectangular, triangular, or trapezoidal. Open-channel flow may also occur in conduits (such as pipes of circular cross section) if the conduit is not flowing full. Sewer pipe systems normally flow less than full and are therefore designed as open channels.

STEADY UNIFORM FLOW

Steady uniform flow encompasses two conditions of flow. *Steady flow*, as defined under closed-conduit flow, refers to the condition in which the flow characteristics at any point do not change with time ($\partial V/\partial t = 0, \partial y/\partial t = 0$, etc.). *Uniform flow* refers to the condition in which the depth, slope, velocity, and cross section remain constant over a given length of channel ($\partial y/\partial L = 0, \partial V/\partial L = 0$, etc.).

In the special case of steady uniform flow, the energy grade line, hydraulic grade line, and channel bottom are all parallel (i.e., their slopes are equal). This is not true for steady nonuniform flow.

NONUNIFORM FLOW

Nonuniform flow occurs when the depth of flow changes along the length of the open channel, or $\partial y/\partial L \neq 0$. Nonuniform flow may be steady or unsteady. It may also be classified as tranquil, rapid, or critical.

LAMINAR FLOW

Laminar flow will generally occur in open-channel flow for values of Reynolds number Re of 2000 or less. The flow *may* be laminar up to Re = 10,000. For open-channel flow, $\text{Re} = 4RV/\nu$, where R is the hydraulic radius (area divided by wetted perimeter), V = velocity of flow, and ν = kinematic viscosity.

THE CHEZY FORMULA for *steady uniform flow*, developed in Problem 10.1, is

$$V = C\sqrt{RS} \tag{1}$$

where V = average velocity
C = coefficient
R = hydraulic radius
S = slope of the energy grade line

COEFFICIENT C can be obtained by using one of the following expressions.

$$C = \sqrt{\frac{8g}{f}} \qquad \text{(See Problem 10.1.)} \qquad (2)$$

$$C = \frac{41.65 + \dfrac{0.00281}{S} + \dfrac{1.811}{n}}{1 + \dfrac{n}{\sqrt{R}}\left(41.65 + \dfrac{0.00281}{S}\right)} \qquad \text{(Kutter)} \qquad (3)$$

$$C = \frac{1.486}{n} R^{1/6} \qquad \text{(Manning)} \qquad (4)$$

$$C = \frac{157.6}{1 + m/\sqrt{R}} \qquad \text{(Bazin)} \qquad (5)$$

$$C = -42 \log\left(\frac{C}{Re} + \frac{\epsilon}{R}\right) \qquad \text{(Powell)} \qquad (6)$$

In expressions (*3*) through (*5*), n and m are roughness factors determined by experiments using water only. Some values are given in Table 9 of the Appendix. In general, the Manning formula is preferred for open-channel flow. The Powell formula will be discussed in Problems 10.9 and 10.11.

DISCHARGE (Q) for steady uniform flow, in terms of the Manning formula, is

$$Q = AV = A\left(\frac{1.486}{n}\right) R^{2/3} S^{1/2} \qquad (7a)$$

This formula results from substituting expression (*4*) into formula (*1*). It gives discharge (Q) in ft^3/sec if area (A) is in ft^2 and hydraulic radius (R) is in ft. [Slope (S) is dimensionless.] For SI units, the Manning formula is

$$Q = AV = A\left(\frac{1.0}{n}\right) R^{2/3} S^{1/2} \qquad (7b)$$

It gives Q in m^3/s if A is in m^2 and R in m.

The conditions associated with steady uniform flow are called normal—thus the terms normal depth and normal slope.

LOST HEAD (h_L) expressed in terms of the Manning formula, is

$$h_L = \left[\frac{Vn}{1.486R^{2/3}}\right]^2 L, \qquad \text{using } S = h_L/L \qquad (8a)$$

for use with fps units, or

$$h_L = \left[\frac{Vn}{1.0R^{2/3}}\right]^2 L \qquad (8b)$$

for use with SI units.

For nonuniform (varied) flow, mean values of V and R can be used with reasonable accuracy. For a long channel, short lengths should be employed in which the changes in depth are about the same magnitude.

VERTICAL DISTRIBUTION OF VELOCITY

Vertical distribution of velocity in an open channel may be assumed parabolic for laminar flow and logarithmic for turbulent flow.

For uniform *laminar* flow in wide open channels of mean depth y_m, the velocity distribution can be expressed as

$$v = \frac{gS}{\nu}\left(yy_m - \tfrac{1}{2}y^2\right) \quad \text{or} \quad v = \frac{\gamma S}{\mu}\left(yy_m - \tfrac{1}{2}y^2\right) \tag{9}$$

The mean velocity V, derived from this equation in Problem 10.3, becomes

$$V = \frac{gSy_m^2}{3\nu} \quad \text{or} \quad V = \frac{\gamma S y_m^2}{3\mu} \tag{10}$$

For uniform *turbulent* flow in wide open channels, the velocity distribution (developed in Problem 10.4) can be expressed as

$$v = 2.5\sqrt{\tau_o/\rho}\,\ln(y/y_o) \quad \text{or} \quad v = 5.75\sqrt{\tau_o/\rho}\,\log(y/y_o) \tag{11}$$

SPECIFIC ENERGY

Specific energy (E) is defined as the energy per unit weight (ft-lb/lb or N·m/N) relative to the bottom of the channel or

$$E = \text{depth} + \text{velocity head} = y + V^2/2g \tag{12a}$$

A more exact expression of the kinetic energy term would be $\alpha V^2/2g$. See Chapter 7 for discussion of the kinetic energy correction factor α.

In terms of the flow rate q per unit of channel width b (i.e., $q = Q/b$),

$$E = y + (1/2g)(q/y)^2 \quad \text{or} \quad q = \sqrt{2g(y^2E - y^3)} \tag{12b}$$

For uniform flow, specific energy remains constant from section to section. For nonuniform flow, the specific energy along the channel may increase or decrease.

CRITICAL DEPTH

Critical depth (y_c) for a constant unit flow q in a rectangular channel occurs when the specific energy is a minimum. As shown in Problems 10.33 and 10.34,

$$y_c = \sqrt[3]{q^2/g} = \tfrac{2}{3}E_c = V_c^2/g \tag{13}$$

This expression can be rearranged to give

$$V_c = \sqrt{gy_c} \quad \text{or} \quad V_c/\sqrt{gy_c} = 1 \text{ for critical flow} \tag{14}$$

Thus, if the Froude number $\text{Fr} = V_c/\sqrt{gy_c} = 1$, critical flow occurs. If $\text{Fr} > 1$, supercritical flow (rapid flow) occurs; and if $\text{Fr} < 1$, subcritical flow (tranquil flow) occurs.

MAXIMUM UNIT FLOW

Maximum unit flow (q_{max}) in a rectangular channel for any given specific energy E, as developed in Problem 10.34, is

$$q_{max} = \sqrt{gy_c^3} = \sqrt{g\left(\tfrac{2}{3}E\right)^3} \tag{15}$$

FOR CRITICAL FLOW IN NONRECTANGULAR CHANNELS, as developed in Problem 10.33,

$$\frac{Q^2}{g} = \frac{A_c^3}{b'} \quad \text{or} \quad \frac{Q^2 b'}{g A_c^3} = 1 \tag{16}$$

where b' is the width of the water surface. We may rearrange equation (*16*) by dividing by A_c^2, as follows:

$$V_c^2/g = A_c/b' \quad \text{or} \quad V_c = \sqrt{gA_c/b'} = \sqrt{gy_m} \tag{17}$$

where term A_c/b' is called the mean depth y_m.

NONUNIFORM FLOW

For nonuniform flow, an open channel is usually divided into lengths L, called *reaches*, for study. To compute backwater curves, the energy equation (see Problem 10.46) yields

$$L = \frac{(V_2^2/2g + y_2) - (V_1^2/2g + y_1)}{S_0 - S} = \frac{E_2 - E_1}{S_0 - S} = \frac{E_1 - E_2}{S - S_0} \tag{18}$$

where S_0 = slope of the channel bottom and S = slope of the energy grade line.

For successive reaches where changes in depth are about the same, energy gradient S can be written

$$S = \left(\frac{nV_{\text{mean}}}{1.486R_{\text{mean}}^{2/3}}\right)^2 \quad \text{or} \quad \frac{V_{\text{mean}}^2}{C^2 R_{\text{mean}}} \tag{19}$$

(Substitute 1.0 for 1.486 for SI units.)

Surface profiles for gradually varied flow conditions in wide rectangular channels can be analyzed by using the expression

$$\frac{dy}{dL} = \frac{S_0 - S}{(1 - V^2/gy)} \tag{20}$$

The term dy/dL represents the slope of the water surface relative to the channel bottom. Thus if dy/dL is positive, the depth is increasing downstream. Problems 10.52 and 10.53 will develop the equation and a system of classifying surface profiles.

HYDRAULIC JUMP

Hydraulic jump occurs when a supercritical flow changes to a subcritical flow. In such cases, the elevation of the liquid surface increases suddenly in the direction of flow. For a constant flow in a rectangular channel, as derived in Problem 10.54,

$$\frac{q^2}{g} = y_1 y_2 \left(\frac{y_1 + y_2}{2}\right) \tag{21}$$

Problems 10.55 through 10.59 further illustrate the phenomenon of hydraulic jump.

OPEN-CHANNEL FLOW IN CIRCULAR CROSS SECTIONS

Problems involving uniform open-channel flow in circular sections can be solved essentially the same way as those involving noncircular sections, the major difference being the generally more difficult task of computing the hydraulic radius for a part of a circular cross section. Problem 10.19 illustrates such a problem.

Computations involving cross sections that are segments of circles, while not sophisticated, are nevertheless tedious. (Problem 10.19 illustrates this, but the computations can be much more tedious if the depth of flow in a circular cross section is the unknown.) Such computations can be facilitated (with some small loss of accuracy) by using the graphical relationship shown in Fig. 10-1, which gives hydraulic elements of a circular section. A separate line is given for each hydraulic element (wetted perimeter, area, discharge, hydraulic radius, and velocity) showing how each of these elements varies with depth of flow. It should be noted that the ordinate represents the ratio (expressed as a percentage) of any particular depth of flow to the depth of flow when the pipe is flowing full (which is, of course, the diameter of the pipe). The abscissa represents similar ratios for all other hydraulic elements. Problems 10.20 and 10.21 illustrate the use of Fig. 10-1 for solving open-channel flow in circular cross sections.

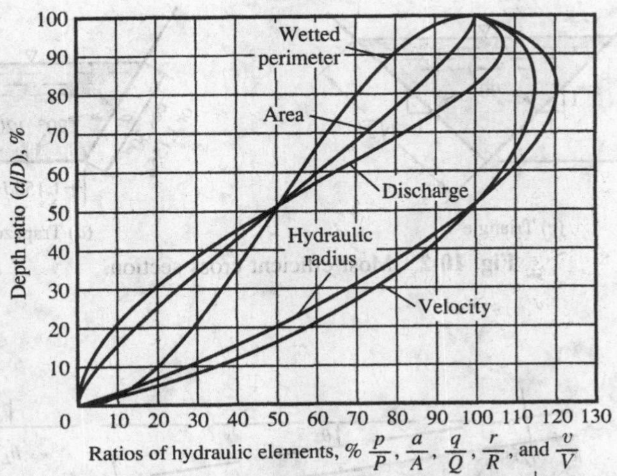

Fig. 10-1 Hydraulic elements of a circular section.

MOST EFFICIENT CROSS SECTIONS

The *most efficient cross section* for an open channel is the one that will have the greatest capacity for a given slope, area, and roughness coefficient. If these parameters remain constant, velocity (and therefore flow rate) will be greatest when the wetted perimeter is smallest. Based on this premise, the most efficient (and therefore most economical) cross section for some common shapes can be determined.

The most efficient of all cross sections is a semicircle because it has the smallest wetted perimeter for a given area. For a rectangular section, the most efficient one has a depth that is half its width. For a triangular section, side slopes of unity give the most efficient section. And for a trapezoidal section, the most efficient one is half a regular hexagon (i.e., three equal sides with interior angles of 120° each). All of these most efficient sections are illustrated in Fig. 10-2. (Also, see Problem 10.30.)

Solved Problems

10.1. Develop the general (Chezy) equation for steady uniform flow in an open channel.

Solution:

In Fig. 10-3, consider volume of liquid $ABCD$ of constant cross section A and length L. The volume may be considered in equilibrium because the flow is steady (zero acceleration). Summing forces acting in

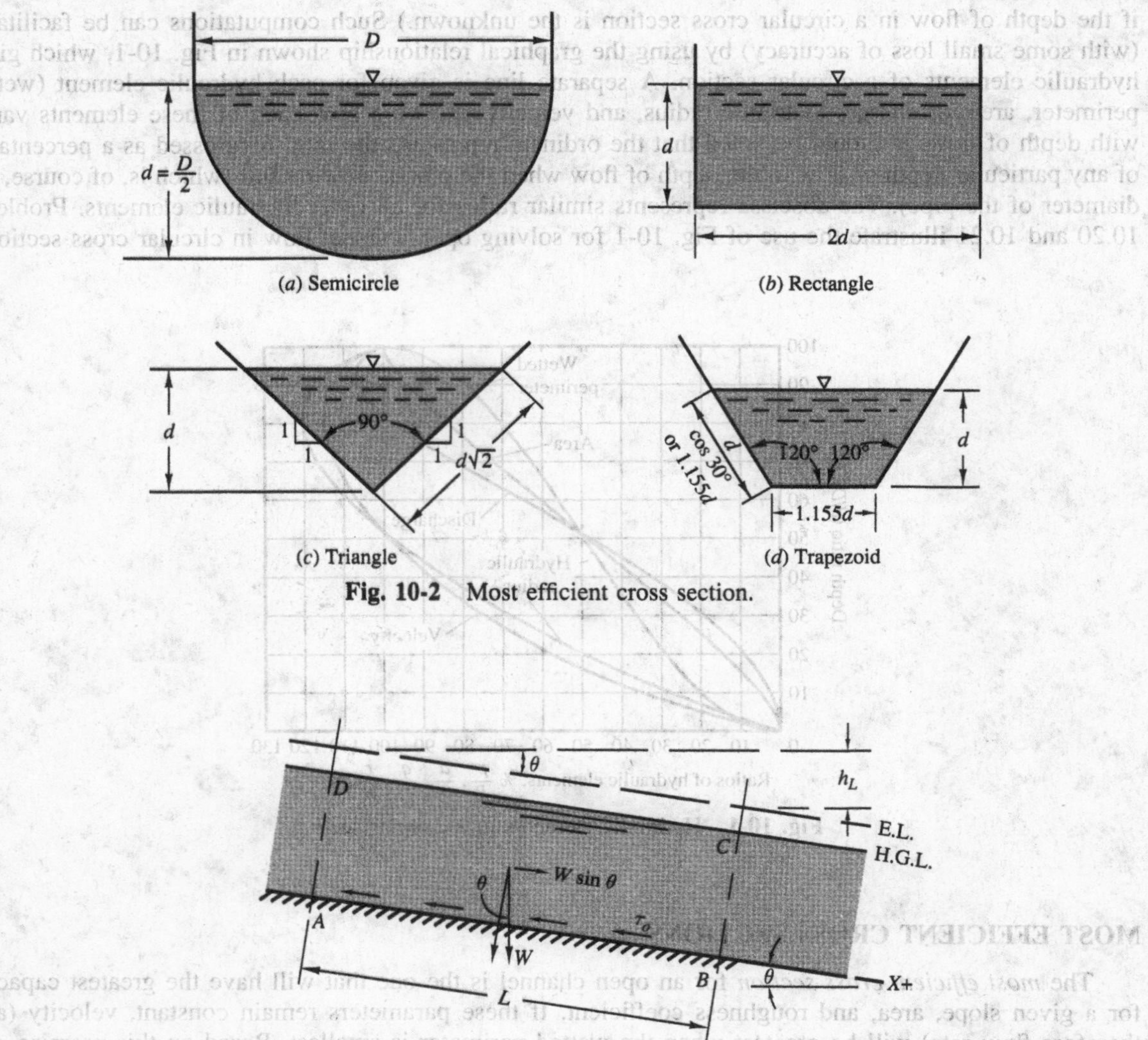

(a) Semicircle (b) Rectangle

(c) Triangle (d) Trapezoid

Fig. 10-2 Most efficient cross section.

Fig. 10-3

the X direction,

$$\text{force on area } AD - \text{ force on area } BC + W \sin\theta - \text{resisting forces} = 0$$

$$\overset{(h_L)}{\gamma \bar{h} A - \gamma \bar{h} A + \gamma A L \sin\theta - \tau_o p L = 0}$$

where τ_o is the boundary shear stress (lb/ft^2 or Pa) acting on an area L ft or m long by wetted perimeter p ft or m wide. Then

$$\gamma A L \sin\theta = \tau_o p L \qquad \text{and} \qquad \tau_o = (\gamma A \sin\theta)/p = \gamma R S \tag{A}$$

since $R = A/p$ and $\sin\theta = \tan\theta = S$ for small values of θ.

It was seen in Chapter 8, Problem 8.5, that $\tau_o = (\gamma/g) f(V^2/8)$. Then

$$\gamma R S = (\gamma/g) f\left(V^2/8\right) \qquad \text{or} \qquad V = \sqrt{(8g/f)RS} = C\sqrt{RS} \tag{B}$$

For *laminar* flow, f may be taken as 64/Re. Hence

$$C = \sqrt{(8g/64)\ \text{Re}} \tag{C}$$

10.2. Show that the vertical distribution of velocity is parabolic in a wide, open channel for uniform laminar flow. (y_m = mean depth of the channel). See Fig. 10-4.

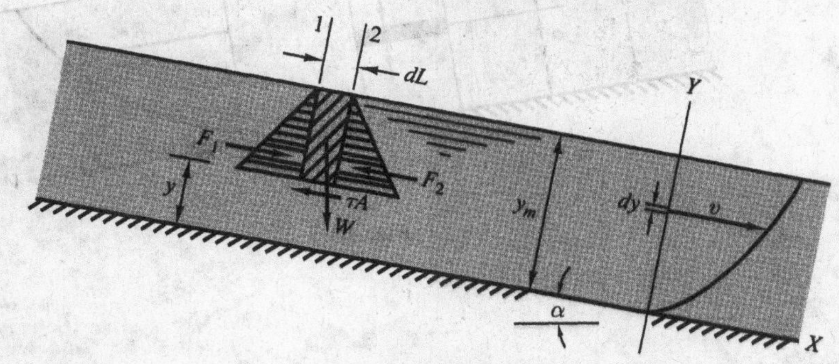

Fig. 10-4

Solution:

When velocity and depth of flow are relatively small, reflecting a Reynolds number <2000, viscosity becomes the dominant flow factor. The resulting flow is laminar. (For open channels, Re is defined as $4RV/\nu$). For the free body shown crosshatched in Fig. 10-4, using $\Sigma F_x = 0$, we obtain

$$F_1 - F_2 + \gamma(y_m - y)dL\ dz \sin\alpha - \tau\ dL\ dz = 0$$

Since $F_1 = F_2$, we obtain

$$\tau = \gamma(y_m - y)\sin\alpha$$

For laminar flow, $\tau = \mu\, dv/dy$, from which we obtain

$$dv = \frac{\gamma}{\mu}(y_m - y)\sin\alpha\ dy = \frac{\gamma S}{\mu}(y_m - y)dy \tag{A}$$

For the small values of angle α associated with the slope of open channels, $\sin\alpha = \tan\alpha$ = slope S. Integrating (A) yields

$$v = \frac{\gamma S}{\mu}\left(yy_m - \tfrac{1}{2}y^2\right) + C \tag{B}$$

Since $v = 0$ when $y = 0$, the value of constant $C = 0$. Equation (B) is a quadratic equation representing a parabola.

10.3. What is the mean velocity V in Problem 10.2?

Solution:

$$\text{mean velocity } V = \frac{Q}{A} = \frac{\int dQ}{\int dA} = \frac{\int v\ da}{\int dA} = \frac{(\gamma S/\mu)\int\left(yy_m - \frac{1}{2}y^2\right)dy\ dz}{\int dy\ dz = y_m\ dz}$$

where dz is a constant (dimension perpendicular to the paper).

$$V = \frac{\gamma S\ dz}{\mu y_m\ dz}\int_o^{y_m}\left(yy_m - \tfrac{1}{2}y^2\right)dy = \frac{\gamma S y_m^2}{3\mu}$$

10.4. For steady uniform flow in wide, open channels, establish a theoretical equation for the mean velocity for smooth surfaces. See Fig. 10-5.

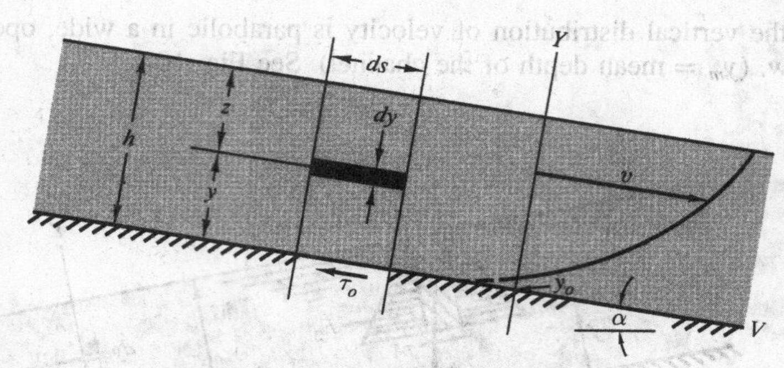

Fig. 10-5

Solution:

For turbulent flow in general, shear stress τ can be expressed as

$$\tau = \rho l^2 (dv/dz)^2$$

where l is the mixing length and a function of z. (See Chapter 8.)

Also, from Problem 10.1, expression (A), $\tau_o = \gamma RS = \gamma hS$, since hydraulic radius R for wide channels approximates the depth.

In the boundary layer, since y is very small, $z \cong h$ and $\tau \cong \tau_o$. Hence we may equate the values of τ_o, or

$$\rho l^2 (dv/dz)^2 = \gamma z S \quad \text{or} \quad dv/dz = \pm\sqrt{gzS/l^2}$$

To integrate this expression, try a value of $l = k(h - z)(z/h)^{1/2}$. Then

$$-\frac{dv}{dz} = \sqrt{gS}\left[\frac{z^{1/2}}{k(h-z)(z/h)^{1/2}}\right] = \frac{\sqrt{gSh}}{k}\left(\frac{1}{h-z}\right)$$

Let $y = (h - z)$ and $dy = -dz$; then

$$+y\left(\frac{dv}{dy}\right) = \frac{\sqrt{gSh}}{k} \quad \text{and} \quad dv = \frac{\sqrt{gSh}}{k}\left(\frac{dy}{y}\right)$$

Since $\tau_o/\rho = \gamma hS/\rho = gSh$,

$$dv = \frac{1}{k}\sqrt{\tau_o/\rho}\left(\frac{dy}{y}\right) \quad \text{or} \quad v = \frac{1}{k}\sqrt{\tau_o/\rho}\ \ln\ y + C$$

For $y = y_o$, $v \cong 0$, then $C = (-1/k)\sqrt{\tau_o/\rho}\ \ln\ y_o$ and

$$v = \frac{1}{k}\sqrt{\tau_o/\rho}\ln(y/y_o) \tag{A}$$

Note: Neglecting the log curve to the left of y_o, while producing an approximation, gives satisfactory results well within the limits of expected accuracy because y_o is very small. See Problem 10.5 for the value of y_o.

In this expression (A), $k \cong 0.40$ and is called the von Karman constant. Inasmuch as the term $\sqrt{\tau_o/\rho}$ has the dimensions of ft/sec or m/s, this term is called the *shear velocity* and is designated by v_*. Thus

$$v = 2.5 v_* \ln(y/y_o) \tag{B}$$

From $Q = AV = (h \times 1)V = \int v(dy \times 1)$, we obtain the value of mean velocity V. Thus

$$V = \frac{\int v(dy \times 1)}{(h \times 1)} = \frac{2.5v_*}{h} \int_0^h (\ln y - \ln y_o)\, dy$$

Using L'Hospital's rule from calculus, the mean velocity for smooth surfaces where a boundary layer exists can be evaluated as

$$V = 2.5v_*[\ln h - \ln y_o - 1] \qquad (C)$$

It will be shown in Problem 10.5 that y_o equals $\nu/9v_*$. Therefore equations (B) and (C) may be written

$$v = 2.5v_* \ln (9v_* y/\nu) \qquad (D)$$

and

$$V = 2.5v_*[\ln h - \ln (\nu/9v_*) - 1] \qquad (E)$$

Frequently the average velocity in an open channel is taken as the velocity observed at a point 0.6 of the depth (measured from the surface). If we accept this value of $\bar{y}$, then we may write, from (B) above, average velocity V as

$$V = 2.5v_* \ln (0.4h/y_o)$$

From Problem 10.5, $y_o = \delta/103$. Then for wide channels this average velocity becomes, since hydraulic radius $R = h$,

$$V = 2.5v_* \ln (41.2R/\delta) \qquad (F)$$

10.5. Determine the value of y_o in the preceding problem.

Solution:

For smooth surfaces, in a boundary (laminar) layer.

$$\tau_o = \mu(dv/dy) = \nu\rho(dv/dy) \quad \text{or} \quad dv/dy = (\tau_o/\rho)/\nu = v_*^2/\nu \text{ (a constant)}$$

Using δ as the thickness of the boundary layer,

$$\int dv = \frac{v_*^2}{\nu} \int_0^\delta dy \quad \text{or} \quad v_\delta = v_*^2\delta/\nu = \text{Re}_* v_* \qquad (A)$$

From experimentation, $\text{Re}_* \cong 11.6$ (fairly constant). Hence,

$$v_*^2\delta/\nu = 11.6v_* \quad \text{or} \quad \delta = 11.6\nu/v_* \qquad (B)$$

Putting $y = \delta$ into equation (B) of Problem 10.4,

$$v_\delta = 2.5v_* \ln (\delta/y_o) \qquad (C)$$

Combining (C) with (A), $\ln \delta/y_o = v_\delta/2.5v_* = \text{Re}_*/2.5 \cong 4.64$,

$$\delta/y_o = e^{4.64} = 103 \quad \text{and} \quad \delta = 103y_o \qquad (D)$$

Then from (B),

$$y_o = \frac{\delta}{103} \cong \frac{11.6\nu}{103v_*} \cong \frac{\nu}{9v_*} \qquad (E)$$

10.6. Water at 60°F flows in a smooth and wide rectangular channel ($n = 0.011$) at a depth of 4 ft and on a slope of 0.0004. Compare the value of C obtained by using Manning's formula with that by using the expression $V = 2.5v_* \ln (41.2R/\delta)$.

Mathcad

Solution:

(*a*) Using Manning's formula, $C = (1.486/n)R^{1/6} = (1.486/0.011)(4^{1/6}) = 170$.

(*b*) Equating the Chezy formula for average velocity V with the given expression,

$$C\sqrt{RS} = 2.5v_* \ln(41.2R/\delta)$$

Substituting $v_* = \sqrt{gSR}$ from Problem 10.4, we obtain

$$C = 2.5\sqrt{g}\ln(41.2R/\delta) \qquad (A)$$

For 60°F water, $\nu = 1.217 \times 10^{-5}$ ft²/sec, and, using $\delta = 11.6\nu/v_*$ from (*B*) of Problem 10.5, we find $C = 177$.

10.7. (*a*) **A wide rectangular channel flows 4 ft deep on a slope of 4 ft in 10,000 ft. Using the theoretical formula for velocity in Problem 10.4, calculate values of theoretical velocities at depth increments of 1/10 depth, assuming the channel is smooth.**

(*b*) **Compare the average of the 0.2 and 0.8 depth values of velocity with the 0.6 depth velocity. Use kinematic viscosity as 1.50×10^{-5} ft²/sec.**

Solution:

(*a*) Since $v_* = \sqrt{\tau_o/\rho} = \sqrt{gRS} = \sqrt{ghS}$ and $y_o = \nu/9v_*$,

$$v = 2.5v_* \ln(y/y_o) = (2.5)(2.303)\sqrt{ghS}\log(9v_* y/\nu)$$

$$= 5.76\sqrt{(32.2)(4)(0.0004)}\ \log\frac{9y\sqrt{(32.2)(4)(0.0004)}}{1.5\times 10^{-5}}$$

$$= 1.307\log(136{,}188y) \qquad (A)$$

Using (*A*) we obtain the following values of velocity v:

Dist. Down (%)	y (ft)	v (ft/sec)
0	4.0	7.50
10.	3.6	7.44
20.	3.2	7.37
30.	2.8	7.29
40.	2.4	7.21
50.	2.0	7.10
60.	1.6	6.98
70.	1.2	6.81
80.	0.8	6.58
90.	0.4	6.19
92.5	0.3	6.03
95.0	0.2	5.80
97.5	0.1	5.40
99.75	0.01	4.10

(*b*) Average of the 0.2 and 0.8 depth values is $V = \frac{1}{2}(7.37 + 6.58) = 6.98$ ft/sec.

The 0.6 depth value is 6.98 ft/sec. Seldom does such exact agreement occur.

10.8. **Assuming the Manning formula for C to be correct, what value of n will satisfy the "smooth" criterion in Problem 10.6?**

Solution:

Equate the values of C, using expression (A) in Problem 10.6, as follows:

$$\frac{1.486R^{1/6}}{n} = 5.75\sqrt{g}\log\left(\frac{41.2R}{\delta}\right) = 5.75\sqrt{g}\log\left(\frac{41.2R\sqrt{gSR}}{11.6\nu}\right)$$

Substituting values and solving, $n = 0.0106$.

10.9. Using the Powell equation, what quantity of liquid will flow in a smooth rectangular channel 2 ft wide, on a slope of 0.010, if the depth is 1.00 ft? (Use $\nu = 0.00042$ ft²/sec.)

Mathcad

Solution:

Equation (6) is $\quad C = -42\log\left(\frac{C}{\text{Re}} + \frac{\epsilon}{R}\right)$

For smooth channels, ϵ/R is small and can be neglected; then

$$C = 42\log \text{Re}/C \qquad (A)$$

From the given data, Re/C can be evaluated using $V = C\sqrt{RS}$:

$$\text{Re} = 4RV/\nu = 4RC\sqrt{RS}/\nu$$

$$\text{Re}/C = 4R^{3/2}S^{1/2}/\nu = 4\left(\tfrac{1}{2}\right)^{3/2}(0.010)^{1/2}/0.00042 = 337$$

Then $C = 42\log 337 = 106$, and

$$Q = CA\sqrt{RS} = (106)(2)\sqrt{(0.50)(0.010)} = 15.0 \text{ cfs}$$

10.10. Water flows in a rectangular concrete open channel that is 12.0 m wide at a depth of 2.5 m. The channel slope is 0.0028. Find the water velocity and flow rate.

Solution:

$$V = \frac{1.0}{n}R^{2/3}S^{1/2}$$

$$R = \frac{(2.5)(12.0)}{2.5 + 12.0 + 2.5} = 1.765 \text{ m} \qquad V = \left(\frac{1.0}{0.013}\right)(1.765)^{2/3}(0.0028)^{1/2} = 5.945 \text{ m/s}$$

$$Q = AV = [(2.5)(12.0)](5.945) = 178 \text{ m}^3/\text{s}$$

10.11. Determine C by the Powell equation for a 2 ft by 1 ft rectangular channel, if $V = 5.50$ ft/sec, $\epsilon/R = 0.0020$, and $\nu = 0.00042$ ft²/sec.

Mathcad

Solution:

First calculate $\text{Re} = 4RV/\nu = (4)(0.50)(5.50)/0.00042 = 26{,}190$. Then

$$C = -42\log\left(\frac{C}{26{,}190} + 0.0020\right)$$

Solving by successive trials, we find that $C = 94$ is satisfactory.

Powell has plotted graphs of C vs Re for various values of relative roughness ϵ/R. The graphs simplify the calculations. They also indicate a close analogy with the Colebrook formula for the flow in pipes.

10.12. (*a*) Show a correlation between roughness factor f and roughness factor n.

(*b*) What is the average shear stress at the sides and bottom of a rectangular flume 3.66 m wide, flowing 1.22 m deep and laid on a slope of 1.60 m/1000 m?

Solution:

(*a*) Taking the Manning formula as a basis of correlation,

$$C = \sqrt{\frac{8g}{f}} = \frac{1.486R^{1/6}}{n}, \qquad \frac{1}{\sqrt{f}} = \frac{1.486R^{1/6}}{n\sqrt{8g}}, \qquad f = \frac{8gn^2}{2.21R^{1/3}}$$

(*b*) From Problem 10.1,

$$\tau_o = \gamma RS = \gamma\left(\frac{\text{area}}{\text{wetted perimeter}}\right)(\text{slope}) = (9.79)\left(\frac{3.66 \times 1.22}{1.22 + 3.66 + 1.22}\right)\left(\frac{1.60}{1000}\right) = 0.0115 \text{ kPa}$$

10.13. What flow can be expected in a 4-ft-wide rectangular cement-lined channel laid on a slope of 4 ft in 10,000 ft, if the water flows 2 ft deep? Use both Kutter's C and Manning's C.

Solution:

(*a*) *Using Kutter's C*: From Table 9, $n = 0.015$. Hydraulic radius $R = (4)(2)/8 = 1.00$ ft.

From Table 10, for $S = 0.0004$, $R = 1.00$, and $n = 0.015$, the value of $C = 98$.

$$Q = AV = AC\sqrt{RS} = (4 \times 2)(98)\sqrt{1.00 \times 0.0004} = 15.7 \text{ cfs}$$

(*b*) *Using Manning's C*:

$$Q = AV = A\frac{1.486}{n}R^{2/3}S^{1/2} = (4 \times 2)\left(\frac{1.486}{0.015}\right)(1.00)^{2/3}(0.0004)^{1/2} = 15.9 \text{ cfs}$$

10.14. Water is to flow at a rate of 30 m³/s in the concrete channel shown in Fig. 10-6. Find the vertical drop of the channel bottom per kilometer of length.

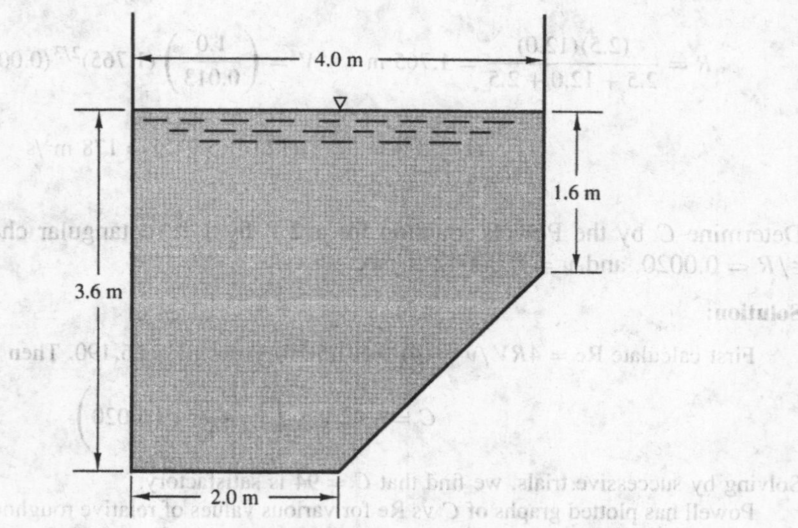

Fig. 10-6

Solution:

$$V = \frac{Q}{A} = \frac{1.0}{n} R^{2/3} S^{1/2}$$

$$A = (3.6)(2.0) + (2.0)\left(\frac{1.6 + 3.6}{2}\right) = 12.40 \text{ m}^2$$

$$R = \frac{12.40}{3.6 + 2.0 + \sqrt{2.0^2 + 2.0^2} + 1.6} = 1.236 \text{ m}$$

$$30/12.40 = \left(\frac{1.0}{0.013}\right)(1.236)^{2/3} S^{1/2}$$

$S = 0.000746$, or 0.000746 m per meter of length, or 0.746 m per kilometer of length

10.15. In a hydraulics laboratory, a flow of 14.56 cfs was measured from a rectangular channel flowing 4 ft wide and 2 ft deep. If the slope of the channel was 0.00040, what is the roughness factor for the lining of the channel?

Solution:

(*a*) Using Kutter's formula,

$$Q = 14.56 = AC\sqrt{RS} = (4 \times 2)C\sqrt{[(4 \times 2)/8](0.0004)} \quad \text{and} \quad C = 91$$

By interpolation in Table 10, $n = 0.016$.

(*b*) Using Manning's formula,

$$Q = 14.56 = A\frac{1.486}{n} R^{2/3} S^{1/2} = (4 \times 2)\left(\frac{1.486}{n}\right)(1)^{2/3}(0.0004)^{1/2}, \; n = 0.0163. \text{ Use } n = 0.016.$$

10.16. On what slope should a 600-mm vitrified sewer pipe be laid in order that 0.17 m³/s will flow when the sewer is half full? What slope if the sewer flows full? (Table 9 gives $n = 0.013$.)

Solution:

$$\text{hydraulic radius } R = \frac{\text{area}}{\text{wetted perimeter}} = \frac{\frac{1}{2}\left(\frac{1}{4}\pi d^2\right)}{\frac{1}{2}(\pi d)} = \frac{1}{4}d$$

(*a*) $Q = 0.17 = A\dfrac{1.0}{n} R^{2/3} S^{1/2} = \dfrac{1}{2}\left(\dfrac{1}{4}\pi\right)(0.600)^2 \times (1.0/0.013)(0.6/4)^{2/3} S^{1/2}$, $\sqrt{S} = 0.0554$ and $S = 0.0031$.

(*b*) $R = \dfrac{1}{4}d = 0.15$ m, as before, and $A = \dfrac{1}{4}\pi(0.600)^2$. Then $\sqrt{S} = 0.0277$ and $S = 0.00077$.

10.17. A trapezoidal channel, bottom width 6 m and side slopes 1/1, flows 1.2 m deep on a slope of 0.0009. For a value of $n = 0.025$, what is the uniform discharge?

Solution:

$$\text{area } A = (6)(1.2) + (2)\left(\tfrac{1}{2}\right)(1.2)(1.2) = 8.64 \text{ m}^2, \; R = 8.64/[6 + (2)(1.2\sqrt{2})] = 0.920 \text{ m}$$

$$Q = (1.0/n)AR^{2/3} S^{1/2} = (1.0/0.025)(8.64)(0.920)^{2/3}(0.0009)^{1/2} = 9.81 \text{ m}^3/\text{s}$$

10.18. Two concrete pipes ($C = 100$) must carry the flow from an open channel of half-square section 6 ft wide and 3 ft deep ($C = 120$). The slope of both structures is 0.00090. (*a*) Determine the diameter of the pipes. (*b*) Find the depth of water in the rectangular channel after it has become stabilized, if the slope is changed to 0.00160, using $C = 120$.

Solution:

(*a*)
$$Q_{\text{channel}} = Q_{\text{pipes}}$$

$$AC\sqrt{RS} = 2AC\sqrt{RS}$$

$$(6 \times 3)(120)\sqrt{\frac{6 \times 3}{12}(0.00090)} = 2\left(\frac{1}{4}\pi d^2\right)(100)\sqrt{\frac{d}{4}(0.00090)}$$

$$79.36 = 2.356 d^{5/2} \quad \text{and} \quad d = 4.08 \text{ ft}$$

(*b*) For depth y, area $A = 6y$ and hydraulic radius $R = \dfrac{6y}{6+2y}$. For the same Q,

$$79.36 = (6y)(120)\sqrt{\left(\frac{6y}{6+2y}\right)(0.00160)}, \qquad 6y\sqrt{\frac{6y}{6+2y}} = 16.53, \qquad y^3 - 2.53y = 7.59$$

Solving by successive trials: Try $y = 2.5$ ft, $(15.6 - 6.3) \neq 7.59$ (decrease y).

Try $y = 2.4$ ft, $(13.8 - 6.1) = 7.7$ (satisfactory).

Thus, depth to nearest 1/10 ft is 2.4 ft.

10.19. An average vitrified sewer pipe is laid on a slope of 0.00020 and is to carry 83.5 cfs when the pipe flows 0.90 full. What size pipe should be used?

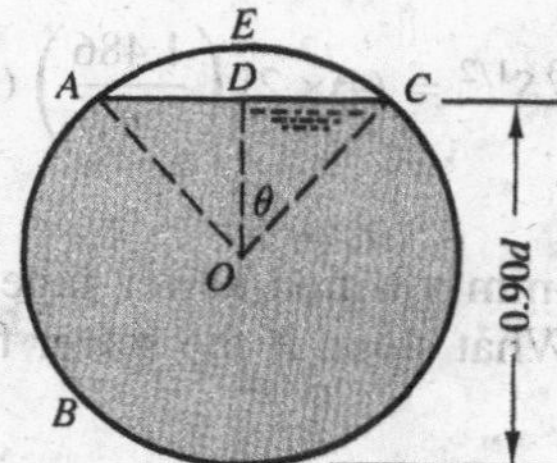

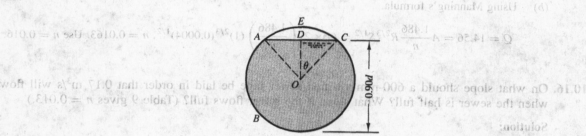

Fig. 10-7

Solution:

From Table 9, $n = 0.015$.

Calculate the hydraulic radius R. (Refer to Fig. 10-7.)

$$R = \frac{A}{p} = \frac{\text{circle} - (\text{sector } AOCE - \text{triangle } AOCD)}{\text{arc } ABC}$$

Angle $\theta = \cos^{-1}(0.40d/0.50d) = \cos^{-1} 0.800$, $\theta = 36°52'$.

Area of sector $AOCE = [(2)(36°52')/360°]\left(\frac{1}{4}\pi d^2\right) = 0.1609d^2$.

Length of arc $ABC = \pi d - [(2)(36°52')/360°](\pi d) = 2.498d$.

Area of triangle $AOCD = (2)\left(\frac{1}{2}\right)(0.40d)(0.40d \tan 36°52') = 0.1200d^2$.

$$R = \frac{\frac{1}{4}\pi d^2 - \left(0.1609d^2 - 0.1200d^2\right)}{2.498d} = \frac{0.7445d^2}{2.498d} = 0.298d$$

(*a*) Using Kutter's C (assumed as 100 for first calculation),

$$Q = CA\sqrt{RS}, \qquad 83.5 = (100)(0.7445d^2)\sqrt{0.298d(0.00020)}, \qquad d^{5/2} = 145.3, \qquad d = 7.33 \text{ ft}$$

Checking on C, $R = 0.298 \times 7.33 = 2.18$ ft. Table 10 gives $C = 112$. Recalculating,

$$d^{5/2} = (145.3)(100/112) = 129.7 \qquad \text{or} \qquad d = 7.00 \text{ ft} \qquad (C \text{ checks satisfactorily.})$$

(*b*) Using Manning's n (and above information),

$$Q = \frac{1.486}{n} AR^{2/3}S^{1/2}$$

$$83.5 = \left(\frac{1.486}{0.015}\right)(0.7445d^2)(0.298d)^{2/3}(0.00020)^{1/2}, \qquad d^{8/3} = 179.4, \qquad d = 7.00 \text{ ft}$$

10.20. Solve Problem 10.19 using the Manning formula and Fig. 10-1.

Solution:

$$d/d_{\text{full}} = 0.90$$

From Fig. 10-1, $Q/Q_{\text{full}} = 106\%$.

$$Q_{\text{full}} = 83.5/1.06 = 78.8 \text{ cfs}$$

$$Q = \frac{1.486}{n} AR^{2/3}S^{1/2}$$

$$78.8 = \left(\frac{1.486}{0.015}\right)(\pi d^2/4)(d/4)^{2/3}(0.00020)^{1/2}$$

$$d = 7.02 \text{ ft}$$

10.21. A 600-mm-diameter concrete pipe on a 1/400 slope carries water at a depth of 240 mm. Find the flow rate.

Solution:

$$Q = \frac{1.0}{n} AR^{2/3}S^{1/2}$$

$$Q_{\text{full}} = (1.0/0.013)[(\pi)(0.600)^2/4](0.600/4)^{2/3}(1/400)^{1/2} = 0.307 \text{ m}^3\text{/s}$$

$$d/d_{\text{full}} = 240/600 = 0.40, \text{ or } 40\%$$

From Fig. 10-1, $Q/Q_{\text{full}} = 32\%$.

$$Q = (0.32)(0.307) = 0.098 \text{ m}^3\text{/s}$$

10.22. How deep will water flow at the rate of 6.80 m³/s in a rectangular channel 6.10 m wide laid on a slope of 0.00010? Use $n = 0.0149$.

Solution:

Employing the Manning formula,

$$Q = \frac{1.0}{n} AR^{2/3}S^{1/2}, \qquad 6.80 = \frac{1.0}{0.0149}(6.10y)\left(\frac{6.10y}{6.10 + 2y}\right)^{2/3}(0.01), \qquad 1.661 = y\left(\frac{6.10y}{6.10 + 2y}\right)^{2/3}$$

Solving by successive trials, we find that $y = 1.60$ m is satisfactory. The water will flow at a depth of 1.60 m, called the normal depth.

10.23. How wide must a rectangular channel be constructed in order to carry 14 m^3/s at a depth of 1.8 m on a slope of 0.00040? Use $n = 0.010$.

Solution:

Using the Manning formula, with $A = 1.8b$ and $R = 1.8b/(b + 3.6)$, and solving by successive trials, we find the required width $b = 4.02$ m.

10.24. After a flood had passed an observation station on a river, an engineer visited the site and, by locating flood marks, performing appropriate surveying, and doing necessary computations, determined that the cross-sectional area, wetted perimeter, and water surface slope at the time of peak flooding were 2960 m^2, 341 m, and 0.00076, respectively. The engineer also noted that the channel bottom was "earth with grass and weeds," for which a handbook gave a Manning n value of 0.030. Estimate the peak flood discharge.

Solution:

$$Q = \frac{1.0}{n} A R^{2/3} S^{1/2}, \qquad Q = \left(\frac{1.0}{0.030}\right)(2960)\left(\frac{2960}{341}\right)^{2/3}(0.00076)^{1/2} = 11{,}500\ \text{m}^3/\text{s}$$

10.25. Develop the Manning equation discharge factors K and K' enumerated in Tables 11 and 12 of the Appendix.

Solution:

The discharge factors to be used in the Manning formula can be evaluated as follows. Any cross-sectional area can be expressed as $A = F_1 y^2$, where F_1 is a dimensionless factor and y^2 is the square of the depth. In similar fashion, hydraulic radius R can be expressed as $R = F_2 y$. Then the Manning formula becomes

$$Q = \frac{1.486}{n}(F_1 y^2)(F_2 y)^{2/3} S^{1/2} \quad \text{or} \quad \frac{Qn}{y^{8/3} S^{1/2}} = 1.486 F_1 F_2^{2/3} = K \tag{1}$$

Similarly, in terms of a base width b, $A = F_3 b^2$, $R = F_4 b$, and

$$\frac{Qn}{b^{8/3} S^{1/2}} = 1.486 F_3 F_4^{2/3} = K' \tag{2}$$

Tables 11 and 12 give values of K and K' for representative trapezoids. Values of K and K' may be calculated for any shape of cross section.

10.26. What are the discharge factors K and K' for a rectangular channel 20 ft wide by 4 ft deep? Compare with the values in Tables 11 and 12.

Solution:

(*a*) $A = F_1 y^2$, $80 = F_1(16)$, $F_1 = 5.0$. $R = F_2 y$, $80/28 = F_2(4)$, $F_2 = 0.714$. $K = 1.486 F_1 F_2^{2/3} = 5.94$.

Table 11 indicates that for $y/b = 4/20 = 0.20$, $K = 5.94$. (Check)

(*b*) $A = F_3 b^2$, $80 = F_3(400)$, $F_3 = 0.20$. $R = F_4 b$, $80/28 = F_4(20)$, $F_4 = 0.143$. $K' = 1.486 F_3 F_4^{2/3} = 0.0813$.

Table 12 indicates that for $y/b = 4/20 = 0.20$, $K' = 0.0812$. (Check)

10.27. After converting given data to fps units ($Q = 240$ cfs, $b = 20$ ft), solve Problem 10.22, using the discharge factors in Table 12.

Solution:

From Problem 10.25, equation (2),

$$\frac{Qn}{b^{8/3}S^{1/2}} = K', \qquad \frac{(240)(0.0149)}{(20)^{8/3}(0.00010)^{1/2}} = 0.121 = K'$$

Table 12 indicates that for trapezoids with vertical sides, a K' of 0.121 represents a depth-to-width ratio (y/b) between 0.26 and 0.28. Interpolating, $y/b = 0.263$. Then $y = (0.263)(20) = 5.26$ ft, (or 1.60 m as calculated in Problem 10.22).

10.28. Solve Problem 10.23 using the discharge factors in Table 11.

Solution:

From Problem 10.25, equation (1),

$$\frac{Qn}{y^{8/3}S^{1/2}} = K, \qquad \frac{(14)(0.010)}{(1.8)^{8/3}(0.00040)^{1/2}} = 1.46 = K$$

Since values of K and K' in Tables 11 and 12 are for the Manning formula utilizing British Engineering System units, values of these parameters must be multiplied by 1.486 when using SI units. Hence, $K = (1.46)(1.486) = 2.17$. By interpolation, the y/b value for $K = 2.17$ is 0.448. Then $b = 1.8/0.448 = 4.02$ m, as found in Problem 10.23.

10.29. A channel with a trapezoidal cross section is to carry 900 cfs. If slope $S = 0.000144$, $n = 0.015$, base width $b = 20$ ft, and the side slopes are 1 vertical to $1\frac{1}{2}$ horizontal, determine the normal depth of flow y_N by formula and by use of tables.

Mathcad

Solution:

(*a*) By formula,

$$900 = \left(\frac{1.486}{0.015}\right)(20y_N + 1.5y_N^2)\left(\frac{20y_N + 1.5y_N^2}{20 + 2y_N\sqrt{3.25}}\right)^{2/3}(0.000144)^{1/2}$$

or

$$757 = \frac{(20y_N + 1.5y_N^2)^{5/3}}{(20 + 2y_N\sqrt{3.25})^{2/3}}$$

Try $y_N = 8.0$ ft: $757 \stackrel{?}{=} \dfrac{(160 + 96)^{5/3}}{(20 + 16\sqrt{3.25})^{2/3}}$ or $757 \neq 772$ (close enough).

The depth of flow can be evaluated by successive trials to the accuracy desired. The normal depth is slightly less than 8.0 ft.

(*b*) Preparatory to using Table 12 in the Appendix,

$$\frac{Qn}{b^{8/3}S^{1/2}} = \frac{(900)(0.015)}{(20)^{8/3}(0.000144)^{1/2}} = 0.382 = K'$$

In Table 12, for $1\frac{1}{2}$ horizontal to 1 vertical,

$$y/b = 0.38,\ K' = 0.353 \qquad \text{and} \qquad y/b = 0.40,\ K' = 0.389$$

Interpolating for $K' = 0.382$ gives $y/b = 0.396$. Then $y_N = (0.396)(20) = 7.92$ ft.

10.30. For a given cross-sectional area, determine the best dimensions for a trapezoidal channel.

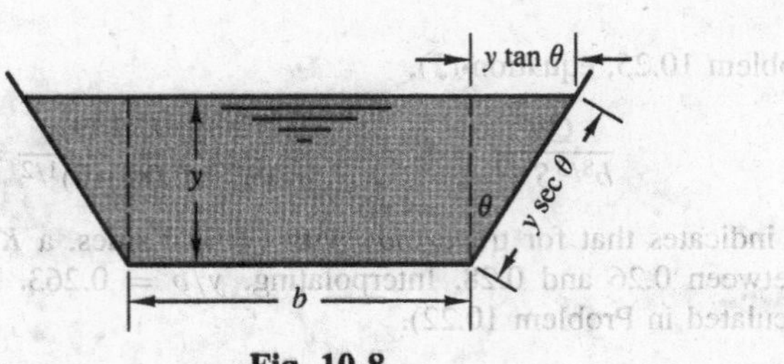

Fig. 10-8

Solution:

Examination of the Chezy equation indicates that for a given cross-sectional area and slope the rate of flow through a channel of given roughness will be a maximum when the hydraulic radius is a maximum. It follows that the hydraulic radius will be a maximum when the wetted perimeter is a minimum. Referring to Fig. 10-8.

$$A = by + (2)\left(\tfrac{1}{2}y\right)(y \tan\theta) \quad \text{or} \quad b = A/y - y\tan\theta$$

$$p = b + 2y\sec\theta \quad \text{or} \quad p = A/y - y\tan\theta + 2y\sec\theta$$

Differentiating p with respect to y and equating to zero,

$$dp/dy = -A/y^2 - \tan\theta + 2\sec\theta = 0 \quad \text{or} \quad A = (2\sec\theta - \tan\theta)y^2$$

$$\text{maximum } R = \frac{A}{p} = \frac{(2\sec\theta - \tan\theta)y^2}{(2\sec\theta - \tan\theta)y^2/y - y\tan\theta + 2y\sec\theta} = \frac{y}{2}$$

Notes:

(1) For all trapezoidal channels, the best hydraulic section is obtained when $R = y/2$. The symmetrical section will be a half-hexagon.

(2) For a rectangular channel (when $\theta = 0°$), $A = 2y^2$ and also $A = by$, yielding $y = b/2$, in addition to $R = y/2$. Thus the best depth is half the width with the hydraulic radius half the depth.

(3) The circle has the least perimeter for a given area. A semicircular open channel will discharge more water than any other shape (for the same area, slope, and factor n).

10.31. (*a*) Determine the most efficient section of a trapezoidal channel, $n = 0.025$, to carry 450 cfs. To prevent scouring, the maximum velocity is to be 3.00 ft/sec and the side slopes of the trapezoidal channel are 1 vertical to 2 horizontal.

(*b*) What slope S of the channel is required? Refer to Fig. 10-8.

Solution:

(*a*)
$$R = \frac{y}{2} = \frac{A}{p} = \frac{by + (2)\left(\frac{1}{2}y\right)(2y)}{b + 2y\sqrt{5}} \quad \text{or} \quad b = 2y\sqrt{5} - 4y \qquad (1)$$

$$A = Q/V = 450/3 = by + 2y^2 \quad \text{or} \quad b = (150 - 2y^2)/y \qquad (2)$$

Equating (*1*) and (*2*), we obtain $y = 7.79$ ft. Substituting in (*2*), $b = 3.68$ ft.

For this trapezoid, $b = 3.68$ ft and $y = 7.79$ ft.

(*b*) $V = (1.486/n)R^{2/3}S^{1/2}$, $\quad 3.00 = (1.486/0.025)(7.79/2)^{2/3}S^{1/2}$, $\quad S = 0.000416$

10.32. An open channel with $n = 0.011$ is to be designed to carry 1.0 m³/s at a slope of 0.0065. Find the most efficient cross section for (*a*) a semicircular section, (*b*) a rectangular section, (*c*) a triangular section, and (*d*) a trapezoidal section.

Solution:

$$v = (1.0/n)R^{2/3}S^{1/2}, \qquad \text{or} \qquad Q/A = (1.0/n)(A/p)^{2/3}S^{1/2}$$

$$A^{5/3}/p^{2/3} = Qn/S^{1/2} = (1.0)(0.011)/0.0065^{1/2} = 0.1364$$

(*a*) For a semicircular section [see Fig. 10-2(*a*)], $A = \pi D^2/8$ and $p = \pi D/2$. Therefore,

$$\frac{(\pi D^2/8)^{5/3}}{(\pi D/2)^{2/3}} = 0.1364 \qquad D = 0.9513 \text{ m} \qquad \text{or} \qquad 951.3 \text{ mm} \qquad \text{or} \qquad d = 951.3/2 = 476 \text{ mm}$$

Note from Fig. 10-2(*a*) that d is the depth of flow, which is, in this case, the radius of the required semicircular section.

(*b*) For a rectangular section [see Fig. 10-2(*b*)], $A = (2d)(d) = 2d^2$ and $p = d+2d+d = 4d$. Therefore,

$$(2d^2)^{5/3}/(4d)^{2/3} = 0.1364$$

$$d = 0.434 \text{ m} \qquad \text{and} \qquad \text{width} = (2)(0.434) = 0.868 \text{ m}$$

(*c*) For a triangular section [see Fig. 10-2(*c*)], $A = (1/2)(d\sqrt{2})(d\sqrt{2}) = d^2$ and $p = (2)(d\sqrt{2}) = 2.828d$. Therefore,

$$(d^2)^{5/3}/(2.828d)^{2/3} = 0.1364$$

$$d = 0.614 \text{ m} \qquad \text{and} \qquad \text{each side} = (0.614)\sqrt{2} = 0.868 \text{ m}$$

(*d*) For a trapezoidal section [see Fig. 10-2(*d*)], $A = (1.155d)(d) + (2)[(d)(d \tan 30°)/2] = 1.732d^2$ and $p = (3)(1.155d) = 3.465d$. Therefore,

$$(1.732d^2)^{5/3}/(3.465d)^{2/3} = 0.1364$$

$$d = 0.459 \text{ m}, \qquad \text{and sides and bottom each} = (1.155)(0.459) = 0.530 \text{ m}.$$

10.33. Develop the expression for critical depth, critical specific energy, and critical velocity for (*a*) rectangular channels and (*b*) any channel.

Mathcad

Solution:

(*a*) **Rectangular Channels.**

As defined,
$$E = y + \frac{V^2}{2g} = y + \frac{1}{2g}\left(\frac{Q/b}{y}\right)^2 = y + \frac{1}{2g}\left(\frac{q}{y}\right)^2 \tag{1}$$

The critical depth for a given flow Q occurs when E is a minimum. Following the usual calculus procedure,

$$\frac{dE}{dy} = \frac{d}{dy}\left[y + \frac{1}{2g}\left(\frac{q}{y}\right)^2\right] = 1 - \frac{q^2}{gy^3} = 0, \qquad q^2 = gy_c^3, \qquad y_c = \sqrt[3]{q^2/g} \tag{2}$$

Eliminating q in (*1*), using values in (*2*),

$$E_c = y_c + \frac{gy_c^3}{2gy_c^2} = \frac{3}{2}y_c \tag{3}$$

Since $q = yV$ (b = unity), expression (2) gives

$$y_c^3 = \frac{q^2}{g} = \frac{y_c^2V_c^2}{g}, \qquad V_c = \sqrt{gy_c}, \qquad \frac{V_c^2}{2g} = \frac{y_c}{2} \tag{4}$$

(*b*) **Any Channel.**

$$E = y + \frac{V^2}{2g} = y + \frac{1}{2g}\left(\frac{Q}{A}\right)^2$$

For a constant Q, and since area A varies with depth y,

$$\frac{dE}{dy} = 1 + \frac{Q^2}{2g}\left(-\frac{2}{A^3}\frac{dA}{dy}\right) = 1 - \frac{Q^2}{A^3 g}\frac{dA}{dy} = 0$$

Area dA is defined as water surface width $b' \times dy$. Substitute in the above equation and obtain

$$\frac{Q^2 b'}{g A_c^3} = 1 \qquad \text{or} \qquad \frac{Q^2}{g} = \frac{A_c^3}{b'} \tag{5}$$

This equation must be satisfied for critical flow conditions. The right-hand side is a function of depth y, and generally a trial-and-error solution is necessary to determine the value of y_c that satisfies equation (*5*).

By dividing Q^2 by A_c^2, or in terms of average velocity, (*5*) may be written

$$V_c^2/g = A_c/b' \qquad \text{or} \qquad V_c = \sqrt{gA_c/b'} \tag{6}$$

Using the mean depth y_m equal to the area A divided by the surface dimension b', equation (*5*) may be written

$$Q = A\sqrt{gA/b'} = A\sqrt{gy_m} \tag{7}$$

Also,

$$V_c = \sqrt{gA_c/b'} = \sqrt{gy_m} \qquad \text{or} \qquad V_c^2/gy_m = 1 \tag{8}$$

The minimum specific energy is, using (*8*),

$$E_{\min} = y_c + V_c^2/2g = y_c + \tfrac{1}{2}y_m \tag{9}$$

For a rectangular channel $A_c = b'y_c$, and (*6*) reduces to equation (*4*) above.

See Fig. 10-9 for equation (*1*) plotted twice, with Q held constant, and with E held constant. When the flow is near the critical, a rippling, unstable surface results. It is not desirable to design channels with slopes near the critical.

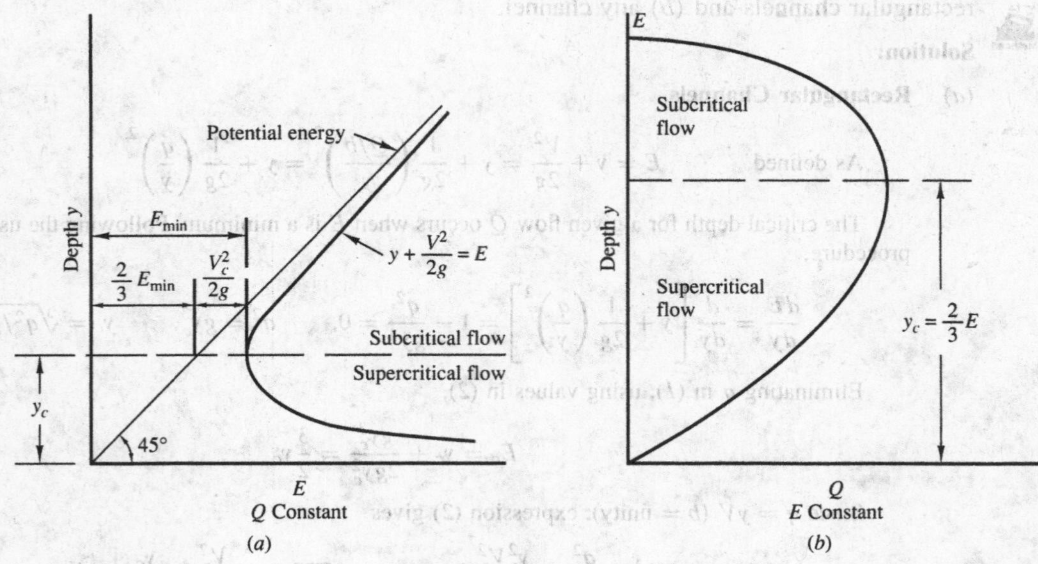

Fig. 10-9

10.34. Derive the expression for the maximum unit flow q in a rectangular channel for a given specific energy E.

Solution:

Solving (*1*) of Problem 10.33 for q gives $q = y\sqrt{2g}(E-y)^{1/2}$. Differentiating with respect to y and equating to zero, we obtain $y_c = \frac{2}{3}E$. Equation (*2*) of Problem 10.33 now becomes

$$q^2_{max} = g\left(\tfrac{2}{3}E_c\right)^3 = gy_c^3 \qquad \text{or} \qquad q_{max} = \sqrt{gy_c^3}$$

Summarizing, for rectangular channels, the characteristics of critical flow are as follows.

(*a*) $E_{min} = \frac{3}{2}\sqrt[3]{q^2/g}$

(*b*) $q_{max} = \sqrt{gy_c^3} = \sqrt{g\left(\frac{2}{3}E_c\right)^3}$

(*c*) $y_c = \frac{2}{3}E_c = V_c^2/g = \sqrt[3]{q^2/g}$

(*d*) $V_c/\sqrt{gy_c} = Fr = 1$

(*e*) Tranquil or subcritical flow occurs when $Fr < 1$ and $y/y_c > 1$.

(*f*) Rapid or supercritical flow occurs when $Fr > 1$ and $y/y_c < 1$.

10.35. A rectangular channel carries 6.0 m³/s. Find the critical depth y_c and critical velocity V_c for a width of (*a*) 4 m and (*b*) 3 m. (*c*) What slope will produce the critical velocity in (*a*) if $n = 0.020$?

Solution:

(*a*) $y_c = \sqrt[3]{q^2/g} = \sqrt[3]{(6.0/4)^2/9.81} = 0.612\text{ m}, \qquad V_c = \sqrt{gy_c} = \sqrt{9.81 \times 0.612} = 2.45\text{ m/s}.$

(*b*) $y_c = \sqrt[3]{q^2/g} = \sqrt[3]{(6.0/3)^2/9.81} = 0.742\text{ m}, \qquad V_c = \sqrt{gy_c} = \sqrt{9.81 \times 0.742} = 2.70\text{ m/s}.$

(*c*) $V_c = \dfrac{1.0}{n}R^{2/3}S^{1/2}, \quad 2.45 = \dfrac{1.0}{0.020}\left(\dfrac{4 \times 0.612}{5.224}\right)^{2/3} S^{1/2}, \; S = 0.00660.$

10.36. A trapezoidal channel with side slopes of 2 horizontal to 1 vertical is to carry a flow of 16.7 m³/s. For a bottom width of 3.6 m, calculate (*a*) the critical depth and (*b*) the critical velocity.

Solution:

(*a*) Area $A_c = 3.6y_c + (2)\left(\frac{1}{2}y_c \times 2y_c\right) = 3.6y_c + 2y_c^2$, and surface width $b' = 3.6 + 4y_c$.

Expression (*5*) of Problem 10.33 yields $\dfrac{(16.7)^2}{9.81} = \dfrac{(3.6y_c + 2y_c^2)^3}{3.6 + 4y_c}$.

Solving this equation by trial, $y_c = 1.06$ m.

(*b*) The critical velocity V_c is determined by using equation (*6*) of Problem 10.33.

$$V_c = \sqrt{\frac{gA_c}{b'}} = \sqrt{\frac{(9.81)(3.82 + 2.25)}{3.6 + 4.24}} = 2.76\text{ m/s}$$

As a check, using $y = y_c = 1.06$, $V_c = Q/A_c = 16.7/\left[(3.6)(1.06) + (2)(1.06)^2\right] = 2.75$ m/s.

10.37. A trapezoidal channel has a bottom width of 20 ft, side slopes of 1 to 1, and flows at a depth of 3.00 ft. For $n = 0.015$ and a discharge of 360 cfs, calculate (*a*) the normal slope, (*b*) the critical slope and critical depth for 360 cfs, and (*c*) the critical slope at the normal depth of 3.00 ft.

Solution:

(a) $$Q = A\frac{1.486}{n}R^{2/3}S_N^{1/2}, \qquad 360 = (60+9)\left(\frac{1.486}{0.015}\right)\left(\frac{69.0}{20+6\sqrt{2}}\right)^{2/3} S_N^{1/2}, \qquad S_N = 0.000853$$

(b) $$V = \frac{Q}{A} = \frac{360}{20y+y^2} \qquad \text{and} \qquad V_c = \sqrt{\frac{gA_c}{b'}} = \sqrt{\frac{(32.2)\,(20y_c+y_c^2)}{20+2y_c}}$$

Equating the velocity terms, squaring, and simplifying, we obtain

$$\frac{[y_c(20+y_c)]^3}{10+y_c} = 8050$$

which when solved by successive trials gives critical depth $y_c = 2.1$ ft.

The critical slope S_c is calculated using the Manning equation:

$$360 = [(20)(2.1)+(2.1)^2]\left(\frac{1.486}{0.015}\right)\left(\frac{(20)(2.1)+(2.1)^2}{20+(2)(2.1\sqrt{2})}\right)^{2/3} S_c^{1/2}, \qquad S_c = 0.00282$$

This slope will maintain uniform critical flow at a critical depth of 2.1 ft and with $Q = 360$ cfs.

(c) From (a), for $y_N = 3.00$ ft, $R = 2.42$ ft and $A = 69.0$ ft^2. Also, using (6) of Problem 10.33,

$$V_c = \sqrt{gA/b'} = \sqrt{(32.2)(69.0)/[20+(2)(3)]} = 9.24 \text{ ft/sec}$$

Substituting these values in the Manning equation for velocity gives

$$9.24 = (1.486/0.015)(2.42)^{2/3}S_c^{1/2}, \qquad S_c = 0.00268$$

This slope will produce uniform critical flow in the trapezoidal channel at a depth of 3.00 ft. Note that in this case the flow $Q = AV = (69.0)(9.24) = 638$ cfs.

10.38. A rectangular channel 9 m wide carries 7.6 m^3/s when flowing 1.00 m deep. (a) What is the specific energy? (b) Is the flow subcritical or supercritical?

Mathcad

Solution:

(a) $$E = y + \frac{V^2}{2g} = y + \left(\frac{1}{2g}\right)\left(\frac{Q}{A}\right)^2 = 1.00 + \left(\frac{1}{19.62}\right)\left(\frac{7.6}{9\times 1}\right)^2 = 1.04 \text{ m.}$$

(b) $y_c = \sqrt[3]{q^2/g} = \sqrt[3]{(7.6/9)^2/9.81} = 0.417$ m.

The flow is subcritical because the depth of flow exceeds the critical depth. (See Problem 10.34.)

10.39. The triangular channel ($n = 0.012$) shown in Fig. 10-10 is to carry water at a flow rate of 10 m^3/s. Find the critical depth, critical velocity, and critical slope of the channel.

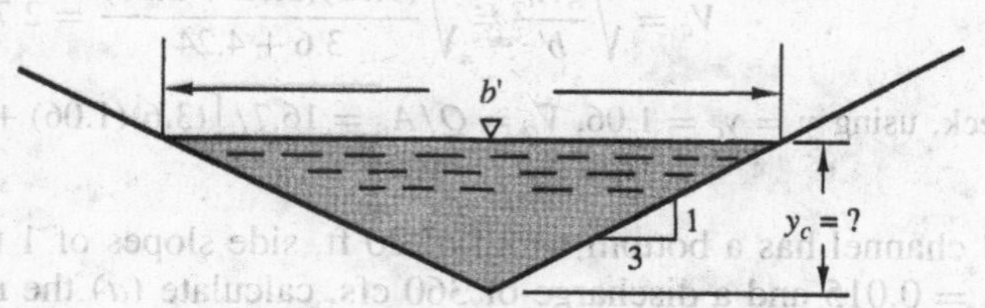

Fig. 10-10

Solution:

From equation (5) of Problem 10.33, $Q^2/g = A_c^3/b'$,

$$A_c = (2)[(y_c)(3y_c)/2] = 3y_c^2; \qquad b' = 6y_c$$

$$10^2/9.81 = (3y_c^2)^3/6y_c, \qquad y_c = 1.178 \text{ m}$$

$$V_c = Q/A = 10/[(3)(1.178)^2] = 2.402 \text{ m/s}$$

$$S_c = \left[(nV_c)/1.0R_c^{2/3}\right]^2$$

$$R_c = A/p = \frac{(3)(1.178)^2}{(2)\sqrt{10}(1.178)} = 0.5588 \text{ m}, \qquad S_c = \left[\frac{(0.012)(2.402)}{(1.0)(0.5588)^{2/3}}\right]^2 = 0.00181$$

10.40. A trapezoidal channel has a bottom width of 6 m and side slopes of 2 horizontal to 1 vertical. When the depth of water is 1.00 m, the flow is 10 m³/s. (*a*) What is the specific energy? (*b*) Is the flow subcritical or supercritical?

Solution:

(*a*) Area $A = (6)(1.00) + (2)(1/2)(1.00)(2.00) = 8.00 \text{ m}^2$.

$$E = y + \left(\frac{1}{2g}\right)\left(\frac{Q}{A}\right)^2 = 1.00 + \left(\frac{1}{19.62}\right)\left(\frac{10}{8.00}\right)^2 = 1.080 \text{ m}$$

(*b*) Using $\dfrac{Q^2}{g} = \dfrac{A_c^3}{b'}, \dfrac{(10)^2}{9.81} = \dfrac{(6y_c + 2y_c^2)^3}{6 + 4y_c}$. Solving by trial, $y_c = 0.611$ m.

Actual depth exceeds critical depth, and flow is subcritical.

10.41. The discharge through a rectangular channel ($n = 0.012$) 4.6 m wide is 11.3 m³/s when the slope is 1 m in 100 m. Is the flow subcritical or supercritical?

Solution:

(1) Investigate critical conditions for the channel.

$$q_{max} = 11.3/4.6 = \sqrt{gy_c^3} \qquad \text{and} \qquad y_c = 0.850 \text{ m}$$

(2) Critical slope for above critical depth can be found by the Chezy-Manning formula.

$$Q = A\frac{1.0}{n}R^{2/3}S_c^{1/2}$$

$$11.3 = (4.6 \times 0.850)\left(\frac{1.0}{0.012}\right)\left(\frac{4.6 \times 0.850}{4.6 + (2)(0.850)}\right)^{2/3} S_c^{1/2}, \qquad S_c = 0.0023$$

Since the stated slope *exceeds* the critical slope, the flow is supercritical.

10.42. A rectangular channel 10 ft wide carries 400 cfs. (*a*) Tabulate (as preliminary to preparing a diagram) depth of flow against specific energy for depths from 1 ft to 8 ft. (*b*) Determine the minimum specific energy. (*c*) What type of flow exists when the depth is 2 ft and when it is 8 ft? (*d*) For $C = 100$, what slopes are necessary to maintain the depths in (*c*)?

Mathcad

Solution:

(*a*) From $E = y + \dfrac{V^2}{2g} = y + \dfrac{(Q/A)^2}{2g}$ we obtain:

$$\begin{aligned}
\text{For } y &= 1 \text{ ft}, & E &= 1.00 + \frac{(400/10)^2}{2g} &&= 25.8 \text{ ft-lb/lb}\\
&= 2 & &= 2.00 + 6.21 &&= 8.21\\
&= 3 & &= 3.00 + 2.76 &&= 5.76\\
&= 4 & &= 4.00 + 1.55 &&= 5.55\\
&= 5 & &= 5.00 + 0.99 &&= 5.99\\
&= 6 & &= 6.00 + 0.69 &&= 6.69\\
&= 7 & &= 7.00 + 0.51 &&= 7.51\\
&= 8 & &= 8.00 + 0.39 &&= 8.39 \text{ ft-lb/lb}
\end{aligned}$$

(*b*) The minimum value of E lies between 5.76 and 5.55 ft-lb/lb.

Using equation (2) of Problem 10.33, $y_c = \sqrt[3]{q^2/g} = \sqrt[3]{(400/10)^2/32.2} = 3.68$ ft.

Then $E_{min} = E_c = \frac{3}{2} y_c = \frac{3}{2}(3.68) = 5.52$ ft-lb/lb.

Note that $E = 8.21$ at 2.00 ft depth and 8.39 at 8.00 ft depth. Figure 10-9(*a*) indicates this fact, i.e., two depths for a given specific energy when flow Q is constant.

(*c*) For 2-ft depth (below critical depth) the flow is supercritical, and for the 8-ft depth the flow is subcritical.

(*d*)
$$Q = CA\sqrt{RS}$$

For $y = 2$ ft, $A = 20$ ft^2 and $R = 20/14 = 1.43$ ft, $400 = (100)(20)\sqrt{1.43S}$ and $S = 0.0280$.

For $y = 8$ ft, $A = 80$ ft^2 and $R = 80/26 = 3.08$ ft, $400 = (100)(80)\sqrt{3.08S}$ and $S = 0.000812$.

10.43. A rectangular flume ($n = 0.012$) is laid on a slope of 0.0036 and carries 580 cfs. For critical flow conditions, what width is required?

Solution:

From Problem 10.34, $q_{max} = \sqrt{g y_c^3}$. Hence $580/b = \sqrt{32.2 y_c^3}$.
Using successive trials, check calculated flow against given flow.

Trial 1. Letting $b = 8.0$ ft, $y_c = \sqrt[3]{(580/8.0)^2/32.2} = 5.47$ ft.
Then $R = A/p = (8.0 \times 5.47)/18.94 = 2.31$ ft.
and $Q = AV = (8.0 \times 5.47)\left[\dfrac{1.486}{0.012}(2.31)^{2/3}(0.0036)^{1/2}\right] = 568$ cfs.

Trial 2. Since the flow must be increased, let $b = 8.50$ ft.
Then $y_c = \sqrt[3]{(580/8.50)^2/32.2} = 5.25$ ft, $R = (8.50 \times 5.25)/19.00 = 2.35$ ft,
and $Q = AV = (8.50 \times 5.25)\left[\dfrac{1.486}{0.012}(2.35)^{2/3}(0.0036)^{1/2}\right] = 586$ cfs.
The result is probably close enough.

10.44. For a constant specific energy of 2.00 N·m/m, what maximum flow can occur in a rectangular channel 3.00 m wide?

Solution:

Critical depth $y_c = \frac{2}{3}E = \frac{2}{3}(2.00) = 1.33$ m.

Critical velocity $V_c = \sqrt{gy_c} = \sqrt{9.81 \times 1.33} = 3.61$ m/s and

$$\text{maximum } Q = AV = (3.00 \times 1.33)(3.61) = 14.4 \text{ m}^3/\text{s}$$

Using $q_{max} = \sqrt{gy_c^3}$, we have

$$\text{maximum} Q = bq_{max} = 3.00\sqrt{(9.81)(1.33)^3} = 14.4\text{m}^3/\text{s}$$

10.45. The triangular channel ($n = 0.013$) shown in Fig. 10-11 is to carry water at a flow rate of 38.5 m^3/s. Find the critical depth, critical velocity, and critical slope of the channel.

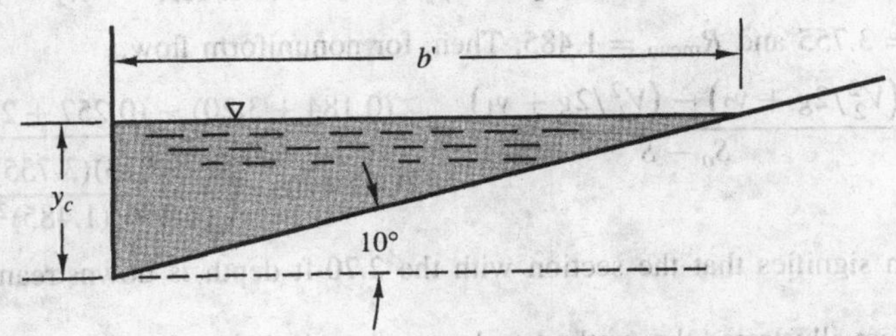

Fig. 10-11

Solution:

From equation (5) of Problem 10.33, $Q^2/g = A_c^3/b'$, and

$$b' = y_c \tan 80° = 5.671y_c$$

$$A_c = y_c b'/2 = (y_c)(5.671y_c)/2 = 2.836y_c^2$$

$$(38.5)^2/9.81 = \left(2.836y_c^2\right)^3/(5.671y_c) \qquad y_c = 2.065 \text{ m}$$

$$V_c = Q/A = 38.5/\left[(2.836)(2.065)^2\right] = 3.184 \text{ m/s}$$

$$R_c = A/p = \frac{(2.836)(2.065)^2}{(2.065) + (2.065/\cos 80°)} = 0.8665 \text{ m}, \qquad S_c = \left[\frac{(0.013)(3.184)}{(1.0)(0.8665)^{2/3}}\right]^2 = 0.00207$$

10.46. Develop a formula for the length–energy–slope relationship for nonuniform flow problems.

Solution:

By applying the energy equation, section 1 to section 2 in the direction of flow, using the datum below the channel bottom, we obtain

$$\text{energy at } 1 - \text{ lost head} = \text{energy at } 2$$
$$\left(z_1 + y_1 + V_1^2/2g\right) - h_L = \left(z_2 + y_2 + V_2^2/2g\right)$$

The slope of the energy line S is h_L/L; then $h_L = SL$. The slope of the channel bottom S_0 is $(z_1 - z_2)/L$; then $z_1 - z_2 = S_oL$. Rearranging and substituting,

$$S_oL + (y_1 - y_2) + \left(V_1^2/2g - V_2^2/2g\right) = SL$$

This expression is usually solved for length L in open-channel studies. Thus

$$L = \frac{(y_1 + V_1^2/2g) - (y_2 + V_2^2/2g)}{S - S_o} = \frac{E_1 - E_2}{S - S_0} \tag{A}$$

The next few problems will illustrate the use of expression (*A*).

10.47. A rectangular flume ($n = 0.013$) is 6 ft wide and carries 66 cfs of water. At a certain section F, the depth is 3.20 ft. If the slope of the channel bed is constant at 0.000400, determine the distance from F where the depth is 2.70 ft. (Use one reach.)

Solution:

Assume the depth is *upstream* from F. Use subscripts 1 and 2, as usual.

$A_1 = (6)(2.70) = 16.20\ \text{ft}^2$, $V_1 = 66/16.20 = 4.07$ ft/sec, $R_1 = 16.20/11.40 = 1.42$ ft

$A_2 = (6)(3.20) = 19.20\ \text{ft}^2$, $V_2 = 66/19.20 = 3.44$ ft/sec, $R_2 = 19.20/12.40 = 1.55$ ft

Hence, $V_{\text{mean}} = 3.755$ and $R_{\text{mean}} = 1.485$. Then, for nonuniform flow,

$$L = \frac{(V_2^2/2g + y_2) - (V_1^2/2g + y_1)}{S_o - S} = \frac{(0.184 + 3.20) - (0.257 + 2.70)}{0.000400 - \left(\dfrac{(0.013)(3.755)}{(1.486)(1.485)^{2/3}}\right)^2} = -1802\ \text{ft}$$

The minus sign signifies that the section with the 2.70-ft depth is downstream from F, not upstream as assumed.

This problem illustrates the method to be employed. A more accurate answer could be obtained by assuming intermediate depths of 3.00 ft and 2.85 ft (or exact depths by interpolating values), calculating ΔL values, and adding these. In this manner a *backwater* curve can be calculated. The backwater curve is not a straight line.

10.48. A rectangular channel 40 ft wide carries 900 cfs of water. The slope of the channel is 0.00283. At section 1 the depth is 4.50 ft, and at section 2, 300 ft downstream, the depth is 5.00 ft. What is the average value of roughness factor n?

Solution:

$A_2 = (40)(5.00) = 200\ \text{ft}^2$, $V_2 = 900/200 = 4.50$ ft/sec, $R_2 = 200/50 = 4.00$ ft

$A_1 = (40)(4.50) = 180\ \text{ft}^2$, $V_1 = 900/180 = 5.00$ ft/sec, $R_2 = 180/49 = 3.67$ ft

Hence, $V_{\text{mean}} = 4.75$ ft/sec and $R_{\text{mean}} = 3.835$ ft. For nonuniform flow,

$$L = \frac{(V_2^2/2g + y_2) - (V_1^2/2g + y_1)}{S_0 - \left(\dfrac{nV}{1.486R^{2/3}}\right)^2}, \qquad 300 = \frac{(0.314 + 5.00) - (0.388 + 4.50)}{0.00283 - \left(\dfrac{n \times 4.75}{(1.486)(3.835)^{2/3}}\right)^2}$$

and $n = 0.0288$.

10.49. A rectangular channel 20 ft wide has a slope of 1 ft per 1000 ft. The depth at section 1 is 8.50 ft, and at section 2, 2000 ft downstream, the depth is 10.25 ft. If $n = 0.011$, determine the probable flow in cfs.

Solution:

Using the stream bed at section 2 as datum,

$$\text{energy at } 1 = y_1 + V_1^2/2g + z_1 = 8.50 + V_1^2/2g + 2.00$$

$$\text{energy at } 2 = y_2 + V_2^2/2g + z_2 = 10.25 + V_2^2/2g + 0$$

The drop in the energy line = energy at 1 − energy at 2. Since the value is unknown, a slope assumption will be made.

$$\text{slope } S = \frac{\text{lost head}}{L} = \frac{(10.50 - 10.25) + \left(V_1^2/2g - V_2^2/2g\right)}{2000} \qquad (1)$$

Assume $S = 0.000144$. Also, the values of A_{mean} and R_{mean} are necessary.

$$A_1 = (20)(8.50) \quad = 170 \text{ ft}^2, \; R_1 = 170/37 \quad = 4.59 \text{ ft}$$

$$A_2 = (20)(10.25) = 205 \text{ ft}^2, \; R_2 = 205/40.5 = 5.06 \text{ ft}$$

Hence, $A_{\text{mean}} = 187.5 \text{ ft}^2$ and $R_{\text{mean}} = 4.825$ ft.

First Approximation.

$$Q = A_m(1.486/n)R_m^{2/3}S^{1/2} = (187.5)(1.486/0.011)(4.825)^{2/3}(0.000144)^{1/2} = 868 \text{ cfs}$$

Check on slope S in equation (*1*) above, as follows:

$$V_1 = 868/170 = 5.11, \; V_1^2/2g = 0.405$$

$$V_2 = 868/205 = 4.23, \; V_2^2/2g = 0.278$$

$$S = \frac{(10.50 - 10.25) + 0.127}{2000} = 0.000188$$

The energy gradient drops 0.376 ft in 2000 ft, which is more than was assumed.

Second Approximation.

Try $S = 0.000210$. Then $Q = 868 \times \left(\dfrac{0.000210}{0.000144}\right)^{1/2} = 1048$ cfs.

Again checking, $\quad V_1 = 1048/170 = 6.16 \text{ ft/sec}, \; V_1^2/2g = 0.589 \text{ ft}$

$$V_2 = 1048/205 = 5.11 \text{ ft/sec}, \; V_2^2/2g = 0.405 \text{ ft}$$

$$S = \frac{(10.50 - 10.25) + 0.184}{2000} = 0.000217$$

This slope checks (reasonably) the assumption made. Hence approximate $Q = 1050$ cfs.

10.50. Water flowing at the normal depth in a rectangular concrete channel that is 12.0 m wide encounters an obstruction, as shown in Fig. 10-12, causing the water level to rise above the normal depth at the obstruction and for some distance upstream. The water discharge is 126 m³/s, and the channel bottom slope is 0.00086. If the depth of water just upstream from the obstruction (y_0) is 4.55 m, find the distance upstream to the point where the water surface is at the normal depth.

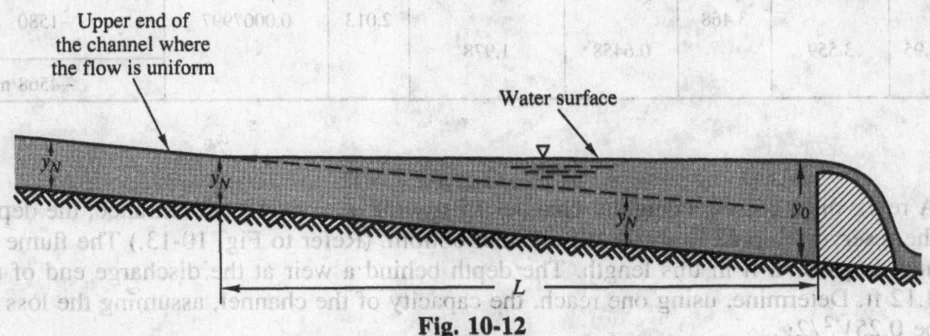

Fig. 10-12

Solution:

$$y_c = \sqrt[3]{(Q/b')^2/g} = \sqrt[3]{(126/12.0)^2/9.81} = 2.24 \text{ m}$$

$$V = Q/A = (1.0/n)R^{2/3}S^{1/2}$$

$$126/(12.0y) = (1.0/0.013)[12.0y/(12.0+2y)]^{2/3}(0.00086)^{1/2}$$

$$(2.256)[12.0y/(12.0+2y)]^{2/3} - 10.5/y = 0$$

A trial-and-error solution of this equation gives a value of y, the normal depth (y_N), of 2.95 m. Since $y_N > y_c$, the flow is subcritical, and computations should proceed upstream. The problem now is to determine the distance from the point where the depth is 4.55 m to the point upstream where the depth is 2.95 m. This will be done (arbitrarily) in 10 equal-depth increments of 0.16 m. The computations are given in the table below; they apply expression (A) of Problem 10.46. The summation of values in column 8 of the table (4568 m) gives the answer to the problem (i.e., the distance upstream from the obstruction to the point where the water surface is at the normal depth).

(1) y(m)	(2) V(m/s) $\frac{126}{12.0\times(1)}$	(3) V_m(m/s)	(4) $V^2/2g$(m) $\frac{(2)^2}{2\times 9.807}$	(5) R(m) $\frac{12.0\times(1)}{12.0+2\times(1)}$	(6) R_m(m)	(7) S $\left[\frac{0.013\times(3)}{(6)^{2/3}}\right]^2$	(8) L(m) $\frac{[(4)+(1)]_2-[(4)+(1)]_1}{0.00086-(7)}$
4.55	2.308		0.2716	2.588			
		2.350			2.562	0.0002662	−236
4.39	2.392		0.2917	2.535			
		2.437			2.508	0.0002946	−243
4.23	2.482		0.3141	2.481			
		2.531			2.453	0.0003272	−253
4.07	2.580		0.3394	2.425			
		2.633			2.396	0.0003654	−266
3.91	2.685		0.3676	2.367			
		2.743			2.338	0.0004098	−284
3.75	2.800		0.3997	2.308			
		2.863			2.277	0.0004626	−311
3.59	2.925		0.4362	2.246			
		2.993			2.214	0.0005246	−353
3.43	3.061		0.4777	2.182			
		3.136			2.150	0.0005989	−429
3.27	3.211		0.5257	2.117			
		3.294			2.083	0.0006893	−613
3.11	3.376		0.5811	2.048			
		3.468			2.013	0.0007997	−1580
2.95	3.559		0.6458	1.978			
							−4568 m

10.51. A reservoir feeds a rectangular channel 15 ft wide, $n = 0.015$. At entrance, the depth of water in the reservoir is 6.22 ft above the channel bottom. (Refer to Fig. 10-13.) The flume is 800 ft long and drops 0.72 ft in this length. The depth behind a weir at the discharge end of the channel is 4.12 ft. Determine, using one reach, the capacity of the channel, assuming the loss at entrance to be $0.25V_1^2/2g$.

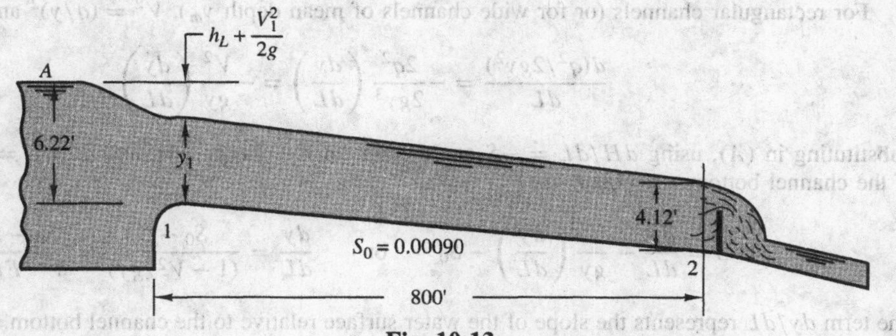

Fig. 10-13

Solution:

The Bernoulli equation, A to 1, datum 1, gives

$$(0 + \text{negl.} + 6.22) - 0.25V_1^2/2g = (0 + V_1^2/2g + y_1) \tag{1}$$

and

$$L = \frac{(V_2^2/2g + y_2) - (V_1^2/2g + y_1)}{S_0 - \left(\dfrac{nV_m}{1.486R_m^{2/3}}\right)^2} \tag{2}$$

Solve these equations by successive trials until L approximates or equals 800 ft.

Try $y_1 = 5.0$ ft; then from (1), $V_1^2/2g = (6.22-5.00)/1.25 = 0.976$ ft, $V_1 = 7.93$ ft/sec, and $q = y_1V_1$ = (5.0) (7.93) = 39.6 cfs/ft, $V_2 = 39.6/4.12 = 9.61$ ft/sec.

$$V_{\text{mean}} = \left(\tfrac{1}{2}\right)(7.93 + 9.61) = 8.77 \text{ ft/sec}$$

and

$$R_{\text{mean}} = \left(\tfrac{1}{2}\right)(R_1 + R_2) = \left(\tfrac{1}{2}\right)[(15 \times 5)/25 + (15 \times 4.12)/23.24] = 2.83 \text{ ft}$$

Substituting in equation (2) above, we find $L = 399$ ft.

Increase the value of y_1 to 5.50 ft and repeat the calculation. Results in tabular form are:

y_1	V_1	q_1	V_2	V_m	R_m	L	*Notes*
5.50	6.09	33.5	8.13	7.11	2.92	2782 ft	∴ decrease y_1
5.20	7.25	37.7	9.15	8.20	2.87	764 ft	close enough

The capacity of the channel = 37.7 × 15 = 566 cfs.

Should greater accuracy be required, start at the lower end and, for unit flow $q = 37.7$ cfs, find the length of the reach to a point where the depth is about 10% more than 4.12 or about 4.50 ft, thence to a depth of 4.90 ft, and so on. If the sum of the lengths exceeds 800 ft, decrease the value of y_1, obtaining an increase in q.

10.52. Derive the expression that gives the slope of the liquid surface in wide rectangular channels for gradually varied flow.

Solution:

The total energy of fluid with respect to an arbitrary datum plane is

$$H = y + V^2/2g + z$$

where the kinetic energy correction factor α is taken as unity. Differentiating this expression with respect to L, the distance along the channel, yields

$$\frac{dH}{dL} = \frac{dy}{dL} + \frac{d(V^2/2g)}{dL} + \frac{dz}{dL} \tag{A}$$

For rectangular channels (or for wide channels of mean depth y_m), $V^2 = (q/y)^2$ and

$$\frac{d(q^2/2gy^2)}{dL} = -\frac{2q^2}{2gy^3}\left(\frac{dy}{dL}\right) = -\frac{V^2}{gy}\left(\frac{dy}{dL}\right)$$

Substituting in (A), using $dH/dL = -S$ or the slope of the energy line, and $dz/dL = -S_0$ or the slope of the channel bottom, we obtain

$$-S = \frac{dy}{dL} - \frac{V^2}{gy}\left(\frac{dy}{dL}\right) - S_0 \quad \text{or} \quad \frac{dy}{dL} = \frac{S_0 - S}{(1 - V^2/gy)} = \frac{S_0 - S}{1 - Fr^2} \tag{B}$$

The term dy/dL represents the slope of the water surface relative to the channel bottom. When the channel slopes down in the direction of flow, S_0 is positive. Similarly, S is positive (always). For uniform flow, $S = S_0$ and $dy/dL = 0$.

Another form of equation (B) can be obtained as follows. Manning's formula is

$$Q = (1.486/n)AR^{2/3}S^{1/2}$$

Solving this equation for the slope of the energy line, using $q = Q/b$, $A = by$, and $R = y$ for wide rectangular channels yields

$$\frac{dH}{dL} = S = \frac{n^2(q^2b^2/b^2y^2)}{(1.486)^2y^{4/3}}$$

Similarly, the slope of the channel bottom, in terms of normal depth y_N and coefficient n_N can be written

$$\frac{dz}{dL} = S_0 = \frac{n_N^2\,(q^2b^2/b^2y_N^2)}{(1.486)^2y_N^{4/3}}$$

Then the first part of equation (B) becomes

$$-\frac{n^2\,(q^2b^2/b^2y^2)}{(1.486)^2y^{4/3}} = \left(1 - \frac{V^2}{gy}\right)\frac{dy}{dL} - \frac{n_N^2\,(q^2b^2/b^2y_N^2)}{(1.486)^2y_N^{4/3}}$$

But $V^2 = q^2/y^2$, $n \cong n_N$, and $q^2/g = y_c^3$. Then

$$\frac{-n^2q^2}{(1.486)^2y^{10/3}} = \frac{dy}{dL}\left(1 - y_c^3/y^3\right) - \frac{n^2q^2}{(1.486)^2\,y_N^{10/3}} \tag{C}$$

$$\frac{dy}{dL} = \frac{(nq/1.486)^2\left[1/y_N^{10/3} - 1/y^{10/3}\right]}{1 - (y_c/y)^3} \tag{D}$$

Using $Q/b = q = y_N\left[(1.486/n)y_N^{2/3}S_0^{1/2}\right]$ or $(nq/1.486)^2 = y_N^{10/3}S_0$, equation (D) becomes

$$\frac{dy}{dL} = S_0\left[\frac{1 - (y_N/y)^{10/3}}{1 - (y_c/y)^3}\right] \tag{E}$$

There are limiting conditions to the surface profiles. For example, as y approaches y_c, the denominator of (E) approaches zero. Thus dy/dL becomes infinite and the curves cross the critical depth line perpendicular to it. Hence surface profiles in the vicinity of $y = y_c$ are only approximate.

Similarly, when y approaches y_N, the numerator approaches zero. Thus the curves approach the normal depth y_N asymptotically.

Finally, as y approaches zero, the surface profile approaches the channel bed perpendicularly, which is impossible under the assumption concerning gradually varied flow.

10.53. Summarize the system for classifying surface profiles for gradually varied flow in wide channels.

Solution:

There are a number of different conditions in a channel that give rise to 12 different types of nonuniform (varied) flow. In expression (E) in Problem 10.52, for positive values of dy/dL, depth y increases downstream along the channel, and for negative values of dy/dL depth y will decrease downstream along the channel.

In Table 10.1 a summary of the 12 different types of varied flow is presented. Several of these will be discussed, and the reader may analyze the remaining types of flow in a similar fashion.

The "mild" classification results from channel slope S_0 being such that normal depth $y_N > y_c$. If depth y is greater than y_N and y_c, the curve is called type 1; if depth y is between y_N and y_c, type 2; and if depth y is smaller than y_N and y_c, type 3.

It will be noticed that, for the type 1 curves, since velocity is decreasing because of increased depth, the water surface must approach a horizontal asymptote (see M_1, C_1, and S_1). Similarly, curves that approach the normal depth line do so asymptotically. As pointed out previously, curves that approach the critical depth line cross it vertically, inasmuch as the denominator of expression (E) in Problem 10.52 becomes zero for such cases. Therefore, curves for critical slopes are exceptions to the preceding statements, as it is impossible to have a water surface curve both tangent and perpendicular to the critical depth line.

On each profile in Table 10.1, the vertical scale is greatly magnified with respect to the horizontal scale. As indicated in the numerical problems for M_1 curves, such profiles may be thousands of feet in extent.

Table 10.1 gives the relations between slopes and depths, the sign of dy/dL, the type of profile, the symbol for the profile, the type of flow, and a sketch representing the form of the profile. The values of y within each profile can be observed to be greater than or less than y_N and/or y_c by studying each sketch.

10.54. For a rectangular channel, develop an expression for the relation between the depths before and after a hydraulic jump. Refer to Fig. 10-14.

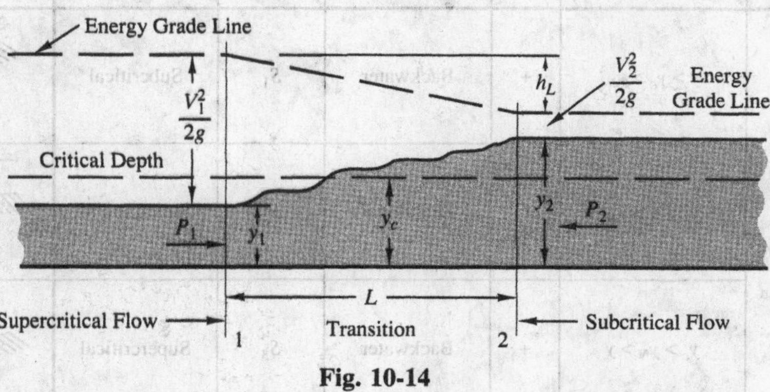

Fig. 10-14

Solution:

For the free body between sections 1 and 2, considering a unit width of channel and unit flow q,

$$P_1 = \gamma \bar{h} A = \gamma \left(\tfrac{1}{2} y_1\right) y_1 = \tfrac{1}{2} \gamma y_1^2, \quad \text{and similarly} \quad P_2 = \tfrac{1}{2} \gamma y_2^2$$

Table 10.1 Types of Varied Flow

Channel Slope	Depth Relations	$\frac{dy}{dL}$	Type of Profile	Symbol	Type of Flow	Form of Profile
Mild $0 < S < S_c$	$y > y_N > y_c$	+	Backwater	M_1	Subcritical	
	$y_N > y > y_c$	−	Dropdown	M_2	Subcritical	
	$y_N > y_c > y$	+	Backwater	M_3	Supercritical	
Horizontal $S = 0$ $y_N = \infty$	$y > y_c$	−	Dropdown	H_2	Subcritical	
	$y_c > y$	+	Backwater	H_3	Supercritical	
Critical $S_N = S_c$ $y_N = y_c$	$y > y_c = y_N$	+	Backwater	C_1	Subcritical	
	$y_c = y = y_N$		Parallel to bed	C_2	Uniform, critical	
	$y_c = y_N > y$	+	Backwater	C_3	Supercritical	
Steep $S > S_c > 0$	$y > y_c > y_N$	+	Backwater	S_1	Subcritical	
	$y_c > y > y_N$	−	Dropdown	S_2	Supercritical	
	$y_c > y_N > y$	+	Backwater	S_3	Supercritical	
Adverse $S < 0$ $y_N = \infty$	$y > y_c$	−	Dropdown	A_2	Subcritical	
	$y_c > y$	+	Backwater	A_3	Supercritical	

From the principle of impulse and momentum,

$$\Delta P_x dt = \Delta \text{ linear momentum } = \frac{W}{g}(\Delta V_x)$$

$$\tfrac{1}{2}\gamma\left(y_2^2 - y_1^2\right) dt = \frac{\gamma q\, dt}{g}(V_1 - V_2)$$

Since $V_2 y_2 = V_1 y_1$ and $V_1 = q/y_1$, the above equation becomes

$$q^2/g = \tfrac{1}{2} y_1 y_2 (y_1 + y_2) \tag{1}$$

Since $q^2/g = y_c^3$, $\quad y_c^3 = \tfrac{1}{2} y_1 y_2 (y_1 + y_2) \quad (2)$

The length of the jump has been found to vary between $4.3 y_2$ and $5.2 y_2$.

The hydraulic jump is an energy dissipator. In designing hydraulic jump stilling basins, knowledge of jump length and depth y_2 is important. Good energy dissipation occurs when $V_1^2/gy_1 = 20$ to 80.

10.55. A rectangular channel 6.1 m wide carries 11.3 m^3/s and discharges onto a 6.1-m-wide apron with no slope at a mean velocity of 6.1 m/s. What is the height of the hydraulic jump? What energy is absorbed (lost) in the jump?

Mathcad

Solution:

(*a*) $V_1 = 6.1$ m/s, $q = 11.3/6.1 = 1.85$ (m^3/s)/m width, and $y_1 = q/V_1 = 1.85/6.1 = 0.303$ m. Then

$$q^2/g = \tfrac{1}{2} y_1 y_2 (y_1 + y_2), \qquad (1.85)^2/9.81 = \tfrac{1}{2}(0.303) y_2 (0.303 + y_2), \qquad 0.349 = 0.0459 y_2 + 0.152 y_2^2$$

from which $y_2 = -1.67$ m, $+1.37$ m. The negative root being extraneous, $y_2 = 1.37$ m and the height of the hydraulic jump is $(1.37 - 0.303) = 1.07$ m.

Note that $y_c = \sqrt[3]{(1.85)^2/9.81}$ or $\sqrt[3]{\tfrac{1}{2} y_1 y_2 (y_1 + y_2)} = 0.70$ m.

Hence the flow at 0.303-m depth is supercritical, and the flow at 1.37-m depth is subcritical.

(*b*) Before the jump, $E_1 = V_1^2/2g + y_1 = (6.1)^2/2g + 0.303 = 2.20$ m·N/N.

After the jump, $E_2 = V_2^2/2g + y_2 = [11.3/(6.1 \times 1.37)]^2/2g + 1.37 = 1.46$ m·N/N.

Lost energy per second $= \gamma Q H = (9.79)(11.3)(2.20 - 1.46) = 81.9$ kW.

10.56. A rectangular channel 16 ft wide carries a flow of 192 cfs. The depth of water on the downstream side of the hydraulic jump is 4.20 ft. (*a*) What is the depth upstream? (*b*) What is the loss of head?

Solution:

(*a*) $q^2/g = \tfrac{1}{2} y_1 y_2 (y_1 + y_2), \qquad (192/16)^2/32.2 = 2.10 y_1 (y_1 + 4.20), \qquad y_1 = 0.457$ ft

(*b*) $A_1 = (16)(0.457) = 7.31$ ft^2, $V_1 = 192/7.31 = 26.3$ ft/sec

$A_2 = (16)(4.20) = 67.2$ ft^2, $V_2 = 192/67.2 = 2.86$ ft/sec

$E_1 = V_1^2/2g + y_1 = (26.3)^2/2g + 0.457 = 11.20$ ft-lb/lb

$E_2 = V_2^2/2g + y_2 = (2.86)^2/2g + 4.20 = 4.33$ ft-lb/lb

Energy lost $= 11.20 - 4.33 = 6.87$ ft-lb/lb or ft.

10.57. Water flows over a concrete spillway into a rectangular channel 9.0 m wide through a hydraulic jump. The depths before and after the jump are 1.55 m and 3.08 m, respectively. Find the rate of flow in the channel.

Solution:

$$q^2/g = \tfrac{1}{2}y_1y_2(y_1+y_2), \qquad q^2/9.81 = \tfrac{1}{2}(1.55)(3.08)(1.55+3.08)$$

$$q = 10.41\ \text{m}^3/(\text{s}\cdot\text{m}), \qquad Q = (10.41)(9.0) = 93.7\ \text{m}^3/\text{s}$$

10.58. After flowing over the concrete spillway of a dam, 9000 cfs then passes over a level concrete apron ($n = 0.013$). The velocity of the water at the bottom of the spillway is 42.0 ft/sec, and the width of the apron is 180 ft. Conditions will produce a hydraulic jump, the depth in the channel below the apron being 10.0 ft. (*a*) In order for the jump to be contained on the apron, how long should the apron be built? (*b*) How much energy is lost from the foot of the spillway to the downstream side of the jump?

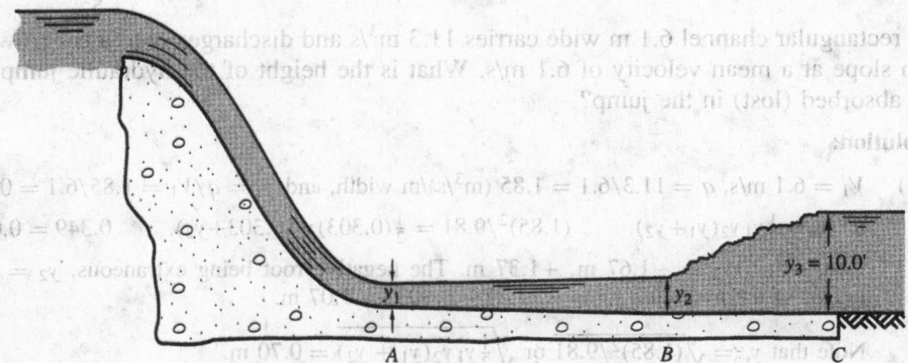

Fig. 10-15

Solution:

(*a*) Referring to Fig. 10-15, first calculate depth y_2 at the upstream end of the jump.

$$q^2/g = \tfrac{1}{2}y_2y_3(y_2+y_3), \qquad (9000/180)^2/32.2 = \tfrac{1}{2}(10y_2)(y_2+10) \qquad y_2 = 1.37\ \text{ft}$$

Also, $$y_1 = q/V_1 = (9000/180)/42.0 = 1.19\ \text{ft}$$

Now calculate length L_{AB} of the retarded flow.

$$V_1 = 42.0\ \text{ft/sec}, \qquad V_1^2/2g = 27.4\ \text{ft}, \qquad R_1 = (180\times 1.19)/182.38 = 1.174\ \text{ft}$$

$$V_2 = \frac{q}{y_2} = \frac{50.0}{1.37} = 36.5\ \text{ft/sec}, \qquad \frac{V_2^2}{2g} = 20.7\ \text{ft} \qquad R_2 = \frac{(180\times 1.37)}{182.74} = 1.349\ \text{ft}$$

Then $V_{\text{mean}} = 39.25$ ft/sec, $R_{\text{mean}} = 1.262$ ft, and

$$L_{AB} = \frac{(V_2^2/2g + y_2) - (V_1^2/2g + y_1)}{S_0 - S} = \frac{(20.7+1.37)-(27.4+1.19)}{0 - \left(\dfrac{0.013\times 39.25}{(1.486)(1.262)^{2/3}}\right)^2} = 75.4\ \text{ft}$$

The length of the jump L_J from B to C is between $4.3y_3$ and $5.2y_3$ ft. Assuming the conservative value of $5.0y_3$,

$$L_J = 5.0\times 10.0 = 50.0\ \text{ft}$$

Hence, total length $ABC = 76 + 50 = 126$ ft (approximately).

(b) Energy at $A = y_1 + V_1^2/2g = 1.19 + 27.4 = 28.59$ ft-lb/lb.

Energy at $C = y_3 + V_3^2/2g = 10.0 + (5.0)^2/2g = 10.39$ ft-lb/lb.

Total energy lost $= \gamma QH = (62.4)(9000)(28.59 - 10.39) = 1.02 \times 10^7$ ft-lb/sec, or 18,600 hp.

10.59. In order that the hydraulic jump below a spillway will not be swept downstream, establish the relationship between the variables indicated in Fig. 10-16. (Prof. E. A. Elevatorski suggested the dimensionless parameters that follow.)

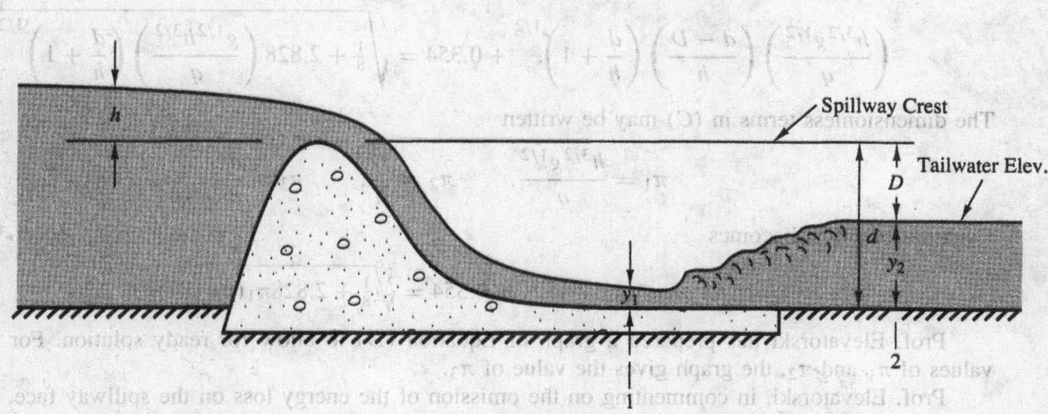

Fig. 10-16

Solution:

The energy equation is applied between a section upstream of the dam where h is measured and section 1, neglecting the velocity head of approach, i.e.,

$$(h + d) + 0 + \text{negl.} \; - \; \text{losses (neglected)} \; = 0 + 0 + V_1^2/2g$$

or $V_1 = \sqrt{2g(h + d)}$.

Since $q = y_1 V_1$,

$$y_1 = \frac{q}{V_1} = \frac{q}{\sqrt{2g(d + h)}}$$

or

$$y_1 = \frac{q}{\sqrt{2g} \;\; (d/h + 1)^{1/2} h^{1/2}} \qquad (A)$$

From Problem 10.54, the hydraulic jump relationship is

$$\frac{y_2^2 - y_1^2}{2} = \frac{qV_1}{g}\left(\frac{y_2 - y_1}{y_2}\right) \quad \text{or} \quad gy_2^2 + gy_1 y_2 = 2qV_1$$

Solving,

$$y_2 = \frac{-y_1 \pm \sqrt{y_1^2 + 8qV_1/g}}{2}$$

Dividing by y_1 produces a dimensionless form

$$\frac{y_2}{y_1} = -\tfrac{1}{2} \pm \tfrac{1}{2}\sqrt{1 + 8qV_1/y_1^2 g} = \tfrac{1}{2}\left[\sqrt{1 + 8q^2/gy_1^3} - 1\right] \qquad (B)$$

Since $y_2 = (d - D)$, $y_2/y_1 = (d - D)/y_1$ is substituted in (B) together with the value of y_1 from (A) to yield

$$\frac{d - D}{y_1} = \frac{1}{2}\left[\sqrt{1 + 8q^2/gy_1^3} - 1\right]$$

$$\frac{2(d - D)\sqrt{2g}\,(d/h + 1)^{1/2}h^{1/2}}{q} + 1 = \sqrt{1 + \frac{8\,(2^{3/2})\,(g^{3/2})\,(d/h + 1)^{3/2}h^{3/2}}{qg}}$$

The equation is put in dimensionless form by multiplying the left-hand side by h/h, dividing through by $\sqrt{8}$, and collecting terms:

$$\left(\frac{h^{3/2}g^{1/2}}{q}\right)\left(\frac{d - D}{h}\right)\left(\frac{d}{h} + 1\right)^{1/2} + 0.354 = \sqrt{\frac{1}{8} + 2.828\left(\frac{g^{1/2}h^{3/2}}{q}\right)\left(\frac{d}{h} + 1\right)^{3/2}} \qquad (C)$$

The dimensionless terms in (C) may be written

$$\pi_1 = \frac{h^{3/2}g^{1/2}}{q}, \qquad \pi_2 = \frac{D}{h}, \qquad \pi_3 = \frac{d}{h}$$

Equation (C) then becomes

$$\pi_1(\pi_3 - \pi_2)(\pi_3 + 1)^{1/2} + 0.354 = \sqrt{\tfrac{1}{8} + 2.828\pi_1(\pi_3 + 1)^{3/2}} \qquad (D)$$

Prof. Elevatorski has prepared a graph of Equation (D) to allow for ready solution. For calculated values of π_1 and π_2, the graph gives the value of π_3.

Prof. Elevatorski, in commenting on the omission of the energy loss on the spillway face, states "by neglecting the loss due to friction, a slight excess of tailwater will be produced in the stilling basin. A slightly-drowned jump produces a better all-around energy dissipator as compared to one designed for the y_2 depth."

10.60. Determine the elevation of the spillway apron if $q = 50$ cfs/ft, $h = 9$ ft, $D = 63$ ft, and the spillway crest is at elevation 200.0 ft.

Solution:

Employing the dimensionless ratios derived in Problem 10.59,

$$\pi_1 = g^{1/2}h^{3/2}/q = (5.67)\,(9^{3/2})\,/50 = 3.06, \qquad \pi_2 = D/h = 63/9 = 7.00, \qquad \pi_3 = d/h = d/9$$

Equation (D) of Problem 10.59 can then be written

$$(3.06)(d/9 - 7.00)(d/9 + 1)^{1/2} + 0.354 = \sqrt{0.125 + (2.828)(3.06)(d/9 + 1)^{3/2}}$$

Solving by successive trials, $d = 77.9$ ft. The elevation of the spillway apron is $(200 - 77.9) = 122.1$ ft above datum.

Supplementary Problems

10.61. Using y_N as the depth in Fig. 10-3, derive an expression for laminar flow along a flat plate of infinite width, considering the free body in Fig. 10-3 to be one unit wide. *Ans.* $y_N^2 = 3\nu V/gS$

10.62. The Darcy friction factor f is usually associated with pipes. However, for Problem 10.61, evaluate Darcy's factor f using the answer given for that problem. *Ans.* 96/Re

10.63. Show that mean velocity V can be expressed as $0.263v_*R^{1/6}/n$.

10.64. Show that Manning's n and Darcy's f are related to each other by $n = 0.093 f^{1/2} R^{1/6}$.

10.65. Calculate the mean velocity in the rectangular channel of Problem 10.7 by summing up the area under the depth–velocity curve. *Ans.* 6.95 ft/sec

10.66. A flow rate of 2.1 m^3/s is to be carried in an open channel at a velocity of 1.3 m/s. Determine the dimensions of the channel cross section and required slope if the cross section is (*a*) rectangular, with depth equal to one-half the width; (*b*) semicircular; and (*c*) trapezoidal, with depth equal to the width of the channel bottom and with side slopes of 1/1. Use $n = 0.020$.
Ans. 0.90 m, 1.80 m, 0.00196, 1.014 m (radius), 0.00167, 0.90 m, 0.00185

10.67. Upon what slope should the flume shown in Fig. 10-17 be laid in order to carry 522.5 cfs? ($C = 100$)
Ans. 0.00373

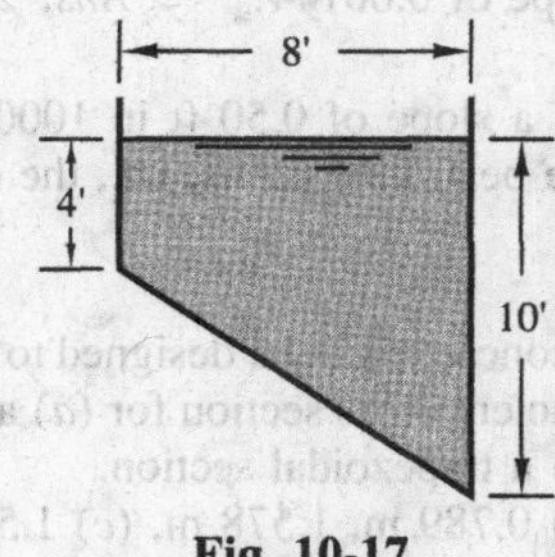

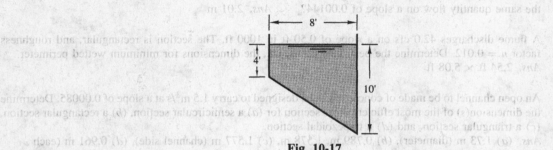

Fig. 10-17

10.68. The canal shown in Fig. 10-18 is laid on a slope of 0.00016. When it reaches a railroad embankment, the flow is to be carried by two concrete pipes ($n = 0.012$) laid on a slope of 2.5 ft in 1000 ft. What size pipes should be used? *Ans.* 4.16 ft

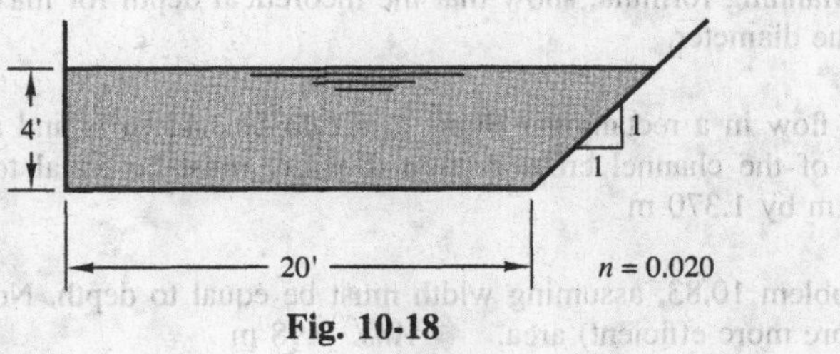

Fig. 10-18

10.69. A flow of 2.22 m^3/s is carried in a flume that is a half-square. The flume is 1220 m long and drops 0.610 m in that length. Using Manning's formula and $n = 0.012$, determine the dimensions.
Ans. 1.95 m × 0.975 m

10.70. Water flows at a depth of 1.90 m in a rectangular canal 2.44 m wide. The average velocity is 0.579 m/s. What is the probable slope on which the canal is laid if $C = 100$? *Ans.* 0.000148

10.71. A canal cut in rock ($n = 0.030$) is trapezoidal in section with a bottom width of 6.10 m and side slopes of 1 on 1. The allowable average velocity is 0.76 m/s. What slope will produce 5.66 m^3/s? *Ans.* 0.000675

10.72. What is the flow of water in a new 24″ vitrified sewer pipe flowing half-full on a slope of 0.0025?
Ans. 5.66 cfs

10.73. A canal ($n = 0.017$) has a slope of 0.00040 and is 10,000 ft long. Assuming the hydraulic radius is 4.80 ft, what correction in grade must be made to produce the same flow if the roughness factor changes to 0.020? *Ans.* new $S = 0.000552$

10.74. How deep will water flow in a 90° V-shaped flume, $n = 0.013$, laid on a grade of 0.00040 if it is to carry 2.55 m^3/s? *Ans.* $y = 1.57$ m

10.75. Water flows in a 60° V-shaped steel flume at a velocity of 1.2 m/s. If the channel slope is 0.0020, determine the depth of flow. *Ans.* 0.921 m

10.76. A given amount of lumber is to be used to build a V-notch triangular flume. What vertex angle should be used for maximum flow on a given slope? *Ans.* 90°

10.77. Water flows 0.914 m deep in a rectangular canal 6.10 m wide, $n = 0.013$, $S = 0.0144$. How deep would the same quantity flow on a slope of 0.00144? *Ans.* 2.01 m

10.78. A flume discharges 42.0 cfs on a slope of 0.50 ft in 1000 ft. The section is rectangular, and roughness factor $n = 0.012$. Determine the best dimensions, i.e., the dimensions for minimum wetted perimeter. *Ans.* 2.54 ft × 5.08 ft

10.79. An open channel to be made of concrete is to be designed to carry 1.5 m^3/s at a slope of 0.00085. Determine the dimension(s) of the most efficient cross section for (*a*) a semicircular section, (*b*) a rectangular section, (*c*) a triangular section, and (*d*) a trapezoidal section.
Ans. (*a*) 1.73 m (diameter), (*b*) 0.789 m, 1.578 m, (*c*) 1.577 m (channel side), (*d*) 0.961 m (each channel side and bottom)

10.80. A lined rectangular canal 16 ft wide carries a flow of 408 cfs at a depth of 2.83 ft. Find n if the slope of the canal is 3.22 ft in 1600 ft. (Use Manning's formula.) *Ans.* $n = 0.0121$

10.81. Find the average shear stress over the wetted perimeter in Problem 10.80. *Ans.* 0.263 lb/ft^2

10.82. Using the Manning formula, show that the theoretical depth for maximum velocity in a circular conduit is 0.81 of the diameter.

10.83. Water is to flow in a rectangular flume at a rate of 1.42 m^3/s and at a slope of 0.0028. Determine the dimensions of the channel cross section if width must be equal to twice the depth. Use $n = 0.017$. *Ans.* 0.685 m by 1.370 m

10.84. Rework Problem 10.83, assuming width must be equal to depth. Note which solution gives the smaller (and therefore more efficient) area. *Ans.* 0.98 m

10.85. A rectangular channel ($n = 0.011$) 18 m wide is to carry water at a flow rate of 35 m^3/s. The slope of the channel is 0.00078. Determine the depth of flow. *Ans.* 0.885 m

10.86. Design the most efficient trapezoidal channel to carry 600 cfs at a maximum velocity of 3.00 ft/sec. Use $n = 0.025$ and side slopes of 1 vertical on 2 horizontal. *Ans.* $b = 4.22$ ft, $y = 9$ ft

10.87. Calculate the slope of the channel in Problem 10.86. *Ans.* 0.000345

10.88. Which canal structure shown in Fig. 10-19 will carry the greater flow if both are laid on the same slope? *Ans.* (*b*) trapezoidal section

10.89. A square box culvert is 8 ft on a side and is installed as shown in Fig. 10-20. What is the hydraulic radius for a depth of 7.66 ft? *Ans.* 2.33 ft

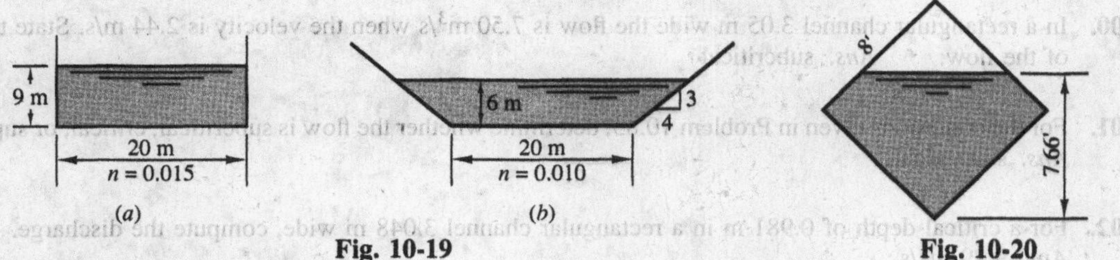

Fig. 10-19 **Fig. 10-20**

10.90. What is the radius of semicircular flume B, shown in Fig. 10-21, if its slope $S = 0.0200$ and $C = 90$?
Ans. $r = 1.80$ ft

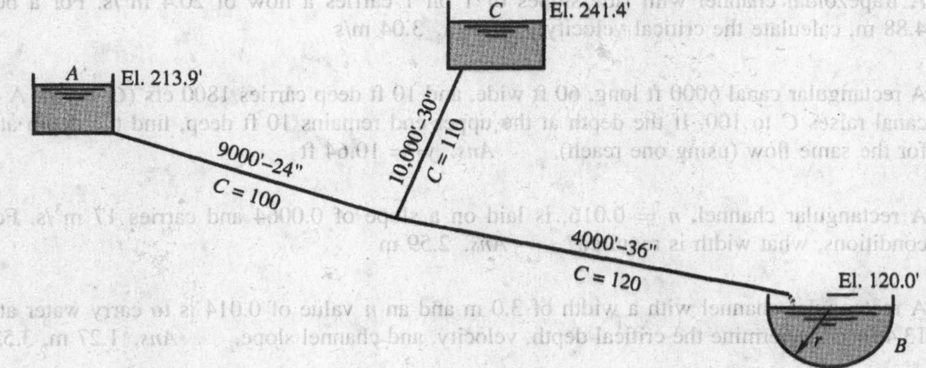

Fig. 10-21

10.91. A concrete pipe of 1.0-m diameter flows half-full at a slope of 0.0012. Determine the flow rate of water in the pipe. *Ans.* 0.415 m^3/s

10.92. A sewer pipe for which $n = 0.014$ is laid on a slope of 0.00018 and is to carry 2.76 m^3/s when the pipe flows at 80% of full depth. Determine the required diameter of pipe. Do not use Fig. 10-1. *Ans.* 2.32 m

10.93. Rework Problem 10.92, using Fig. 10-1. *Ans.* 2.34 m

10.94. A 1.0-m-diameter pipe must carry water at a discharge rate of 0.40 m^3/s and a velocity of 0.80 m/s. Determine the slope and the depth of flow. *Ans.* 0.00059, 0.63 m

10.95. Calculate the specific energy when a flow of 220 cfs is produced in a rectangular channel 10 ft wide at a depth of 3.0 ft. *Ans.* 3.84 ft

10.96. Calculate the specific energy when 8.78 m^3/s flows in a trapezoidal channel, base width 2.44 m with side slopes 1 on 1, at a depth of 1.19 m. *Ans.* 1.40 m

10.97. A sewer pipe 6 ft in diameter carries 80.6 cfs when flowing 4 ft deep. What is the specific energy? *Ans.* 4.25 ft

10.98. At what depths may the water flow at 220 cfs in Problem 10.95 for a specific energy of 5.00 ft-lb/lb? What is the critical depth? *Ans.* 1.46 ft and 4.65 ft, 2.47 ft

10.99. In a rectangular channel 3.05 m wide the flow is 7.50 m^3/s. Is the flow subcritical or supercritical at depths of 0.610 m, 0.914 m, and 1.219 m? *Ans.* supercritical, subcritical, subcritical

10.100. In a rectangular channel 3.05 m wide the flow is 7.50 m^3/s when the velocity is 2.44 m/s. State the nature of the flow. *Ans.* subcritical

10.101. For the conditions given in Problem 10.83, determine whether the flow is subcritical, critical, or supercritical. *Ans.* subcritical

10.102. For a critical depth of 0.981 m in a rectangular channel 3.048 m wide, compute the discharge. *Ans.* 9.29 m^3/s

10.103. Determine the critical slope of a rectangular channel 20 ft wide, $n = 0.012$, when the flow is 990 cfs. *Ans.* 0.00207

10.104. A trapezoidal channel with side slopes of 1 on 1 carries a flow of 20.4 m^3/s. For a bottom width of 4.88 m, calculate the critical velocity. *Ans.* 3.04 m/s

10.105. A rectangular canal 6000 ft long, 60 ft wide, and 10 ft deep carries 1800 cfs ($C = 75$). A cleaning of the canal raises C to 100. If the depth at the upper end remains 10 ft deep, find the depth at the lower end for the same flow (using one reach). *Ans.* $y_2 = 10.64$ ft

10.106. A rectangular channel, $n = 0.016$, is laid on a slope of 0.0064 and carries 17 m^3/s. For critical flow conditions, what width is required? *Ans.* 2.59 m

10.107. A rectangular channel with a width of 3.0 m and an n value of 0.014 is to carry water at a flow rate of 13.4 m^3/s. Determine the critical depth, velocity, and channel slope. *Ans.* 1.27 m, 3.52 m/s, 0.00400

10.108. A rectangular channel, $n = 0.012$, 3.05 m wide laid on a slope of 0.0049, carries 13.6 m^3/s. The channel is to be contracted to produce critical flow. What width of contracted section will accomplish this, neglecting any loss in the gradual reduction in width? *Ans.* 1.37 m

10.109. In a rectangular channel 12 ft wide, $C = 100$, $S = 0.0225$, the flow is 500 cfs. The slope of the channel changes to 0.00250. How far below the point of change in slope is the depth 2.75 ft, using one reach? *Ans.* 104 ft

10.110. A rectangular channel 12.0 m wide is laid on a slope of 0.0028. The depth of flow at one section is 1.5 m, while the depth of flow at another section 500 m downstream is 1.80 m. Determine the probable rate of flow if $n = 0.026$. *Ans.* 44.1 m^3/s

10.111. Using the data in Problem 10.109, calculate (*a*) the critical depth in the flatter channel, (*b*) the depth required for uniform flow in the flatter channel, (*c*) the depth just before the hydraulic jump occurs, using equation from Problem 10.54. (Note that this depth occurs 104 ft from change in grade, from Problem 10.109.)
Ans. (*a*) 3.78 ft, (*b*) 5.05 ft, (*c*) 2.75 ft

10.112. Show that the critical depth in a triangular channel is $2V_c^2/g$.

10.113. Show that the critical depth in a triangular channel can be expressed as 4/5 of the minimum specific energy.

10.114. Show that the critical depth in a parabolic channel is 3/4 of the minimum specific energy if the channel is y_c deep and b' wide at the water surface.

10.115. For a rectangular channel, show that the discharge q per foot of width equals $3.087E_{\min}^{3/2}$.

10.116. For a triangular channel, show that discharge $Q = 1.148(b'/y_c)E_{\min}^{5/2}$.

10.117. For a parabolic channel, show that discharge $Q = 2.005b'E_{\min}^{3/2}$.

10.118. Water flows at a rate of 20.0 m^3/s through a rectangular channel 4.0 m wide from a "steep slope" to a "mild slope," creating a hydraulic jump. The upstream depth of flow is 1.20 m. Determine (*a*) the downstream depth of flow, (*b*) the energy (head) loss in the jump, and (*c*) the upstream and downstream velocities.
Ans. (*a*) 1.55 m, (*b*) 0.006 m of water, (*c*) 4.17 and 3.23 m/s

Chapter 11

Flow of Compressible Fluids

INTRODUCTION

In previous chapters most flow problems considered fluids that are virtually incompressible, primarily water. Many flow problems, however, involve fluids that are compressible, such as air. As a general rule, gases are compressible whereas liquids are more or less incompressible. Analysis of compressible flow is often more complicated than that of incompressible flow because mass density varies with applied pressure in the case of compressible flow.

ISOTHERMAL FLOW

Isothermal means at the same (or constant) temperature. Isothermal flow of a compressible fluid in a conduit occurs if heat transferred out of the fluid (through the conduit walls) and energy converted to heat by friction offset each other so that the temperature of the fluid is constant. This condition can occur in an uninsulated conduit with the same temperature inside and outside the conduit and with low velocities of fluid flowing in the conduit. Compressible fluid flow in long conduits can often be analyzed as isothermal flow.

Isothermal flow in a pipe can be analyzed by using the following formula:

$$p_1^2 - p_2^2 = \frac{G^2RT}{gA^2}\left[f\frac{L}{D} + 2\ln\frac{p_1}{p_2}\right] \tag{1}$$

where p_1 = pressure at point 1
p_2 = pressure at point 2
G = weight flow rate
R = gas constant
T = absolute temperature of the fluid
g = acceleration of gravity
A = cross-sectional area of the conduit
f = friction factor
L = length of segment from point 1 to point 2
D = diameter of pipe

Equation (*1*) has some limitations that need to be pointed out. First of all, the equation should not be used for large pressure drops. Also, the pipe diameter must be constant, and pressure changes due to differences in elevation are considered negligible. Finally, it is limited to ordinary pressure ranges.

ISENTROPIC FLOW

Flow that is both adiabatic and frictionless is called *isentropic flow*. (Adiabatic means that no heat is transferred to the system from its surroundings or vice versa.) Isentropic flow exhibits no change in entropy. Isentropic flow is approximated in practice if the flow occurs very quickly (so there is little chance for heat transfer) and with a small amount of friction. Hence, isentropic analysis can be applied to high-velocity gas flow over short distances wherein friction and heat transfer should be relatively small.

Isentropic flow can be analyzed by using the equations

$$\frac{v_2^2 - v_1^2}{2g} = \frac{p_1}{\gamma_1}\left(\frac{k}{k-1}\right)\left[1 - \left(\frac{p_2}{p_1}\right)^{(k-1)/k}\right] \tag{2}$$

or

$$\frac{v_2^2 - v_1^2}{2g} = \frac{p_2}{\gamma_2}\left(\frac{k}{k-1}\right)\left[\left(\frac{p_1}{p_2}\right)^{(k-1)/k} - 1\right] \tag{3}$$

where v_2 = velocity at point 2
v_1 = velocity at point 1
g = acceleration of gravity
p_2 = pressure at point 2
p_1 = pressure at point 1
k = specific heat ratio
γ_1 = specific weight at point 1
γ_2 = specific weight at point 2

THE CONVERGENT NOZZLE

Isentropic flow of compressible fluid from a large tank through a convergent nozzle, as shown in Fig. 11-1 is discussed in this section. Indicated on Fig. 11-1 are the pressure, mass density, and temperature (p_1, ρ_1, and T_1) at a point within the tank. Since the tank is "large," velocity here is assumed to be near zero. Also indicated on Fig. 11-1 are the same parameters as well as the velocity of flow and area of the nozzle (p_2, ρ_2, T_2, v_2, and A_2) at the exit of the nozzle. Also indicated is p_2', the pressure outside the tank.

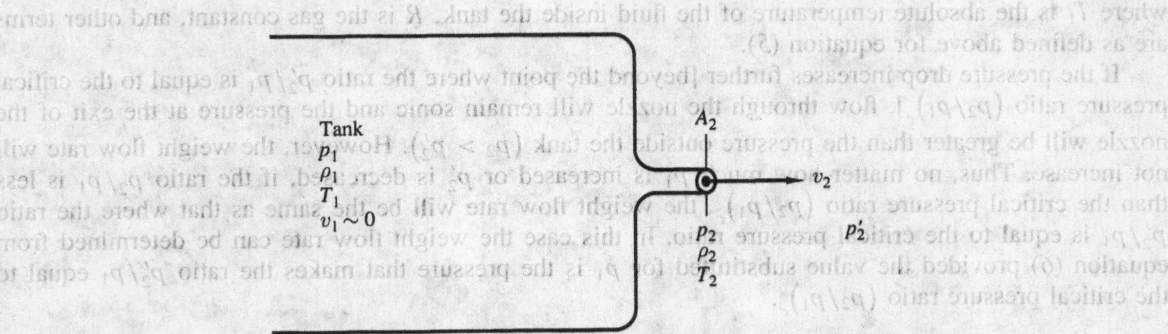

Fig. 11-1 Convergent nozzle.

In a convergent nozzle, flow through the nozzle's throat will be either sonic or subsonic. If flow is sonic, the Mach number is equal to unity, and the ratio p_2/p_1 must be equal to the "critical pressure ratio" as defined by

$$\left(\frac{p_2}{p_1}\right)_c = \left(\frac{2}{k+1}\right)^{k/(k-1)} \tag{4}$$

where $(p_2/p_1)_c$ = critical pressure ratio
k = specific heat ratio

If flow through the throat is subsonic, the ratio p_2/p_1 will be larger than $(p_2/p_1)_c$.

Obviously, in order to have appreciable flow from the tank through the nozzle out of the tank, pressure inside the tank must be greater than pressure outside the tank (that is, $p_1 > p_2'$). If the pressure drop is small $[(p_2'/p_1) > (p_2/p_1)_c]$, flow through the nozzle will be subsonic and the pressure at the exit of the nozzle will be the same as the pressure outside the tank $(p_2 = p_2')$. In this case, the weight flow rate can be determined from the equation

$$G = A_2\sqrt{\frac{2gk}{k-1}p_1\gamma_1\left[\left(\frac{p_2}{p_1}\right)^{2/k} - \left(\frac{p_2}{p_1}\right)^{(k+1)/k}\right]} \tag{5}$$

where G = weight flow rate
A_2 = throat area
g = acceleration of gravity
k = specific heat ratio
p_1 = pressure inside the tank
γ_1 = specific weight of fluid inside the tank
p_2 = pressure of the jet at the exit of the nozzle

If the pressure drop increases (either by increasing p_1 or decreasing p_2', or both), flow through the nozzle will remain subsonic until the point is reached where the ratio p_2'/p_1 is equal to the critical pressure ratio $(p_2/p_1)_c$. At this point, flow through the nozzle will be sonic and the pressure at the exit of the nozzle will be the same as the pressure outside the tank $(p_2 = p_2')$. In this case, the weight flow rate can be determined from the equation

$$G = \frac{A_2 p_1}{\sqrt{T_1}}\sqrt{\frac{gk}{R}\left(\frac{2}{k+1}\right)^{(k+1)/(k-1)}} \tag{6}$$

where T_1 is the absolute temperature of the fluid inside the tank, R is the gas constant, and other terms are as defined above for equation (5).

If the pressure drop increases further [beyond the point where the ratio p_2'/p_1 is equal to the critical pressure ratio $(p_2/p_1)_c$], flow through the nozzle will remain sonic and the pressure at the exit of the nozzle will be greater than the pressure outside the tank $(p_2 > p_2')$. However, the weight flow rate will not increase. Thus, no matter how much p_1 is increased or p_2' is decreased, if the ratio p_2'/p_1 is less than the critical pressure ratio $(p_2/p_1)_c$, the weight flow rate will be the same as that where the ratio p_2'/p_1 is equal to the critical pressure ratio. In this case the weight flow rate can be determined from equation (6) provided the value substituted for p_1 is the pressure that makes the ratio p_2'/p_1 equal to the critical pressure ratio $(p_2/p_1)_c$.

COMPRESSIBLE FLOW THROUGH A CONSTRICTION

Figure 11-2 depicts compressible fluid flow through a constriction in a conduit. The various parameters are shown at point 1 in the larger section and point 2 in the smaller section. This configuration is similar to the convergent nozzle (Fig. 11-1), but it differs because the cross section at point 1 is generally not large enough to neglect velocity v_1.

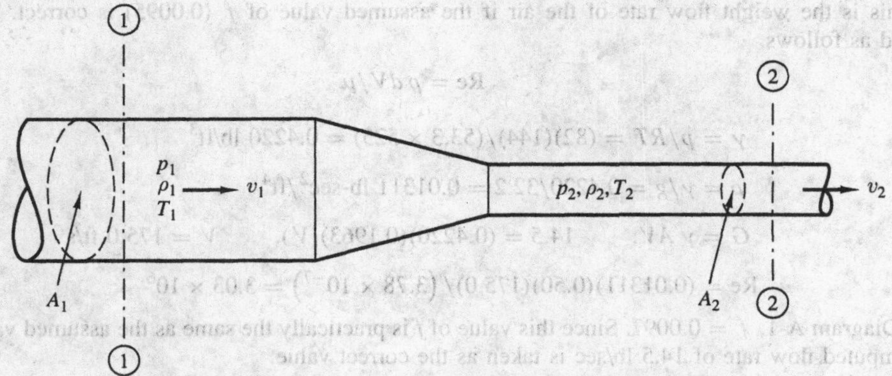

Fig. 11-2 Compressible flow through a constriction.

The weight flow rate for compressible flow through a constriction can be determined from the equation

$$G = \frac{A_2}{\sqrt{1-(p_2/p_1)^{2/k}(A_2/A_1)^2}}\sqrt{\frac{2gk}{k-1}p_1\,\gamma_1\left[\left(\frac{p_2}{p_1}\right)^{2/k}-\left(\frac{p_2}{p_1}\right)^{(k+1)/k}\right]} \tag{7}$$

where the terms are similar to those of equation (5). (Also, refer to Fig. 11-2.) This equation is similar to the one for convergent nozzles [equation (5)]; and, in fact, as A_1 gets larger compared to A_2, it approaches equation (5).

Solved Problems

11.1. Air at 65°F flows isothermally through a 6-in-diameter pipe. The pressure at one section is 82 psia, and that at a section 550 ft downstream is 65 psia. The pipe surface is "smooth." Find the weight flow rate of air.

Solution:

$$p_1^2 - p_2^2 = \frac{G^2RT}{gA^2}\left[f\frac{L}{D} + 2\ln\frac{p_1}{p_2}\right]$$

From Table 1 in the Appendix, $R = 53.3$ ft/°R.

$$A = (\pi)(0.5)^2/4 = 0.1963 \text{ ft}^2$$

To determine the value of f, the Reynolds number must first be evaluated. However, to evaluate it, the velocity must be known. Since it is not known and cannot be determined until the flow rate is known, a value of $f = 0.0095$ will be assumed initially. Therefore

$$(82^2 - 65^2)(144)^2 = \frac{(G^2)(53.3)(525)}{(32.2)(0.1963)^2}\left[(0.0095)\left(\frac{550}{0.50}\right) + (2)\left(\ln\frac{82}{65}\right)\right]$$

$$G = 14.5 \text{ lb/sec}$$

This is the weight flow rate of the air if the assumed value of f (0.0095) is correct. This must be checked as follows.

$$\text{Re} = \rho d V/\mu$$

$$\gamma = p/RT = (82)(144)/(53.3 \times 525) = 0.4220 \text{ lb/ft}^3$$

$$\rho = \gamma/g = 0.4220/32.2 = 0.01311 \text{ lb-sec}^2/\text{ft}^4$$

$$G = \gamma AV, \qquad 14.5 = (0.4220)(0.1963)(V), \qquad V = 175.0 \text{ ft/s}$$

$$\text{Re} = (0.01311)(0.50)(175.0)/\left(3.78 \times 10^{-7}\right) = 3.03 \times 10^6$$

From Diagram A-1, $f = 0.0097$. Since this value of f is practically the same as the assumed value (0.0095), the computed flow rate of 14.5 lb/sec is taken as the correct value.

11.2. Air at 18°C flows isothermally through a 300-mm-diameter pipe at a flow rate of 0.450 kN/s. The pipe is "smooth." If the pressure at one section is 550 kPa, find the pressure at a section 200 m downstream.

Solution:

$$p_1^2 - p_2^2 = \frac{G^2RT}{gA^2}\left[f\frac{L}{D} + 2\ln\frac{p_1}{p_2}\right]$$

From Table 1 in the Appendix, $R = 29.3$ m/K.

$$A = (\pi)(0.30)^2/4 = 0.07069 \text{ m}^2$$

To determine the value of f, the Reynolds number must first be evaluated. However, to evaluate it, values of mass density and velocity must be determined, as follows.

$$\gamma = p/RT = 550/[(29.3)(291)] = 0.06451 \text{ kN/m}^3$$

$$\gamma = \rho g, \qquad 0.06451 = (\rho)(9.807), \qquad \rho = 0.006578 \text{ kN}\cdot\text{s}^2/\text{m}^4$$

$$G = \gamma AV, \qquad 0.450 = (0.06451)(0.07069)(V), \qquad V = 98.68 \text{ m/s}$$

$$\text{Re} = \rho DV/\mu = (0.006578)(0.30)(98.68)/\left(1.81 \times 10^{-8}\right) = 1.08 \times 10^7$$

From Diagram A-1, using the "smooth pipe" line, $f = 0.0080$. Substituting into equation (*1*) and neglecting temporarily the second term inside the brackets,

$$550^2 - p_2^2 = \frac{(0.450)^2(29.3)(291)}{(9.807)(0.07069)^2}[(0.0080)(200/0.30)], \qquad p_2 = 339 \text{ kPa}$$

Substituting this value of p_2 into the term that was neglected and solving for the p_2 on the left-hand side of the equation,

$$550^2 - p_2^2 = \frac{(0.450)^2(29.3)(291)}{(9.807)(0.07069)^2} \times \left[\frac{(0.0080)(200)}{(0.30)} + (2)\left(\ln\frac{550}{339}\right)\right], \qquad p_2 = 284 \text{ kPa}$$

Again, substituting this value of p_2 into the term that was originally neglected and solving for the p_2 on the left-hand side of the equation yields a value of $p_2 = 261$ kPa. Several more similar iterations (not shown) yield a final value of 233 kPa for p_2.

11.3. Isentropic flow of nitrogen occurs in a 2-in-diameter pipe. At one point the velocity of flow, pressure, and unit weight are 409 ft/sec, 85 psia, and 0.655 lb/ft³, respectively. Find the velocity at a second point a short distance away, where the pressure is 83 psia.

Mathcad

Solution:

$$\frac{v_2^2 - v_1^2}{2g} = \frac{p_1}{\gamma_1}\left(\frac{k}{k-1}\right)\left[1 - \left(\frac{p_2}{p_1}\right)^{(k-1)/k}\right]$$

From Table 1 in the Appendix, $k = 1.40$.

$$\frac{v_2^2 - 409^2}{(2)(32.2)} = [(85)(144)/0.655]\left(\frac{1.40}{1.40-1}\right)\left[1-(83/85)^{(1.40-1)/1.40}\right]$$

$$v_2 = 443 \text{ ft/sec}$$

11.4. At one point on a streamline in an isentropic airflow, the velocity, pressure, and unit weight are 30.5 m/s, 350 kPa (absolute), and 0.028 kN/m^3, respectively. Find the pressure at a second point on the streamline, where the velocity is 150 m/s.

Solution:

$$\frac{v_2^2 - v_1^2}{2g} = \frac{p_1}{\gamma_1}\left(\frac{k}{k-1}\right)\left[1-\left(\frac{p_2}{p_1}\right)^{(k-1)/k}\right]$$

From Table 1 in the Appendix, $k = 1.40$.

$$\frac{150^2 - 30.5^2}{(2)(9.807)} = (350/0.028)\left(\frac{1.40}{1.40-1}\right)\left[1-(p_2/350)^{(1.40-1)/1.40}\right]$$

$$p_2 = 320 \text{ kPa}$$

11.5. Air at 28°C flows from a large tank through a convergent nozzle that has an exit diameter of 10 mm. Discharge is to the atmosphere, where atmospheric pressure is 96.5 kPa. Air pressure inside the tank is 40.0 kPa (gage). What is the weight flow rate through the nozzle?

Solution:

$$\left(\frac{p_2}{p_1}\right)_c = \left(\frac{2}{k+1}\right)^{k/(k-1)}$$

$$\left(\frac{p_2}{p_1}\right)_c = \left(\frac{2}{1.40+1}\right)^{1.40/(1.40-1)} = 0.5283, \qquad \frac{p_2'}{p_1} = \frac{96.5}{40+96.5} = 0.7070$$

Since the value of the ratio p_2'/p_1, 0.7070, is greater than the value of the critical pressure ratio, 0.5283, flow through the nozzle will be subsonic and equation (5) applies.

$$G = A_2\sqrt{\frac{2gk}{k-1}p_1\gamma_1\left[\left(\frac{p_2}{p_1}\right)^{2/k} - \left(\frac{p_2}{p_1}\right)^{(k+1)/k}\right]}$$

$$A_2 = (\pi)(0.010)^2/4 = 0.00007854 \text{ m}^2$$

$$\gamma = p/RT = 136.5/[(29.3)(301)] = 0.01548 \text{ kN/m}^3$$

$$G = (0.00007854) \times \sqrt{\left(\frac{(2)(9.807)(1.40)}{1.40-1}\right)(136.5)(0.01548)\sqrt{\left(\frac{96.5}{136.5}\right)^{2/1.40} - \left(\frac{96.5}{136.5}\right)^{(1.40+1)/1.40}}}$$

$$G = 0.000227 \text{ kN/s}, \qquad \text{or } 0.227 \text{ N/s}$$

11.6. Air at 75°F flows from a large tank through a convergent nozzle that has an exit diameter of 1.5 in. The discharge is to the atmosphere, where atmospheric pressure is 14.0 psia. The air pressure inside the tank is 25.0 psig. Find the weight flow rate of air from the nozzle.

Solution:

$$\left(\frac{p_2}{p_1}\right)_c = \left(\frac{2}{k+1}\right)^{k/(k-1)}, \qquad \left(\frac{p_2}{p_1}\right)_c = \left(\frac{2}{1.40+1}\right)^{1.40/(1.40-1)} = 0.5283$$

$$p_2'/p_1 = 14.0/(14.0 + 25.0) = 0.3590$$

Since the value of the ratio p_2'/p_1 of 0.3590 is less than the value of the critical pressure ratio of 0.5283, flow through the nozzle will be sonic, and equation (*6*) applies.

$$G = \frac{A_2 p_1}{\sqrt{T_1}} \sqrt{\frac{gk}{R}\left(\frac{2}{k+1}\right)^{(k+1)/(k-1)}}$$

$$A_2 = (\pi)(1.5)^2/4 = 1.767 \text{ in}^2, \text{ or } 0.01227 \text{ ft}^2$$

$$p_1 = p_2'/(p_2/p_1)_c = 14.0/0.5283 = 26.5 \text{ psia}$$

$$G = \frac{(0.01227)(26.5 \times 144)}{\sqrt{535}} \sqrt{(32.2)\left(\frac{1.40}{53.3}\right)\left(\frac{2}{1.40+1}\right)^{(1.40+1)/(1.40-1)}} = 1.08 \text{ lb/sec}$$

11.7. Air flows through a 2-in constriction in a 3-in-diameter pipeline. Pressure and temperature of the air in the pipeline are 108 psig and 105°F, respectively, and the pressure in the constriction is 81 psig. Barometric pressure is 14.5 psia. What is the weight flow rate of the air in the pipeline?

Solution:

$$G = \frac{A_2}{\sqrt{1-(p_2/p_1)^{2/k}(A_2/A_1)^2}} \sqrt{\frac{2gk}{k-1} p_1 \gamma_1 \left[\left(\frac{p_2}{p_1}\right)^{2/k} - \left(\frac{p_2}{p_1}\right)^{(k+1)/k}\right]}$$

$$A_2 = (\pi)(2)^2/4 = 3.142 \text{ in}^2, \text{ or } 0.02182 \text{ ft}^2$$

$$A_1 = (\pi)(3)^2/4 = 7.069 \text{ in}^2, \text{ or } 0.04909 \text{ ft}^2$$

$$\gamma = p/RT = (122.5 \times 144)/(53.3 \times 565) = 0.5858 \text{ lb/ft}^3$$

$$G = \frac{0.02182}{\sqrt{1-(95.5/122.5)^{2/1.40}(0.02182/0.04909)^2}} \sqrt{\left[\frac{(2)(32.2)(1.40)}{(1.40-1)}\right](122.5 \times 144)(0.5858)}$$

$$\times \sqrt{\left[\left(\frac{95.5}{122.5}\right)^{2/1.40} - \left(\frac{95.5}{122.5}\right)^{(1.40+1)/1.40}\right]} = 7.87 \text{ lb/sec}$$

Supplementary Problems

11.8. Air at 100°F flows isothermally through a 4-in-diameter pipe. Pressures at sections 1 and 2 are 120 and 80 psia, respectively. Section 2 is located 400 ft downstream from section 1. Determine the weight flow rate of the air. Assume the pipe has a "smooth" surface. *Ans.* 10.56 lb/sec

11.9. Air at 85°F flows isothermally through a 6-in-diameter pipe at a flow rate of 10 lb/sec. The pipe surface is very smooth. If the pressure at one section is 70 psia, determine the pressure at a section 600 ft downstream from the first section. *Ans.* 58.9 lb/in^2

11.10. Oxygen flows isentropically through a 100-mm-diameter pipe. At one point, the velocity of flow, pressure, and unit weight are 125 m/s, 450 kPa, and 0.058 kN/m^3, respectively. At another point a short distance away from the first, the pressure is 360 kPa. Determine the velocity at the second point. *Ans.* 220 m/s

11.11. At one point in a streamline of isentropic airflow, the velocity, pressure, and unit weight are 80 m/s, 405 kPa, and 0.046 kN/m^3, respectively. The velocity of flow at a second point in the streamline is 165 m/s. Determine the pressure at the second point. *Ans.* 358 kPa

11.12. Air at 30°C flows from a large tank through a convergent nozzle that has an exit diameter of 20 mm. Discharge of air is to the atmosphere where atmospheric pressure is 954.0 kPa (absolute). Air pressure inside the tank is 50 kPa (gage). Determine the weight flow rate of air through the nozzle. *Ans.* 1.00 N/s

11.13. Air at 60°F flows from a large tank through a convergent nozzle that has an exit diameter of 1.0 in. Discharge of air is to the atmosphere where atmospheric pressure is 14.5 psia. Air pressure inside the tank is 30.0 psig. Determine the weight flow rate of the air through the nozzle. *Ans.* 0.503 lb/sec

11.14. Air flows through a 1-in-diameter constriction in a 2-in-diameter pipeline. The pressure and temperature of air in the pipeline are 100 psig and 102°F, respectively. Barometric pressure is 14.7 psia. Calculate the weight flow rate of the air in the pipeline if the pressure in the constriction is 78 psig. *Ans.* 1.67 lb/sec

Chapter 12

Measurement of Flow of Fluids

INTRODUCTION

Numerous devices are used in engineering practice to measure the flow of fluids. Velocity measurements are made with Pitot tubes, current meters, and rotating and hot-wire anemometers. In model studies, photographic methods are often used. Quantity measurements are accomplished by means of orifices, tubes, nozzles, Venturi meters and flumes, elbow meters, weirs, numerous modifications of the foregoing, and various patented meters. In order to apply the hydraulic devices intelligently, use of the Bernoulli equation and additional knowledge of the characteristics and coefficients of each device are imperative. In the absence of reliable values of coefficients, a device should be calibrated for the expected operating conditions.

Formulas developed for incompressible fluids may be used for compressible fluids where the pressure differential is small relative to the total pressure. In many practical cases such small differentials occur. However, where compressibility must be considered, special formulas are required.

PITOT TUBE

The pitot tube measures the velocity at a point by virtue of the fact that the tube measures the stagnation pressure, which exceeds the local static pressure by $\gamma\left(V^2/2g\right)$. In an open stream of fluid, since the local pressure is zero gage, the height to which the liquid rises in the tube measures the velocity head. Problems 12.1 and 12.5 develop expressions for the flow of incompressible and compressible fluids, respectively.

COEFFICIENT OF VELOCITY

The coefficient of velocity (c_v) is the ratio of the actual mean velocity in the cross section of a stream (jet) to the theoretical mean velocity that would occur without friction. Thus

$$c_v = \frac{\text{actual mean velocity}}{\text{theoretical mean velocity}} \tag{1}$$

COEFFICIENT OF CONTRACTION

The coefficient of contraction (c_c) is the ratio of the area of the contracted section of a stream (jet) to the area of the opening through which the fluid flows. Thus

$$c_c = \frac{\text{area of stream (jet)}}{\text{area of opening}} = \frac{A_{\text{jet}}}{A_o} \tag{2}$$

COEFFICIENT OF DISCHARGE

The coefficient of discharge (c) is the ratio of the actual discharge through the device to the theoretical discharge. This coefficient can be expressed as

$$c = \frac{\text{actual flow } Q}{\text{theoretical flow } Q} \tag{3}$$

More practically, when the coefficient of discharge c has been determined experimentally,

$$Q = cA\sqrt{2gH} \tag{4}$$

Where A = cross-sectional area of device
H = total head causing flow

The coefficient of discharge can also be written in terms of the coefficient of velocity and the coefficient of contraction, i.e.,

$$c = c_v \times c_c \tag{5}$$

The coefficient of discharge is not constant. For a given device, it varies with Reynolds number. In the Appendix the following information will be found:

(1) Table 7 contains coefficients of discharge for circular orifices discharging water at about 60°F into the atmosphere. Few authoritative data are available for all fluids throughout wide ranges of Reynolds number.

(2) Diagram C indicates the variation of c' with Reynolds number for three pipe orifice ratios. No authoritative data are available below a Reynolds number of about 10,000.

(3) Diagram D shows the variation of c with Reynolds number for three long-radius flow nozzle ratios (pipeline nozzles).

(4) Diagram E indicates the variation of c with Reynolds number for five sizes of Venturi meters of diameter ratios of 0.500.

LOST HEAD

The lost head in orifices, tubes, nozzles, and Venturi meters is expressed as

$$\text{lost head of fluid} = \left(\frac{1}{c_v^2} - 1\right)\frac{V_{\text{jet}}^2}{2g} \tag{6}$$

When this expression is applied to a Venturi meter, V_{jet} = throat velocity and $c_v = c$.

WEIRS

Weirs measure the flow of liquids, usually water, in open channels. A number of empirical formulas are available in engineering literature, each with its limitations. Only a few will be listed here. Most weirs are rectangular: the *suppressed* weir with no end contractions and generally used for larger flows, and the *contracted* weir for smaller flows. Other weirs are triangular, trapezoidal, parabolic, and proportional flow. For accurate results, a weir should be calibrated in place under the conditions under which it will be used.

THEORETICAL WEIR FORMULA

The theoretical weir formula for rectangular weirs, developed in Problem 12.29, is

$$Q = \frac{2}{3}cb\sqrt{2g}\left[\left(H + \frac{V^2}{2g}\right)^{3/2} - \left(\frac{V^2}{2g}\right)^{3/2}\right] \tag{7}$$

where Q = flow in cfs (or m^3/s)
c = coefficient (to be determined experimentally)
b = length of weir crest in feet (or meters)
H = head on weir in feet (or meters) (height of level liquid surface above crest)
V = average velocity of approach in ft/sec or m/s.

FRANCIS FORMULA

The Francis formula, based upon experiments on rectangular weirs from 3.5 ft (1.1 m) to 17 ft (5.2 m) long under heads from 0.6 ft (0.2 m) to 1.6 ft (0.5 m), is

$$Q = 3.33^* \left(b - \frac{nH}{10}\right)\left[\left(H + \frac{V^2}{2g}\right)^{3/2} - \left(\frac{V^2}{2g}\right)^{3/2}\right] \tag{8}$$

where the notation is the same as above and

$n = 0$ for a suppressed weir

$n = 1$ for a weir with one contraction

$n = 2$ for a fully contracted weir

BAZIN FORMULA

The Bazin formula (lengths from 1.64 ft to 6.56 ft under heads from 0.164 ft to 1.969 ft) is

$$Q = \left(3.25 + \frac{0.0789}{H}\right)\left[1 + 0.55\left(\frac{H}{H+Z}\right)^2\right] bH^{3/2} \tag{9}$$

where Z = height of the weir crest above the channel bottom.
The bracketed term becomes negligible for low velocities of approach.

FTELEY AND STEARNS FORMULA

The Fteley and Stearns formula (lengths 5 ft and 19 ft under heads from 0.07 ft to 1.63 ft) for suppressed weirs is

$$Q = 3.31b\left(H + \alpha\frac{V^2}{2g}\right)^{3/2} + 0.007b \tag{10}$$

where α = factor dependent upon crest height Z (tables of values required).

THE TRIANGULAR WEIR FORMULA (developed in Problem 12.30) is

$$Q = \frac{8}{15} c \tan\frac{\theta}{2}\sqrt{2g}\, H^{5/2} \tag{11}$$

or, for a given weir,

$$Q = mH^{5/2} \tag{12}$$

THE TRAPEZOIDAL WEIR FORMULA (of Cipolletti) is

$$Q = 3.367bH^{3/2} \tag{13}$$

This weir has side (end) slopes of 1 horizontal to 4 vertical.

FOR DAMS USED AS WEIRS the expression for approximate flow is

$$Q = mbH^{3/2} \tag{14}$$

where m = experimental factor, usually from model studies.

Nonuniform flow over broad-crested weirs is discussed in Problem 12.37.

TIME TO EMPTY TANKS by means of an orifice is (see Problem 12.40)

$$t = \frac{2A_T}{cA_o\sqrt{2g}}\left(h_1^{1/2} - h_2^{1/2}\right) \qquad \text{(constant cross section, no inflow)} \tag{15}$$

$$t = \int_{h_1}^{h_2} \frac{-A_T\,dh}{Q_{\text{out}} - Q_{\text{in}}} \qquad \text{(inflow < outflow, constant cross section)} \tag{16}$$

For a tank whose cross section is not constant, see Problem 12.43.

TIME TO EMPTY TANKS by means of weirs is calculated by using (see Problem 12.45)

$$t = \frac{2A_T}{mL}\left(H_2^{-1/2} - H_1^{-1/2}\right) \tag{17}$$

TIME TO ESTABLISH FLOW in a pipeline is (see Problem 12.47)

$$t = \frac{LV_f}{2gH}\ln\left(\frac{V_f + V}{V_f - V}\right) \tag{18}$$

Solved Problems

12.1. A Pitot tube having a coefficient of 0.98 is used to measure the velocity of water at the center of a pipe. The stagnation pressure head is 5.67 m and the static pressure head in the pipe is 4.73 m. What is the velocity?

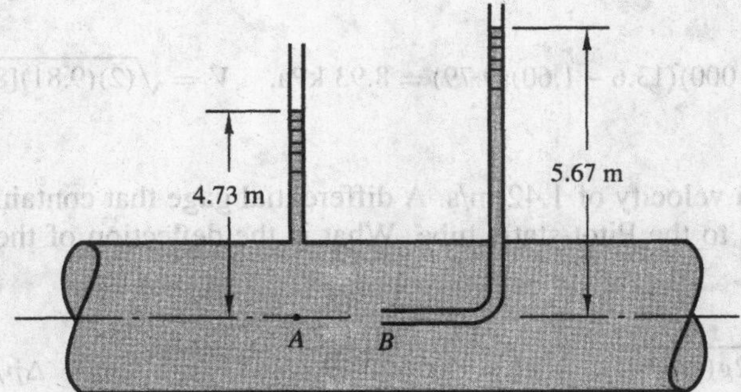

Fig. 12-1

Solution:

If the tube is shaped and positioned properly, a point of zero velocity (stagnation point) is developed at B in front of the open end of the tube (see Fig. 12-1). Applying the Bernoulli theorem from A in the undisturbed liquid to B yields

$$\left(\frac{p_A}{\gamma}+\frac{V_A^2}{2g}+0\right)-\underset{\text{(assumed)}}{\text{no loss}}=\left(\frac{p_B}{\gamma}+0+0\right) \tag{1}$$

Then, for an ideal "frictionless" fluid,

$$\frac{V_A^2}{2g}=\frac{p_B}{\gamma}-\frac{p_A}{\gamma} \quad \text{or} \quad V_A=\sqrt{2g\left(\frac{p_B}{\gamma}-\frac{p_A}{\gamma}\right)} \tag{2}$$

For the actual tube, a coefficient c, which depends upon the design of the tube, must be introduced. The actual velocity for the problem above would be

$$V_A=c\sqrt{2g(p_B/\gamma-p_A/\gamma)}=0.98\sqrt{2g(5.67-4.73)}=4.21 \text{ m/s}$$

The above equation will apply to all incompressible fluids. The value of c may be taken as unity in most engineering problems. Solving (*1*) for the stagnation pressure at B gives

$$p_B=p_A+\frac{1}{2}\rho V_A^2, \quad \text{where } \rho=\gamma/g \tag{3}$$

12.2. Air flows through a duct, and the Pitot-static tube measuring the velocity is attached to a differential gage containing water. If the deflection of the gage is 4 in, calculate the air velocity, assuming the specific weight of air is constant at 0.0761 lb/ft^3 and the coefficient of the tube is 0.98.

Mathcad

Solution:

For the differential gage, $(p_B-p_A)/\gamma=(4/12)(62.4)/0.0761=273$ ft air. Then

$$V=0.98\sqrt{(64.4)(273)}=130 \text{ ft/sec}$$

(See Problems 12.26 through 12.28 for acoustic velocity considerations.)

12.3. Carbon tetrachloride (sp gr 1.60) flows through a pipe. The differential gage attached to the Pitot-static tube shows a 76-mm deflection of mercury. Assuming $c=1.00$, find the velocity.

Solution:

$$p_B-p_A=(76/1000)(13.6-1.60)(9.79)=8.93 \text{ kPa}, \quad V=\sqrt{(2)(9.81)[8.93/(1.60\times 9.79)]}=3.34 \text{ m/s}$$

12.4. Water flows at a velocity of 1.42 m/s. A differential gage that contains a liquid of specific gravity 1.25 is attached to the Pitot-static tube. What is the deflection of the gage fluid?

Solution:

$$V=c\sqrt{2g(\Delta p/\gamma)}, \quad 1.42=1.00\sqrt{(2)(9.81)(\Delta p/\gamma)}, \quad \text{and} \quad \Delta p/\gamma=0.103 \text{ m of water}$$

Applying differential gage principles, $0.103=(1.25-1)h$ and $h=0.412$ m deflection.

12.5. Develop the expression for measuring the flow of a gas with a Pitot tube.

Solution:

The flow from A to B in Fig. 12-1 may be considered adiabatic and with negligible loss. Using the Bernoulli equation (D) in Problem 7.21 of Chapter 7, A to B, we obtain

$$\left[\left(\frac{k}{k-1}\right)\left(\frac{p_A}{\gamma_A}\right)+\frac{V_A^2}{2g}+0\right]-\text{negligible loss}=\left[\left(\frac{k}{k-1}\right)\left(\frac{p_A}{\gamma_A}\right)\left(\frac{p_B}{p_A}\right)^{(k-1)/k}+0+0\right]$$

or

$$\frac{V_A^2}{2g}=\left(\frac{k}{k-1}\right)\left(\frac{p_A}{\gamma_A}\right)\left[\left(\frac{p_B}{p_A}\right)^{(k-1)/k}-1\right] \tag{1}$$

The term p_B is the stagnation pressure. This expression (*1*) is usually rearranged, introducing the ratio of the velocity at A to the acoustic velocity c of the undisturbed fluid.

From Chapter 1, the acoustic velocity $c=\sqrt{E/\rho}=\sqrt{kp/\rho}=\sqrt{kpg/\gamma}$. Combining with equation (*1*) above,

$$\frac{V_A^2}{2}=\left(\frac{c^2}{k-1}\right)\left[\left(\frac{p_B}{p_A}\right)^{(k-1)/k}-1\right],\quad\text{or}\quad\frac{p_B}{p_A}=\left[1+\left(\frac{k-1}{2}\right)\left(\frac{V_A}{c}\right)^2\right]^{k/(k-1)} \tag{2}$$

Expanding by the binomial theorem,

$$\frac{p_B}{p_A}=1+\left(\frac{k}{2}\right)\left(\frac{V_A}{c}\right)^2\left[1+\left(\frac{1}{4}\right)\left(\frac{V_A}{c}\right)^2-\left(\frac{k-2}{24}\right)\left(\frac{V_A}{c}\right)^4+\cdots\right] \tag{3}$$

In order to compare this expression with formula (*3*) of Problem 12.1, multiply through by p_A and replace kp_A/c^2 by ρ_A, obtaining

$$p_B=p_A+\frac{1}{2}\rho_A V_A^2\left[1+\left(\frac{1}{4}\right)\left(\frac{V_A}{c}\right)^2-\left(\frac{k-2}{24}\right)\left(\frac{V_A}{c}\right)^4\cdots\right] \tag{4}$$

The above expressions apply to all compressible fluids for ratios of V/c less than unity. For ratios over unity, shockwave and other phenomena occur, the adiabatic assumption is not sufficiently accurate, and the derivation no longer applies. The ratio V/c is called the *Mach number*.

The bracketed term in (*4*) is greater than unity, and the first two terms provide sufficient accuracy. The effect of compressibility is to increase the stagnation-point pressure over that of an incompressible fluid [see expression (*3*) of Problem 12.1].

Acoustic velocities will be discussed in Problems 12.26 through 12.28.

12.6. Air flowing under atmospheric conditions ($\gamma = 12.0\ \text{N/m}^3$ at 15°C) at a velocity of 90 m/s is measured by a Pitot tube. Calculate the error in the stagnation pressure by assuming the air to be incompressible.

Solution:

Using formula (*3*) of Problem 12.1,

$$p_B=p_A+\tfrac{1}{2}\rho V^2=101{,}400+\left(\tfrac{1}{2}\right)(12.0/9.81)(90)^2=106{,}350\ \text{Pa}=106.35\ \text{kPa}$$

Using formula (*4*) of Problem 12.5 and $c=\sqrt{kgRT}=\sqrt{(1.4)(9.81)(29.3)(288)}=340$ m/s,

$$p_B=101{,}400+\left(\tfrac{1}{2}\right)(12.0/9.81)(90)^2\left[1+\left(\tfrac{1}{4}\right)(90/340)^2+\cdots\right]$$

$$=106{,}440\ \text{Pa}=106.44\ \text{kPa absolute}$$

The error in the stagnation pressure is 0.1%, and the error in (p_B-p_A) is about 2.0%.

12.7. The difference between the stagnation pressure and the static pressure measured by a Pitot-static device is 412 lb/ft². The static pressure is 14.5 psi absolute and the temperature in the airstream is 60°F. What is the velocity of the air, assuming air is (*a*) compressible and (*b*) incompressible?

Solution:

(*a*) $p_A = (14.5)(144) = 2088$ psf absolute and $c = \sqrt{kgRT} = \sqrt{(1.4)(32.2)(53.3)(520)} = 1118$ ft/sec.

From equation (2) of Problem 12.5,
$$\frac{p_B}{p_A} = \left[1 + \left(\frac{k-1}{2}\right)\left(\frac{V_A}{c}\right)^2\right]^{k/(k-1)}$$

$$\frac{2088 + 412}{2088} = \left[1 + \left(\frac{1.4-1}{2}\right)\left(\frac{V_A}{1118}\right)^2\right]^{1.4/0.4}, \quad V_A = 574 \text{ ft/sec}$$

(*b*) $\gamma = \dfrac{(14.5)(144)}{(53.3)(520)} = 0.0753$ lb/ft³ and $V = \sqrt{2g(p_B/\gamma - p_A/\gamma)} = \sqrt{2g(412/0.0753)}$ $= 594$ ft/sec.

12.8. Air flows at 800 ft/sec through a duct. At standard barometric pressure, the stagnation pressure is −5.70 ft of water, gage. The stagnation temperature is 145°F. What is the static pressure in the duct?

Solution:

With two unknowns in equation (2) of Problem 12.5, assume a V/c ratio (Mach number) of 0.72. Then
$$(-5.70 + 34.0)(62.4) = p_A\left[1 + \left(\tfrac{1}{2}\right)(1.4-1)(0.72)^2\right]^{1.4/0.4}$$

and $p_A = (62.4)(28.3)/1.412 = 1251$ lb/ft² absolute.

Checking the assumption, using the adiabatic relation
$$\frac{T_B}{T_A} = \left(\frac{p_B}{p_A}\right)^{(k-1)/k}, \quad \frac{460+145}{T_A} = \left(\frac{28.3 \times 62.4}{1251}\right)^{0.4/1.4}, \quad T_A = 548°\text{R}$$

Also, $c = \sqrt{kgRT} = \sqrt{(1.4)(32.2)(53.3)(548)} = 1147$ ft/sec.

Then $V/c = 800/1147 = 0.697$ and $p_A = \dfrac{62.4 \times 28.3}{\left[1 + (0.2)(0.697)^2\right]^{1.4/0.4}} = 1277$ lb/ft² absolute.

No further refinement is necessary.

12.9. A 100-mm-diameter standard orifice discharges water under a 6.1-m head. What is the flow?

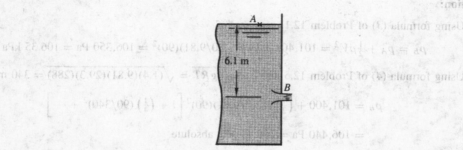

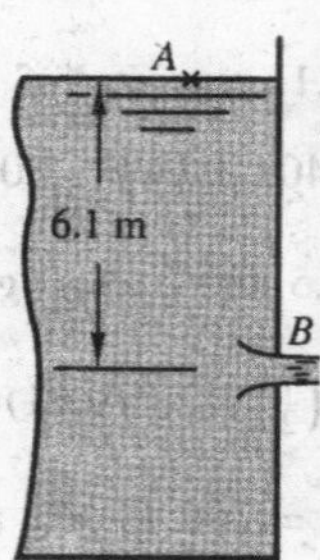

Fig. 12-2

Solution:

Applying the Bernoulli equation, A to B in Fig. 12-2, datum B,

$$(0+0+6.1) - \left(\frac{1}{c_v^2} - 1\right)\left(\frac{V_{\text{jet}}^2}{2g}\right) = \left(\frac{V_{\text{jet}}^2}{2g} + \frac{p_B}{\gamma} + 0\right)$$

But the pressure head at B is zero (as discussed in Chap. 5, Problem 5.6). Then

$$V_{\text{jet}} = c_v\sqrt{2g \times 6.1}$$

Also, $Q = A_{\text{jet}}\,V_{\text{jet}}$, which, using the definitions of the coefficients, becomes

$$Q = (c_c A_o)c_v\sqrt{2g \times 6.1} = cA_o\sqrt{2g \times 6.1}$$

From Table 7, $c = 0.594$ for $D = 100$ mm and $h = 6.1$ m. Hence

$$Q = (0.594)\left[\tfrac{1}{4}\pi(0.100)^2\right]\sqrt{2g \times 6.1} = 0.0510\ \text{m}^3/\text{s}.$$

12.10. The actual velocity in the contracted section of a jet of liquid flowing from a 50-mm-diameter orifice is 8.53 m/s under a head of 4.57 m. (*a*) What is the value of the coefficient of velocity? (*b*) If the measured discharge is 0.0114 m³/s, determine the coefficients of contraction and discharge.

Solution:

(*a*) Actual velocity $= c_v\sqrt{2gH}$, $\quad 8.53 = c_v\sqrt{(2)(9.81) \times 4.57}$, $\quad c_v = 0.901$.

(*b*) Actual $Q = cA\sqrt{2gH}$, $\quad 0.0114 = c\left[\tfrac{1}{4}\pi(50/1000)^2\right]\sqrt{(2)(9.81) \times 4.57}$, $\quad c = 0.613$.

From $c = c_v \times c_c$, $\quad c_c = 0.613/0.901 = 0.680$.

12.11. Oil flows through a standard 25-mm-diameter orifice under a 5.49 m head at the rate of 0.00314 m³/s. The jet strikes a wall 1.52 m away and 0.119 m vertically below the centerline of the contracted section of the jet. Compute the coefficients.

Solution:

(*a*) $Q = cA\sqrt{2gH}$, $\quad 0.00314 = c\left[\dfrac{1}{4}\pi\left(\dfrac{25}{1000}\right)^2\right]\sqrt{2g(5.49)}$, $\quad c = 0.616$

(*b*) From kinematic mechanics, $x = Vt$ and $y = \tfrac{1}{2}gt^2$. Here x and y represent the coordinates of the jet, as measured.

Eliminate t and obtain $\quad x^2 = (2V^2/g)y$.

Substituting, $\quad (1.52)^2 = (2V^2/9.81)(0.119)$, and actual $V = 9.76$ m/s in jet.

Then $9.76 = c_v\sqrt{2g(5.49)}$ and $c_v = 0.940$. Finally, $c_c = c/c_v = 0.616/0.940 = 0.655$.

12.12. The tank in Problem 12.9 is closed and the air space above the water is under pressure, causing the flow to increase to 0.075 m³/s. Find the pressure in the air space in psi.

Solution:

$$Q = cA_o\sqrt{2gH} \quad \text{or} \quad 0.075 = c\left[\tfrac{1}{4}\pi(0.100)^2\right]\sqrt{2g(6.1 + p/\gamma)}$$

Table 7 indicates that c does not change appreciably at the range of head under consideration. Using $c = 0.593$ and solving, $p/\gamma = 7.12$ m water (the assumed c checks for total head H). Then

$$p' = \gamma h = (9.79)(7.12) = 69.7 \text{ kPa}$$

12.13. Oil of specific gravity 0.720 flows through a 3″ diameter orifice whose coefficients of velocity and contraction are 0.950 and 0.650, respectively. What must be the reading of gage A in Fig. 12-3 in order for the power in the jet C to be 8.00 hp?

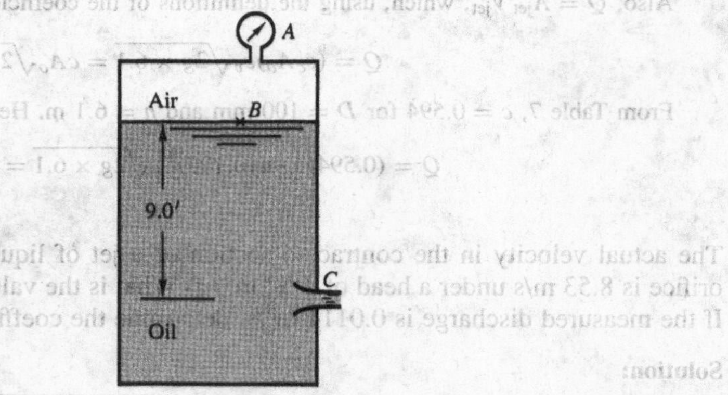

Fig. 12-3

Solution:

The velocity in the jet can be calculated from the value of the power in the jet:

$$\text{horsepower in jet} = \frac{\gamma Q H_{\text{jet}}}{550} = \frac{\gamma(c_c A_o V_{\text{jet}})\left(0 + V_{\text{jet}}^2/2g + 0\right)}{550}$$

$$8.00 = \frac{(0.720 \times 62.4)(0.650)\left[\frac{1}{4}\pi\left(\frac{1}{4}\right)^2\right] V_{\text{jet}}^3/2g}{550}$$

Solving, $V_{\text{jet}}^3 = 197{,}669$ and $V_{\text{jet}} = 58.3$ ft/sec.
Applying the Bernoulli equation, B to C, datum C,

$$\left(\frac{p_A}{\gamma} + \text{negl.} + 9.0\right) - \left[\frac{1}{(0.95)^2} - 1\right]\frac{(58.3)^2}{2g} = \left(0 + \frac{(58.3)^2}{2g} + 0\right)$$

and $p_A/\gamma = 49.5$ ft of oil. Then $p_A' = \gamma h/144 = (0.720 \times 62.4)(49.5)/144 = 15.4$ psi.
Note: The reader should not confuse the total head H causing flow with the value of H_{jet} in the horsepower expression. They are *not* the same.

12.14. For the 4″-diameter short tube shown in Fig. 12-4, (*a*) what flow of water at 75°F will occur under a head of 30 ft? (*b*) What is the pressure head at section B? (*c*) What maximum head can be used if the tube is to flow full at exit? (Use $c_v = 0.82$.)

Solution:

For a standard short tube, the stream contracts at B to about 0.62 of the area of the tube. The lost head from A to B has been measured at about 0.042 times the velocity head at B.

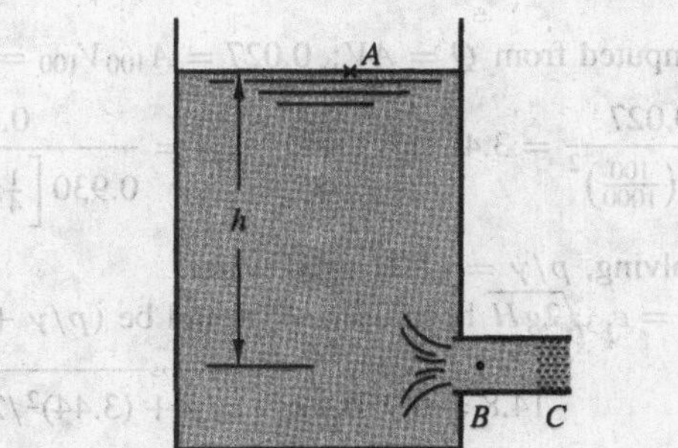

Fig. 12-4

(*a*) Applying the Bernoulli equation, *A* to *C*, datum *C*,

$$(0 + \text{negl.} + 30) - \left[\frac{1}{(0.82)^2} - 1\right]\frac{V_{\text{jet}}^2}{2g} = \left(0 + \frac{V_{\text{jet}}^2}{2g} + 0\right)$$

and $V_{\text{jet}} = 36.0$ ft/sec. Then $Q = A_{\text{jet}} V_{\text{jet}} = \left[1.00 \times \frac{1}{4}\pi\left(\frac{1}{3}\right)^2\right](36.0) = 3.14$ cfs.

(*b*) Now the Bernoulli equation, *A* to *B*, datum *B*, gives

$$(0 + \text{negl.} + 30) - 0.042\frac{V_B^2}{2g} = \left(\frac{p_B}{\gamma} + \frac{V_B^2}{2g} + 0\right) \qquad (A)$$

Also, $Q = A_B V_B = A_C V_C$ or $c_c A V_B = A V_c$ or $V_B = V_{\text{jet}}/c_c = 36.0/0.62 = 58.1$ ft/sec.

Substituting in equation (*A*), $30 = \frac{p_B}{\gamma} + 1.042\frac{(58.1)^2}{2g}$ and $\frac{p_B}{\gamma} = -24.6$ ft of water.

(*c*) As the head causing flow through the short tube is increased, the pressure head at *B* will become less and less. For steady flow (and with the tube full at exit), the pressure head at *B* must not be less than the vapor pressure head for the liquid at the particular temperature. From Table 1 in the Appendix, for water at 75°F this value is 0.43 psia or about 1.0 ft absolute at sea level (−33.0 ft gage).

From (*A*) above, $$h = \frac{p_B}{\gamma} + 1.042\frac{V_B^2}{2g} = -33.0 + 1.042\frac{V_B^2}{2g} \qquad (B)$$

Also, $$c_c A V_B = A V_C = A c_v\sqrt{2gh}$$

Thus $$V_B = \frac{c_v}{c_c}\sqrt{2gh} \quad \text{or} \quad \frac{V_B^2}{2g} = \left(\frac{c_v}{c_c}\right)^2 h = \left(\frac{0.82}{0.62}\right)^2 h = 1.75h$$

Substituting in (*B*), $h = -33.0 + (1.042)(1.75h)$ and $h = 40.1$ ft of water (75°F).

Any head over 40 ft will cause the stream to spring free of the sides of the tube. The tube will then function as an orifice.

Cavitation may result at vapor pressure conditions (see Chapter 14).

12.15. Water flows through a 100-mm pipe at the rate of 0.027 m³/s and thence through a nozzle attached to the end of the pipe. The nozzle tip is 50 mm in diameter, and the coefficients of velocity and contraction for the nozzle are 0.950 and 0.930, respectively. What pressure head must be maintained at the base of the nozzle if atmospheric pressure surrounds the jet?

Solution:

Apply the Bernoulli equation, base of nozzle to jet.

$$\left(\frac{p}{\gamma} + \frac{V_{100}^2}{2g} + 0\right) - \left[\frac{1}{(0.950)^2} - 1\right]\frac{V_{\text{jet}}^2}{2g} = \left(0 + \frac{V_{\text{jet}}^2}{2g} + 0\right)$$

and the velocities are computed from $Q = AV$: $0.027 = A_{100}V_{100} = A_{\text{jet}}V_{\text{jet}} = (c_c A_{50})V_{\text{jet}}$. Thus

$$V_{100} = \frac{0.027}{\frac{1}{4}\pi\left(\frac{100}{1000}\right)^2} = 3.44 \text{ m/s} \quad \text{and} \quad V_{\text{jet}} = \frac{0.027}{0.930\left[\frac{1}{4}\pi\left(\frac{50}{1000}\right)^2\right]} = 14.8 \text{ m/s}.$$

Substituting above and solving, $p/\gamma = 11.77$ m of water.

Had the formula $V_{\text{jet}} = c_v\sqrt{2gH}$ been used, H would be $(p/\gamma + V_{100}^2/2g)$ or

$$14.8 = 0.950\sqrt{2g\left[p/\gamma + (3.44)^2/2g\right]}$$

from which $p/\gamma = 11.77$ m of water, as before.

12.16. A 100-mm base diameter by 50-mm tip diameter nozzle points downward, and the pressure head at the base of the nozzle is 7.92 m of water. The base of the nozzle is 0.914 m above the tip, and the coefficient of velocity is 0.962. Determine the power in the jet of water.

Solution:

For a nozzle, unless c_c is given, it may be taken as unity. Therefore $V_{\text{jet}} = V_{50\text{ mm}}$.

Before the power can be calculated, both V and Q must be found. Using the Bernoulli equation, base to tip, datum tip, gives

$$\left(7.92 + \frac{V_{100}^2}{2g} + 0.914\right) - \left[\frac{1}{(0.962)^2} - 1\right]\frac{V_{50}^2}{2g} = \left(0 + \frac{V_{50}^2}{2g} + 0\right)$$

and $A_{100}V_{100} = A_{50}V_{50}$ or $V_{100}^2 = (50/100)^4 V_{50}^2 = \frac{1}{16}V_{50}^2$. Solving, $V_{50} = 13.0$ m/s.

$$\text{power in jet} = \gamma Q H_{\text{jet}} = (9.79)\left[\tfrac{1}{4}\pi(50/1000)^2(13.0)\right]\left[0 + (13.0)^2/2g + 0\right] = 2.15 \text{ kW}$$

12.17. Water flows through a 12″ × 6″ Venturi meter at the rate of 1.49 cfs, and the differential gage is deflected 3.50 ft, as shown in Fig. 12-5. The specific gravity of the gage liquid is 1.25. Determine the coefficient of the meter.

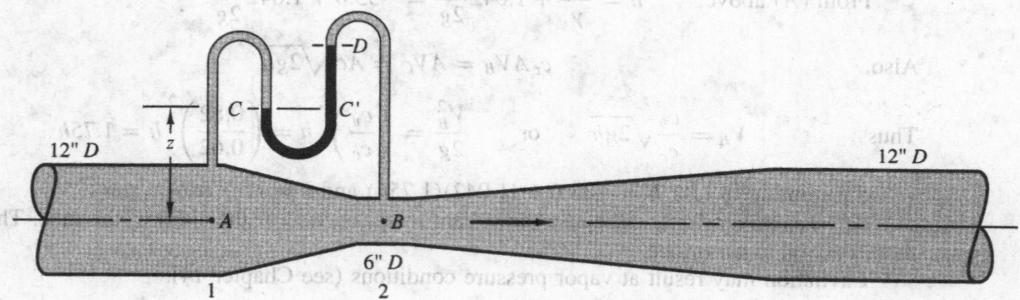

Fig. 12-5

Solution:

The coefficient of a Venturi meter is the same as the coefficient of discharge ($c_c = 1.00$ and thus $c = c_v$). Flow coefficient K should not be confused with meter coefficient c. Clarification will be made at the end of this problem.

Applying the Bernoulli equation, A to B, ideal case, yields

$$\left(\frac{p_A}{\gamma} + \frac{V_{12}^2}{2g} + 0\right) - \text{no lost head} = \left(\frac{p_B}{\gamma} + \frac{V_6^2}{2g} + 0\right)$$

and $V_{12}^2 = (A_6/A_{12})^2 V_6^2$. Solving, $V_6 = \sqrt{\dfrac{2g(p_A/\gamma - p_B/\gamma)}{1-(A_6/A_{12})^2}}$ (no lost head).

The true velocity (and hence the true value of flow Q) will be obtained by multiplying the ideal value by the coefficient c of the meter. Thus

$$Q = A_6 V_6 = A_6\, c\sqrt{\frac{2g(p_A/\gamma - p_B/\gamma)}{1-(A_6/A_{12})^2}} \qquad (1)$$

To obtain the differential pressure head indicated above, the principles of the differential gage must be used.

$$p_C = p_{C'}$$

$$(p_A/\gamma - z) = p_B/\gamma - (z + 3.50) + (1.25)(3.50) \quad \text{or} \quad (p_A/\gamma - p_B/\gamma) = 0.875 \text{ ft}$$

Substituting in (*1*), $1.49 = \frac{1}{4}\pi\left(\frac{1}{2}\right)^2 c\sqrt{2g(0.875)/(1-1/16)}$ and $c = 0.979$ (use 0.98).

Note: Equation (*1*) is sometimes written $Q = KA_2\sqrt{2g(\Delta p/\gamma)}$, where K is called the *flow coefficient*. It is apparent that

$$K = \frac{c}{\sqrt{1-(A_2/A_1)^2}} \quad \text{or} \quad \frac{c}{\sqrt{1-(D_2/D_1)^4}}$$

Tables or charts that give K can readily be used to obtain c if it is so desired. Diagrams in this book give values of c. The conversion factors to obtain the values of K for certain diameter-ratio devices are indicated on the several diagrams in the Appendix.

12.18. Water flows upward through a vertical 300 mm by 150 mm Venturi meter whose coefficient is 0.980. The differential gage deflection is 1.18 m of liquid of specific gravity 1.25, as shown in Fig. 12-6. Determine the flow in cfs.

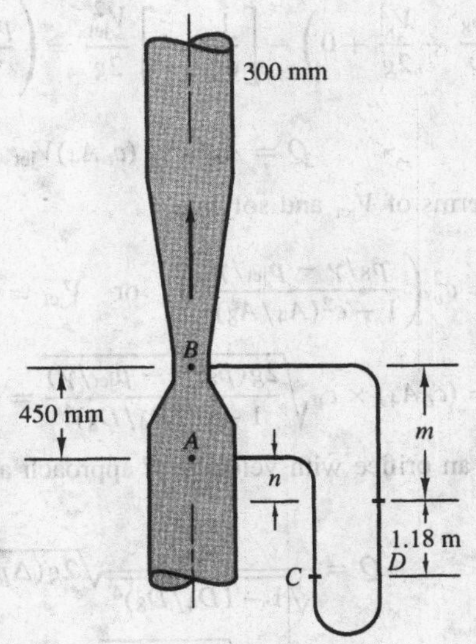

Fig. 12-6

Solution:

Reference to the Bernoulli equation in Problem 12.17 indicates that, for this problem, $z_A = 0$ and $z_B = 0.450$ m. Then

$$Q = cA_{150}\sqrt{\frac{2g[(p_A/\gamma - p_B/\gamma) - 0.450]}{1-(1/2)^4}}$$

Using the principles of the differential gage to obtain $\Delta p/\gamma$.

$$p_C/\gamma = p_D/\gamma \text{ (m of water units)}$$

$$p_A/\gamma + (n+1.18) = p_B/\gamma + m + (1.25)(1.18)$$

$$[(p_A/\gamma - p_B/\gamma) - (m-n)] = (1.18)(1.25 - 1.00)$$

$$[(p_A/\gamma - p_B/\gamma) - 0.450] = 0.295 \text{ m of water}$$

Substituting into the equation for flow, $Q = (0.980)\left(\frac{1}{4}\pi\right)\left(\frac{150}{1000}\right)^2\sqrt{2g(0.295)/(1-1/16)} = 0.0430 \text{ m}^3/\text{s}$.

12.19. Water at 100°F flows at the rate of 0.525 cfs through a 4″-diameter orifice used in an 8″ pipe. What is the difference in pressure head between the upstream section and the contracted section (vena contracta section)?

Mathcad

Solution:

In Diagram C of the Appendix, it is observed that c' varies with Reynolds number. Note that Reynolds number must be calculated for the orifice cross section, not for the contracted section of the jet or for the pipe section. This value is

$$\text{Re} = \frac{V_o D_o}{\nu} = \frac{\left(4Q/\pi D_o^2\right) D_o}{\nu} = \frac{4Q}{\nu\pi D_o} = \frac{(4)(0.525)}{(\pi)(0.00000739)(4/12)} = 271{,}000$$

Diagram C for $\beta = 0.500$ gives $c' = 0.604$.

Applying the Bernoulli theorem, pipe section to jet section, produces the general equation for incompressible fluids, as follows:

$$\left(\frac{p_8}{\gamma} + \frac{V_8^2}{2g} + 0\right) - \left[\frac{1}{c_v^2} - 1\right]\frac{V_{\text{jet}}^2}{2g} = \left(\frac{p_{\text{jet}}}{\gamma} + \frac{V_{\text{jet}}^2}{2g} + 0\right)$$

and

$$Q = A_8 V_8 = (c_c A_4) V_{\text{jet}}$$

Substituting for V_8 in terms of V_{jet} and solving,

$$\frac{V_{\text{jet}}^2}{2g} = c_v^2\left(\frac{p_8/\gamma - p_{\text{jet}}/\gamma}{1 - c^2(A_4/A_8)^2}\right) \quad \text{or} \quad V_{\text{jet}} = c_v\sqrt{\frac{2g(p_8/\gamma - p_{\text{jet}}/\gamma)}{1 - c^2(D_4/D_8)^4}}$$

Then

$$Q = A_{\text{jet}} V_{\text{jet}} = (c_c A_4) \times c_v\sqrt{\frac{2g(p_8/\gamma - p_{\text{jet}}/\gamma)}{1 - c^2(D_4/D_8)^4}} = cA_4\sqrt{\frac{2g(p_8/\gamma - p_{\text{jet}}/\gamma)}{1 - c^2(D_4/D_8)^4}}.$$

More conveniently, for an orifice with velocity of approach and a contracted jet, the equation can be written

$$Q = \frac{c' A_4}{\sqrt{1-(D_4/D_8)^4}}\sqrt{2g(\Delta p/\gamma)} \qquad (1)$$

or

$$Q = K A_4\sqrt{2g(\Delta p/\gamma)} \qquad (2)$$

where K is called the *flow coefficient*. The meter coefficient c' can be determined experimentally for a given ratio of diameter of orifice to diameter of pipe, or the flow coefficient K may be preferred.

Proceeding with the solution by substituting in the above expression (*1*),

$$0.525 = \frac{0.604 \times \frac{1}{4}\pi(4/12)^2}{\sqrt{1-(1/2)^4}}\sqrt{2g(\Delta p/\gamma)} \text{ and } \Delta p/\gamma = (p_8/\gamma - p_{\text{jet}}/\gamma) = 1.44 \text{ ft water}$$

12.20. For the pipe orifice in Problem 12.19, what pressure difference in psi would cause the same quantity of turpentine at 68°F to flow? (See Appendix for sp gr and ν.)

Solution:

$$\text{Re} = \frac{4Q}{\pi\nu D_o} = \frac{(4)(0.525)}{(\pi)(0.0000186)(4/12)} = 108{,}000. \text{ From Diagram } C, \text{ for } \beta = 0.500, c' = 0.607.$$

$$\text{Then} \quad 0.525 = \frac{0.607 \times \frac{1}{4}\pi(4/12)^2}{\sqrt{1-(1/2)^4}}\sqrt{2g(\Delta p/\gamma)}, \quad \text{from which}$$

$$\Delta\frac{p}{\gamma} = \left(\frac{p_8}{\gamma} - \frac{p_{\text{jet}}}{\gamma}\right) = 1.43 \text{ ft turpentine} \quad \text{and} \quad \Delta p' = \frac{\gamma h}{144} = \frac{(0.862 \times 62.4)(1.43)}{144} = 0.534 \text{ psi.}$$

12.21. Determine the flow of water at 20°C through a 150-mm orifice installed in a 250-mm pipeline if the pressure head differential for vena contracta taps is 1.10 m of water.

Mathcad

Solution:

This type of problem was met in the flow of fluids in pipes. The value of c' cannot be found inasmuch as Reynolds number cannot be computed. Referring to Diagram C, for $\beta = 0.600$, a value of c' of 0.610 will be assumed. Using this assumed value,

$$Q = \frac{0.610 \times \frac{1}{4}\pi(150/1000)^2}{\sqrt{1-(0.60)^4}}\sqrt{(2)(9.81)(1.10)} = 0.0537 \text{ m}^3/\text{s}$$

$$\text{Then} \quad \text{Re} = \frac{(4)(0.0537)}{\pi\,(9.84 \times 10^{-7})\,(150/1000)} = 463{,}000 \text{ (trial value)}$$

From Diagram C, $\beta = 0.600$, $c' = 0.609$. Recalculation of the flow using $c' = 0.609$ gives $Q = 0.0536$ m^3/s. (Reynolds number is unaffected.)

Special Note: Professor R. C. Binder of Purdue University suggests on pages 132–133 of his fluid mechanics text (second edition) that this type of problem need not be a "cut and try" proposition. He proposes that special lines be drawn on the coefficient–Reynolds number chart. In the case of the pipe orifice, equation (*1*) of Problem 12.19 can be written

$$\frac{Q}{A_4} = \frac{c'\sqrt{2g(\Delta p/\gamma)}}{\sqrt{1-(D_4/D_8)^4}} = V_4 \quad \text{since } Q = AV$$

$$\text{But} \quad \text{Re} = \frac{V_4 D_4}{\nu} = \frac{c'\sqrt{2g(\Delta p/\gamma)} \times D_4}{\nu\sqrt{1-(4/8)^4}} \quad \text{or} \quad \frac{\text{Re}}{c'} = \frac{D_4\sqrt{2g(\Delta p/\gamma)}}{\nu\sqrt{1-(4/8)^4}}$$

$$\text{or, in general,} \quad \frac{\text{Re}}{c'} = \frac{D_o\sqrt{2g(\Delta p/\gamma)}}{\nu\sqrt{1-(D_o/D_p)^4}}$$

Two straight lines called T-lines have been drawn on Diagram C, one for $\text{Re}/c' = 700{,}000$ and one for $\text{Re}/c' = 800{,}000$. For Problem 12.21, the calculated

$$\frac{\text{Re}}{c'} = \frac{(150/1000)\sqrt{(2)(9.81)(1.10)}}{(9.84 \times 10^{-7})\sqrt{1-(0.60)^4}} = 759{,}000$$

As near as can be read, the 759,000 line cuts the $\beta = 0.600$ curve at $c' = 0.609$. The flow Q is then calculated readily.

12.22. A nozzle with a 4″-diameter tip is installed in a 10″ pipe. Medium fuel oil at 80°F flows through the nozzle at the rate of 3.49 cfs. Assume the calibration of the nozzle is represented by curve $\beta = 0.40$ on Diagram D. Calculate the differential gage reading if a liquid of specific gravity 13.6 is the gage liquid.

Solution:

The Bernoulli equation, pipe section to jet, yields the same equation as was obtained in Problem 12.17 for the Venturi meter, since the nozzle is designed for a coefficient of contraction of unity.

$$Q = A_4 V_4 = A_4 c \sqrt{\frac{2g(p_A/\gamma - p_B/\gamma)}{1-(4/10)^4}} \qquad (1)$$

Diagram D indicates that c varies with Reynolds number.

$$V_4 = \frac{Q}{A_4} = \frac{3.49}{\frac{1}{4}\pi(4/12)^2} = 40.0 \text{ ft/sec} \quad \text{and} \quad \text{Re} = \frac{40.0 \times 4/12}{3.65 \times 10^{-5}} = 365{,}000$$

The curve for $\beta = 0.40$ gives $c = 0.993$. Thus

$$3.49 = \frac{1}{4}\pi(4/12)^2 \times 0.993 \sqrt{\frac{2g(p_A/\gamma - p_B/\gamma)}{1-(4/10)^4}}$$

and $(p_A/\gamma - p_B/\gamma) = 24.5$ ft of fuel oil.

Differential gage principles produce, using sp gr of the oil = 0.851 from the Appendix,

$$24.5 = h(13.6/0.851 - 1) \quad \text{and} \quad h = 1.64 \text{ ft (gage reading)}$$

Had the gage differential reading been given, the procedure used in the preceding problem would be utilized, e.g., a value of c assumed, Q calculated, Reynolds number obtained, and c read from the appropriate curve on Diagram D. If c differs from the assumed value, the calculation is repeated until the coefficient checks in.

12.23. Derive an expression for the flow of a compressible fluid through a nozzle flow meter and a Venturi meter.

Solution:

Since the change in velocity takes place in a very short period of time, little heat can escape and adiabatic conditions will be assumed. The Bernoulli theorem for compressible flow was shown in Chapter 7, equation (D) of Problem 7.21, to give

$$\left[\left(\frac{k}{k-1}\right)\frac{p_1}{\gamma_1} + \frac{V_1^2}{2g} + z_1\right] - H_L = \left[\left(\frac{k}{k-1}\right)\left(\frac{p_1}{\gamma_1}\right)\left(\frac{p_2}{p_1}\right)^{(k-1)/k} + \frac{V_2^2}{2g} + z_2\right]$$

For a nozzle meter and for a horizontal Venturi meter, $z_1 = z_2$ and the lost head will be taken care of by means of the coefficient of discharge. Also, since $c_c = 1.00$,

$$W = \gamma_1 A_1 V_1 = \gamma_2 A_2 V_2$$

Then upstream $V_1 = W/\gamma_1 A_1$, downstream $V_2 = W/\gamma_2 A_2$. Substituting and solving for W,

$$\frac{W^2}{\gamma_2^2 A_2^2} - \frac{W^2}{\gamma_1^2 A_1^2} = 2g\left(\frac{k}{k-1}\right)\left(\frac{p_1}{\gamma_1}\right)\left[1-\left(\frac{p_2}{p_1}\right)^{(k-1)/k}\right]$$

or (ideal) $$W = \frac{\gamma_2 A_2}{\sqrt{1-(\gamma_2/\gamma_1)^2 (A_2/A_1)^2}}\sqrt{\frac{2gk}{k-1}(p_1/\gamma_1)\times[1-(p_2/p_1)^{(k-1)/k}]}$$

It may be more practical to eliminate γ_2 under the radical. Since $\gamma_2/\gamma_1 = (p_2/p_1)^{1/k}$,

$$\text{(ideal) } W = \gamma_2 A_2 \sqrt{\frac{\dfrac{2gk}{k-1}(p_1/\gamma_1)\times\left[1-(p_2/p_1)^{(k-1)/k}\right]}{1-(A_2/A_1)^2\,(p_2/p_1)^{2/k}}} \qquad (1)$$

The true value of W is obtained by multiplying the right-hand side of the equation by coefficient c.

For comparison, equation (*1*) of Problem 12.17 and equation (*1*) of Problem 12.22 (for incompressible fluids) can be written

$$W = \gamma Q = \frac{\gamma A_2 c}{\sqrt{1-(A_2/A_1)^2}}\sqrt{2g(\Delta p/\gamma)}$$

or $$W = \gamma K A_2 \sqrt{2g(\Delta p/\gamma)}$$

The above equation can be expressed more generally so that it will apply to both compressible and incompressible fluids. An expansion (adiabatic) factor Y is introduced, and the value of γ_1 at the inlet is specified. The fundamental relation is then

$$W = \gamma_1 K A_2 Y \sqrt{2g(\Delta p/\gamma_1)} \qquad (2)$$

For incompressible fluids, $Y = 1$. For compressible fluids, equate expressions (*1*) and (*2*) and solve for Y. Thus,

$$Y = \sqrt{\frac{1-(A_2/A_1)^2}{1-(A_2/A_1)^2(p_2/p_1)^{2/k}} \times \frac{[k/(k-1)]\left[1-(p_2/p_1)^{k-1)/k}\right](p_2/p_1)^{2/k}}{1-p_2/p_1}}$$

This expansion factor Y is a function of three dimensionless ratios. Table 8 lists some typical values for nozzle flow meters and for Venturi meters.

Note: Values of Y' for orifices and for orifice meters should be determined experimentally. The values differ from the above value of Y because the coefficient of contraction is not unity nor is it a constant. Knowing Y', solutions are identical to those that follow for flow nozzles and Venturi meters. The reader is referred to experiments by H. B. Reynolds and J. A. Perry as two sources of material.

12.24. Air at a temperature of 80°F flows through a 4″ pipe and through a 2″ flow nozzle. The pressure differential is 0.522 ft of oil, sp gr 0.910. The pressure upstream from the nozzle is 28.3 psi gage. How many pounds per second are flowing for a barometric reading of 14.7 psi, (*a*) assuming the air has constant density and (*b*) assuming adiabatic conditions?

Mathcad

Solution:

(*a*) $$\gamma_1 = \frac{(28.3+14.7)(144)}{(53.3)(460+80)} = 0.215 \text{ lb/ft}^3$$

From differential gage principles, using pressure heads in feet of air,

$$\frac{\Delta p}{\gamma_1} = 0.522\left(\frac{\gamma_{oil}}{\gamma_{air}} - 1\right) = 0.522\left(\frac{0.910 \times 62.4}{0.215} - 1\right) = 137 \text{ ft of air}$$

Assuming $c = 0.980$ and using equation (*1*) of Problem 12.22 after multiplying by γ_1, we have

$$W = \gamma_1 Q = 0.215 \times \frac{1}{4}\pi(2/12)^2(0.980)\sqrt{\frac{2g(137)}{1-(2/4)^4}} = 0.446 \text{ lb/sec}$$

To check the value of c, find the Reynolds number and use the appropriate curve on Diagram *D*. (Here $\gamma_1 = \gamma_2$ and $\nu = 16.9 \times 10^{-5}$ ft²/sec at standard atmosphere from Table 1*B*.)

$$V_2 = \frac{W}{A_2\gamma_2} = \frac{W}{(\pi d_2^2/4)\gamma_2}$$

Then $\text{Re} = \dfrac{V_2 d_2}{\nu} = \dfrac{4W}{\pi d_2 \nu \gamma_2} = \dfrac{(4)(0.446)}{(\pi)(2/12)(16.9 \times 14.7/43.0)10^{-5}(0.215)} = 274{,}000$

From Diagram *D*, $c = 0.986$. Recalculating, $W = 0.449$ lb/sec.

Further refinement in calculation is not warranted inasmuch as the Reynolds number will not be changed materially nor will the value of c read from Diagram *D*.

(*b*) Calculate pressures and specific weights first.

$$p_1 = (28.3 + 14.7)(144) = 6192 \text{ lb/ft}^2, \quad p_2 = (6192 - 137 \times 0.215) = 6163 \text{ lb/ft}^2$$

$\dfrac{p_2}{p_1} = \dfrac{6163}{6192} = 0.995$ and $\left(\dfrac{\gamma_2}{\gamma_1}\right)^k = 0.994$ (see Chapter 1). Then $\gamma_2 = 0.214$ lb/ft³.

Table 8 gives some values of expansion factor Y referred to in Problem 12.23. Interpolation may be used, in this case between the pressure ratio of 0.95 and 1.00 to obtain Y for $p_2/p_1 = 0.994$. For $k = 1.40$ and $d_2/d_1 = 0.50$, we obtain $Y = 0.997$.

Assuming $c = 0.980$, from examination of Diagram *D* and noting that $K = 1.032c$, equation (*2*) of Problem 12.23 becomes

$$W = \gamma_1 K A_2 Y\sqrt{2g(\Delta p/\gamma_1)}$$

$$= (0.215)(1.032 \times 0.980) \times \tfrac{1}{4}\pi(2/12)^2 \times 0.997\sqrt{(64.4)(137)} = 0.444 \text{ lb/sec}$$

Checking c, $\quad \text{Re} = \dfrac{4W}{\pi d_2 \nu \gamma_2} = \dfrac{(4)(0.444)}{(\pi)(2/12)(16.9 \times 14.7/43.0)10^{-5}(0.214)} = 274{,}000$

and $c = 0.986$ (Diagram *D*, curve $\beta = 0.50$).

Recalculating, $W = 0.447$ lb/sec. Further refinement is not essential. Note that little error is introduced in part (*a*) by assuming a constant density of air.

12.25. An 8″ × 4″ Venturi meter is used to measure the flow of carbon dioxide at 68°F. The deflection of the water column in the differential gage is 71.8 in, and the barometer reads 30.0 in of mercury. For a pressure at entrance of 18.0 psi absolute, calculate the weight flow.

Mathcad

Solution:

The absolute pressure at entrance $= p_1 = 18.0 \times 144 = 2592$ psf absolute, and the specific weight γ_1 of the carbon dioxide is

$$\gamma_1 = \frac{2592}{(34.9)(460 + 68)} = 0.141 \text{ lb/ft}^3$$

The pressure difference = (71.8/12)(62.4 − 0.141) = 373 psf, and hence the absolute pressure at throat = p_2 = 2592 − 373 = 2219 psf absolute.

To obtain the specific weight γ_2, we use $\dfrac{p_2}{p_1} = \dfrac{2219}{2592} = 0.856$ and $\dfrac{\gamma_2}{\gamma_1} = (0.856)^{1/k}$ (see Chap. 1). Thus $\gamma_2 = (0.141)(0.856)^{1/1.30} = 0.1251$ lb/ft³.

$$W = \gamma_1 K A_2 Y \sqrt{2g(\Delta p/\gamma_1)} \quad \text{lb/sec}$$

Using $k = 1.30$, $d_2/d_1 = 0.50$, and $p_2/p_1 = 0.856$, Y (Table 8) = 0.909 by interpolation. Assuming $c = 0.985$, from Diagram *E*, and noting that $K = 1.032c$, we have

$$W = (0.141)(1.032 \times 0.985) \times \frac{1}{4}\pi(4/12)^2 \times 0.909\sqrt{2g(373/0.141)} = 4.69 \text{ lb/sec}$$

To check the assumed value of c, determine the Reynolds number and use the appropriate curve on Diagram *E*. From Problem 12.24,

$$\text{Re} = \frac{4W}{\pi d_2 \nu \gamma_2} = \frac{(4)(4.69)}{(\pi)(4/12)(9.1 \times 14.7/18.0 \times 10^{-5})(0.1251)} = 1.93 \times 10^6$$

From Diagram *E*, $c = 0.984$. Recalculating, $W = 4.69$ lb/sec.

12.26. Establish the relationship that limits the velocity of a compressible fluid in convergent passages (acoustic velocity).

Solution:

Neglecting the velocity of approach in the Bernoulli equation (*D*) of Problem 7.21, Chapter 7, for an ideal fluid we obtain

$$\frac{V_2^2}{2g} = \frac{k}{k-1}\left(\frac{p_1}{\gamma_1}\right)\left[1 - \left(\frac{p_2}{p_1}\right)^{(k-1)/k}\right] \tag{1}$$

Also, if $(p_2/\gamma_2)^{1/k}$ had been substituted for $(p_1/\gamma_1)^{1/k}$ before the integration that produced equation (*D*), the velocity head would have been

$$\frac{V_2^2}{2g} = \frac{k}{k-1}\left(\frac{p_2}{\gamma_2}\right)\left[\left(\frac{p_1}{p_2}\right)^{(k-1)/k} - 1\right] \tag{2}$$

If the fluid attains acoustic velocity c_2 at section 2, then $V_2 = c_2$ and $V_2^2 = c_2^2 = kp_2 g/\gamma_2$, (see Chap. 1). Substituting in equation (2),

$$\frac{kp_2 g}{2g\gamma_2} = \frac{k}{k-1}\left(\frac{p_2}{\gamma_2}\right)\left[\left(\frac{p_1}{p_2}\right)^{(k-1)/k} - 1\right]$$

which simplifies to

$$\frac{p_2}{p_1} = \left(\frac{2}{k+1}\right)^{k/(k-1)} \tag{3}$$

This ratio p_2/p_1 is called the *critical pressure ratio* and depends upon the fluid flowing. For values of p_2/p_1 equal to or less than the critical pressure ratio, a gas will flow at the acoustic velocity. The pressure in a free jet flowing at the acoustic velocity will *equal or exceed* the pressure that surrounds it.

12.27. Carbon dioxide discharges through a 12.5-mm hole in the wall of a tank in which the pressure is 758 kPa gage and the temperature is 20°C. What is the velocity in the jet (standard barometer)?

Solution:

From Table 1*A*, $R = 19.2$ m/K and $k = 1.30$.

$$\gamma_1 = \frac{p_1}{RT_1} = \frac{758 + 101}{(19.2)(293)} = 0.153 \text{ kN/m}^3$$

$$\text{critical } \frac{p_2}{p_1} = \left(\frac{2}{k+1}\right)^{k/(k-1)} = \left(\frac{2}{2.30}\right)^{1.30/0.30} = 0.546$$

$$\text{ratio } \left(\frac{\text{atmosphere}}{\text{tank pressure}}\right) = \frac{101}{859} = 0.118$$

Since this ratio is less than the critical pressure ratio, the pressure of the escaping gas $= 0.546 \times p_1$. Hence $p_2 = 0.546 \times 859 = 469$ kPa absolute.

$$V_2 = c_2 = \sqrt{1.3 \times 9.81 \times 19.3 \times T_2} = \sqrt{246T_2}$$

where $T_2/T_1 = (p_2/p_1)^{(k-1)/k} = (0.546)^{0.30/1.30} = 0.870$, $T_2 = 255$ K. Then $V_2 = \sqrt{246 \times 255} = 250$ m/s.

12.28. Nitrogen flows through a duct in which changes in cross section occur. At a particular cross section the velocity is 366 m/s, the pressure is 83 kPa absolute, and the temperature is 30°C. Assuming no friction losses and adiabatic conditions, (*a*) what is the velocity at a section where the pressure is 124 kPa absolute and (*b*) what is the Mach number at this section?

Solution:

For nitrogen, $R = 30.3$ m/K and $k = 1.40$, from Table 1*A* of the Appendix.

(*a*) From Problem 7.21 of Chapter 7, equation (*D*) for adiabatic conditions may be written

$$\frac{V_2^2}{2g} - \frac{V_1^2}{2g} = \frac{k}{k-1}\left(\frac{p_1}{\gamma_1}\right)\left[1 - \left(\frac{p_2}{p_1}\right)^{(k-1)/k}\right]$$

where no lost head is considered and $z_1 = z_2$.

Calculate the specific weight of nitrogen at cross section 1.

$$\gamma_1 = \frac{p_1}{RT_1} = \frac{83{,}000}{(30.3)(30+273)} = 9.04 \text{ N/m}^3 \quad (\text{or use } p_1/\gamma_1 = RT_1)$$

Then $\dfrac{V_2^2}{2g} - \dfrac{(366)^2}{2g} = \left(\dfrac{1.40}{0.40}\right)\left(\dfrac{83{,}000}{9.04}\right)\left[1 - \left(\dfrac{124}{83}\right)^{0.40/1.40}\right]$, from which $V_2 = 239$ m/s.

(*b*) Mach number $= \dfrac{V_2}{c_2} = \dfrac{239}{\sqrt{kgRT_2}}$ where $\dfrac{T_2}{T_1} = \left(\dfrac{p_2}{p_1}\right)^{(k-1)/k}$ or $\dfrac{T_2}{303} = \left(\dfrac{124}{83}\right)^{0.286} = 1.122.$

Then $T_2 = 340$ K, and Mach number $= \dfrac{239}{\sqrt{1.40 \times 9.81 \times 30.3 \times 340}} = 0.635.$

12.29. Develop the theoretical formula for flow over a rectangular weir.

Solution:

Consider the rectangular opening in Fig. 12-7 to extend the full width W of the channel ($b = W$). With the liquid surface in the dashed position, application of the Bernoulli theorem between A and an elemental strip of height dy in the jet produces, for ideal conditions,

$$(0 + V_A^2/2g + y) - \text{no losses} = (0 + V_{\text{jet}}^2/2g + 0)$$

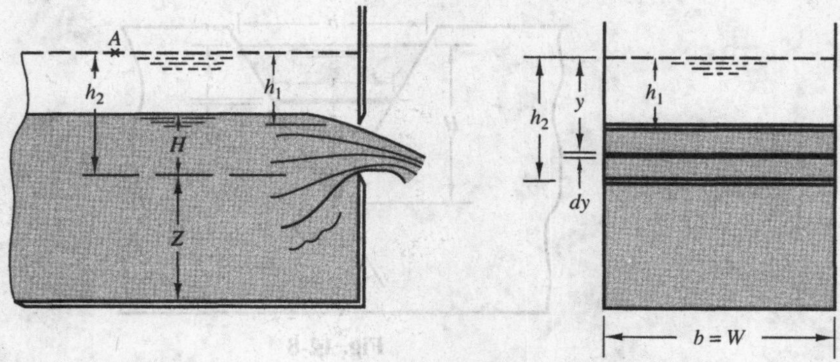

Fig. 12-7

where V_A represents the average velocity of the particles approaching the opening. Thus,

$$\text{ideal } V_{\text{jet}} = \sqrt{2g\left(y + V_A^2/2g\right)}$$

and

$$\text{ideal } dQ = dA\, V_{\text{jet}} = (b\, dy)V_{\text{jet}} = b\sqrt{2g}\left(y + V_A^2/2g\right)^{1/2} dy$$

$$\text{ideal } Q = b\sqrt{2g}\int_{h_1}^{h_2}\left(y + V_A^2/2g\right)^{1/2} dy$$

A weir exists when $h_1 = 0$. Let H replace h_2, and introduce a coefficient of discharge c to obtain the actual flow. Then

$$Q = cb\sqrt{2g}\int_0^H\left(y + V_A^2/2g\right)^{1/2} dy$$

$$= \frac{2}{3}cb\sqrt{2g}\left[\left(H + V_A^2/2g\right)^{3/2} - \left(V_A^2/2g\right)^{3/2}\right]$$

$$= mb\left[\left(H + V_A^2/2g\right)^{3/2} - \left(V_A^2/2g\right)^{3/2}\right] \tag{1}$$

Notes:

(1) For a fully contracted rectangular weir, the end contractions cause a reduction in flow. Length b is corrected to recognize this condition, and the formula becomes

$$Q = m\left(b - \frac{2}{10}H\right)\left[\left(H + V_A^2/2g\right)^{3/2} - \left(V_A^2/2g\right)^{3/2}\right] \tag{2}$$

(2) For high weirs and most contracted weirs, the velocity head of approach is negligible and

$$Q = m\left(b - \frac{2}{10}H\right)H^{3/2} \qquad \text{for contracted weirs} \tag{3}$$

or

$$Q = mbH^{3/2} \qquad \text{for suppressed weirs} \tag{4}$$

(3) Coefficient of discharge c is not constant. It embraces the many complexities not included in the derivation, such as surface tension, viscosity, density, nonuniform velocity distribution, secondary flows, and possibly others.

12.30. Derive the theoretical formula for flow through a triangular-notched weir. Refer to Fig. 12-8.

Solution:

From Problem 12.29,

$$V_{\text{jet}} = \sqrt{2g(y + \text{negligible } V^2/2g)} \quad \text{and} \quad \text{ideal } dQ = dA\, V_{\text{jet}} = x\, dy\sqrt{2gy}$$

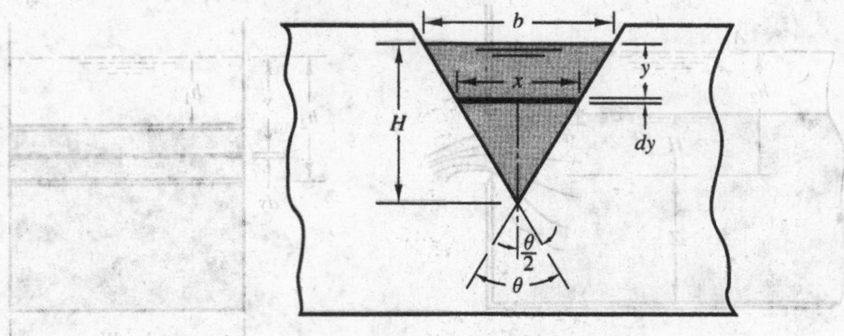

Fig. 12-8

By similar triangles,

$$\frac{x}{b} = \frac{H - y}{H} \quad \text{and} \quad b = 2H \tan\frac{\theta}{2}$$

Then actual $Q = (b/H)c\sqrt{2g}\int_0^H (H - y)y^{1/2}dy$.

Integrating and substituting, $$Q = \frac{8}{15}c\sqrt{2g}H^{5/2}\tan\frac{1}{2}\theta \qquad (1)$$

12.31. During a test on an 8-ft suppressed weir that was 3 ft high, the head was maintained constant at 1.000 ft. In 38.0 sec, 7600 gallons of water were collected. Find weir factor m in equations (*1*) and (*4*) of Problem 12.29.

Solution:

(*a*) Change the measured flow to cfs. $Q = 7600/(7.48 \times 38.0) = 26.7$ cfs.

(*b*) Check the velocity of approach. $V = Q/A = 26.7/(8 \times 4) = 0.834$ ft/sec. Then

$$V^2/2g = (0.834)^2/2g = 0.0108 \text{ ft}$$

(*c*) Using (*1*), $$Q = mb\left[\left(H + V^2/2g\right)^{3/2} - \left(V^2/2g\right)^{3/2}\right]$$

or $$26.7 = m \times 8\left[(1.000 + 0.0108)^{3/2} - (0.0108)^{3/2}\right]$$

and $m = 3.29$.

Using (*4*), $$Q = 26.7 = mbH^{3/2} = m \times 8 \times (1.000)^{3/2}$$

and $m = 3.34$ (about 1.5% higher neglecting the velocity of approach terms).

12.32. Determine the flow over a suppressed weir 3.00 m long and 1.20 m high under a head 0.914 m. The value of m is 1.91.

Mathcad

Solution:

Since the velocity head term cannot be calculated, an approximate flow is

$$Q = mbH^{3/2} = (1.91)(3.00)(0.914)^{3/2} = 5.01 \text{ m}^3/\text{s}$$

For this flow, $V = 5.01/(3.00 \times 2.114) = 0.790$ m/s and $V^2/2g = 0.032$ m. Using equation (*1*) of Problem 12.29,

$$Q = (1.91)(3.00)\left[(0.914 + 0.032)^{3/2} - (0.032)^{3/2}\right] = 5.24 \text{ m}^3/\text{s}$$

This second calculation shows an increase of 0.23 m^3/s or about 4.6% over the first calculation. Further calculation will generally produce an unwarranted refinement, i.e., beyond the accuracy of the formula itself.

However, to illustrate, the revised velocity of approach would be

$$V = 5.24/(3.00 \times 2.114) = 0.826 \text{ m/s}, \qquad \text{and} \qquad V^2/2g = 0.035 \text{ m}$$

and
$$Q = (1.91)(3.00)\left[(0.914 + 0.035)^{3/2} - (0.035)^{3/2}\right] = 5.26 \text{ m}^3/\text{s}$$

12.33. A suppressed weir 25.0 ft long is to discharge 375.0 cfs into a channel. The weir factor $m = 3.42$. To what height Z (nearest 1/100 ft) can the weir be built, if the water behind the weir must not exceed 6 ft in depth?

Solution:

Velocity of approach $V = Q/A = 375.0/(25 \times 6) = 2.50$ ft/sec.

Then
$$375.0 = 3.42 \times 25.0\left[\left(H + \frac{(2.50)^2}{2g}\right)^{3/2} - \left(\frac{(2.50)^2}{2g}\right)^{3/2}\right] \quad \text{and} \quad H = 2.59 \text{ ft}$$

Height of weir is $Z = 6.00 - 2.59 = 3.41$ ft.

12.34. A contracted weir, 1.25 m high, is to be installed in a channel 2.5 m wide. The maximum flow over the weir is 1.70 m³/s when the total depth back of the weir is 2.00 m. What length of weir should be installed if $m = 1.88$?

Solution:

Velocity of approach $V = Q/A = 1.70/(2.5 \times 2.00) = 0.340$ m/s. It appears that the velocity head is negligible in this case.

$$Q = m\left(b - \tfrac{2}{10}H\right)(H)^{3/2}, \qquad 1.70 = (1.88)\left(b - \tfrac{2}{10} \times 0.75\right)(0.75)^{3/2}, \qquad b = 1.54 \text{ m long.}$$

12.35. The discharge from a 6″-diameter orifice, under a 10.0-ft head, $c = 0.600$, flows into a rectangular weir channel and over a contracted weir. The channel is 6 ft wide, and for the weir, $Z = 5.00$ ft and $b = 1.00$ ft. Determine the depth of water in the channel if $m = 3.35$.

Solution:

The discharge through the orifice is

$$Q = cA\sqrt{2gh} = 0.600 \times \tfrac{1}{4}\pi\left(\tfrac{1}{2}\right)^2\sqrt{2g(10.0)} = 2.99 \text{ cfs}$$

For the weir,
$$Q = m(b - \tfrac{2}{10}H)H^{3/2} \quad \text{(velocity head neglected)}$$

or
$$2.99 = (3.35)(1.00 - 0.20H)H^{3/2} \quad \text{and} \quad H^{3/2} - 0.20H^{5/2} = 0.893$$

By successive trials, $H = 1.09$ ft, and the depth $= Z + H = 5.00 + 1.09 = 6.09$ ft.

12.36. The discharge of water over a 45° triangular weir is 0.0212 m³/s. For $c = 0.580$, determine the head on the weir.

Solution:

$$Q = \tfrac{8}{15}c\sqrt{2g}\left(\tan\tfrac{1}{2}\theta\right)H^{5/2}, \qquad 0.0212 = \tfrac{8}{15}(0.580)\sqrt{2g}\left(\tan 22\tfrac{1}{2}^\circ\right)H^{5/2}, \qquad H = 0.268 \text{ m}$$

12.37. Establish the equation for flow over a broad-crested weir assuming no lost head (Fig. 12-9).

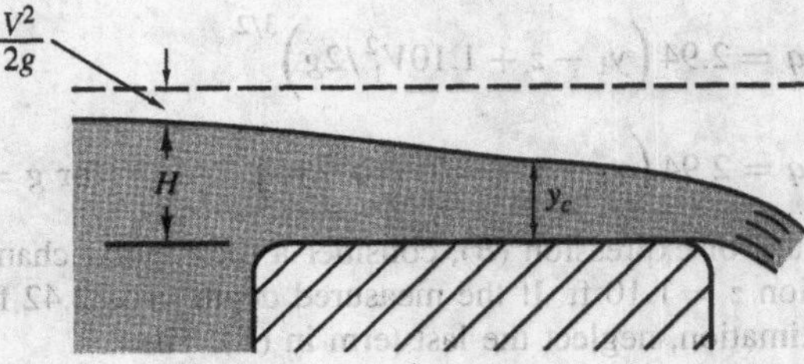

Fig. 12-9

Solution:

At the section where critical flow occurs, $q = V_c y_c$. But $y_c = V_c^2/g = \frac{2}{3}E$ and $V_c = \sqrt{g\left(\frac{2}{3}E_c\right)}$. Hence the theoretical value of flow q becomes

$$q = \sqrt{g\left(\tfrac{2}{3}E_c\right)} \times \tfrac{2}{3}E_c$$

However, the value of E_c is difficult to measure accurately, because the critical depth is difficult to locate. The practical equation becomes

$$q = CH^{3/2}$$

The weir should be calibrated in place to obtain accurate results.

12.38. Develop an expression for a critical flow meter and illustrate the use of the formula.

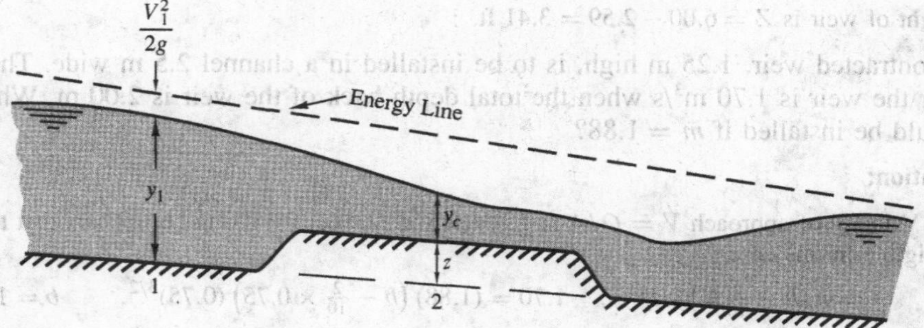

Fig. 12-10

Solution:

An excellent method of measuring flow in open channels is by means of a constriction (Fig. 12-10). The measurement of the critical depth is not required. The depth y_1 is measured a short distance upstream from the constriction. The raised floor should be about $3y_c$ long and of such height as to have the critical velocity occur on it.

For a rectangular channel of constant width, the Bernoulli equation is applied between sections 1 and 2, in which the lost head in accelerated flow is taken as 1/10 of the difference in velocity heads, i.e.,

$$y_1 + \frac{V_1^2}{2g} - \left(\frac{1}{10}\right)\left(\frac{V_c^2}{2g} - \frac{V_1^2}{2g}\right) = \left(y_c + \frac{V_c^2}{2g} + z\right)$$

This equation neglects the slight drop in the channel bed between 1 and 2. Recognizing that $E_c = y_c + V_c^2/2g$, we rearrange as follows:

$$\left(y_1 + 1.10V_1^2/2g\right) = \left[z + 1.0E_c + \left(\tfrac{1}{10}\right)\left(\tfrac{1}{3}E_c\right)\right]$$

$$\left(y_1 - z + 1.10V_1^2/2g\right) = 1.033E_c = (1.033)\left(\tfrac{3}{2}\sqrt[3]{q^2/g}\right)$$

or $$q = 2.94\left(y_1 - z + 1.10V_1^2/2g\right)^{3/2} \qquad (A)$$

Since $q = V_1 y_1$, $$q = 2.94\left(y_1 - z + 0.0171q^2/y_1^2\right)^{3/2} \quad (\text{for } g = 32.2 \text{ ft/sec}^2) \qquad (B)$$

To illustrate the use of expression (B), consider a rectangular channel 10 ft wide with a critical depth meter having dimension $z = 1.10$ ft. If the measured depth y_1 is 2.42 ft, what is the discharge Q?

As a first approximation, neglect the last term in (B). Then

$$q = (2.94)(2.42 - 1.10)^{3/2} = 4.46 \text{ cfs/ft width}$$

Now, using the entire equation (*B*), by successive trials we find $q = 4.80$. Hence

$$Q = q(10) = (4.80)(10) = 48.0 \text{ cfs}$$

12.39. What length of trapezoidal (Cipolletti) weir should be constructed so that the head will be 1.54 ft when the discharge is 122 cfs?

Solution:

$$Q = 3.367bH^{3/2}, \qquad 122 = 3.367b(1.54)^{3/2}, \qquad b = 19.0 \text{ ft}$$

12.40. Establish the formula to determine the time to lower the liquid level in a tank of constant cross section by means of an orifice. Refer to Fig. 12-11.

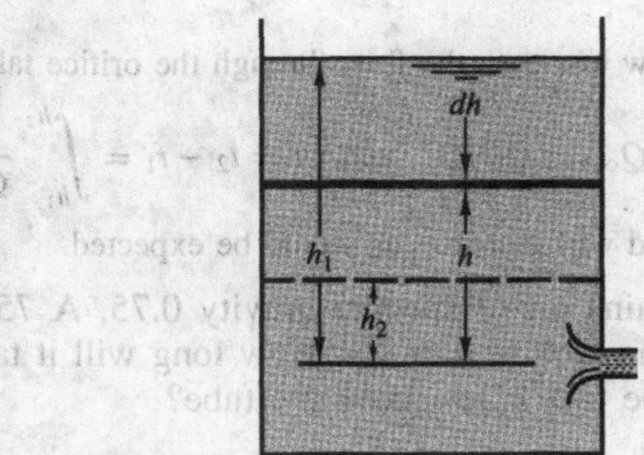

Fig. 12-11

Solution:

Inasmuch as the head is changing with time, we know that $\partial V/\partial t \neq 0$, i.e., we do not have steady flow. This means that the energy equation should be amended to include an acceleration term, which complicates the solution materially. As long as the head does not change too rapidly, no appreciable error will be introduced by assuming steady flow, thus neglecting the acceleration-head term. An approximate check on the error introduced is given in Problem 12.41.

Case A.

With *no inflow* taking place, the instantaneous flow will be

$$Q = cA_o\sqrt{2gh}$$

In time interval dt, the small volume dV discharged will be $Q\,dt$. In the same time interval, the head will decrease dh and the volume discharged will be the area of the tank A_T times dh. Equating these values,

$$\left(c\,A_o\sqrt{2gh}\right)dt = -A_T\,dh$$

where the negative sign signifies that h decreases as t increases. Solving for t yields

$$t = \int_{t_1}^{t_2} dt = \frac{-A_T}{cA_o\sqrt{2g}}\int_{h_1}^{h_2} h^{-1/2}dh$$

or

$$t = t_2 - t_1 = \frac{2A_T}{cA_o\sqrt{2g}}\left(h_1^{1/2} - h_2^{1/2}\right) \qquad (1)$$

In using this expression, an average value of coefficient of discharge c can be used without producing significant error in the result. As h_2 approaches zero, a vortex will form and the orifice will cease to flow full. However, using $h_2 = 0$ will not produce serious error in most cases.

Equation (1) can be rewritten by multiplying and dividing by $\left(h_1^{1/2}+h_2^{1/2}\right)$. There results

$$t = t_2 - t_1 = \frac{A_T(h_1 - h_2)}{\frac{1}{2}\left(cA_o\sqrt{2gh_1}+cA_o\sqrt{2gh_2}\right)} \tag{2}$$

Noting that the volume discharged in time $(t_2 - t_1)$ is $A_T(h_1 - h_2)$, this equation simplifies to

$$t = t_2 - t_1 = \frac{\text{volume discharged}}{\frac{1}{2}(Q_1+Q_2)} = \frac{\text{volume discharged}}{\text{average flow } Q} \tag{3}$$

Problem 12.43 will illustrate a case where tank cross section is not constant but can be expressed as a function of h. Other cases, such as reservoirs emptying, are beyond the scope of this book (see water supply engineering texts).

Case B.

With a constant rate of inflow less than the flow through the orifice taking place,

$$-A_T\,dh = (Q_{out} - Q_{in})dt \quad \text{and} \quad t = t_2 - t_1 = \int_{h_1}^{h_2} \frac{-A_T\,dh}{Q_{out}-Q_{in}}$$

If Q_{in} exceeds Q_{out}, the head will increase, as would be expected.

12.41. A 1.22-m-diameter tank contains oil of specific gravity 0.75. A 75-mm-diameter short tube is installed near the bottom of the tank ($c = 0.85$). How long will it take to lower the level of the oil from 1.83 m above the tube to 1.22 m above the tube?

Mathcad

Solution:

$$t = t_2 - t_1 = \frac{2A_T}{cA_o\sqrt{2g}}\left(h_1^{1/2} - h_2^{1/2}\right) = \frac{2\times\frac{1}{4}\pi(1.22)^2}{0.85\times\frac{1}{4}\pi(0.075)^2\sqrt{2g}}\left(1.83^{1/2} - 1.22^{1/2}\right) = 35 \text{ s.}$$

In order to evaluate the approximate effect of assuming steady flow, the change of velocity with time t is estimated as

$$\frac{\partial V}{\partial t} \cong \frac{\Delta V}{\Delta t} = \frac{\sqrt{2g(1.83)} - \sqrt{2g(1.22)}}{35} = 0.0314 \text{ m/s}^2$$

This is about $\frac{1}{3}$% of acceleration g, a negligible addition to the acceleration g. Such an accuracy is not warranted in these illustrations of unsteady flow, particularly as orifice coefficients are not known to such a degree of accuracy.

12.42. The initial head on an orifice was 2.75 m, and when the flow was terminated the head was measured at 1.22 m. Under what constant head H would the same orifice discharge the same volume of water in the same time interval? Assume coefficient c constant.

Solution:

Volume under falling head = volume under constant head

$$\tfrac{1}{2}cA_o\sqrt{2g}\left(h_1^{1/2}+h_2^{1/2}\right)\times t = cA_o\sqrt{2gH}\times t$$

Substituting and solving, $\frac{1}{2}(\sqrt{2.75}+\sqrt{1.22}) = \sqrt{H}$, and $H = 1.91$ m.

12.43. A tank has the form of a frustum of a cone, with the 8-ft diameter uppermost and the 4-ft diameter at the bottom. The bottom contains an orifice whose average coefficient of discharge may be taken as 0.60. What size orifice will empty the tank in 6 min if the depth full is 10.0 ft? Refer to Fig. 12-12.

Solution:

From Problem 12.40,

$$Q\,dt = -A_T\,dh \quad \text{and} \quad cA_o\sqrt{2gh}\,dt = -\pi x^2\,dh$$

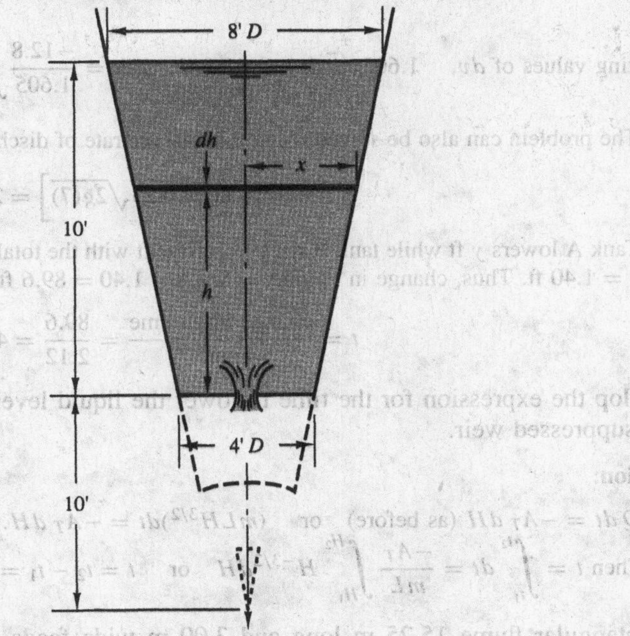

Fig. 12-12

and, by similar triangles, $x/4 = (10+h)/20$. Then

$$\left(0.60 \times \tfrac{1}{4}\pi d_o^2\sqrt{2g}\right) dt = -\pi \frac{(10+h)^2}{25} h^{-1/2} dh$$

$$d_o^2 \int dt = \frac{-4\pi}{25\pi \times 0.60\sqrt{2g}} \int_{10}^{0} (10+h)^2 h^{-1/2}\, dh$$

Since $\int dt = 360$ sec,

$$d_o^2 = \frac{+4}{360 \times 25 \times 0.60\sqrt{2g}} \int_0^{10} (100h^{-1/2} + 20h^{1/2} + h^{3/2}) dh$$

Integrating and solving, we obtain $d^2 = 0.109$ and $d = 0.33$ ft. Use $d = 4''$ orifice.

12.44. Two square tanks have a common wall in which an orifice, area = 0.25 ft^2 and coefficient = 0.80, is located. Tank *A* is 8 ft on a side, and the initial depth above the orifice is 10.0 ft. Tank *B* is 4 ft on a side, and the initial depth above the orifice is 3.0 ft. How long will it take for the water surfaces to be at the same elevation?

Solution:

At any instant the difference in level of the surfaces may be taken as head *h*. Then

$$Q = 0.80 \times 0.25\sqrt{2gh}$$

and the change in volume $dv = Q\,dt = 1.605\sqrt{h}\,dt$.

In this interval of time *dt* the change in head is *dh*. Consider the level in tank A to have fallen *dy*; then the corresponding rise in level in tank B will be the ratio of the areas times *dy*, or $(64/16)\,dy$. The change in head is thus

$$dh = dy + (64/16)dy = 5\,dy$$

The change in volume $dv = 8 \times 8 \times dy \left[= 4 \times 4 \times (64/16)dy \text{ also}\right]$

or, using *dh*, $$dv = (64/5)dh = 12.8dh$$

Equating values of dv, $1.605\sqrt{h}\,dt = -12.8dh$, $dt = \dfrac{-12.8}{1.605}\displaystyle\int_7^0 h^{-1/2}dh$, $t = 42.2$ sec.

The problem can also be solved using the average rate of discharge expressed in (3) of Problem 12.40.

$$Q_{\text{av}} = \tfrac{1}{2}\left[0.80 \times 0.25\sqrt{2g(7)}\right] = 2.12 \text{ cfs}$$

Tank A lowers y ft while tank B rises $(64/16)y$ ft with the total change in level of 7 ft; then $y + 4y = 7$ and $y = 1.40$ ft. Thus, change in volume $= 8 \times 8 \times 1.40 = 89.6$ ft^3 and

$$t = \frac{\text{change in volume}}{\text{average } Q} = \frac{89.6}{2.12} = 42.3 \text{ sec}$$

12.45. Develop the expression for the time to lower the liquid level in a tank, lock, or canal by means of a suppressed weir.

Solution:

$Q\,dt = -A_T\,dH$ (as before) or $(mLH^{3/2})dt = -A_T\,dH$.

Then $t = \displaystyle\int_{t_1}^{t_2} dt = \frac{-A_T}{mL}\int_{H_1}^{H_2} H^{-3/2}dH$ or $t = t_2 - t_1 = \dfrac{2A_T}{mL}\left(H_2^{-1/2} - H_1^{-1/2}\right)$.

12.46. A rectangular flume 15.25 m long and 3.00 m wide feeds a suppressed weir under a head of 0.30 m. If the supply to the flume is cut off, how long will it take for the head on the weir to decrease to 100 mm? Use $m = 1.84$.

Solution:

From Problem 12.45, $t = \dfrac{2(15.25 \times 3.00)}{1.84 \times 3.00}\left[\dfrac{1}{\sqrt{0.100}} - \dfrac{1}{\sqrt{0.30}}\right] = 22.2$ s

12.47. Determine the time required to establish flow in a pipeline of length L under a constant head H discharging into the atmosphere, assuming an inelastic pipe, an incompressible fluid, and constant friction factor f.

Solution:

The final velocity V_f can be determined from the Bernoulli equation, as follows:

$$H - f\left(\frac{L}{d}\right)\left(\frac{V_f^2}{2g}\right) - k\left(\frac{V_f^2}{2g}\right) = \left(0 + \frac{V_f^2}{2g} + 0\right)$$

In this equation, the minor losses are represented by the $kV_f^2/2g$ term, and the energy in the jet at the end of the pipe is kinetic energy represented by $V_f^2/2g$. This equation can be written in the form

$$\left[H - f\left(\frac{L_E}{d}\right)\left(\frac{V_f^2}{2g}\right)\right] = 0 \qquad (1)$$

where L_E is the equivalent length of pipe for the system (see Problem 9.3, Chapter 9).

From Newton's equation of motion, at any instant

$$\gamma(AH_e) = M\frac{dV}{dt} = \frac{\gamma}{g}(AL)\frac{dV}{dt}$$

where H_e is the effective head at any instant and V is a function of time and not length. Rearranging the equation,

$$dt = \left(\frac{\gamma AL}{g\gamma AH_e}\right)dV \quad \text{or} \quad dt = \frac{L\,dV}{gH_e} \qquad (2)$$

In equation (*1*), for all intermediate values of V the term in the brackets is not zero but is the effective head available to cause the acceleration of the liquid. Hence expression (*2*) may be written

$$\int dt = \int \frac{L\,dV}{g\left[H - f\left(\frac{L_E}{d}\right)\left(\frac{V^2}{2g}\right)\right]} = \int \frac{L\,dV}{g\left[f\left(\frac{L_E}{d}\right)\left(\frac{V_f^2}{2g}\right) - f\left(\frac{L_E}{d}\right)\left(\frac{V^2}{2g}\right)\right]} \qquad (3A)$$

Since from (*1*), $\dfrac{fL_E}{2gd} = \dfrac{H}{V_f^2}$, $\qquad \displaystyle\int dt = \int \frac{L\,dV}{g\left(H - HV^2/V_f^2\right)} \qquad (3B)$

or

$$\int_o^t dt = \frac{L}{gH}\int_0^{V_f} \frac{V_f^2}{V_f^2 - V^2}\,dV$$

Integrating,

$$t = \frac{LV_f}{2gH}\ln\left(\frac{V_f + V}{V_f - V}\right) \qquad (4)$$

It will be noted that as V approaches final velocity V_f, $(V_f - V)$ approaches zero. Thus, mathematically, time t approaches infinity.

Equation (*3B*) may be rearranged, using symbol ϕ for the ratio V/V_f. Then

$$\frac{dV}{dt} = \frac{gH}{L}\left(1 - V^2/V_f^2\right) = \frac{gH}{L}\left(1 - \phi^2\right) \qquad (5)$$

Using $V = V_f\phi$ and $\dfrac{dV}{dt} = V_f\dfrac{d\phi}{dt}$, we obtain

$$\frac{d\phi}{1 - \phi^2} = \frac{gH\,dt}{V_f L}$$

Integrating,

$$\frac{1}{2}\ln\left(\frac{1+\phi}{1-\phi}\right) = \frac{gHt}{V_f L} + C$$

and when $t = 0$, $C = 0$. Then

$$\frac{1+\phi}{1-\phi} = e^{2gHt/V_f L}$$

Using hyperbolic functions, $\phi = \tanh(gHt/V_f L)$, and since $\phi = V/V_f$,

$$V = V_f \tanh\frac{gHt}{V_f L} \qquad (6)$$

The advantage of expression (*6*) is that the value of velocity V in terms of the final velocity V_f can be calculated for any chosen time.

12.48. Simplify equation (*4*) in Problem 12.47 to give the time to establish flow such that velocity V equals (*a*) 0.75, (*b*) 0.90, and (*c*) 0.99 times final velocity V_f.

Solution:

(*a*) $\quad t = \dfrac{LV_f}{2gH}\ln\left[\dfrac{V_f + 0.75V_f}{V_f - 0.75V_f}\right] = \left(\dfrac{LV_f}{2gH}\right)(2.3026)\log\dfrac{1.75}{0.25} = 0.973\dfrac{LV_f}{gH}$

(*b*) $\quad t = \dfrac{LV_f}{2gH}\ln\dfrac{1.90}{0.10} = \left(\dfrac{LV_f}{2gH}\right)(2.3026)\log\dfrac{1.90}{0.10} = 1.472\dfrac{LV_f}{gH}$

(*c*) $\quad t = \dfrac{LV_f}{2gH}\ln\dfrac{1.99}{0.01} = \left(\dfrac{LV_f}{2gH}\right)(2.3026)\log\dfrac{1.99}{0.01} = 2.647\dfrac{LV_f}{gH}$

12.49. Water is discharged from a tank through 2000 ft of 12″ pipe ($f = 0.020$). The head is constant at 20 ft. Valves and fittings in the line produce losses of $21\left(V^2/2g\right)$. After a valve is opened, how long will it take to attain a velocity of 0.90 of the final velocity?

Solution:

Bernoulli's equation, tank surface to end of pipe, will yield

$$(0+0+H) - [f(L/d) + 21.0]V^2/2g = (0 + V^2/2g + 0)$$

or $H = [0.020(2000/1) + 22.0]V^2/2g = 62.0(V^2/2g)$. Then using the procedure in Problem 9.3 of Chapter 9,

$$62.0\,(V^2/2g) = 0.020(L_E/1)\,(V^2/2g) \qquad \text{or} \qquad L_E = 3100 \text{ ft}$$

Inasmuch as equation (4) in Problem 12.47 does not contain L_E, the final velocity must be calculated, as follows:

$$H = f\left(\frac{L_E}{d}\right)\left(\frac{V_f^2}{2g}\right) \qquad \text{or} \qquad V_f = \sqrt{\frac{2gdH}{fL_E}} = \sqrt{\frac{(64.4)(1)(20.0)}{(0.020)(3100)}} = 4.56 \text{ ft/sec}$$

Substituting in (b) of Problem 12.48 yields $t = (1.472)\dfrac{(2000)(4.56)}{(32.2)(20.0)} = 20.8$ sec.

12.50. In Problem 12.49, what velocity will be attained in 10 sec and in 15 sec?

Solution:

In equation (6) of Problem 12.47, evaluate gHt/V_fL.

$$\text{For 10 sec, } \frac{32.2 \times 20 \times 10}{4.56 \times 2000} = 0.706. \quad \text{For 15 sec, } \frac{32.2 \times 20 \times 15}{4.56 \times 2000} = 1.059.$$

Using a table of hyperbolic functions and equation (6), $V = V_f \tanh\left(gHt/V_fL\right)$, we obtain

For 10 sec, $\quad V = 4.56 \tanh 0.706 = 4.56 \times 0.6082 = 2.77$ ft/sec

For 15 sec, $\quad V = 4.56 \tanh 1.059 = 4.56 \times 0.7853 = 3.58$ ft/sec

It will be noted that the value V/V_f is represented by the value of the hyperbolic tangent. In the solution above, 61% and 79% of the final velocity are attained in the 10 and 15 sec, respectively.

12.51. A rectangular channel 20 ft wide, $n = 0.025$, flows 5 ft deep on a slope of 14.7 ft in 10,000 ft. A suppressed weir C, 2.45 ft high, is built across the channel ($m = 3.45$). Taking the elevation of the bottom of the channel just upstream from the weir to be 100.00′, estimate (using one reach) the elevation of the water surface at a point A, 1000 ft upstream from the weir.

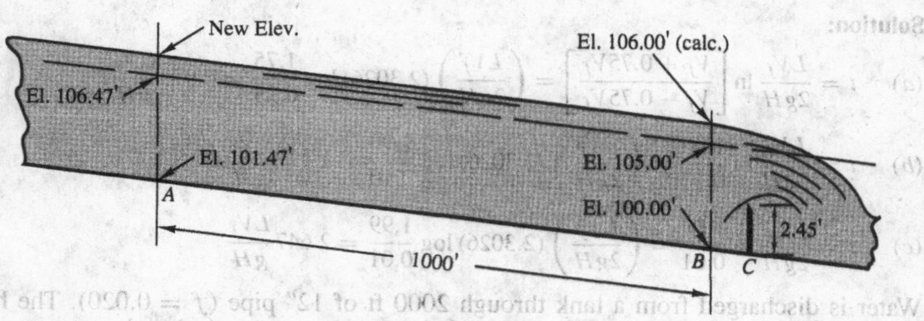

Fig. 12-13

Solution:

Calculate the new elevation of water surface at B in Fig. 12-13 (before drop-down). Note that the flow is nonuniform, as the depths, velocities, and areas are not constant after the weir is installed.

$$Q = (20 \times 5)(1.486/0.025)(100/30)^{2/3}(0.00147)^{1/2} = 509 \text{ cfs}$$

For an estimated depth of 6 ft just upstream from the weir,

$$\text{velocity of approach } V = Q/A = 509/(20 \times 6) = 4.24 \text{ ft/sec}$$

The weir formula gives $509 = 3.45 \times 20 \left[\left(H + \frac{(4.24)^2}{2g} \right)^{3/2} - \left(\frac{(4.24)^2}{2g} \right)^{3/2} \right]$

$$H = 3.56 \text{ ft}$$
$$\text{height } Z = \underline{2.45 \text{ ft}}$$
$$\text{depth } y = 6.01 \text{ ft (assumption checks)}$$

New elevation at A must lie between 106.47 and 107.47. Try an elevation of 106.90 (and check in the Bernoulli equation).

$$\text{new area at } A = (20)(106.90 - 101.47) = 108.6 \text{ ft}^2, \text{ and } V = 509/108.6 = 4.69 \text{ ft/sec}$$

$$\text{mean velocity} = \tfrac{1}{2}(4.24 + 4.69) = 4.46 \text{ ft/sec}$$

$$\text{mean hydraulic radius } R = \left(\tfrac{1}{2}\right)(120.0 + 108.6)/\left[\tfrac{1}{2}(32.0 + 30.9)\right] = 3.63$$

$$\text{lost head } h_L = \left(\frac{Vn}{1.486R^{2/3}} \right)^2 L = \left(\frac{4.46 \times 0.025}{(1.486)(3.63)^{2/3}} \right)^2 (1000) = 1.01 \text{ ft}$$

Now apply the Bernoulli equation, A to B, datum B,

$$\left[106.90 + (4.69)^2/2g\right] - 1.01 = \left[106.00 + (4.24)^2/2g\right]$$

which reduces to $106.23 = 106.28$ (approximately)

The difference of 0.05 ft is within the error in roughness factor n alone. Further refinement does not seem justified. Use elevation of 106.90 ft.

Supplementary Problems

12.52. Turpentine at 20°C flows through a pipe in which a Pitot-static tube having a coefficient of 0.97 is centered. The differential gage containing mercury shows a deflection of 102 mm. What is the center velocity?
Ans. 5.27 m/s

12.53. Air at 120°F flows by a Pitot-static tube at a velocity of 60.0 ft/sec. If the coefficient of the tube is 0.95, what water differential reading is expected, assuming constant specific weight of the air at atmospheric pressure? *Ans.* 0.816 in

12.54. The lost head through a 2″-diameter orifice under a certain head is 0.540 ft, and the velocity of the water in the jet is 22.5 ft/sec. If the coefficient of discharge is 0.61, determine the head causing flow, the diameter of the jet, and the coefficient of velocity. *Ans.* 8.39 ft, 1.59″, 0.97

12.55. What size standard orifice is required to discharge 0.016 m³/s of water under a head of 8.69 m?
Ans. 50 mm

12.56. A sharp-edged orifice has a diameter of 1″ and coefficients of velocity and contraction of 0.98 and 0.62, respectively. If the jet drops 3.08 ft in a horizontal distance of 8.19 ft, determine the flow in cfs and the head on the orifice. *Ans.* 0.0632 cfs, 5.66 ft

12.57. Oil of specific gravity 0.800 flows from a closed tank through a 75-mm-diameter orifice at the rate of 0.026 m^3/s. The diameter of the jet is 58.5 mm. The level of the oil is 7.47 m above the orifice, and the air pressure is equivalent to −152 mm of mercury. Determine the three coefficients of the orifice. *Ans.* 0.580, 0.590, 0.982

12.58. Refer to Fig. 12-14. The 3″-diameter orifice has coefficients of velocity and contraction of 0.950 and 0.632, respectively. Determine (*a*) the flow for the deflection of the mercury indicated and (*b*) the power in the jet. *Ans.* (*a*) 1.04 cfs, (*b*) 2.06 hp

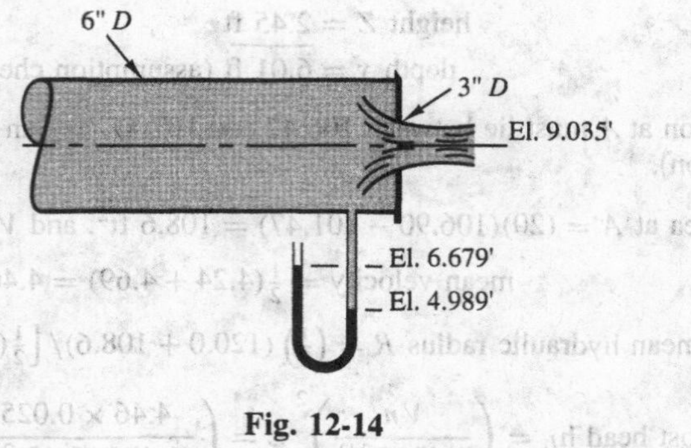

Fig. 12-14

12.59. Refer to Fig. 12-15. Heavy fuel oil at 60°F flows through the 3″-diameter end-of-pipe orifice, causing the deflection of the mercury in the U-tube gage. Determine the power in the jet. *Ans.* 2.90 hp

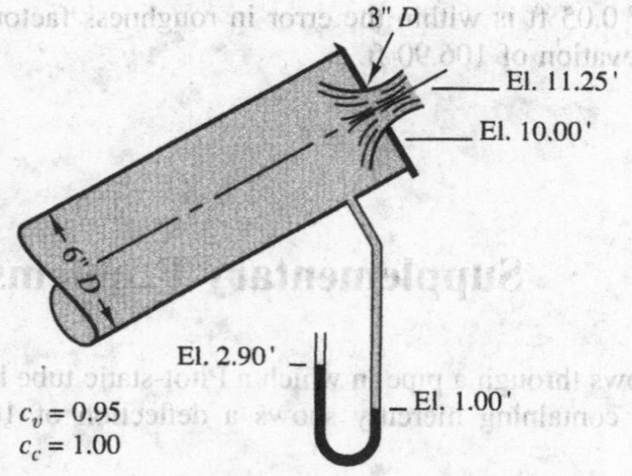

Fig. 12-15

12.60. Steam locomotives sometimes take on water by means of a scoop that dips into a long, narrow tank between the rails. If the lift into the tank is 2.74 m, at what velocity must the train travel (neglecting friction)? *Ans.* 26.4 km/h

12.61. Air at 15°C flows through a large duct and thence through a 75-mm-diameter hole in the thin metal (c = 0.62). A U-tube gage containing water registers 31.7 mm. Considering the specific weight of air constant, what is the flow through the opening? *Ans.* 46 N/min

12.62. An oil having specific gravity 0.926 and viscosity 350 Saybolt seconds flows through a 3″-diameter pipe orifice placed in a 5″-diameter pipe. The differential gage registers a pressure drop of 21.5 psi. Determine flow Q. *Ans.* 1.93 cfs

12.63. A nozzle with a 50-mm-diameter tip is attached at the end of a 2000-mm horizontal pipeline. The coefficients of velocity and contraction are respectively 0.976 and 0.909. A pressure gage attached at the base of the nozzle and located 2.16 m above its centerline reads 221 kPa. Determine the flow of water in m³/s. *Ans.* 0.040 m³/s

12.64. When the flow of water through a horizontal 300 mm by 150 mm Venturi meter ($c = 0.95$) is 0.111 m³/s, find the deflection of the mercury in the differential gage attached to the meter. *Ans.* 157 mm

12.65. When 4.20 cfs of water flows through a 12″ by 6″ Venturi meter, the differential gage records a difference of pressure heads of 7.20 ft. What is the indicated coefficient of discharge of the meter? *Ans.* 0.964

12.66. The loss of head from entrance to throat of a 250 mm by 125 mm Venturi meter is 1/16 times the throat velocity head. When the mercury in the differential gage attached to the meter deflects 102 mm, what is the indicated flow of water? *Ans.* 0.063 m³/s

12.67. A 300 mm by 150 mm Venturi meter ($c = 0.985$) carries 0.0566 m³/s of water with a differential gage reading of 0.634 m. What is the specific gravity of the gage liquid? *Ans.* 1.75

12.68. Methane flows at a rate of 16.5 lb/sec through a 12″ by 6″ Venturi meter at a temperature of 60°F. The pressure at the meter inlet is 49.5 psi absolute. Using $k = 1.31$, $R = 96.3$ ft/°R, $\nu = 1.94 \times 10^{-4}$ ft²/sec at 1 atm, and $\gamma = 0.0416$ lb/ft³ at 68°F and 1 atm, calculate the expected deflection of the mercury in the differential gage. *Ans.* 1.03 ft

12.69. Water flows through a 150-mm pipe in which a long-radius flow nozzle 75 mm in diameter is installed. For a deflection of 152 mm of mercury in the differential gage, calculate the rate of flow. (Assume $c = 0.98$ from Diagram D.) *Ans.* 0.028 m³/s

12.70. Water at 30°C flows at the rate of 0.046 m³/s through the nozzle in Problem 12.69. What is the deflection of the mercury in the differential gage? (Use Diagram D.) *Ans.* 393 mm

12.71. If a dust proofing oil at 30°C had been flowing at 0.046 m³/s in Problem 12.70, what would have been the deflection of the mercury? *Ans.* 372 mm

12.72. If air at 20°C flows through the same pipe and nozzle as in Problem 12.69, how many newtons of air per second will flow if the absolute pressures in pipe and jet are 207 kPa and 172 kPa, respectively? *Ans.* 17 N/s

12.73. What depth of water must exist behind a rectangular sharp-crested suppressed weir 1.52 m long and 1.22 m high when a flow of 0.283 m³/s passes over it? (Use the Francis formula.) *Ans.* 1.44 m

12.74. A flow of 30 cfs occurs in a rectangular flume 4 ft deep and 6 ft wide. Find the height at which the crest of a sharp-edged suppressed weir should be placed in order that water will not overflow the sides of the flume, ($m = 3.33$) *Ans.* 2.72 ft

12.75. A flow of 10.9 m^3/s passes over a suppressed weir which is 4.88 m long. The total depth upstream from the weir must not exceed 2.44 m. Determine the height to which the crest should be placed to carry this flow. ($m = 1.85$) *Ans.* 1.34 m

12.76. A suppressed weir ($m = 3.33$) under a constant head of 0.300 ft feeds a tank containing a 3″-diameter orifice. The weir, which is 2 ft long and 2.70 ft high, is located in a rectangular channel. The lost head through the orifice is 2.00 ft, and $c_c = 0.65$. Determine the head to which the water will rise in the tank and the coefficient of velocity for the orifice. *Ans.* $h = 20.3$ ft, $c_v = 0.95$

12.77. A contracted weir 1.22 m long is placed in a rectangular channel 2.74 m wide. The height of weir crest is 1.00 m and the head is 381 mm. Determine the flow, using $m = 1.88$. *Ans.* 0.504 m^3/s

12.78. A triangular weir has a 90° notch. What head will produce 1200 gpm? ($m = 2.50$) *Ans.* 1.027 ft

12.79. A 36″ pipeline containing a 36″ by 12″ Venturi meter supplies water to a rectangular canal. The pressure is 30.0 psi at the inlet of the Venturi and 8.67 psi at the throat. A suppressed weir ($m = 3.33$), 3 ft high, placed in the canal, discharges under a 9″ head. What is the probable width of the canal? *Ans.* 20.0 ft

12.80. Water flows over a suppressed weir ($m = 1.85$) that is 3.66 m long and 0.610 m high. For a head of 0.366 m, find the flow. *Ans.* 1.54 m^3/s

12.81. A tank 12 ft long and 4 ft wide contains 4 ft of water. How long will it take to lower the water to a 1-ft depth if a 3″ diameter orifice ($c = 0.60$) is opened in the bottom of the tank? *Ans.* 406 sec

12.82. A rectangular tank 4.88 m by 1.22 m contains 1.22 m of oil, specific gravity 0.75. If it takes 10 min and 10 s to empty the tank through a 100 mm diameter orifice in the bottom, determine the average value of the coefficient of discharge. *Ans.* 0.60

12.83. In Problem 12.82, for a coefficient of discharge of 0.60, what will be the depth after the orifice has been flowing for 5 min.? *Ans.* 0.314 m

12.84. A tank with a trapezoidal cross section has a constant length of 5 ft. When the water is 8 ft deep above the 2-in-diameter orifice ($c = 0.65$), the width of the water surface is 6 ft and, at a 3-ft-head depth, the width of water surface is 4 ft. How long will it take to lower the water from the 8-ft depth to the 3-ft depth? *Ans.* 482 sec

12.85. A suppressed weir is located in the end of a tank that is 3.05 m square. If the initial head on the weir is 0.610 m, how long will it take for 3.54 m^3 of water to run out of the tank? ($m = 1.85$) *Ans.* 2.68 s

12.86. A rectangular channel 60 ft long by 10 ft wide is discharging its flow over a 10-ft-long suppressed weir under a head of 1.00 ft. If the supply is suddenly cut off, what will be the head on the weir in 36 sec? ($m = 3.33$) *Ans.* 0.25 ft

12.87. Two orifices in the side of a tank are one above the other vertically and are 1.83 m apart. The total depth of water in the tank is 4.27 m, and the head on the upper orifice is 1.22 m. For the same values of c_v, show that the jets will strike the horizontal plane on which the tank rests at the same point.

12.88. A 6″-diameter orifice discharges 12.00 cfs of water under a head of 144 ft. This water flows into a 12-ft-wide rectangular channel at a depth of 3 ft, and then it flows over a contracted weir. The head on the weir is 1.000 ft. What are the length of the weir and the coefficient of the orifice? *Ans.* 3.80 ft, $c = 0.635$

12.89. The head on a suppressed weir G that is 12 ft long is 1.105 ft, and the velocity of approach can be neglected. For the system shown in Fig. 12-16, what is the pressure head at B? Sketch the hydraulic grade lines. *Ans.* 193.2 ft

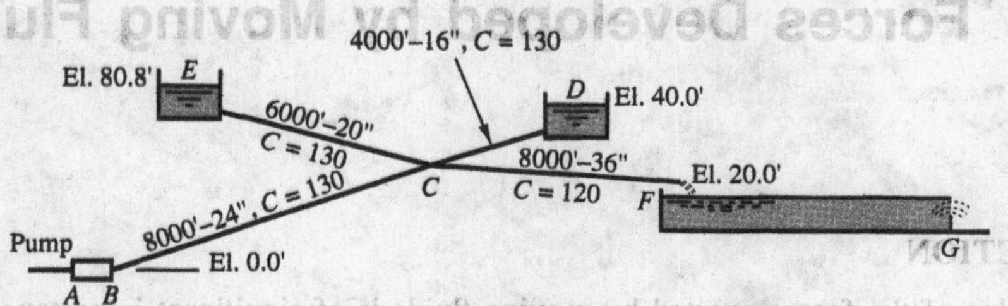

Fig. 12-16

12.90. In Fig. 12-17 the elevation of the hydraulic grade line at B is 50.0′ and pipes BC and BD are arranged so that the flow from B divides equally. What is the elevation of the end of the pipe at D, and what is the head that will be maintained on the 4″-diameter orifice E? *Ans.* El. 23.8′, $h = 22.5$ ft

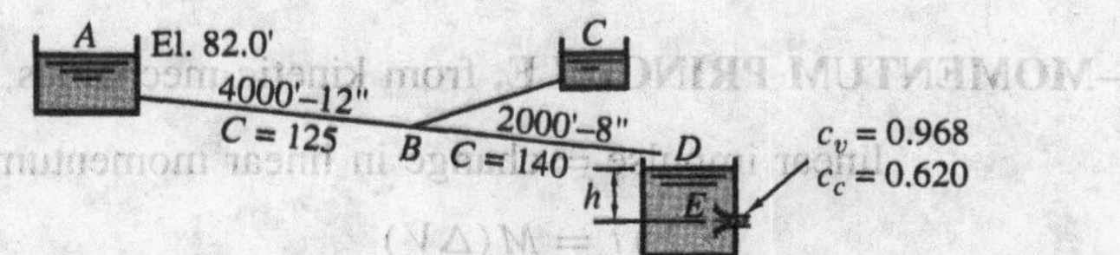

Fig. 12-17

12.91. For the tank shown in Fig. 12-18, using an average coefficient of discharge of 0.65 for the 2″ diameter orifice, how long will it take to lower the tank level 4 ft. *Ans.* 390 sec

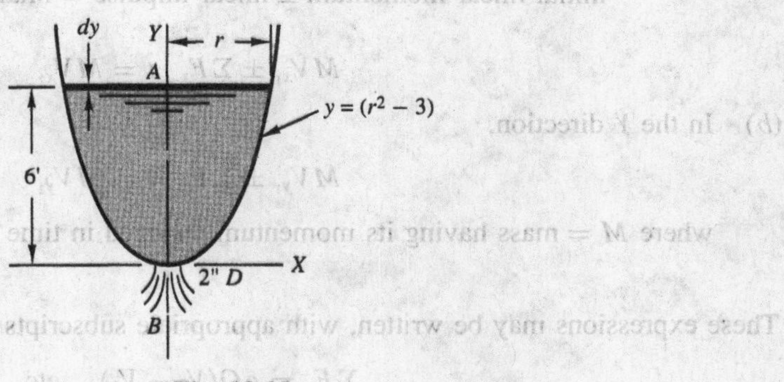

Fig. 12-18

12.92. A broad-crested weir is 1.25 ft high above the bottom of a rectangular channel 10 ft wide. The measured head above the weir crest is 1.95 ft. Determine the approximate flow in the channel. (Use $c = 0.92$) *Ans.* 83.5 cfs

Chapter 13

Forces Developed by Moving Fluids

INTRODUCTION

Knowledge of the forces exerted by moving fluids is of significant importance in the analysis and design of such objects as pumps, turbines, airplanes, rockets, spacecraft, propellers, ships, automobile bodies, buildings, and many hydraulic devices. The energy relationship is not sufficient to solve most of these problems. One additional tool of mechanics, the momentum principle, is most important. The boundary layer theory provides a further basis for analysis. Extensive and continuing experiments add data regarding laws of variation of fundamental coefficients.

THE IMPULSE–MOMENTUM PRINCIPLE, from kinetic mechanics, states that

$$\text{linear impulse} = \text{change in linear momentum}$$

or

$$(\Sigma F)\,t = M(\Delta V)$$

The quantities in the equation are vector quantities and must be added and subtracted accordingly. Components are usually most convenient, and to avoid possible mistakes in sign the following forms are suggested:

(*a*) In the *X* direction,

$$\text{initial linear momentum} \pm \text{linear impulse} = \text{final linear momentum}$$

$$MV_{x_1} \pm \Sigma F_x \cdot t = MV_{x_2} \tag{1}$$

(*b*) In the *Y* direction,

$$MV_{y_1} \pm \Sigma F_y \cdot t = MV_{y_2} \tag{2}$$

where M = mass having its momentum changed in time t.

These expressions may be written, with appropriate subscripts x, y, or z, in the following form:

$$\Sigma F_x = \rho Q(V_2 - V_1)_x, \text{ etc.} \tag{3}$$

THE MOMENTUM CORRECTION FACTOR β, evaluated in Problem 13.1, is

$$\beta = \frac{1}{A}\int_A (v/V)^2\, dA \tag{4}$$

For laminar flow in pipes, $\beta = 1.33$. For turbulent flow in pipes, β varies from 1.01 to 1.07. In most cases β can be considered as unity.

DRAG

Drag is the component of the resultant force exerted by a fluid on a body *parallel* to the relative motion of the fluid. The usual equation is

$$\text{drag in lb (or N)} = C_D \rho A \frac{V^2}{2} \tag{5}$$

LIFT

Lift is the component of the resultant force exerted by a fluid on a body *perpendicular* to the relative motion of the fluid. The usual equation is

$$\text{lift in lb (or N)} = C_L \rho A \frac{V^2}{2} \tag{6}$$

where C_D = the drag coefficient, which is dimensionless
C_L = the lift coefficient, which is dimensionless
ρ = the density of the fluid, in slugs/ft^3 (or kg/m^3)
A = some characteristic area in ft^2 (or m^2), usually the area projected on a plane perpendicular to the relative motion of the fluid
V = relative velocity of the fluid with respect to the body, in ft/sec (or m/s)

TOTAL DRAG FORCE

Total drag force consists of friction drag and pressure drag. However, seldom are both these effects of appreciable magnitude simultaneously. For objects that exhibit no lift, profile drag is synonymous with total drag. The following tabulation will illustrate.

Object	Friction drag		Pressure drag		Total drag
1. Spheres	negligible	+	Pressure drag	=	total drag
2. Cylinders (axis perpendicular to velocity)	negligible	+	Pressure drag	=	total drag
3. Disks and thin plates (perpendicular to velocity)	zero	+	Pressure drag	=	total drag
4. Thin plates (parallel to velocity)	Friction drag	+	negligible to zero	=	total drag
5. Well-streamlined objects	Friction drag	+	small to negligible	=	total drag

DRAG COEFFICIENTS

Drag coefficients are dependent upon Reynolds number at low and intermediate velocities but are independent at high velocity. However, at high velocities the drag coefficient is related to the Mach number, which has negligible effect at low velocities. Diagrams *F*, *G*, and *H* illustrate the variations for certain geometric shapes. Problems 13.24 and 13.40 discuss these relations.

For flat plates and airfoils, the drag coefficients are usually tabulated for the plate area and for the chord-length product, respectively.

LIFT COEFFICIENTS

Kutta gives theoretical maximum values of lift coefficients for thin flat plates not normal to the relative velocity of the fluid as

$$C_L = 2\pi \sin \alpha \tag{7}$$

where α = the angle of attack or the angle the plate makes with the relative velocity of the fluid. In normal range of operation, airfoil sections have values about 90% of this theoretical maximum. Angle α should not exceed 25°.

MACH NUMBER

Mach number is the dimensionless ratio of the velocity of the fluid to the acoustic velocity (sometimes called celerity).

$$\text{Mach number} = \text{Ma} = \frac{V}{c} = \frac{V}{\sqrt{E/\rho}} \tag{8}$$

For gases, $c = \sqrt{kgRT}$ (see Chapter 1).

Values of V/c up to the critical value of 1.0 indicate subsonic flow; at 1.0, sonic flow; and above 1.0, supersonic flow (see Diagram *H*).

BOUNDARY LAYER THEORY

Boundary layer theory was first developed by Prandtl. He showed that for a moving fluid all friction losses occur within a thin layer adjacent to a solid boundary (called the boundary layer) and that flow outside this layer can be considered frictionless. The velocity near the boundary is affected by boundary shear. In general, the boundary layer is very thin at the upstream boundaries of an immersed object but increases in thickness due to the continual action of shear stress.

At low Reynolds numbers, the entire boundary layer is governed by viscous forces and laminar flow occurs therein. For intermediate values of Reynolds number the boundary layer is laminar near the boundary surface and turbulent beyond. For high Reynolds numbers the entire boundary layer is turbulent.

FLAT PLATES

For flat plates of length L, held parallel to the relative motion of a fluid, the following equations are applicable.

1. Boundary Layer Laminar (up to Reynolds number about 500,000).

(*a*) Mean drag coefficient (C_D) $= \dfrac{1.328}{\sqrt{\text{Re}}} = \dfrac{1.328}{\sqrt{VL/\nu}}$ (*9*)

(*b*) Boundary layer thickness δ at any distance x is given by

$$\frac{\delta}{x} = \frac{5.20}{\sqrt{\text{Re}_x}} = \frac{5.20}{\sqrt{Vx/\nu}} \tag{10}$$

(*c*) Shear stress τ_o is estimated by

$$\tau_o = 0.33\rho V^{3/2}\sqrt{\nu/x} = 0.33(\mu V/x)\sqrt{\text{Re}_x} = \frac{0.33\rho V^2}{\sqrt{\text{Re}_x}} \tag{11}$$

where V = velocity of the fluid approaching the boundary (ambient velocity)
x = distance from the leading edge
L = total length of plate
Re_x = local Reynolds number, for distance x

It can be seen that the thickness of the boundary layer will increase as the square root of dimension x increases and also as the square root of the kinematic viscosity increases, while δ

will decrease as the square root of the velocity increases. Similarly, the boundary shear τ_0 will increase as ρ and μ increase, will decrease as the square root of x increases, and will increase as the power of V increases.

2. **Boundary Layer Turbulent** (smooth boundary)

(*a*) Mean drag coefficient

$$(C_D) = \frac{0.074}{\text{Re}^{0.20}} \qquad \text{for } 2 \times 10^5 < \text{Re} < 10^7 \tag{12}$$

$$= \frac{0.455}{(\log_{10}\text{Re})^{2.58}} \qquad \text{for } 10^6 < \text{Re} < 10^9 \tag{13}$$

For a rough boundary, the drag coefficient varies with the relative roughness ϵ/L and not with Reynolds number.

K. E. Schoenherr suggests the formula $1/\sqrt{C_D} = 4.13\log(C_D\text{Re}_x)$, which is considered more accurate than expressions (*12*) and (*13*), particularly for Reynolds numbers greater than 2×10^7.

(*b*) Boundary layer thickness δ is estimated by

$$\frac{\delta}{x} = \frac{0.38}{\text{Re}_x^{0.20}} \qquad \text{for } 5 \times 10^4 < \text{Re} < 10^6 \tag{14}$$

$$= \frac{0.22}{\text{Re}_x^{0.167}} \qquad \text{for } 10^6 < \text{Re} < 5 \times 10^8 \tag{15}$$

(*c*) Shear stress is estimated by

$$\tau_o = \frac{0.023\rho V^2}{(\delta V/\nu)^{1/4}} = 0.0587\frac{V^2}{2}\rho\left(\frac{\nu}{xV}\right)^{1/5} \tag{16}$$

3. **Boundary Layer in Transition** from laminar to turbulent on plate surface (Re from about 500,000 to about 20,000,000).

(*a*) Mean drag coefficient

$$(C_D) = \frac{0.455}{(\log_{10}\text{Re})^{2.58}} - \frac{1700}{\text{Re}} \tag{17}$$

Diagram *G* illustrates the variation of C_D with Reynolds number for these three conditions of flow.

WATER HAMMER

Water hammer is the term used to express the resulting shock caused by the sudden decrease in the motion (velocity) of a fluid. In a pipeline the time of travel of the pressure wave up and back (round trip) is given by

$$\text{time (round trip)} = 2 \times \frac{\text{length of pipe}}{\text{celerity of pressure wave}}$$

or

$$T = \frac{2L}{c} \tag{18}$$

The increase in pressure caused by the sudden closing of a valve is calculated by

$$\text{change in pressure} = \text{density} \times \text{celerity} \times \text{change in velocity}$$

or

$$dp = \rho c\, dV \quad \text{or} \quad dh = c\, dV/g \tag{19}$$

where dh is the change in pressure head.

For rigid pipes, the velocity (celerity) of the pressure wave is

$$c = \sqrt{\frac{\text{bulk modulus of fluid}}{\text{density of fluid}}} = \sqrt{\frac{E_B}{\rho}} \tag{20}$$

For nonrigid pipes the expression is

$$c = \sqrt{\frac{E_B}{\rho[1 + (E_B/E)(d/t)]}} \tag{21}$$

where E = modulus of elasticity of pipe walls,
d = inner diameter of pipe
t = thickness of wall of pipe

SUPERSONIC SPEEDS

Supersonic speeds completely change the nature of the flow. The drag coefficient is related to Mach number Ma (see Diagram H) since viscosity has a small effect upon the drag. The pressure disturbance created forms a cone with the apex at the nose of the body or projectile. The cone represents the wave front or *shock wave*, which may be photographed. The cone angle or *Mach angle* α is given by

$$\sin\alpha = \frac{\text{celerity}}{\text{velocity}} = \frac{1}{V/c} = \frac{1}{\text{Ma}} \tag{22}$$

Solved Problems

13.1. Determine the momentum correction factor β that should be applied when using average velocity V in the momentum calculation (two-dimensional flow).

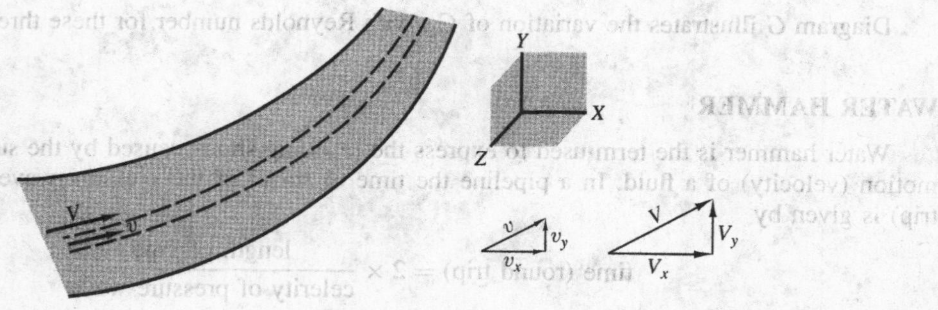

Fig. 13-1

Solution:

The mass discharge dM through the streamtube shown in Fig. 13-1 is $\rho\,dQ$. The correct momentum in the X direction is

$$\text{Mom}_x = \int dM\,v_x = \int \rho\,dQ\,v_x = \int \rho v_x(v\,dA)$$

Using the average velocity for the cross section, the correct momentum would be

$$\text{Mom}_x = \beta\,(MV_x) = \beta\,(\rho QV_x) = \beta\rho(AV)V_x$$

Equating the two values of correct momentum produces

$$\beta = \frac{\int \rho v_x (v\,dA)}{\rho A V(V_x)} = \frac{1}{A}\int_A (v/V)^2 dA$$

since, from the velocity vector diagrams, $v_x/V_x = v/V$.

13.2. Evaluate the momentum correction factor if the velocity profile satisfies the equation $v = v_{max}\left[\left(r_o^2 - r^2\right)/r_o^2\right]$. (See Chap. 7, Problem 7.18 for sketch.)

Solution:

From Problem 7.18, Chap. 7, the average velocity was found to be $\frac{1}{2}v_{max}$. Using this value for average velocity V, we obtain

$$\beta = \frac{1}{A}\int_A \left(\frac{v}{V}\right)^2 dA = \frac{1}{\pi r_o^2}\int_0^{r_o}\left[\frac{v_{max}\left(r_o^2 - r^2\right)/r_o^2}{\frac{1}{2}v_{max}}\right]^2 (2\pi r\,dr)$$

$$= \frac{8}{r_o^6}\left(\tfrac{1}{2}r_o^6 - \tfrac{1}{2}r_o^6 + \tfrac{1}{6}r_o^6\right) = \tfrac{4}{3} = 1.33$$

13.3. A jet of water 3″ in diameter and moving to the right impinges on a flat plate held normal to its axis. (*a*) For a velocity of 80.0 ft/sec, what force will keep the plate in equilibrium? (*b*) Compare the average dynamic pressure on the plate with the maximum (stagnation) pressure if the plate is 20 times the area of the jet.

Solution:

(*a*) Let the X axis lie along the jet's path. Thus the plate destroys the initial momentum of the water in the X direction. Taking M as the mass of water that has its momentum reduced to zero in dt seconds and F_x to the left as the force exerted by the plate on the water, we have

$$\text{initial linear momentum} - \text{linear impulse} = \text{final linear momentum}$$

$$M(80.0) - F_x dt = M(0)$$

$$\frac{\gamma Q}{g}dt\,(80.0) - F_x dt = 0$$

and
$$F_x = \frac{(62.4)\overbrace{\left[(\pi/4)(3/12)^2\right]}^{A}\overbrace{(80.0)}^{V}\times \overbrace{80.0}^{V}}{32.2} = 609 \text{ lb (to the left for equilibrium).}$$

There is no Y component of force involved in this problem, the Y components on the plate balancing (and canceling) each other. Note that time dt cancels and might well have been chosen as 1 sec.

For the plate it is well to recognize that this impulse-momentum expression can be arranged as follows:

$$F = MV = \frac{\gamma Q}{g}V = \frac{\gamma}{g}\left(AV\right)V = \rho A V^2 \qquad (1)$$

(*b*) For the average pressure, divide the total dynamic force by the area over which it is acting.

$$\text{average pressure} = \frac{\text{force}}{\text{area}} = \frac{\rho A V^2}{20A} = \frac{\rho V^2}{20} = \frac{\gamma}{10}\left(\frac{V^2}{2g}\right)$$

From Problems 12.1 and 12.5 of Chapter 12, the stagnation pressure $= p_s = \gamma\left(V^2/2g\right)$. Thus the average pressure is 1/10 the stagnation pressure for this case.

13.4. A curved plate deflects a 76-mm-diameter stream of water through an angle of 45°. For a velocity in the jet of 40 m/s to the right, compute the value of the components of the force developed against the curved plate (assuming no friction).

Solution:

The components will be chosen along the jet's initial path and normal thereto. The water has its momentum changed by the force exerted by the plate.

(*a*) For the *X* direction, taking + to the right, assuming F_x positive,

initial linear momentum + linear impulse = final linear momentum.

$$MV_{x_1} + F_x\,dt = MV_{x_2}$$

$$\frac{\gamma Q\,dt}{g}V_{x_1} + F_x\,dt = \frac{\gamma Q\,dt}{g}V_{x_2}$$

Rearranging and noting that $V_{x_2} = +V_{x_1}\cos 45°$, we obtain

$$F_x = \frac{(9.79)\left[(\pi/4)(76/1000)^2\right](40)}{9.81}(40 \times \cos 45° - 40) = -2.12 \text{ kN}$$

where the negative sign indicates that F_x is to the left (assumed to the right). Had F_x been assumed to the left, the solution would yield +2.12 kN, the sign signifying that the assumption was correct.

The effect of the water on the plate is opposite and equal to the effect of the plate on the water. Hence, *X* component on plate = 2.12 kN to the right.

(*b*) For the *Y* direction, taking *up* positive,

$$MV_{y_1} + F_y\,dt = MV_{y_2}$$

$$0 + F_y\,dt = \frac{(9.79)(0.00454)(40)\,dt}{9.81}(\sin 45° \times 40)$$

and $F_y = +5.13$ kN upward on water. Hence, *Y* component on plate = 5.13 kN downward.

13.5. The force exerted by a 25-mm-diameter stream of water against a flat plate held normal to the stream's axis is 645 N. What is the flow?

Solution:

From equation (*1*) of Problem 13.3,

$$F_x = \frac{9{,}790QV}{9.81} = \rho AV^2$$

$$645 = \frac{9{,}790\left[(\pi/4)(25/1000)^2\right]V^2}{9.81} \quad \text{and} \quad V = 36.3 \text{ m/s}$$

Then $Q = AV = \left[(\pi/4)(25/1000)^2\right](36.3) = 0.0178 \text{ m}^3/\text{s}$.

13.6. If the plate in Problem 13.3 were moving to the right with a velocity of 30.0 ft/sec, what force would the jet exert on the plate?

Solution:

Using $t = 1$ sec, initial $MV_{x_1} + F_x(1) =$ final MV_{x_2}.

In this case the mass of water having its momentum changed is not identical with the mass for the stationary plate. For the fixed plate, in 1 sec a mass of

$$(\gamma/g)(\text{volume}) = (\gamma/g)(A \times 80.0)$$

has its momentum changed. For the moving plate, in 1 sec the mass striking the plate is

$$M = (\gamma/g)[A(80.0 - 30.0)]$$

where (80.0 − 30.0) is the relative velocity of the water with respect to the plate.

$$\text{Then } F_x = \frac{(62.4)\left[(\pi/4)(3/12)^2\right](80.0-30.0)}{32.2}(30.0-80.0)$$

and F_x = force of plate on water = −238 lb, to the left. Hence the force of the water on the plate is 238 lb to the right.

Had the plate been moving to the left at 30.0 ft/sec, more water would have had its momentum changed in any time t. The value for V_{x_2} is −30.0 ft/sec. The magnitude of the force would be

$$F_x = \frac{(62.4)(0.0491)[80.0-(-30.0)]}{32.2}(-30.0-80.0) = -1150 \text{ lb to the left on the water}$$

13.7. The fixed surface shown in Fig. 13-2 divides the jet so that 1.00 cfs goes in each direction. For an initial velocity of 48.0 ft/sec, find the values of the X and Y components to keep the surface in equilibrium (assuming no friction).

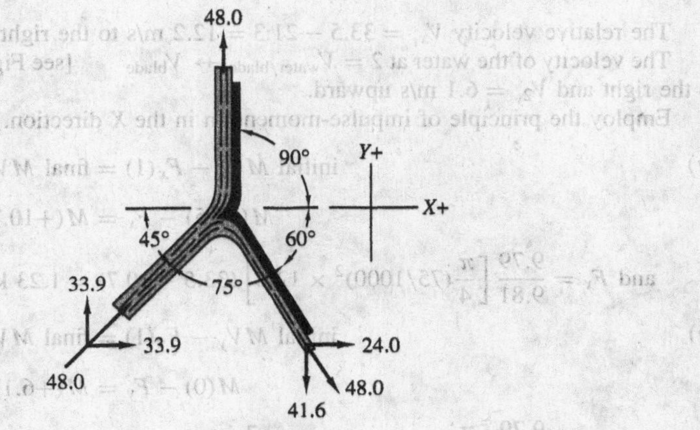

Fig. 13-2

Solution:

(*a*) In the X direction, using $t = 1$ sec,

$$MV_{x_1} - F_x(1) = \tfrac{1}{2}MV_{x_2} + \tfrac{1}{2}MV'_{x_2}$$

$$\frac{(62.4)(2.00)}{32.2}(33.9) - F_x = \frac{62.4}{32.2}\left(\frac{2.00}{2}\right)(0+24.0)$$

and $F_x = +84.9$ lb to the left.

(*b*) In the Y direction,

$$MV_{y_1} - F_y(1) = \tfrac{1}{2}MV_{y_2} - \tfrac{1}{2}MV'_{y_2}$$

$$\frac{(62.4)(2.00)}{32.2}(33.9) - F_y = \frac{62.4}{32.2}\left(\frac{2.00}{2}\right)(+48.0-41.6)$$

and $F_y = 119.0$ lb downward.

13.8. A 75-mm-diameter jet has a velocity of 33.5 m/s. It strikes a blade moving in the same direction at 21.3 m/s. The deflection angle of the blade is 150°. Assuming no friction, calculate the X and Y components of the force exerted by the water on the blade. [Refer to Fig. 13-3(*a*).]

Mathcad

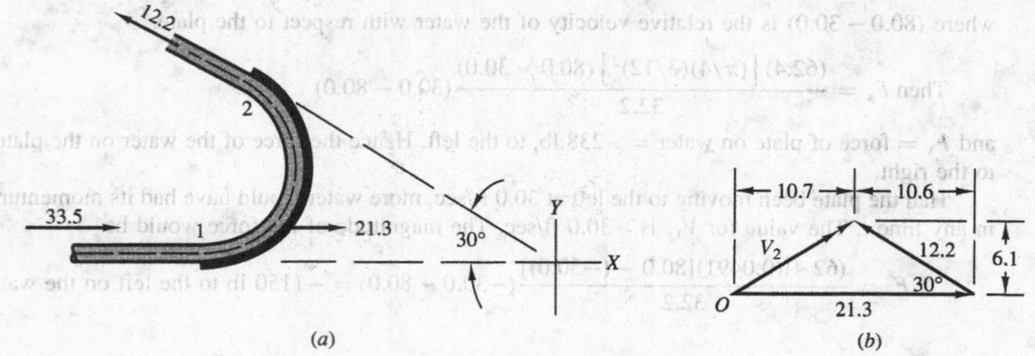

Fig. 13-3

Solution:

The relative velocity $V_{x_1} = 33.5 - 21.3 = 12.2$ m/s to the right.

The velocity of the water at $2 = V_{\text{water/blade}} \nrightarrow V_{\text{blade}}$ [see Fig. 13-3(*b*)], from which $V_{2_x} = 10.7$ m/s to the right and $V_{2_y} = 6.1$ m/s upward.

Employ the principle of impulse-momentum in the X direction.

(*a*)
$$\text{initial } MV_x - F_x(1) = \text{final } MV_x$$
$$M(33.5) - F_x = M(+10.7)$$

and $F_x = \dfrac{9.79}{9.81}\left[\dfrac{\pi}{4}(75/1000)^2 \times 12.2\right](33.5 - 10.7) = 1.23$ kN to the left on the water.

(*b*)
$$\text{initial } MV_y - F_y(1) = \text{final } MV_y$$
$$M(0) - F_y = M(+6.1)$$

and $F_y = \dfrac{9.79}{9.81}\left[\dfrac{\pi}{4}(75/1000)^2 \times 12.2\right](0 - 6.1) = -0.328$ kN, upward on the water.

The components of the force exerted by the water on the blade are 1.23 kN to the right and 0.328 kN downward.

13.9. If friction in Problem 13.8 reduced the velocity of the water with respect to the blade from 12.2 m/s to 10.7 m/s, what would be (*a*) the components of the force exerted by the blade on the water and (*b*) the final absolute velocity of the water?

Solution:

The components of the absolute velocity at 2 are found by solving a vector triangle similar to Fig. 13.3(*b*) using 21.3 horizontally and 10.7 upward to the left at 30°. Thus

$$V_{2_x} = 12.1 \text{ m/s to the right} \quad \text{and} \quad V_{2_y} = 5.33 \text{ m/s upward}$$

(*a*) Then $F_x = \dfrac{9.79}{9.81}\left[\dfrac{\pi}{4}(75/1000)^2 \times 12.2\right](33.5 - 12.1) = 1.15$ kN, to the left on the water

$$F_y = \frac{9.79}{9.81}\left[\frac{\pi}{4}(75/1000)^2 \times 12.2\right](0 - 5.33) = -0.287 \text{ kN, upward on the water}$$

(*b*) From the components shown above, the absolute velocity of the water leaving the blade is

$$V_2 = \sqrt{(12.1)^2 + (5.33)^2} = 13.2 \text{ m/s upward and to the right at an angle}$$
$$\theta_x = \tan^{-1} 5.33/12.1 = 23.8°.$$

13.10. For a given velocity of jet, determine the conditions that will produce the maximum amount of work (or power) on a series of moving blades (neglecting friction along blades).

Mathcad

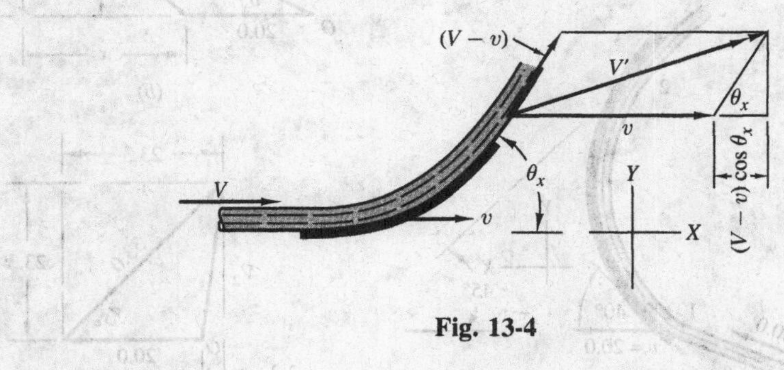

Fig. 13-4

Solution:

Consider first the velocity of the blades v that will produce maximum power. Referring to Fig. 13-4, the expression for power in the X direction will be developed, taking motion of the blade along the X axis. Since the entire jet strikes either one blade or several blades, the mass flowing per second is having its momentum changed or $M = (\gamma/g)AV$.

power = work done per sec = force × distance traveled per sec in direction of force

(1) Determine the force by the momentum principle. The final absolute velocity in the X direction is

$$V'_x = v + (V - v)\cos\theta_x$$

and initial momentum − linear impulse = final momentum

$$MV - F_x(1) = M[v + (V - v)\cos\theta_x]$$

or
$$F_x = (\gamma AV/g)[(V - v)(1 - \cos\theta_x)]$$

Then
$$\text{power } P = (\gamma AV/g)[(V - v)(1 - \cos\theta_x)]v \qquad (1)$$

Since $(V - v)v$ is the variable that must attain its maximum value for maximum power, equating the first derivative to zero we obtain

$$dP/dv = (\gamma AV/g)(1 - \cos\theta_x)(V - 2v) = 0$$

Thus $v = V/2$, or the blades should have a velocity equal to half the jet velocity.

(2) Examination of expression (*1*), above, for given values of V and v indicates that maximum power results when $\theta_x = 180°$ (by inspection). Since this angle is impractical, an angle of about 170° has proven to be satisfactory. The reduction in power is small percentagewise.

(3) In the Y direction, the unbalanced force is made equal to zero by using cusped blades, diverting one-half the mass of water toward each end of the Y axis.

13.11. (*a*) Referring to Fig. 13-5, at what angle should a jet of water moving at 50.0 ft/sec impinge upon a series of blades moving at 20.0 ft/sec in order that the water will strike tangent to the blades (no shock)?

(*b*) What power is developed if the flow is 2.86 cfs?

(*c*) What is the efficiency of the blades?

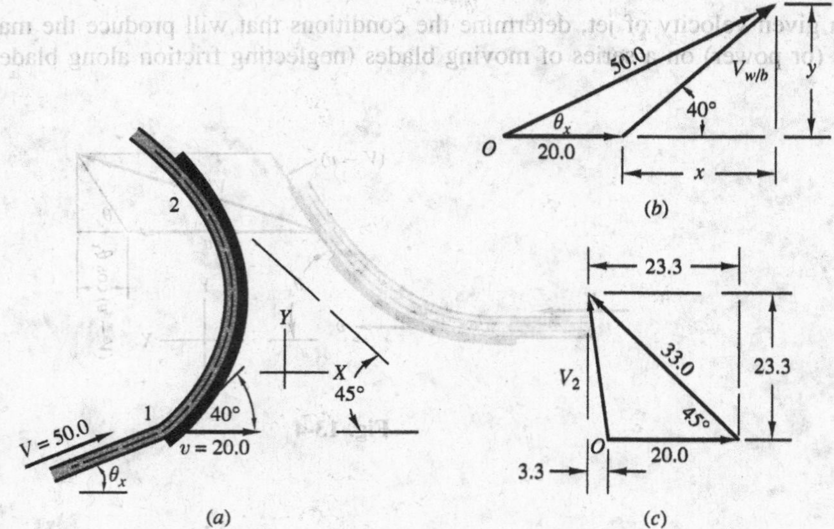

Fig. 13-5

Solution:

(*a*) velocity of water = velocity of water/blade $\nrightarrow$ velocity of blade

or 50.0 at $\angle\theta_x = ?$ at 40° $\nrightarrow$ 20.0 →

From the vector diagram of Fig. 13-5(*b*), $50\cos\theta_x = 20.0 + x$, $50\sin\theta_x = y$, and $\tan 40° = y/x = 0.8391$. Solving these equations, $\theta_x = 25°05'$.

(*b*) Solving Fig. 13-5(*b*) for the velocity of the water with respect to the blades,

$$y = 50\sin\theta_x = 50\sin 25°05' = 21.20 \text{ ft/sec} \quad \text{and} \quad V_{w/b} = y/(\sin 40°) = 33.0 \text{ ft/sec}$$

Also, absolute $V_{x_2} = 3.3$ ft/sec to the left, from Fig. 13-5(*c*). Then

$$\text{force } F_x = \frac{62.4 \times 2.86}{32.2}[50\times0.906-(-3.3)] = 269 \text{ lb} \quad \text{and} \quad \text{power } E_x = 269\times20 = 5380 \text{ ft-lb/sec}$$

(*c*) $$\text{Efficiency} = \frac{5380}{\frac{1}{2}M(50)^2} = \frac{5380}{6928} = 77.7\%.$$

13.12. A 600-mm pipe carrying 0.889 m³/s of oil (sp gr 0.85) has a 90° bend in a horizontal plane. The loss of head in the bend is 1.07 m of oil, and the pressure at the entrance is 293 kPa. Determine the resultant force exerted by the oil on the bend.

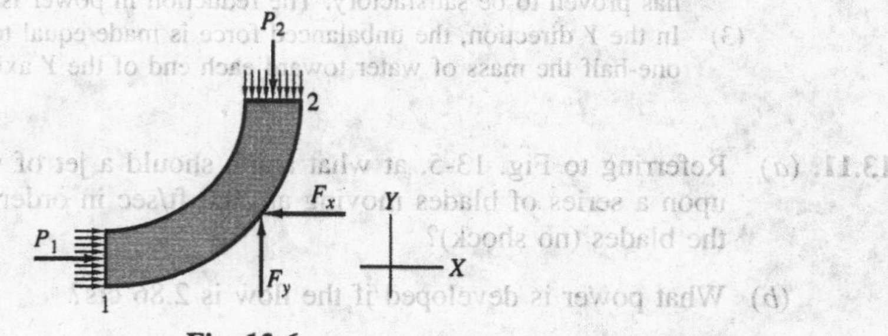

Fig. 13-6

Solution:

Referring to Fig. 13-6, the free-body diagram indicates the static and dynamic forces acting on the mass of oil in the bend. These forces are calculated as follows.

(*a*) $P_1 = p_1 A = 293 \times \frac{1}{4}\pi(0.600)^2 = 82.8$ kN.

(*b*) $P_2 = p_2 A$, where $p_2 = p_1 -$ loss, from the Bernoulli equation since $z_1 = z_2$ and $V_1 = V_2$.

Then $$P_2 = (293 - 0.85 \times 9.79 \times 1.07) \times \tfrac{1}{4}\pi(0.600)^2 = 80.3 \text{ kN}$$

(*c*) Using the impulse-momentum principle, and knowing that $V_1 = V_2 = Q/A = 3.14$ m/s,

$$MV_{x_1} + \Sigma(\text{forces in } X \text{ direction}) \times 1 = MV_{x_2}$$

$$82.8 - F_x = (0.85 \times 9.79 \times 0.889/9.81)(0 - 3.14) = -2.4 \text{ kN}$$

and $$F_x = +85.2 \text{ kN, to the left on the oil}$$

(*d*) Similarly, for $t = 1$ sec

$$MV_{y_1} + \Sigma(\text{forces in } Y \text{ direction}) \times 1 = MV_{y_2}$$

$$F_y - 80.3 = (0.85 \times 9.79 \times 0.889/9.81)(3.14 - 0) = +2.4 \text{ kN}$$

and $$F_y = +82.7 \text{ kN, upward on the oil.}$$

On the pipe bend, the resultant force R acts to the right and downward and is

$$R = \sqrt{(85.2)^2 + (82.7)^2} = 118.7 \text{ kN at } \theta_x = \tan^{-1} 82.7/85.2 = 44.1°$$

13.13. The 24″ pipe of Problem 13.12 is connected to a 12″ pipe by a standard *reducer* fitting (see Fig. 13-7). For the same flow of 31.4 cfs of oil, and a pressure of 40.0 psi, what force is exerted by the oil on the reducer, neglecting any lost head?

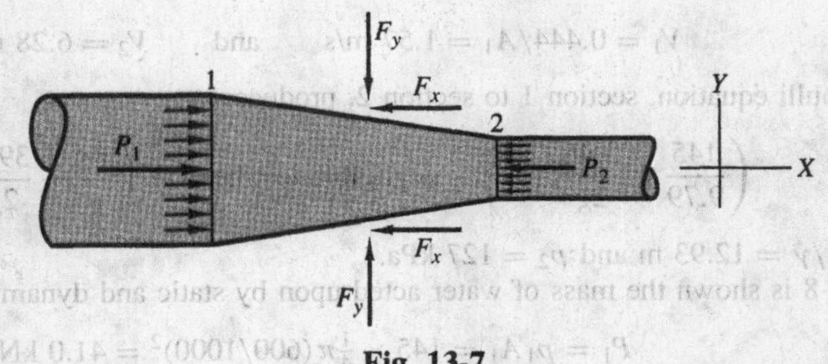

Fig. 13-7

Solution:

Since $V_1 = 10.0$ ft/sec, $V_2 = (2/1)^2 \times 10.0 = 40.0$ ft/sec. Also, the Bernoulli equation between section 1 at entrance and section 2 at exit yields

$$\left(\frac{p_1}{\gamma} + \frac{(10)^2}{2g} + 0\right) - \text{negligible lost head} = \left(\frac{p_2}{\gamma} + \frac{(40)^2}{2g} + 0\right)$$

Solving, $\dfrac{p_2}{\gamma} = \dfrac{40.0 \times 144}{0.85 \times 62.4} + \dfrac{100}{2g} - \dfrac{1600}{2g} = 85.3$ ft of oil and $p_2 = 31.4$ psi.

Figure 13-7 represents the forces acting on the mass of oil in the reducer.

$$P_1 = p_1 A_1 = 40.0 \times \tfrac{1}{4}\pi(24)^2 = 18{,}096 \text{ lb (to the right)}$$

$$P_2 = p_2 A_2 = 31.4 \times \tfrac{1}{4}\pi(12)^2 = 3551 \text{ lb (to the left)}$$

In the X direction the momentum of the oil is changed. Then

$$MV_{x_1} + \Sigma \text{ (forces in } X \text{ direction)} \times 1 = MV_{x_2}$$

$$(18{,}096 - 3551 - F_x)1 = (0.85 \times 62.4 \times 31.4/32.2)(40.0 - 10.0)$$

and $F_x = 13{,}000$ lb acting to the left on the oil.

The forces in the Y direction will balance each other and $F_y = 0$. Hence the force exerted by the oil on the reducer is 13,000 lb to the right.

13.14. A 45° reducing bend, 600-mm diameter upstream, 300-mm diameter downstream, has water flowing through it at the rate of 0.444 m³/s under a pressure of 145 kPa. Neglecting any loss in the bend, calculate the force exerted by the water on the reducing bend.

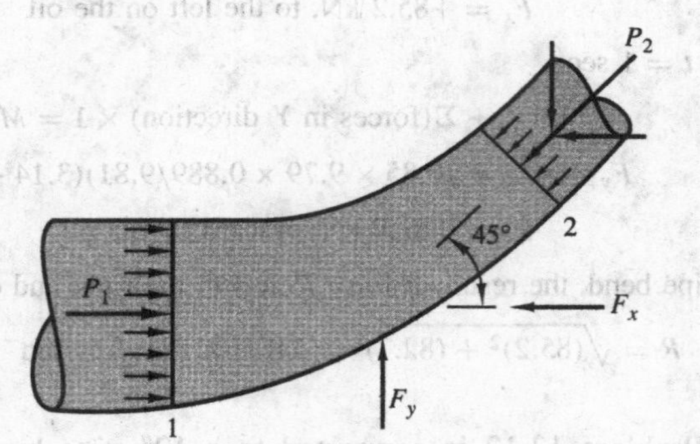

Fig. 13-8

Solution:

$$V_1 = 0.444/A_1 = 1.57 \text{ m/s} \qquad \text{and} \qquad V_2 = 6.28 \text{ m/s}$$

The Bernoulli equation, section 1 to section 2, produces

$$\left(\frac{145}{9.79} + \frac{2.46}{2g} + 0\right) - \text{negligible lost head} = \left(\frac{p_2}{\gamma} + \frac{39.4}{2g} + 0\right)$$

from which $p_2/\gamma = 12.93$ m and $p_2 = 127$ kPa.

In Fig. 13-8 is shown the mass of water acted upon by static and dynamic forces.

$$P_1 = p_1 A_1 = 145 \times \tfrac{1}{4}\pi(600/1000)^2 = 41.0 \text{ kN}$$

$$P_2 = p_2 A_2 = 127 \times \tfrac{1}{4}\pi(300/1000)^2 = 8.98 \text{ kN}$$

$$P_{2_x} = P_{2_y} = 8.98 \times 0.707 = 6.35 \text{ kN}$$

In the X direction,

$$MV_{x_1} + \Sigma(\text{forces in } X \text{ direction}) \times 1 = MV_{x_2}$$

$$(41.0 - 6.35 - F_x)(1) = [(9.79 \times 0.444)/9.81][(6.28 \times 0.707) - 1.57]$$

and $F_x = 33.4$ kN to the left.

In the Y direction,

$$(+F_y - 6.35)1 = [(9.79 \times 0.444)/9.81][(6.28 \times 0.707) - 0]$$

and $F_y = 8.32$ kN upward.

The force exerted by the water on the reducing bend is $F = \sqrt{(33.4)^2 + (8.32)^2} = 34.4$ kN to the right and downward, at an angle $\theta_x = \tan^{-1}(8.32/33.4) = 13°59'$.

13.15. Referring to Fig 13-9, a 2″-diameter stream of water strikes a 4-ft-square door that is at an angle of 30° with the stream's direction. The velocity of the water in the stream is 60.0 ft/sec, and the jet strikes the door at its center of gravity. Neglecting friction, what normal force applied at the edge of the door will maintain equilibrium?

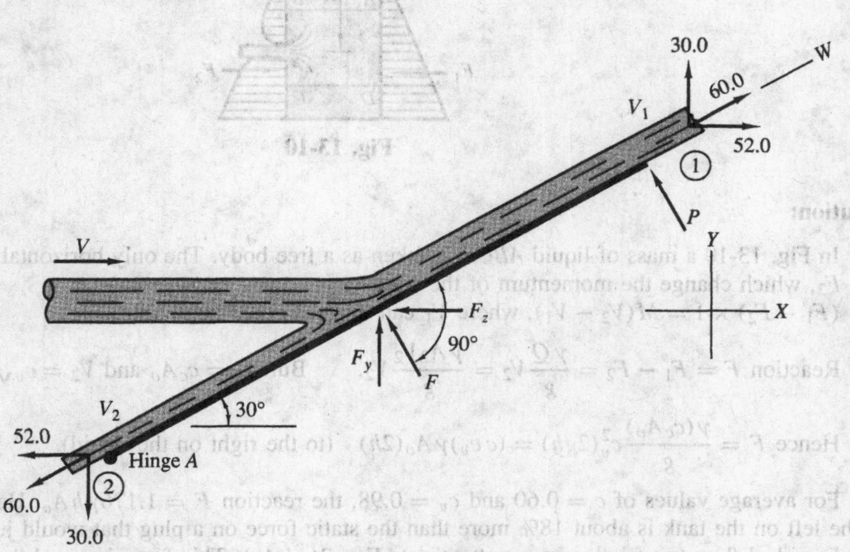

Fig. 13-9

Solution:

The force exerted by the door on the water will be normal to the door (no friction). Hence, since no forces act in the W direction in the figure, there will be no change in momentum in that direction. Thus, using W components,

$$\text{initial momentum} \pm 0 = \text{final momentum}$$

$$+M(V\cos 30°) = +M_1V_1 - M_2V_2$$

$$(\gamma/g)(A_{\text{jet}}V)(V\cos 30°) = (\gamma/g)(A_1V_1)V_1 - (\gamma/g)(A_2V_2)V_2$$

But $V = V_1 = V_2$ (friction neglected). Then $A_{\text{jet}}\cos 30° = A_1 - A_2$ and, from the equation of continuity, $A_{\text{jet}} = A_1 + A_2$. Solving,

$$A_1 = A_{\text{jet}}(1+\cos 30°)/2 = A_{\text{jet}} \times 0.933 \quad \text{and} \quad A_2 = A_{\text{jet}}(1-\cos 30°)/2 = A_{\text{jet}} \times 0.067$$

The stream divides as indicated, and the momentum equation produces, for the X direction,

$$\left[\frac{62.4}{32.2}\left(\tfrac{1}{4}\pi\right)\left(\tfrac{1}{6}\right)^2(60)\right](60) - F_x(1) = \left[\frac{62.4}{32.2}\left(\tfrac{1}{4}\pi\right)\left(\tfrac{1}{6}\right)^2(0.933)(60)\right](52.0) + \left[\frac{62.4}{32.2}\left(\tfrac{1}{4}\pi\right)\left(\tfrac{1}{6}\right)^2(0.067)(60)\right](-52.0)$$

and $F_x = 38.0$ lb.

Similarly, in the Y direction,

$$M(0) + F_y(1) = \left[\frac{62.4}{32.2}(0.0218)(0.933)(60)\right](30.0) + \left[\frac{62.4}{32.2}(0.0218)(0.067)(60)\right](-30.0)$$

and $F_y = 65.9$ lb.

For the door as the free body, $\Sigma M_{\text{hinge}} = 0$ and

$$+(38.0)(1) + (65.9)(2 \times 0.866) - P(4) = 0 \quad \text{or} \quad P = 38.0 \text{ lb}$$

13.16. **Determine the reaction of a jet flowing through an orifice on the containing tank.**

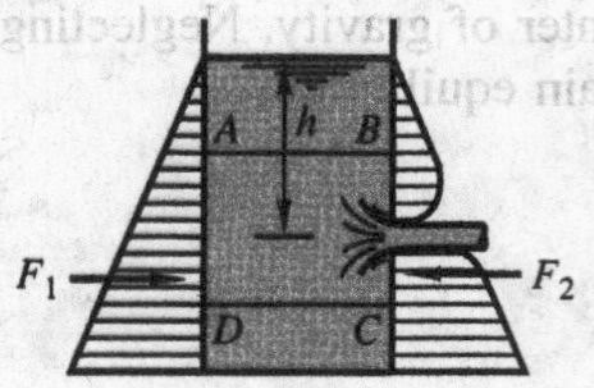

Fig. 13-10

Solution:

In Fig. 13-10 a mass of liquid $ABCD$ is taken as a free body. The only horizontal forces acting are F_1 and F_2, which change the momentum of the water.

$(F_1 - F_2) \times 1 = M(V_2 - V_1)$, where V_1 can be considered negligible.

Reaction $F = F_1 - F_2 = \dfrac{\gamma Q}{g} V_2 = \dfrac{\gamma A_2 V_2}{g} V_2$. But $A_2 = c_c A_o$ and $V_2 = c_v\sqrt{2gh}$.

Hence $F = \dfrac{\gamma(c_c A_o)}{g} c_v^2(2gh) = (c\,c_v)\gamma A_o(2h)$ (to the right on the liquid).

For average values of $c = 0.60$ and $c_v = 0.98$, the reaction $F = 1.176\gamma h A_o$. Hence the force acting to the left on the tank is about 18% more than the static force on a plug that would just fill the orifice.

For ideal flow (no friction, no contraction), $F = 2(\gamma h A_o)$. This force is equal to twice the force on a plug that would just fill the orifice.

For a nozzle ($c_c = 1.00$), the reaction $F = c_v^2 \gamma A(2h)$, where h would be the effective head causing the flow.

13.17. **The jets from a garden sprinkler are 25 mm in diameter and are normal to the 0.6-m radius (see Fig. 13-11). If the pressure at the base of the nozzles is 350 kPa, what force must be applied on each sprinkler pipe, 0.3 m from the center of rotation, to maintain equilibrium? (Use $c_v = 0.80$ and $c_c = 1.00$.)**

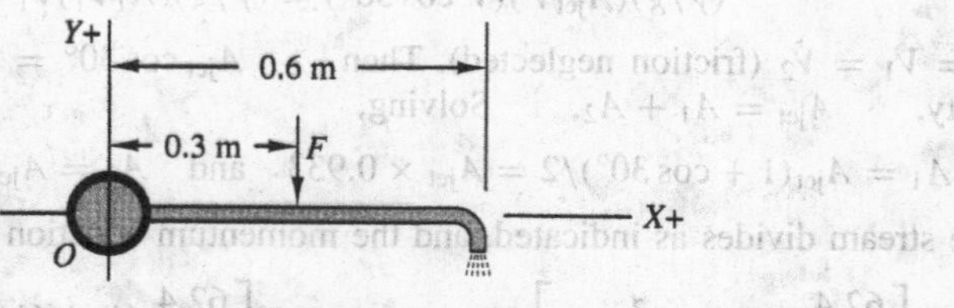

Fig. 13-11

Solution:

The reaction of the sprinkler jet can be calculated from the momentum principle. Inasmuch as the force that causes a change in momentum in the X direction acts along the X axis, no torque is exerted. We are interested, therefore, in the change in momentum in the Y direction. But the initial momentum in the Y direction is zero. The jet velocity

$$V_Y = c_v\sqrt{2gh} = 0.80\sqrt{2g(350/9.79 + \text{ negligible velocity head})} = 21.2 \text{ m/s}$$

Thus
$$F_Y\,dt = M(V_Y) = \left[\frac{9.79}{9.81} \times \tfrac{1}{4}\pi(25/1000)^2 \times 21.2\,dt\right](-21.2)$$

or $F_Y = -0.220$ kN downward on the water. Hence the force of the jet on the sprinkler is 0.220 kN upward. Then

$$\Sigma M_O = 0, \qquad F(0.3) - (0.220)(0.6) = 0, \qquad F = 0.440 \text{ kN for equilibrium}$$

13.18. Develop basic thrust equations for propulsive devices.

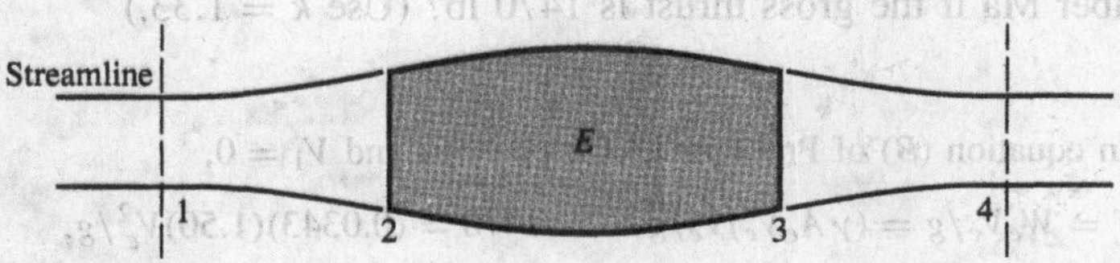

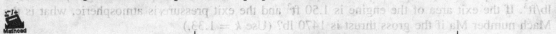

Fig. 13-12

Solution:

In Fig. 13-12, consider the air-breathing engine E that uses W lb of air per second. At section 1, the velocity V_1 of the air entering the engine is taken as the flight velocity. Also the air enters at atmospheric pressure (where no shock waves occur). In the engine E the air is compressed and heated by combustion. The air leaves the nozzle at section 3 at a high exit velocity and hence with greatly increased momentum.

In most air-breathing engines, the weight of air per second at exit is greater than the weight of air per second entering the engine due to the addition of fuel. This increase is about 2%. The weight of air at exit is usually measured at section 3.

The thrust can be evaluated in terms of the change in momentum, as follows:

$$\text{thrust } F = \frac{W_{\text{exit}} V_4}{g} - \frac{W_1 V_1}{g} \tag{A}$$

In those cases where the pressure at section 3 may be greater than atmospheric pressure, an additional acceleration of the gas is provided. The additional force is the difference in pressure times the area at section 3. Thus, for change in momentum from sections 1 to 3, we obtain

$$F = \frac{W_{\text{exit}} V_3}{g} + A_3(p_3 - p_4) - \frac{W_1 V_1}{g} \tag{B}$$

If the effective exhaust velocity is required, equations (A) and (B) can be solved simultaneously to produce

$$V_4 = V_3 + \frac{gA_3}{W_3}(p_3 - p_4) \tag{C}$$

Note that if $p_3 = p_4$, $V_4 = V_3$.

The term $W_1 V_1 / g$ is called the *negative thrust* or *ram drag*. The gross thrust (produced by the nozzle) is $W_3 V_4 / g$ in (A) and $W_3 V_3 / g + A_3(p_3 - p_4)$ in (B).

For a rocket, equation (A) is used to evaluate the thrust, recognizing that $V_1 = 0$ for such a device.

13.19. A jet engine is being tested in the laboratory. The engine consumes 50 lb of air per sec and 0.5 lb of fuel per sec. If the exit velocity of the gases is 1500 ft/sec, what is the thrust?

Solution:

Using formula (A) in Problem 13.18, thrust $F = (50.5 \times 1500 - 50 \times 0)/32.2 = 2350$ lb.

13.20. A jet engine operates at 180 m/s and consumes air at the rate of 0.25 kN/s. At what velocity should the air be discharged in order to develop a thrust of 6.7 kN?

Solution:

Thrust $F = 6.7 = (0.25/9.81)(V_{\text{exit}} - 180)$, from which $V_{\text{exit}} = 443$ m/s.

13.21. A turbojet engine is tested in the laboratory under conditions simulating an altitude where atmospheric pressure $p = 785.3$ psfa, temperature $T = 429.5°\text{R}$, and specific weight $\gamma = 0.0343$ lb/ft^3. If the exit area of the engine is 1.50 ft^2 and the exit pressure is atmospheric, what is the Mach number Ma if the gross thrust is 1470 lb? (Use $k = 1.33$.)

Solution:

Since in equation (*B*) of Problem 13.18, $p_3 = p_4$ and $V_1 = 0$,

$$F = W_e V_e/g = (\gamma A_e V_e)V_e/g, \qquad 1470 = (0.0343)(1.50)V_e^2/g, \qquad V_e = 959 \text{ ft/sec}$$

$$\text{Ma} = V_e/c = V_e/\sqrt{kgRT} = 959/\sqrt{(1.33)(32.2)(53.3)(429.5)} = 0.97$$

13.22. In Problem 13.21, what would be the gross thrust if the exit pressure becomes 10 psia and the Mach number is 1.00? (Use $k = 1.33$.)

Solution:

In order to evaluate the exit velocity for the new exit conditions, the temperature at exit must be calculated from

$$T_e/429.5 = (10 \times 144/785.3)^{(k-1)/k} \quad \text{from which} \quad T_e = 499°\text{R}$$

Then

$$V_e = \text{Ma}\ c = \text{Ma}\sqrt{kgRT} = 1.00\sqrt{(1.33)(32.2)(53.3)(499)} = 1067 \text{ ft/sec}$$

Furthermore, the specific weight at exit must be calculated from

$$(\gamma_1/\gamma_2)^k = p_1/p_2, \quad (\gamma_e/0.0343)^{1.33} = 10 \times 144/785.3, \quad \gamma_e = 0.0541 \text{ lb/ft}^3$$

Using (*B*) of Problem 13.18, $F = (0.0541)(1.50)(1067)^2/32.2 + (1.50)(1440 - 785.3) - 0 = 3851$ lb.

13.23. A rocket device burns its propellant at a rate of 0.0676 kN/s. The exhaust gases leave the rocket at a relative velocity of 980 m/s and at atmospheric pressure. The exhaust nozzle has an area of 0.032 m^2, and the gross weight of the rocket is 2.2 kN. At the given instant, 2000 kW is developed by the rocket engine. What is the rocket velocity?

Solution:

For a rocket, no air enters the device and the section 1 terms in formula (*B*) of Problem 13.18 are zero. Also, since the exit pressure is atmospheric, $p_3 = p_4$. Thus, the thrust

$$F_T = (W_e/g)V_e = (0.0676/9.81)(980) = 6.75 \text{ kN}$$

and since 2000 kW $= F_T V_{\text{rocket}}$, $V_{\text{rocket}} = 296$ m/s.

13.24. Assuming that the drag force is a function of density, viscosity, elasticity, and velocity of the fluid and a characteristic area, show that the drag force is a function of the Mach number and the Reynolds number (see also Chapter 6, Problems 6.9 and 6.16).

Solution:

As illustrated in Chapter 6, a dimensional analysis study will provide the desired relation, as follows.

$$F_D = f_1(\rho, \mu, E, V, A) \quad \text{or} \quad F_D = C\,\rho^a\,\mu^b\,E^c\,V^d\,L^{2e}$$

Then, dimensionally, $F^1L^0T^0 = (F^aT^{2a}L^{-4a})(F^bT^bL^{-2b})(F^cL^{-2c})(L^dT^{-d})L^{2e}$

and $$1 = a + b + c, \quad 0 = -4a - 2b - 2c + d + 2e, \quad 0 = 2a + b - d$$

Solving in terms of b and c yields

$$a = 1 - b - c, \quad d = 2 - b - 2c, \quad e = 1 - b/2$$

Substituting, $F_D = C\rho^{1-b-c}\mu^b E^c V^{2-b-2c} L^{2-b}$.
Expressing this equation in the commonly used form produces

$$F = CA_\rho V^2 \left(\frac{\mu}{L_\rho V}\right)^b \left(\frac{E}{\rho V^2}\right)^c$$

or $$F = A\rho V^2 f_2(\text{Re, Ma})$$

This equation indicates that the drag coefficient of objects of given configuration and alignment will depend upon their Reynolds and Mach numbers only.

For incompressible fluids, the Reynolds number is dominant, and the effect of the Mach number is small to negligible; thus the drag coefficient becomes a function of Reynolds number only. (See Diagrams F and G in the Appendix.) Indeed, for small values of Ma, a fluid may be considered as incompressible insofar as the coefficient of drag is concerned.

When the Mach number is equal to or greater than 1.0 (with velocities of the fluid equal to or greater than the velocity of sound) the drag coefficient is a function of Ma only. (See Diagram H in the Appendix.) However, there are often instances where the coefficient of drag depends on both Re and Ma.

A similar derivation may be presented regarding the coefficient of lift, and the conclusions stated above are equally applicable to the coefficient of lift. Use of the Buckingham pi theorem is suggested.

13.25. A 50-mph wind strikes a 6 ft by 8 ft sign normal to its surface. For a standard barometer, what force acts against the sign? $(\gamma = 0.0752 \text{ lb/ft}^3)$

Solution:

For a small jet of fluid striking a large plate at rest, we have seen that the force exerted by the fluid is

$$\text{force}_x = \Delta(MV_x) = (\gamma/g)(AV_x)V_x = \rho AV_x^2$$

The stationary plate under consideration affects a large amount of air. The momentum is not reduced to zero in the X direction as was the case for the jet of water. Tests on plates moving through fluids at different velocities indicate that the drag coefficient varies with the length-to-width ratio and that its value is essentially constant above Re = 1000. (See Diagram F, Appendix.) It is immaterial whether an object moves in the fluid at rest or the fluid moves past a stationary object; drag coefficients and drag forces are the same for each case. It is the relative velocity that is important.

The coefficient C_D is employed in the following equation: force $F = C_D\rho A\dfrac{V^2}{2}$

This is sometimes written to include velocity head, as follows: force $F = C_D\gamma A\dfrac{V^2}{2g}$

Using $C_D = 1.20$ from Diagram F,

$$\text{force } F = 1.20\left(\frac{0.0752}{32.2}\right)(48)\frac{(50 \times 5280/3600)^2}{2} = 362 \text{ lb}$$

13.26. A flat plate, 1.2 m by 1.2 m, moves at 6.7 m/s normal to its plane. At standard pressure and 20°C air temperature, determine the resistance of the plate (a) moving through air ($\gamma = 11.8 \text{ N/m}^3$) and ($b$) moving through water at 16°C.

Solution:

(*a*) Diagram *F* indicates $C_D = 1.16$ for length/width = 1.0

$$\text{drag force} = C_D \rho A \frac{V^2}{2} = (1.16)\left(\frac{11.8}{9.81}\right)(1.2 \times 1.2)\frac{(6.7)^2}{2} = 45.1 \text{ N}$$

(*b*) $$\text{drag force} = C_D \rho A \frac{V^2}{2} = (1.16)(1000)(1.2 \times 1.2)\frac{(6.7)^2}{2} = 37{,}500 \text{ N, or } 37.5 \text{ kN.}$$

13.27. A long copper wire, 10 mm in diameter, is stretched taut and is exposed to a wind at a velocity of 27.0 m/s normal to the wire. Compute the drag force per meter of length.

Solution:

For air at 20°C, Table 1 gives $\rho = 1.2 \text{ kg/m}^3$ and $\nu = 1.49 \times 10^{-5} \text{ m}^2/\text{s}$. Then

$$\text{Re} = \frac{Vd}{\nu} = \frac{27.0 \times 10/1000}{1.49} \times 10^5 = 18{,}100$$

From Diagram *F*, $C_D = 1.30$. Then

$$\text{drag force} = C_D \rho A \frac{V^2}{2} = (1.30)(1.2)\left(1 \times \frac{10}{1000}\right)\frac{(27.0)^2}{2} = 5.69 \text{ N per m of length}$$

13.28. A 3 ft by 4 ft plate moves at 44 ft/sec in still air at an angle of 12° with the horizontal. Using a coefficient of drag $C_D = 0.17$ and a coefficient of lift $C_L = 0.72$, determine (*a*) the resultant force exerted by the air on the plate, (*b*) the frictional force, and (*c*) the horsepower required to keep the plate moving. (Use $\gamma = 0.0752 \text{ lb/ft}^3$.)

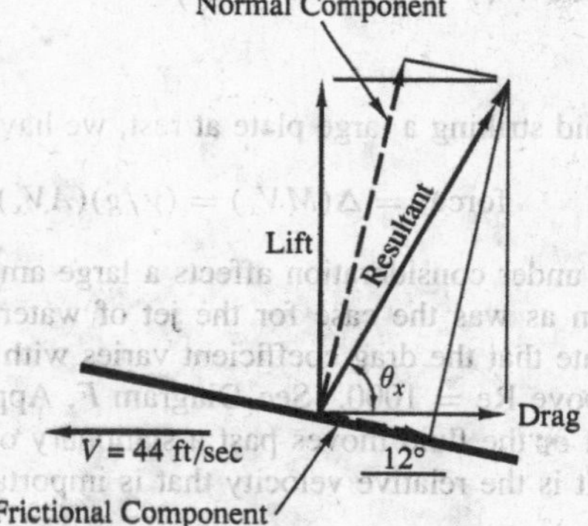

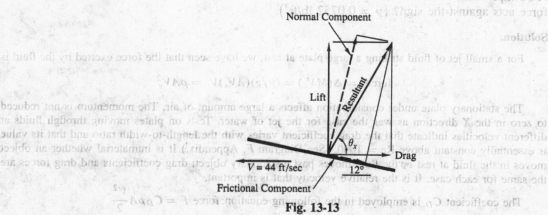

Fig. 13-13

Solution:

(*a*) $$\text{Drag force} = C_D\left(\frac{\gamma}{g}\right)A\frac{V^2}{2} = (0.17)\left(\frac{0.0752}{32.2}\right)(12)\frac{(44)^2}{2} = 4.61 \text{ lb.}$$

$$\text{Lift force} = C_L\left(\frac{\gamma}{g}\right)A\frac{V^2}{2} = (0.72)\left(\frac{0.0752}{32.2}\right)(12)\frac{(44)^2}{2} = 19.5 \text{ lb.}$$

Referring to Fig. 13-13, the resultant of the drag and lift components is

$$R = \sqrt{(4.61)^2 + (19.5)^2} = 20.0 \text{ lb} \quad \text{acting on the plate at } \theta_x = \tan^{-1}(19.5/4.61) = 76°42'$$

(*b*) The resultant force might also have been resolved into a normal component and a frictional component (shown dotted in the figure). From the vector triangle,

$$\text{frictional component} = R\cos\left(\theta_x + 12°\right) = (20.0)(0.0227) = 0.45 \text{ lb}$$

(*c*) Horsepower = (force in direction of motion × velocity)/550 = (4.61 × 44)/550 = 0.369.

13.29. If an airplane weighs 17.8 kN and has a wing area of 28 m², what *angle of attack* must the wings make with the horizontal at a speed of 160 km/h? Assume the coefficient of lift varies linearly from 0.35 at 0° to 0.80 at 6° and use $\gamma = 11.8 \text{ N/m}^3$ for air.

Mathcad

Solution:

For equilibrium in the vertical direction, $\Sigma Y = 0$. Hence, lift − weight = 0, or

$$\text{weight} = C_L \gamma A \frac{V^2}{2g}, \qquad 17{,}800 = C_L(11.8)(28)\frac{(160 \times 1000/3600)^2}{2g}, \qquad C_L = 0.535$$

By interpolation between 0° and 6°, angle of attack = 2.5°.

13.30. What wing area is required to support a 22.2-kN plane flying at an *angle of attack* of 5° at 27 m/s? Use coefficients given in Problem 13.29.

Solution:

From given data (or from a curve), $C_L = 0.725$ for 5° angle by interpolation. As in Problem 13.29,

$$\text{weight} = \text{lift}, \quad 22{,}200 = (0.725)(11.8/9.81)(A)(27)^2/2, \quad A = 69.8 \text{ m}^2$$

13.31. An airfoil of 400 ft² area has an angle of attack of 6° and is traveling at 80 ft/sec. If the coefficient of drag varies linearly from 0.040 at 4° to 0.120 at 14°, what horsepower is required to maintain the velocity in air at 40°F and 13.0 psi absolute?

Solution:

$$\gamma = \frac{p}{RT} = \frac{13.0 \times 144}{(53.3)(460 + 40)} = 0.0702 \text{ lb/ft}^3 \text{ for air}$$

For 6° angle of attack $C_D = 0.056$, by interpolation.

$$\text{drag force} = C_D \rho A V^2/2 = (0.056)(0.0702/32.2)(400)(80)^2/2 = 156.3 \text{ lb}$$

$$\text{power} = (156.3 \text{ lb})(80 \text{ ft/sec})/550 = 22.7 \text{ horsepower}$$

13.32. In Problem 13.31, for a coefficient of lift of 0.70 and a chord length of 5 ft, determine (*a*) the lift force and (*b*) the Reynolds and Mach numbers.

Solution:

(*a*) Lift force $F_L = C_L \rho A V^2/2 = (0.70)(0.0702/g)(400)(80)^2/2 = 1950$ lb.

(*b*) The characteristic length in the Reynolds number is the chord length. Then

$$\text{Re} = \frac{VL\rho}{\mu} = \frac{80 \times 5 \times 0.0702}{\left(3.62 \times 10^{-7}\right)(32.2)} = 2{,}410{,}000$$

It will be remembered that the absolute coefficient of viscosity does not change with pressure variations.

$$\text{Ma} = V/\sqrt{E/\rho} = V/\sqrt{kgRT} = 80/\sqrt{(1.4)(32.2)(53.3)(500)} = 0.073$$

13.33. An air wing of 25-m^2 area moves at 26 m/s. If 10 kW is required to keep the wing in motion, what angle of attack is indicated using the same variation of the coefficient of drag as in Problem 13.31? Use $\gamma = 11.0\ \text{N/m}^3$.

Solution:

$$10{,}000 = (\text{force})(26), \qquad (\text{force}) = 385\ \text{N}$$

$$\text{force} = C_D \rho A V^2/2, \quad 385 = C_D(11.0/9.81)(25)(26)^2/2, \quad C_D = 0.0406$$

Using the data regarding angle of attack and C_D, interpolation yields 4.1° angle of attack.

13.34. Consider the area on one side of a moving van to be 600 ft^2. Determine the resultant force acting on the side of the van when the wind is blowing at 10 mph normal to the area when the van is (*a*) at rest and (*b*) moving at 30 mph normal to the direction of the wind. In (*a*) use $C_D = 1.30$, and in (*b*) use $C_D = 0.25$ and $C_L = 0.60$. ($\rho = 0.00237$ slug/ft^3)

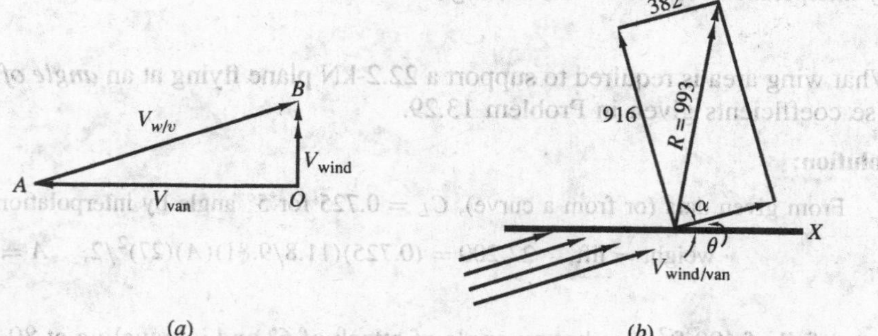

Fig. 13-14

Solution:

(*a*) The force acting normal to the area $= C_D(\rho/2)AV^2$. Then

$$\text{resultant force} = (1.30)(0.00237/2)(600)(10 \times 5280/3600)^2 = 199\ \text{lb normal to area}$$

(*b*) It will be necessary to calculate the relative velocity of the wind with respect to the van. From kinetic mechanics,

$$V_{\text{wind}} = V_{\text{wind/van}} \nrightarrow V_{\text{van}}$$

Figure 13-14(*a*) indicates this vector relationship, i.e.,

$$OB = OA \nrightarrow AB = 30.0 \nrightarrow V_{w/v}$$

Thus the relative velocity $= \sqrt{(30)^2 + (10)^2} = 31.6$ mph to the right and upward at an angle $\theta = \tan^{-1}(10/30) = 18.4°$.

The component of the resultant force normal to the relative velocity of wind with respect to van is

$$\text{lift force} = C_L(\rho/2)AV^2 = (0.60)(0.00237/2)(600)(31.6 \times 5280/3600)^2$$

$$= 916\ \text{lb normal to the relative velocity}$$

The component of the resultant force parallel to the relative motion of wind to van is

$$\text{drag force} = C_D(\rho/2)AV^2 = (0.25)(0.00237/2)(600)(31.6 \times 5280/3600)^2$$

$$= 382\ \text{lb parallel to the relative velocity}$$

Referring to Fig. 13-14*(b)*, the resultant force $= \sqrt{(916)^2 + (382)^2} = 992$ lb at an angle $\alpha = \tan^{-1}(916/382) = 67.4°$. Hence the angle with the longitudinal axis (*X* axis) is $18.4° + 67.4° = 85.8°$.

13.35. A kite weighs 2.50 lb and has an area of 8.00 ft². The tension in the kite string is 6.60 lb when the string makes an angle of 45° with the horizontal. For a wind of 20 mph, what are the coefficients of lift and drag if the kite assumes an angle of 8° with the horizontal? Consider the kite essentially a flat plate and $\gamma_{air} = 0.0752$ lb/ft³.

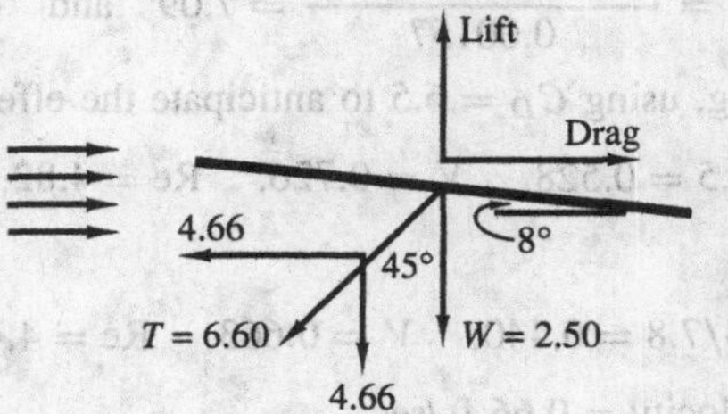

Fig. 13-15

Solution:

Figure 13-15 indicates the forces acting on the kite taken as a free body. The components of the tension are 4.66 lb each.

From $\Sigma X = 0$,

$$\text{drag} = 4.66 \text{ lb}$$

From $\Sigma Y = 0$,

$$\text{lift} = 4.66 + 2.50 = 7.16 \text{ lb}$$

$$\text{drag force} = C_D\rho AV^2/2, \quad 4.66 = C_D(0.0752/32.2)(8.00)(20 \times 5280/3600)^2/2, \quad C_D = 0.58$$

$$\text{lift force} = C_L\rho AV^2/2, \quad 7.16 = C_L(0.0752/32.2)(8.00)(20 \times 5280/3600)^2/2, \quad C_L = 0.89$$

13.36. A man weighing 756 N is descending from an airplane using a 5.5-m-diameter parachute. Assuming a drag coefficient of 1.00 and neglecting the weight of the parachute, what maximum terminal velocity will be attained?

Solution:

The forces on the parachute are the weight down and the drag force up. For equilibrium, $\Sigma Y = 0$ (velocity constant).

$$W = C_D\rho AV^2/2, \quad 756 = (1.00)(11.8/9.81)(\pi \times 5.5^2/4)V^2/2, \quad V = 7.27 \text{ m/s}$$

13.37. A steel ball $\frac{1}{8}$ inch in diameter and weighing 0.284 lb/in³ is falling in an oil of specific gravity 0.908 and kinematic viscosity 0.00157 ft²/sec. What is the terminal velocity of the ball?

Mathcad

Solution:

The forces acting on the steel ball are the weight down, the buoyant force up, and the drag force up. For constant velocity, $\Sigma Y = 0$, and, transposing,

$$\text{weight of sphere} - \text{buoyant force} = \text{drag force}$$

or

$$\gamma_s(\text{volume}) - \gamma_o(\text{volume}) = C_D\rho AV^2/2$$

Using lb/in^3 × in^3 = weight,

$$\frac{4}{3}\pi\left(\frac{1}{16}\right)^3\left(0.284-\frac{0.908\times 62.4}{1728}\right)=C_D\left(\frac{0.908\times 62.4}{32.2}\right)\pi\left(\frac{1}{12\times 16}\right)^2\left(\frac{V^2}{2}\right)$$

Assuming a value of $C_D = 3.00$ (see Diagram F, spheres) and solving,

$$V^2 = 3.43/C_D = 1.143 \quad \text{and} \quad V = 1.069 \text{ ft/sec}$$

Check C_D assumed by calculating Reynolds number and using Diagram F.

$$\text{Re} = \frac{Vd}{\nu} = \frac{1.069\times 1/(8\times 12)}{0.00157} = 7.09 \quad \text{and} \quad C_D = 5.6 \text{ (increase } C_D)$$

Recalculating and checking, using $C_D = 6.5$ to anticipate the effect of increased value of C_D,

$$V^2 = 3.43/6.5 = 0.528, \quad V = 0.726, \quad \text{Re} = 4.82, \quad C_D = 7.2 \text{ (increase } C_D)$$

Trying $C_D = 7.8$,

$$V^2 = 3.43/7.8 = 0.440, \quad V = 0.663, \quad \text{Re} = 4.40, \quad C_D = 7.8 \text{ (checks)}$$

Therefore, the terminal velocity = 0.66 ft/sec.

Had the Reynolds number been less than 0.60, the equation for the drag force could be written

$$C_D\rho AV^2/2 = (24/\text{Re})\rho AV^2/2 = (24\nu/Vd)\rho(\pi d^2/4)V^2/2.$$

Since $\mu = \rho\nu$, drag force $= 3\pi\mu dV$.

13.38. A 25-mm-diameter sphere of lead, weighing 111 kN/m^3, is moving downward in an oil at a constant velocity of 0.357 m/s. Calculate the absolute viscosity of the oil if the specific gravity is 0.93.

Solution:

As in Problem 13.37, but using kN/m^3 × m^3 = weight,

$$(\gamma_s - \gamma_o)\,(\text{volume}) = C_D\rho AV^2/2$$

Then $(111 - 0.93\times 9.79)(4\pi/3)(0.0125)^3 = C_D(0.93\times 9.79/9.81)\pi(0.0125)^2(0.357)^2/2$ and $C_D = 28.7$.

From Diagram F, for $C_D = 28.7$, Re = 0.85 and

$$0.85 = Vd/\nu = (0.357)(0.025)/\nu, \quad \nu = 0.0105 \text{ m}^2/\text{s}.$$

Thus $$\mu = \nu\rho = (0.0105)(0.93\times 9.79)/9.81 = 9.75\times 10^{-3} \text{ kN}\cdot\text{s/m}^2$$

13.39. A sphere 12.5 mm in diameter rises in an oil at the maximum velocity of 0.037 m/s. What is the specific weight of the sphere if the density of the oil is 917 kg/m^3 and the absolute viscosity is 0.034 N · s/m^2?

Solution:

For constant velocity upward, $\Sigma Y = 0$ and

$$\text{buoyant force} - \text{weight} - \text{drag} = 0$$

$$(4\pi/3)(6.25/1000)^3(917\times 9.81 - \gamma_s) = C_D(917)\pi(6.25/1000)^2(0.037)^2/2 \qquad (1)$$

$$(8996 - \gamma_s) = 75.32C_D$$

The coefficient of drag can be evaluated using Diagram F and Reynolds number.

$$\text{Reynolds number} = \frac{Vd\rho}{\mu} = \frac{(0.037)(12.5/1000)(917)}{0.034} = 12.5$$

Then from Diagram F, $C_D = 3.9$ (for sphere) and, from (1),

$$\gamma_s = 8996 - 75.32 \times 3.9 = 8700 \text{ N/m}^3 = 8.70 \text{ kN/m}^3$$

13.40. For laminar flow at low Reynolds numbers, show that the coefficient of drag for a sphere is equal to 24/Re (shown graphically on Diagram F in the Appendix).

Solution:

The drag force $F_D = C_D \rho A V^2/2$, as previously noted. For laminar flow the drag force depends upon the viscosity and the velocity of the fluid and the diameter d of the sphere. Thus,

$$F_D = f(\mu, V, d) = C\,\mu^a\,V^b\,d^c$$

Then
$$F^1\,L^0\,T^0 = \left(F^a\,T^a\,L^{-2a}\right)\left(L^b\,T^{-b}\right)\left(L^c\right)$$

and
$$1 = a, \quad 0 = -2a + b + c, \quad 0 = a - b$$

from which $a = 1, b = 1$, and $c = 1$. Thus, drag force $F_D = C(\mu V d)$. G. G. Stokes has shown mathematically that $C = 3\pi$, a fact that has been confirmed by many experiments.

We may now equate the two expressions for drag force, substitute $\frac{1}{4}\pi d^2$ for projected area A, and then solve for C_D.

$$3\pi\mu V d = C_D\rho\left(\tfrac{1}{4}\pi d^2\right)V^2/2 \quad \text{and} \quad C_D = \frac{24\mu}{V d \rho} = \frac{24}{\text{Re}}$$

13.41. For laminar flow of a fluid past a thin plate, develop an expression for the thickness δ of the boundary layer, assuming that the velocity distribution equation is $v = V\left(\dfrac{2y}{\delta} - \dfrac{y^2}{\delta^2}\right)$.

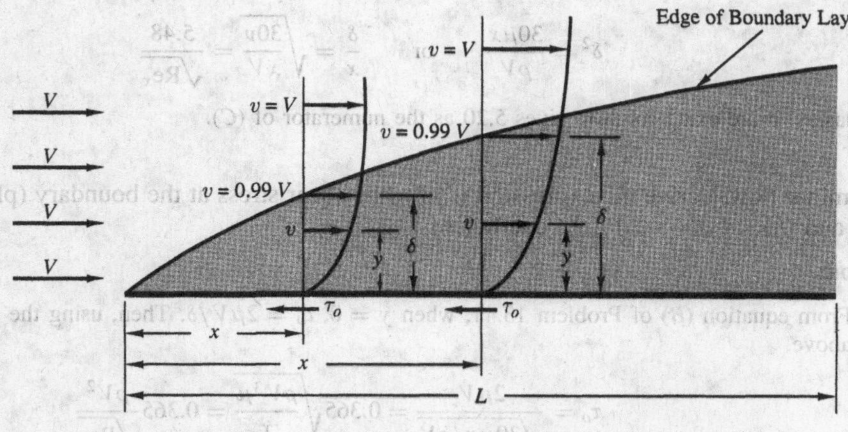

Fig. 13-16

Solution:

It is assumed that steady flow occurs ($\partial v/\partial t = 0$), that the velocity outside the boundary layer is uniformly V, that δ is very small with respect to distance x, and that $dp/dy = 0 = dp/dx$, both outside and within the boundary layer. Furthermore, by definition the edge of the boundary layer is considered as the locus of points where the velocity is 0.99 of the undisturbed velocity V.

The mass passing through any section of the boundary layer per unit width is $\int_0^\delta \rho v(dy \times 1)$, and the change in velocity at any point is $(V - v)$. Inasmuch as the pressure forces on the section cancel each other and thus do not contribute to the change in momentum, the change in momentum is caused by the shearing force $\tau_o dA$ or $\tau_o(dx \times 1)$. From the above, the change in momentum in unit time is

$$\int_0^\delta \rho(V - v)v(dy \times 1)$$

This expression is equal to the shearing force acting for a unit time, or

$$\text{drag force/unit width, } F'_D = \int_0^x \tau_o(dx \times 1) = \int_0^\delta \rho(V - v)v(dy \times 1)$$

Substituting the parabolic velocity relation into the equation gives

$$F'_D = \int_0^\delta \rho\left(V - \frac{2yV}{\delta} + \frac{y^2V}{\delta^2}\right)(V)\left(\frac{2y}{\delta} - \frac{y^2}{\delta^2}\right)dy$$

$$= \rho V^2 \int_0^\delta \left(1 - \frac{2y}{\delta} + \frac{y^2}{\delta^2}\right)\left(\frac{2y}{\delta} - \frac{y^2}{\delta^2}\right)dy = \frac{2}{15}\rho V^2 \delta \qquad (A)$$

To obtain a useful expression for δ, consider that $\tau_o dx =$ the differential unit drag force dF'_D and that the flow is laminar. Then, in $\tau_o = \mu(dv/dy)_o$, the term

$$\left(\frac{dv}{dy}\right)_0 = \frac{d}{dy}\left[V\left(\frac{2y}{\delta} - \frac{y^2}{\delta^2}\right)\right] = \frac{2V}{\delta}\left(1 - \frac{y}{\delta}\right) \qquad (B)$$

Substituting the values above in $\mu(dv/dy)_0 = dF'_D/dx$ and realizing that the shear stress is τ_0 when $y = 0$, we obtain $\mu(2V/\delta) = \frac{2}{15}\rho V^2(d\delta/dx)$ or

$$\int_0^\delta \delta\, d\delta = \frac{15\mu}{\rho V}\int_0^x dx$$

which yields

$$\delta^2 = \frac{30\mu x}{\rho V} \qquad \text{or} \qquad \frac{\delta}{x} = \sqrt{\frac{30\nu}{xV}} = \frac{5.48}{\sqrt{\text{Re}_x}} \qquad (C)$$

Blasius' more exact solution gives 5.20 as the numerator of (*C*).

13.42. For laminar flow, derive the expression (*a*) for the shear stress at the boundary (plate) in Problem 13.41 and (*b*) for the local drag coefficient C_D.

Solution:

(*a*) From equation (*B*) of Problem 13.41, when $y = 0$, $\tau_o = 2\mu V/\delta$. Then, using the value of δ in (*C*) above,

$$\tau_o = \frac{2\mu V}{\sqrt{30\mu x/\rho V}} = 0.365\sqrt{\frac{\rho V^3 \mu}{x}} = 0.365\frac{\rho V^2}{\sqrt{\text{Re}_x}} \qquad (A)$$

Results of experimentation give the more exact formula as

$$\tau_o = 0.33\sqrt{\frac{\rho V^3 \mu}{x}} = 0.33\frac{\rho V^2}{\sqrt{\text{Re}_x}} \qquad (B)$$

(*b*) The local drag coefficient C_{D_x} is obtained by equating $\tau_o A$ and the local drag force, i.e.,

$$F_D = \tau_o A = C_{D_x}\rho A V^2/2$$

or

$$C_{D_x} = \frac{2\tau_o}{\rho V^2} = \frac{0.66\rho V^2}{\rho V^2\sqrt{\text{Re}_x}} = \frac{0.66}{\sqrt{\text{Re}_x}} \qquad (C)$$

It can be seen that the total drag force on one side of a plate is the sum of all the ($\tau_o\, dA$) terms or

$$F_D = \int_0^L \tau_o(dx \cdot 1) = \int_0^L 0.33\sqrt{\rho V^3 \mu}\left(x^{-1/2}\, dx\right) = 0.33\left(2L^{1/2}\right)\sqrt{\rho V^3 \mu}$$

In the usual form, $F_D = C_D \rho A V^2/2$. For this case $A = L \times 1$; hence

$$C_D \rho L V^2/2 = 0.33(2)\sqrt{\rho V^3 \mu L} \quad \text{and} \quad C_D = 1.32\sqrt{\frac{\mu}{\rho V L}} = \frac{1.32}{\sqrt{\text{Re}}} \qquad (D)$$

13.43. A 4 ft by 4 ft thin plate is held parallel to a stream of air moving at 10 ft/sec (standard conditions). Calculate (*a*) the surface drag of the plate, (*b*) the thickness of the boundary layer at the trailing edge, and (*c*) the shear stress at the trailing edge.

Mathcad

Solution:

(*a*) Since the "skin friction" drag coefficient varies with Reynolds number, Re should be found.

$$\text{Re} = VL/\nu = (10)(4)/\left(16.0 \times 10^{-5}\right) = 250{,}000 \text{ (laminar range)}$$

Assuming laminar boundary layer conditions over the entire plate,

$$\text{coefficient } C_D = 1.328/\sqrt{\text{Re}} = 1.328/\sqrt{250{,}000} = 0.00266$$

$$\text{drag force } D \text{ (two sides)} = 2C_D \rho A V^2/2 = (0.00266)(0.0752/32.2)(4 \times 4)(10)^2 = 0.0099 \text{ lb}$$

(*b*) $\dfrac{\delta}{x} = \dfrac{5.20}{\sqrt{\text{Re}_x}}$ and $\delta = \dfrac{(5.20)(4)}{\sqrt{250{,}000}} = 0.0416$ ft.

(*c*) $\tau = 0.33\dfrac{\mu V}{x}\sqrt{\text{Re}_x} = (0.33)\dfrac{\left(3.75 \times 10^{-7}\right)(10)}{4}\sqrt{250{,}000} = 0.000155 \text{ lb/ft}^2$.

13.44. A smooth plate, 10 ft by 4 ft, moves through air (60°F) at a relative velocity of 4 ft/sec parallel to the plate surface and to its length. Calculate the drag force on one side of the plate (*a*) assuming laminar conditions and (*b*) assuming turbulent conditions over the entire plate. (*c*) For laminar conditions, estimate the thickness of the boundary layer at the middle of the plate and at the trailing edge.

Solution:

(*a*) Calculate Reynolds number: $\text{Re} = VL/\nu = (4)(10)/\left(15.8 \times 10^{-5}\right) = 253{,}000$.

For laminar conditions, $C_D = \dfrac{1.328}{\sqrt{\text{Re}}} = \dfrac{1.328}{\sqrt{253{,}000}} = 0.00264$ (also see Diagram *G*).

$$\text{drag force} = C_D \rho A V^2/2 = (0.00264)(0.00237)(10 \times 4)\left(4^2\right)/2 = 0.00200 \text{ lb}$$

(*b*) For turbulent conditions, with $\text{Re} < 10^7$, $C_D = \dfrac{0.074}{\text{Re}^{0.20}}$ [see equation (*12*).]

Then

$$C_D = \frac{0.074}{(253{,}000)^{0.20}} = 0.00615 \text{ (also see Diagram } G\text{)}.$$

$$\text{drag force} = C_D \rho A V^2/2 = (0.00615)(0.00237)(10 \times 4)\left(4^2\right)/2 = 0.00464 \text{ lb}$$

(*c*) For $x = 5$ ft, $\text{Re}_x = (4)(5)/\left(15.8 \times 10^{-5}\right) = 126{,}600$

Note that Reynolds number is calculated for $L = x$ ft. This Reynolds number value is referred to as the *local Reynolds number*. Then

$$\delta = \frac{5.20x}{\sqrt{\text{Re}_x}} = \frac{(5.20)(5)}{\sqrt{126{,}600}} = 0.073 \text{ ft}$$

For $x = 10$ ft, $\text{Re}_x = 253{,}000$ and $\delta = \dfrac{5.20x}{\sqrt{\text{Re}_x}} = \dfrac{(5.20)(10)}{\sqrt{253{,}000}} = 0.103$ ft

13.45. A smooth rectangular plate 1.2 m wide by 24.4 m long moves through 20°C water in the direction of its length. The drag force on the plate (two sides) is 8.0 kN. Find (*a*) the velocity of the plate, (*b*) the thickness of the boundary layer at the trailing edge, and (*c*) the length x_c of the laminar boundary layer if laminar conditions obtain at the leading edge.

Solution:

(*a*) For the length of plate and for water the fluid, turbulent flow can be assumed with some degree of confidence. Examining Diagram *G*, assume $C_D = 0.002$.

$$\text{drag force} = 2C_D \rho A V^2/2, \qquad 8000 = (C_D)(1000)(1.2 \times 24.4)V^2$$

$$V^2 = \frac{0.273}{C_D} = \frac{0.273}{0.002}, \qquad V = 11.7 \text{ m/s}$$

Reynolds number $\text{Re} = (11.7)(24.4)/\left(9.84 \times 10^{-7}\right) = 2.90 \times 10^8$. Thus the boundary layer is turbulent, as assumed. Continuing,

$$C_D = \frac{0.455}{\left[\log\left(2.90 \times 10^8\right)\right]^{2.58}} = 0.00184, \qquad V^2 = \frac{0.273}{0.00184} = 148.4, \qquad V = 12.2 \text{ m/s}$$

Recalculation of Reynolds number gives 3.03×10^8; then

$$C_D = \frac{0.455}{\left[\log\left(3.03 \times 10^8\right)\right]^{2.58}} = 0.00183 \quad \text{and} \quad V = 12.2 \text{ m/s}$$

This value is within the accuracy anticipated.

(*b*) The thickness of the boundary layer for turbulent flow is estimated by using equation (*15*).

$$\frac{\delta}{x} = \frac{0.22}{\text{Re}^{0.167}} \quad \text{and} \quad \delta = \frac{(0.22)(24.4)}{\left(3.03 \times 10^8\right)^{0.167}} = 0.206 \text{ m}$$

(*c*) Assume the critical Reynolds number is about 500,000, i.e, the transition range's lower limit.

$$\text{Re}_c = \frac{Vx_c}{\nu}, \qquad 500{,}000 = \frac{12.2x_c}{9.84 \times 10^{-7}}, \qquad x_c = 0.040 \text{ m}$$

13.46. The 10 ft by 4 ft plate of Problem 13.44 is held in water (50°F) moving at 4 ft/sec parallel to its length. Assuming laminar conditions in the boundary layer at the leading edge of the plate, (*a*) locate where the boundary layer flow changes from laminar to turbulent, (*b*) estimate the thickness of the boundary layer at this point, and (*c*) compute the friction drag of the plate.

Solution:

(*a*) Reynolds number $\text{Re} = VL/\nu = (4)(10)/\left(1.410 \times 10^{-5}\right) = 2{,}840{,}000$. This value of Reynolds number indicates that the boundary layer flow is in the transition range. Assuming the critical Reynolds number is 500,000, the location of the end of the laminar flow conditions can be estimated by using

$$\frac{x_c}{L} = \frac{\text{critical Re}}{\text{full plate Re}} \qquad \text{or} \qquad x_c = (10)\left(\frac{500{,}000}{2{,}840{,}000}\right) = 1.76 \text{ ft}$$

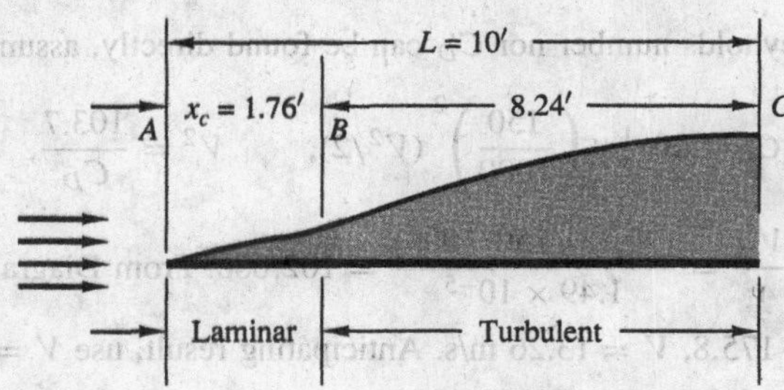

Fig. 13-17

(*b*) The thickness of the boundary layer at this point is estimated to be

$$\delta_c = \frac{5.20x_c}{\sqrt{\text{Re}_c}} = \frac{(5.20)(1.76)}{\sqrt{500{,}000}} = 0.0129 \text{ ft}$$

(*c*) The friction drag can be computed by adding to the drag from the laminar boundary layer up to x_c (see Figure 13-17) the drag from the turbulent boundary layer, *B* to *C*. The latter value of drag is obtained by computing the drag for turbulent flow along the entire plate (*A* to *C*) less the drag of the fictitious turbulent layer from *A* to *B*.

(1) Laminar drag, *A* to *B*, on one side.

$$\text{drag force} = C_D \rho A \frac{V^2}{2} = \frac{1.328}{\sqrt{\text{Re}_c}} \rho A \frac{V^2}{2} = \frac{1.328}{\sqrt{500{,}000}}(1.94)(4 \times 1.76)\frac{4^2}{2} = 0.205 \text{ lb}$$

(2) Turbulent drag, *A* to *C*, if conditions were turbulent for the entire length of the plate.

$$\text{drag force} = C_D \rho A \frac{V^2}{2} \quad \text{(one side)}$$

$$= \frac{0.074}{\text{Re}^{0.20}} \rho A \frac{V^2}{2} = \frac{0.074}{(2{,}840{,}000)^{0.20}}(1.94)(4 \times 10)\frac{4^2}{2} = 2.352 \text{ lb}$$

(3) Fictitious turbulent drag, *A* to *B*.

$$\text{drag force} = C_D \rho A \frac{V^2}{2} \quad \text{(one side)}$$

$$\frac{0.074}{\text{Re}_c^{0.20}} \rho A \frac{V^2}{2} = \frac{0.074}{(500{,}000)^{0.20}}(1.94)(4 \times 1.76)\frac{4^2}{2} = 0.586 \text{ lb}$$

$$\text{Total drag force (two sides)} = (2)[0.205 + (2.352 - 0.586)] = 3.94 \text{ lb.}$$

Had the Reynolds number for the entire plate been over 10^7, equation (*13*) at the beginning of this chapter should be used in part (2) above.

A weighted value of C'_D might be obtained for the entire plate by equating the above total drag force to the drag force expression, as follows.

$$\text{total drag force} = 2C'_D \rho A \frac{V^2}{2}, \qquad 3.94 = 2C'_D(1.94)(4 \times 10)\frac{4^2}{2}, \qquad C'_D = 0.00317$$

13.47. Measurements on a smooth sphere 150 mm in diameter in an air stream 20°C gave a force for equilibrium equal to 1.1 N. At what velocity was the air moving?

Solution:

Total drag $= C_D \rho A V^2/2$, where $C_D =$ overall drag coefficient.

Since neither Reynolds number nor C_D can be found directly, assume $C_D = 1.00$. Then

$$1.1 = C_D(1.20)\tfrac{1}{4}\pi\left(\frac{150}{1000}\right)^2(V^2/2), \qquad V^2 = \frac{103.7}{C_D}, \qquad V = 10.18 \text{ m/s}$$

Calculate $\text{Re} = \dfrac{Vd}{\nu} = \dfrac{(10.18)(150/1000)}{1.49 \times 10^{-5}} = 102{,}000$. From Diagram F, $C_D = 0.59$ (for spheres).

Then $V^2 = \dfrac{103.7}{0.59} = 175.8$, $V = 13.26$ m/s. Anticipating result, use $V = 13.6$ m/s.

Recalculate $\text{Re} = \dfrac{Vd}{\nu} = \dfrac{(13.6)(150/1000)}{1.49 \times 10^{-5}} = 137{,}000$. From Diagram F, $C_D = 0.56$.

Then $V^2 = 103.7/0.56 = 185.2$, $V = 13.61$ m/s (satisfactory accuracy).

13.48. For instantaneous closure of a valve in a pipeline, determine the increase in the pressure produced.

Solution:

Let p' equal the change in pressure due to the closure of the valve. The impulse-momentum equation can be employed to evaluate the change in momentum, i.e.,

$$F_x = \frac{\gamma Q}{g}(V_2 - V_1) \quad \text{in the } x \text{ direction} \tag{A}$$

Neglecting friction, the unbalanced force that causes a change in momentum of the liquid in the pipeline is $p'A$. Equation (A) then becomes

$$-p'A = \frac{\gamma(Ac)}{g}(0 - V_1) \tag{B}$$

where $\gamma Ac/g$ represents the mass of liquid having its momentum changed and c represents the celerity of the pressure wave. This pressure wave reduces the velocity to zero as it passes each section. Then

$$p' = \rho c V_1 \tag{C}$$

Equation (C) can be written in terms of pressure head h', i.e.,

$$h' = \frac{cV_1}{g} \tag{D}$$

13.49. What is the formula for the celerity of the pressure wave due to rapid closing of a valve in a pipeline, considering the pipe to be rigid?

Solution:

By "rapid closure" is meant any time $\leq 2L/c$. To establish an expression for celerity c, the principles of work and energy and of momentum must be used.

The kinetic energy of the water will be converted to elastic energy, thereby compressing the water. The kinetic energy is $\frac{1}{2}MV_1^2 = \frac{1}{2}(\gamma AL/g)V_1^2$, where A is the cross-sectional area of the pipe and L is the length of the pipe.

The bulk modulus of elasticity of the water is $E_B = \dfrac{-\Delta p}{(\Delta \text{ volume})/(\text{original volume})}$

Thus the volume reduction is $\qquad \Delta \text{ volume} = \dfrac{(\text{volume})(\Delta p)}{E_B} = \dfrac{(AL)(\gamma h)}{E_B}.$

The work of compression = average pressure intensity times the volume reduction, i.e.,

$$\tfrac{1}{2}(\gamma AL/g)V_1^2 = \tfrac{1}{2}\gamma h(AL\gamma h/E_B) \tag{A}$$

or

$$h^2 = V_1^2 E_B/g\gamma \tag{B}$$

The momentum principle yields (neglecting friction),

$$MV_1 - \Sigma(F_x dt) = MV_2, \qquad -\gamma hA = (\gamma Q/g)(0 - V_1), \qquad \gamma hA = (\gamma/g)(Ac)V_1$$

or

$$h = cV_1/g \tag{C}$$

Substituting in (*B*), we obtain $c^2V_1^2/g^2 = V_1^2 E_B/g\gamma$, from which

$$c = \sqrt{E_B/\rho} \tag{D}$$

13.50. Develop the formula for the celerity of the pressure wave due to rapid closing of a valve in a pipeline, considering the pipe to be nonrigid.

Solution:

In this solution, the elasticity of the walls of the pipe must be considered, in addition to the factors included in the solution of Problem 13.49.

For the pipe, the work done in stretching the pipe walls is equal to the average force exerted in the pipe walls times the strain. From a free-body diagram of one half of the pipe cross section, using $\Sigma Y = 0$, $2T = pdL = \gamma hdL$. Also, unit strain $\epsilon = \sigma/E$, where $\sigma = pr/t = \gamma hr/t$. (See hoop tension in Chapter 3.) In this derivation, head h represents the pressure head above normal, caused by the rapid closing of the valve.

$$\text{work} = \text{average force} \times \text{strain} = \tfrac{1}{2}\left(\tfrac{1}{2}\gamma hdL\right)(2\pi r\epsilon)$$
$$= \tfrac{1}{4}\gamma hdL(2\pi r)(\gamma hr/tE)$$

Adding this value to equation (*A*) of Problem 13.49 gives

$$\tfrac{1}{2}(\gamma AL/g)V_1^2 = \tfrac{1}{2}\gamma h(AL\gamma h/E_B) + \tfrac{1}{4}\gamma hdL\,(2\pi\gamma hr^2/tE)$$

which, after substituting $h = cV_1/g$ from (*C*) of Problem 13.49, gives

$$\frac{V_1^2}{g} = \frac{c^2V_1^2}{g^2}\left(\frac{\gamma}{E_B} + \frac{\gamma d}{tE}\right)$$

$$\text{celerity } c = \sqrt{\frac{1}{\rho(1/E_B + d/Et)}} = \sqrt{\frac{E_B}{\rho(1 + E_B d/Et)}}$$

13.51. Compare the velocities of the pressure waves traveling along a rigid pipe containing (*a*) water at 60°F, (*b*) glycerin at 68°F, and (*c*) oil of sp gr 0.800. Use values of bulk modulus for glycerin and oil of 630,000 and 200,000 psi, respectively.

Solution:

$$c = \sqrt{\frac{\text{bulk modulus in psf}}{\text{density of fluid}}}$$

(*a*) $$c = \sqrt{\frac{313{,}000 \times 144}{1.94}} = 4820 \text{ ft/sec}$$

(*b*) $$c = \sqrt{\frac{630{,}000 \times 144}{1.262 \times 62.4/32.2}} = 6090 \text{ ft/sec}$$

(*c*) $$c = \sqrt{\frac{200{,}000 \times 144}{0.800 \times 62.4/32.2}} = 4310 \text{ ft/sec}$$

13.52. In Problem 13.51, if these liquids were flowing in a 12″ pipe at 4.0 ft/sec and the flow was stopped suddenly, what increase in pressure could be expected, assuming the pipe to be rigid?

Solution:

$$\text{change (increase) in pressure} = \rho c \times \text{change in velocity}$$

(*a*) Increase in pressure $= (1.94)(4820)(4-0) = 37{,}400\ \text{lb/ft}^2 = 260$ psi.

(*b*) Increase in pressure $= (2.45)(6090)(4) = 59{,}700\ \text{lb/ft}^2 = 414$ psi.

(*c*) Increase in pressure $= (1.55)(4310)(4) = 26{,}700\ \text{lb/ft}^2 = 186$ psi.

13.53. A 48″ steel pipe, $\frac{3}{8}''$ thick, carries water at 60°F at a velocity of 6 ft/sec. If the pipe line is 10,000 ft long and a valve at the discharge end is shut in 2.50 sec, what increase in stress in the walls of the pipe can be expected?

Solution:

The pressure wave would travel from valve to inlet end and back again in

$$\text{time (round trip)} = 2\left(\frac{\text{length of pipeline}}{\text{celerity of pressure wave}}\right)$$

The celerity of the pressure wave for a nonrigid pipe is given by

$$c = \sqrt{\frac{E_B}{\rho[1 + (E_B/E)(d/t)]}}$$

where the two ratios are dimensionless when proper units are used.

Taking E for steel $= 30 \times 10^6$ psi, $c = \sqrt{\dfrac{313{,}000 \times 144}{(1.94)\left[1 + \dfrac{313{,}000}{30 \times 10^6}\left(\dfrac{48}{3/8}\right)\right]}} = 3150$ ft/sec

and time $= (2)(10{,}000/3150) = 6.35$ sec.

But the valve was closed in 2.50 sec. This is equivalent to a *sudden closure*, since the pressure wave must reverse itself when it reaches the closed valve.

$$\text{increase in pressure} = \rho c(dV) = (1.94)(3150)(6) = 36{,}700\ \text{psf} = 255\ \text{psi}$$

From the hoop tension formula for thin-shelled cylinders,

$$\text{tensile stress } \sigma = \frac{\text{pressure} \times \text{radius}}{\text{thickness}} = \frac{255 \times 24}{3/8} = 16{,}300\ \text{psi increase}$$

This increase in stress added to the design value of 16,000 psi approaches the elastic limit of steel. The closing time of the valve should be increased to at least 6.5 sec, preferably several times 6.29 sec.

13.54. A valve is suddenly closed in a 75-mm pipe carrying glycerin at 20°C. The increase in pressure is 690 kPa. What is the probable flow in m³/s? Use $\rho = 1260\ \text{kg/m}^3$ and $E_B = 4.34$ GPa.

Solution:

The value of the celerity was calculated in Problem 13.51 as 6090 ft/sec, or 1860 m/s.

The change in pressure $= \rho c \times$ change in velocity

$$690{,}000 = (1260)(1860)V \quad \text{from which} \quad V = 0.294\ \text{m/s}.$$

Thus $Q = AV = \frac{1}{4}\pi(75/1000)^2 \times 0.294 = 0.0013\ \text{m}^3/\text{s}$.

13.55. Air at 27°C flows with a velocity of 6 m/s through a 1.5-m-square ventilation duct. If the control devices are suddenly closed, what force will be exerted on the 1.5 m × 1.5 m area of closure?

Solution:

For air at 27°C, $\rho = 1.17\ \text{kg/m}^3$, and the celerity

$$c = \sqrt{kgRT} = \sqrt{(1.4)(9.81)(29.3)(273+27)} = 347\ \text{m/s}$$

Then, using $\Delta p = \rho c V$, the force is

$$F = \Delta p \times \text{ area} = (\rho c V)A = (1.17)(347)(6)(1.5 \times 1.5) = 5480\ \text{N} = 5.48\ \text{kN}$$

13.56. A sonar transmitter operates at 2 impulses per second. If the device is held at the surface of fresh water at 40°F and the echo is received midway between impulses, how deep is the water? (The depth is known to be less than 2000 ft.)

Solution:

The celerity of sound in water at 40°F is computed by using

$$c = \sqrt{\frac{\text{bulk modulus}}{\text{density of fluid}}} = \sqrt{\frac{296{,}000 \times 144}{1.94}} = 4690\ \text{ft/sec}$$

(*a*) The distance traveled by the sound wave (down and back again) in $\frac{1}{2}$ of $\frac{1}{2}$ sec, or $\frac{1}{4}$ sec (one-half impulse time), is

$$2 \times \text{depth} = \text{velocity} \times \text{time}$$
$$= 4690 \times \tfrac{1}{4} \quad \text{and} \quad \text{depth} = 586\ \text{ft (least depth)}$$

(*b*) Had the depth exceeded 586 ft, for the echo to be heard midway between impulses the sound would have traveled for 3/2 of 1/2 sec, or 3/4 sec. Then

$$\text{depth} = \left(\tfrac{1}{2}\right)(4690) \times \tfrac{3}{4} = 1760\ \text{ft}$$

(*c*) For depths beyond the 2000-ft limit suggested, we obtain

$$\text{depth} = \left(\tfrac{1}{2}\right)(4690) \times \tfrac{5}{4} = 2930\ \text{ft}$$

$$\text{depth} = \left(\tfrac{1}{2}\right)(4690) \times \tfrac{7}{4} = 4100\ \text{ft, and so on}$$

13.57. A projectile moves at 670 m/s through still air at 38°C and 100 kPa. Determine (*a*) the Mach number, (*b*) the Mach angle, and (*c*) the drag force for shape *B* on Diagram *H*, assuming a diameter of 200 mm.

Solution:

(*a*) Celerity $c = \sqrt{kgRT} = \sqrt{(1.4)(9.81)(29.3)(273+38)} = 354$ m/s.

$$\text{Mach number Ma} = \frac{V}{c} = \frac{670}{354} = 1.89$$

(*b*) Mach angle $\alpha = \sin^{-1}\dfrac{1}{\text{Ma}} = \sin^{-1}\dfrac{1}{1.89} = 31.9°$.

(*c*) From Diagram *H*, shape *B*, for Mach number of 1.89 the coefficient of drag is 0.60.

The specific weight of the air is $\gamma = \dfrac{p}{RT} = \dfrac{100{,}000}{(29.3)(273+38)} = 11.0\ \text{N/m}^3$.

$$\text{drag force} = C_D \rho A V^2/2 = (0.60)(11.0/9.81) \times \tfrac{1}{4}\pi(200/1000)^2 \times (670)^2/2 = 4740\ \text{N} = 4.74\ \text{kN}$$

13.58. From a photograph the Mach angle for a projectile moving through air was 40°. Calculate the speed of the bullet for air conditions in Problem 13.57. (Celerity $c = 354$ m/s.)

Solution:

$$\sin\alpha = \frac{c}{V} = \frac{1}{\text{Ma}}. \text{ Then } \sin 40^\circ = \frac{354}{V}, \text{ and } V = 551 \text{ m/s.}$$

13.59. What should be the diameter of a sphere, sp gr 2.50, in order for its freely falling velocity to attain the acoustic velocity? Use $\rho = 1.22$ kg/m^3 for air.

Solution:

For freely falling body, drag force − weight = 0, and, from Diagram H, $C_D = 0.80$.

For air at 16°C, $c = \sqrt{kgRT} = \sqrt{(1.4)(9.81)(29.3)(289)} = 341$ m/s.

Since weight = drag force,

$$(2.50 \times 9790) \times (4\pi/3)(d/2)^3 = 0.80 \times 1.22 \times (\pi d^2/4) \times (341)^2/2, \quad d = 3.48 \text{ m}$$

Supplementary Problems

13.60. Show that the momentum correction factor β in Problem 7.85 of Chapter 7 is 1.20.

13.61. Show that the momentum correction factor β in Problem 7.83 of Chapter 7 is 1.02.

13.62. Determine the momentum correction factor β for Problem 7.92 in Chapter 7. *Ans.* $\dfrac{(K+1)^2(K+2)^2}{2(2K+1)(2K+2)}$

13.63. Show that the momentum correction factor β for Problem 8.86 in Chapter 8 is 1.12.

13.64. A jet of oil 2″ in diameter strikes a flat plate held normal to the stream's path. For a velocity in the jet of 80 ft/sec, calculate the force exerted on the plate by the oil, sp gr 0.85. *Ans.* 230 lb

13.65. If the plate in Problem 13.64 were moving at 30 ft/sec in the same direction as the fluid, what force would be exerted on the plate by the oil? At 30 ft/sec in the opposite direction, what force would be exerted? *Ans.* 89.5 lb, 434 lb

13.66. A jet of water 50 mm in diameter exerts a force of 2.67 kN on a flat plate held normal to the jet's path. What is the rate of discharge? *Ans.* 0.0736 m^3/s

13.67. Water flowing at the rate of 0.034 m^3/s strikes a flat plate held normal to its path. If the force exerted on the plate is 721 N, calculate the diameter of the stream. *Ans.* 45.2 mm

13.68. A jet of water 2″ in diameter strikes a curved blade at rest and is deflected 135° from its original direction. Neglecting friction along the blade, find the resultant force exerted on the blade if the jet velocity is 90 ft/sec. *Ans.* 630 lb, $\theta_x = -22.6°$

13.69. If the blade in Problem 13.68 were moving at 20 ft/sec against the direction of the water, what resultant force would be exerted on the blade and what power would be required to maintain the motion? *Ans.* 943 lb, 31.7 hp

13.70. A stationary blade deflects a 50-mm-diameter jet moving at 35.1 m/s through 180°. What force does the blade exert on the water? *Ans.* 5.0 kN

13.71. A horizontal 300-mm pipe contracts to a 150-mm diameter. If the flow is 0.127 m^3/s of oil, sp gr 0.88, and the pressure in the smaller pipe is 265 kPa, what resultant force is exerted on the contraction, neglecting friction? *Ans.* 15.4 kPa.

13.72. A vertical reducing bend (see Fig. 13-18) carries 12.6 cfs of oil, sp gr 0.85, at a pressure of 20.5 psi entering the bend at A. The diameters at A and B are 16″ and 12″, respectively, and the volume between A and B is 3.75 cu ft. Neglecting friction, find the force on the bend. *Ans.* 5180 lb, $\theta_x = -76.1°$

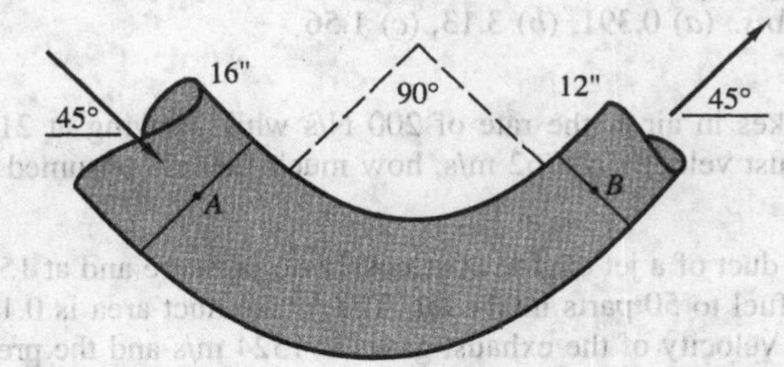

Fig. 13-18

13.73. A model of a motorboat is driven at 4.57 m/s by means of a jet of water 25 mm in diameter, ejected directly astern. The velocity of the jet relative to the model is 35.1 m/s. What is the driving force? *Ans.* 543 N

13.74. A 50-mm-diameter nozzle, $c_v = 0.97$, is attached to a tank and discharges a stream of oil, sp gr 0.80, horizontally under a head of 11 m. What horizontal force is exerted on the tank? *Ans.* 328 N

13.75. A toy balloon weighing 0.23 lb is filled with air having a density of 0.00250 slug/ft^3. The small filling tube, $\frac{1}{4}''$ in diameter, is pointed downward, and the balloon is released. If the air escapes at the initial rate of 0.30 cfs, what is the instantaneous acceleration (neglecting friction)? *Ans.* 60.2 ft/sec^2

13.76. A jet propulsion device is moving upstream in a river with absolute velocity of 8.69 m/s. The stream is flowing at 2.29 m/s. A jet of water issues from the device at a relative velocity of 18.29 m/s. If the flow in the jet is 1.416 m^3/s, what thrust is developed on the jet propulsion device? *Ans.* 10.4 kN

13.77. What weight will be supported by a wing area of 500 ft^2 at an angle of attack of 4° and an airspeed of 100 ft/sec? Use $C_L = 0.65$ and 60°F air. *Ans.* 3850 lb

13.78. At what speed should a plane having a wing area of 46.45 m^2 and weighing 26.7 kN be flown if the angle of attack is 8°? Use $C_L = 0.90$. *Ans.* 32.3 m/s

13.79. What wing area should a plane weighing 2000 lb have to enable it to land at a speed of 35 mph? Use max. $C_L = 1.50$. *Ans.* 425 ft^2

13.80. If the drag on a wing of 27.87-m^2 area is 3.02 kN, at what speed is the wing moving for an angle of attack of 7°? Use $C_D = 0.05$. *Ans.* 59.7 m/s

13.81. A 30-mph wind blows at an angle of 8° to the plane of a signboard that is 12 ft long by 2 ft wide. Using values of $C_L = 0.52$ and $C_D = 0.09$, calculate the force acting on the signboard at right angles to the wind direction and the force acting parallel to the wind direction. Assume standard air at 60°F. *Ans.* 28.6 lb, 4.95 lb

13.82. Prove that for a given angle of attack the drag force on an airfoil is the same for all altitudes. (For a given angle of attack, C_D will not change with altitude.)

13.83. A wing model of 3-ft span and 4″ chord is tested at a fixed angle of attack in a wind tunnel. The air, at standard pressure and 80°F, has a velocity of 60 mph. The lift and drag are measured at 5.8 lb and 0.50 lb, respectively. Determine the lift and drag coefficients. *Ans.* 0.657, 0.0566

13.84. Calculate the Mach number for (*a*) an airplane moving through standard air at 20°C at 483 km/h, (*b*) a rocket moving through air at 20°C at 3860 km/h, and (*c*) a projectile moving through standard air at 20°C at 1930 km/h. *Ans.* (*a*) 0.391, (*b*) 3.13, (*c*) 1.56

13.85. A turbojet engine takes in air at the rate of 200 N/s while moving at 213 m/s. If the thrust developed is 12.0 kN for an exhaust velocity of 762 m/s, how much fuel is consumed per second? *Ans.* 10.6 N/s

13.86. Air enters the intake duct of a jet engine at atmospheric pressure and at 152 m/s. Fuel is used by the engine at the rate of 1 part fuel to 50 parts intake air. The intake duct area is 0.139 m^2, and the density of the air is 1.24 kg/m^3. If the velocity of the exhaust gases is 1524 m/s and the pressure is atmospheric, what thrust is developed? *Ans.* 37 kN

13.87. An automobile has a projected area of 32.0 ft^2 and moves at 50 mph in still air at 80°F. If $C_D = 0.45$, what horsepower is necessary to overcome air resistance? *Ans.* 11.8 hp

13.88. A train travels at 121 km/h through standard air at 16°C and is 152 m long. Consider the surfaces of the train to be 1393 m^2 in area and equivalent to a smooth flat plate. For a turbulent leading edge, what is the skin friction drag? *Ans.* 1.67 kN on one side

13.89. A cylinder 610 mm in diameter and 4.57 m long moves at 48.3 km/h through 16°C water (parallel to its length). What coefficient of drag is indicated if the skin friction drag is 1.60 kN? *Ans.* $C_D = 0.00204$

13.90. Calculate the friction drag on a plate 1 ft wide by 3 ft long placed longitudinally (*a*) in a stream of water at 70°F flowing at 1.0 ft/sec and (*b*) in a stream of heavy fuel oil at 70°F at 1.0 ft/sec. *Ans.* (*a*) 0.0144 lb, (*b*) 0.161 lb

13.91. A balloon 4 ft in diameter weighs 4.0 lb and is acted upon by a buoyant force averaging 5.0 lb. Using $\rho = 0.00228$ slug/ft^3 and $\nu = 17.0 \times 10^{-5}$ ft^2/sec, estimate the velocity with which it will rise. *Ans.* 18.7 ft/sec

13.92. Estimate the terminal velocity of a hailstone 13 mm in diameter using air temperature = 4°C and specific gravity of the hailstone = 0.90. *Ans.* 16.8 m/s

13.93. An object having a projected area of 0.557 m^2 moves at 48.3 km/h. If the drag coefficient is 0.30, compute the drag force in 16°C water and in 16°C standard air. *Ans.* 15.0 kN, 18.3 N

13.94. A body travels through 60° standard air at 60 mph, and 5.5 hp is required to maintain this speed. If the projected area is 13.5 ft^2, find the drag coefficient. *Ans.* 0.28

13.95. A smooth rectangular plate 0.61 m wide by 24.38 m long moves at 12.2 m/s in the direction of its length through oil. (*a*) Calculate the drag force on the plate and the thickness of the boundary layer at the trailing edge. (*b*) How long is the laminar boundary layer? Use kinematic viscosity = 1.49×10^{-5} m^2/s and $\gamma = 8.33$ kN/m^3. *Ans.* (*a*) 5.0 kN, 326 mm; (*b*) 610 mm

13.96. Assuming a 24″ steel pipe to be rigid, what pressure increase occurs when a flow of 20.0 cfs of oil, specific gravity 0.85 and bulk modulus 250,000 psi, is stopped suddenly? *Ans.* 341 psi

13.97. If the pipeline in Problem 13.96 is 8000 ft long, how much time should be allowed for closing a valve to avoid water hammer? *Ans.* more than 3.42 sec

13.98. If a 24″ steel pipe 8000 ft long is designed for a stress of 15,000 psi under a maximum static head of 1085 ft of water, how much will the stress in the walls of the pipe increase when a quick-closing valve stops a flow of 30.0 cfs? ($E_B = 300{,}000$ psi) *Ans.* 472 psi

13.99. Calculate the Mach angle for a bullet moving at 518 m/s through air at 98.6 kPa and 16°C. *Ans.* 41°13′

13.100. What is the drag force of a projectile (shape *A* in Diagram *H*) 102 mm in diameter when it moves at 579 m/s through air at 10°C and 98.6 kPa? *Ans.* 854 N

Chapter 14

Fluid Machinery

FLUID MACHINERY

Consideration is given here to fundamental principles upon which the design of pumps, blowers, turbines, and propellers is based. The essential tools are the principles of impulse-momentum (Chapter 13) and of the forced vortex (Chapter 5) and the laws of similarity (Chapter 6). Modern hydraulic turbines and centrifugal pumps are highly efficient machines with few differences in their characteristics. For each design there is a definite relationship between the speed of rotation N, discharge or flow Q, head H, diameter D of the rotating element, and power P.

For ROTATING CHANNELS, the torque and power produced are evaluated by

$$\text{torque } T = \frac{\gamma Q}{g}(V_2 r_2 \cos\alpha_2 - V_1 r_1 \cos\alpha_1) \tag{1}$$

and

$$\text{power } P = \frac{\gamma Q}{g}(V_2 u_2 \cos\alpha_2 - V_1 u_1 \cos\alpha_1) \tag{2}$$

The development and notation are explained in Problem 14.1.

WATER WHEELS, TURBINES, PUMPS, AND BLOWERS

These have certain constants that are commonly evaluated. Details are developed in Problem 14.5.

(1) The *speed factor* ϕ is defined as

$$\phi = \frac{\text{peripheral velocity of rotating element}}{\sqrt{2gH}} = \frac{u}{\sqrt{2gH}} \tag{3}$$

where u = radius of rotating element in ft × angular velocity in radians/sec $= r\omega$ ft/sec.

This factor is also expressed as

$$\phi = \frac{\text{diameter in inches} \times \text{rpm}}{1840\sqrt{H}} = \frac{D_1 N}{1840\sqrt{H}} \tag{4}$$

(2) (*a*) The *speed relation* may be expressed as

$$\frac{\text{diameter } D \text{ in ft} \times \text{speed } N \text{ in rpm}}{\sqrt{g \times \text{head } H \text{ in ft}}} = \text{constant } C'_N \tag{5a}$$

Also,

$$H = \frac{D^2 N^2}{C_N^2} \tag{5b}$$

in which g is incorporated in the C_N coefficient.

(*b*) The *unit speed* is defined as the speed of a geometrically similar (homologous) rotating element having a diameter of 1 in, operating under a head of 1 ft. This unit speed (N_u

in rpm) is usually expressed in terms of D_1 in inches and N in rpm. Thus

$$N_u = \frac{D\text{ (in)} \times \text{rpm}}{\sqrt{H}} = \frac{D_1 N}{\sqrt{H}} \tag{6a}$$

Also,

$$N = N_u \frac{\sqrt{H}}{D_1} \tag{6b}$$

(3) (*a*) The *discharge relation* may be expressed as

$$\frac{\text{discharge } Q \text{ in cfs}}{(\text{diameter } D \text{ in ft})^2 \sqrt{\text{head } H \text{ in ft}}} = \text{constant } C_Q \tag{7a}$$

Also,

$$Q = C_Q D^2 \sqrt{H} = C_Q D^2 \left(\frac{DN}{C_N}\right) = C'_Q D^3 N \tag{7b}$$

The coefficient C_Q may also be expressed in terms of gpm flow units. In taking coefficients from texts or handbooks, units should be checked or mistakes will occur.

If C_Q is made the same for two homologous units, then C_N, C_P, and the efficiency will be the same, unless very viscous fluids are involved.

(*b*) The *unit discharge* is defined as the discharge of an homologous rotating element 1 inch in diameter, operating under a head of 1 ft. The unit flow Q_u in cfs is written

$$Q_u = \frac{\text{flow } Q \text{ in cfs}}{(\text{diameter } D, \text{ in})^2 \sqrt{\text{head } H, \text{ ft}}} = \frac{Q}{D_1^2 \sqrt{H}} \tag{8a}$$

Also,

$$Q = Q_u D_1^2 \sqrt{H} \tag{8b}$$

(4) (*a*) The *power relation*, obtained by using values of Q and H in equations (*7b*) and (*5a*) is

$$P = \frac{\gamma Q H}{550e} = \frac{\gamma \left(C_Q D^2 \sqrt{H}\right) H}{550e} = C_P D^2 H^{3/2} \quad \text{(hp)} \tag{9a}$$

Also

$$P = \frac{\gamma \left(C'_Q D^3 N\right)}{550e} \times \frac{D^2 N^2}{g\left(C'_N\right)^2} = C'_P \rho D^5 N^3 \quad \text{(hp)} \tag{9b}$$

Note: Delete 550 for power in watts.

(*b*) *Unit power* is defined as the power developed by an homologous rotating element 1 inch in diameter, operating under a head of 1 ft. The unit power P_u is

$$P_u = \frac{P}{D_1^2 H^{3/2}} \quad \text{and} \quad P = P_u D_1^2 H^{3/2} \tag{10}$$

SPECIFIC SPEED

Specific speed is defined as the speed of a homologous rotating element of such diameter that it develops unit power for a unit of head. The specific speed N_S may be expressed in two forms, as follows:

(1) *For turbines*, the general equation is

$$N_S = \frac{N\sqrt{P}}{\sqrt{\rho}(gH)^{5/4}} \tag{11a}$$

Also, $$N_S = N_u\sqrt{P_u} = \frac{N\sqrt{P}}{H^{5/4}} \qquad (11b)$$

has common application for water turbines.

(2) *For pumps and blowers*, the general equation is

$$N_S = \frac{N\sqrt{Q}}{(gH)^{3/4}} \qquad (12a)$$

Also $$N_S = N_u\sqrt{Q_u} = \frac{N\sqrt{Q}}{H^{3/4}} \qquad (12b)$$

has common application.

EFFICIENCY

Efficiency is expressed as a ratio. It will vary with speed and discharge.

For turbines,

$$\text{overall efficiency } e = \frac{\text{power derived from shaft}}{\text{power supplied by water}} \qquad (13)$$

$$\text{hydraulic efficiency } e_h = \frac{\text{power utilized by unit}}{\text{power supplied by water}}$$

For pumps,

$$\text{efficiency } e = \frac{\text{power output}}{\text{power input}} = \frac{\gamma QH}{\text{power input}} \qquad (14)$$

CAVITATION

Cavitation causes the rapid disintegration of metal in pump impellers, turbine runners and blades, Venturi meters, and, on occasion, pipelines. It occurs when the pressure in the liquid falls below the vapor pressure of that liquid.

PROPULSION BY PROPELLERS

Propulsion by propellers has been the motive power of aircraft and marine craft for some time. Also, propellers are used as fans and as means of producing power from the wind. Propeller design will not be attempted here, but expressions for thrust and power are matters of fluid mechanics. The following expressions are developed in Problem 14.23.

$$\text{thrust } F = \frac{\gamma Q}{g}(V_{\text{final}} - V_{\text{initial}}) \qquad (15)$$

$$\text{power output } P_o = \frac{\gamma Q}{g}(V_{\text{final}} - V_{\text{initial}})\,V_{\text{initial}} \qquad (16)$$

$$\text{power input } P_i = \frac{\gamma Q}{g}\left(\frac{V_{\text{final}}^2 - V_{\text{initial}}^2}{2}\right) \qquad (17)$$

$$\text{efficiency } e = \frac{\text{power output}}{\text{power input}} = \frac{2V_{\text{initial}}}{V_{\text{final}} + V_{\text{initial}}} \qquad (18)$$

PROPELLER COEFFICIENTS involve thrust, torque, and power. They may be expressed as follows.

$$\text{thrust coefficient } C_F = \frac{\text{thrust } F}{\rho N^2 D^4} \tag{19}$$

High values of C_F produce good propulsion.

$$\text{torque coefficient } C_T = \frac{\text{torque } T}{\rho N^2 D^5} \tag{20}$$

High values of C_T are common for turbines and windmills.

$$\text{power coefficient } C_P = \frac{\text{power } P}{\rho N^3 D^5} \tag{21}$$

This last coefficient is in the same form as equation (*9b*).

All three coefficients are dimensionless if N is in revolutions per second.

Solved Problems

14.1. Determine the torque and power developed by a rotating element (such as a pump impeller or turbine runner) under steady flow conditions.

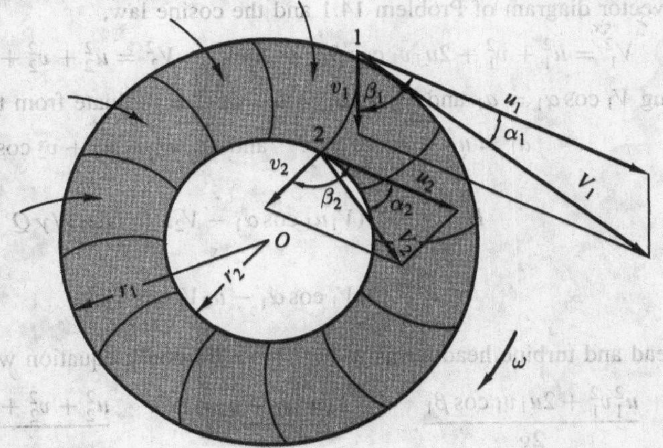

Fig. 14-1

Solution:

Let Fig. 14-1 represent water entering the curved channels formed by the rotating element at radius r_1 and leaving at radius r_2. The relative velocities of the water with respect to a blade are shown as v_1 entering at 1 and v_2 at exit 2. The linear velocity of the blade is u_1 at 1 and u_2 at 2. The vector diagrams show the absolute velocities of the water (V_1 and V_2).

For the elementary mass of water flowing in dt seconds the change in angular momentum is caused by the angular impulse exerted by the runner. That is,

$$\text{initial angular momentum} + \text{angular impulse} = \text{final angular momentum}$$

or $$(dM)V_1 \times r_1 \cos\alpha_1 + \text{torque} \times dt = (dM)V_2 \times r_2 \cos\alpha_2$$

But $dM = (\gamma/g)Q\,dt$. Substituting and solving for the torque exerted on the water, we obtain

$$\text{torque } T = \frac{\gamma}{g} Q\,(V_2 r_2 \cos\alpha_2 - V_1 r_1 \cos\alpha_1)$$

Thus the torque exerted by the fluid on the rotating part is

$$T = \frac{\gamma}{g} Q\,(V_1 r_1 \cos\alpha_1 - V_2 r_2 \cos\alpha_2)$$

Power is equal to torque times angular velocity:

$$P = T\omega = \frac{\gamma}{g} Q\,(V_1 r_1 \cos\alpha_1 - V_2 r_2 \cos\alpha_2)\,\omega$$

Since $u_1 = r_1\omega$ and $u_2 = r_2\omega$, this expression becomes

$$P = \frac{\gamma}{g} Q\,(V_1 u_1 \cos\alpha_1 - V_2 u_2 \cos\alpha_2) \tag{1}$$

The expressions developed here are applicable to pumps and turbines alike. The important point is that, in the development, point 1 was *upstream* and point 2 was *downstream.*

14.2. Establish the Bernoulli equation for a rotating turbine runner.

Solution:

Writing the Bernoulli equation, point 1 to point 2 in Fig. 14-1, yields

$$\left(\frac{V_1^2}{2g} + \frac{p_1}{\gamma} + z_1\right) - H_T - \text{lost head } H_L = \left(\frac{V_2^2}{2g} + \frac{p_2}{\gamma} + z_2\right)$$

From the vector diagram of Problem 14.1 and the cosine law,

$$V_1^2 = u_1^2 + v_1^2 + 2u_1 v_1 \cos\beta_1 \qquad \text{and} \qquad V_2^2 = u_2^2 + v_2^2 + 2u_2 v_2 \cos\beta_2$$

Also, letting $V_1 \cos\alpha_1 = a_1$ and $V_2 \cos\alpha_2 = a_2$, we can evaluate from the vector diagram

$$a_1 = u_1 + v_1 \cos\beta_1 \qquad \text{and} \qquad a_2 = u_2 + v_2 \cos\beta_2$$

Furthermore,

$$H_T = \frac{\gamma Q}{g}(V_1 u_1 \cos\alpha_1 - V_2 u_2 \cos\alpha_2)/\gamma Q$$

$$= \frac{1}{g}(u_1 V_1 \cos\alpha_1 - u_2 V_2 \cos\alpha_2) \tag{1}$$

The velocity head and turbine head terms in the above Bernoulli equation would then become

$$\frac{u_1^2 v_1^2 + 2u_1 v_1 \cos\beta_1}{2g}, \qquad \frac{2\,(u_1 a_1 - u_2 a_2)}{2g}, \qquad \frac{u_2^2 + v_2^2 + 2u_2 v_2 \cos\beta_2}{2g}$$

Simplifying and incorporating these terms in the Bernoulli equation produces

$$\left(\frac{v_1^2}{2g} - \frac{u_1^2}{2g} + \frac{p_1}{\gamma} + z_1\right) + \frac{u_2^2}{2g} - H_L = \left(\frac{v_2^2}{2g} + \frac{p_2}{\gamma} + z_2\right)$$

or

$$\left(\frac{v_1^2}{2g} + \frac{p_1}{\gamma} + z_1\right) - \left(\frac{u_1^2 - u_2^2}{2g}\right) - H_L = \left(\frac{v_2^2}{2g} + \frac{p_2}{\gamma} + z_2\right) \tag{2}$$

in which velocities v are relative values and the term in the second set of parentheses is designated as the head created by the forced vortex or the centrifugal head.

14.3. A turbine rotates at 100 rpm and discharges 28.6 cfs. The pressure head at exit is 1.00 ft, and the hydraulic efficiency under these conditions is 78.5%. The physical data are $r_1 = 1.50$ ft, $r_2 = 0.70$ ft, $\alpha_1 = 15°$, $\beta_2 = 135°$, $A_1 = 1.25$ ft², $A_2 = 0.818$ ft², $z_1 = z_2$. Assuming a lost head

Mathcad

of 4.00 ft, determine (*a*) the power delivered to the turbine, (*b*) the total available head and the head utilized, and (*c*) the pressure at entrance.

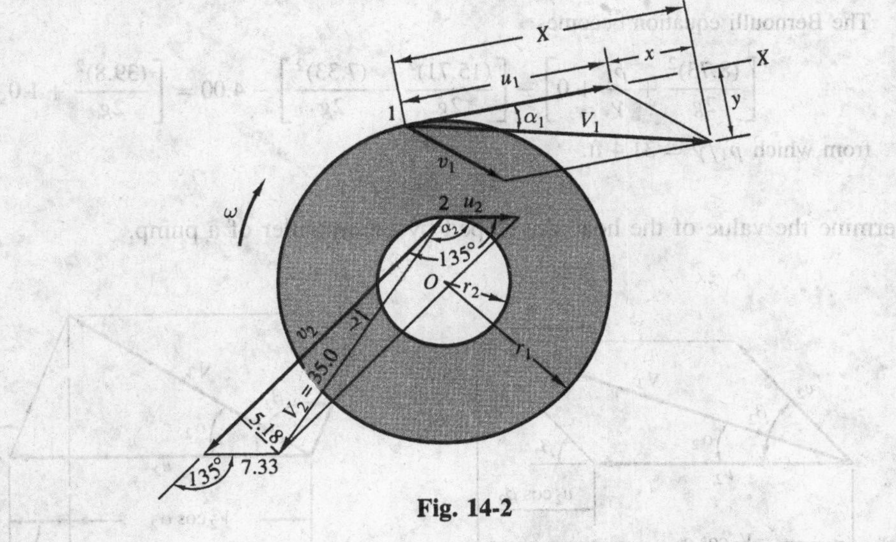

Fig. 14-2

Solution:

(*a*) Preliminary calculations must be made before substituting in the power equation [equation (*1*) of Problem 14.1].

$$V_1 = Q/A_1 = 28.6/1.25 = 22.9 \text{ ft/sec}, \qquad V_2 = 28.6/0.818 = 35.0 \text{ ft/sec}$$

$$V_1 \cos\alpha_1 = 22.9 \times \cos 15° = 22.12 \text{ ft/sec}$$

$$u_1 = (1.50)(2\pi)(100/60) = 15.71 \text{ ft/sec}, \qquad u_2 = (0.70)(2\pi)(100/60) = 7.33 \text{ ft/sec}$$

From the vector diagram in Fig. 14-2, where $\gamma = \sin^{-1} 5.18/35.0 = 8°31'$, we have

$$\alpha_2 = 135° - \gamma = 126°29' \qquad \text{and} \qquad V_2\cos\alpha_2 = (35.0)(-0.595) = -20.8 \text{ ft/sec}$$

Then

$$\text{power } P = \frac{62.4 \times 28.6}{550 \times 32.2}[(15.71)(22.12) - (7.33)(-20.8)] = 50.4 \text{ hp}$$

(*b*) $\text{Efficiency} = \dfrac{\text{output}}{\text{input}} = \dfrac{\text{head utilized}}{\text{head available}}$, but

$$\text{head utilized} = \frac{\text{horsepower utilized} \times 550}{\gamma Q}, \qquad \text{or} \qquad H_T = \frac{50.4 \times 550}{62.4 \times 28.6} = 15.53 \text{ ft}$$

Thus,

$$\text{head available} = 15.53/0.785 = 19.8 \text{ ft}$$

(*c*) In order to use equation (2) of Problem 14.2, we must calculate the two relative velocities. Referring again to the vector diagram of Fig. 14-2, we obtain

$$X = 22.9\cos 15° = 22.12 \text{ ft/sec [as in } (a)]$$

$$y = 22.9\sin 15° = 5.93 \text{ ft/sec}$$

$$x = (X - u_1) = 22.12 - 15.71 = 6.41 \text{ ft/sec}$$

$$v_1 = \sqrt{(5.93)^2 + (6.41)^2} = 8.73 \text{ ft/sec}$$

In similar fashion,

$$v_2 = V_2 \cos\gamma + u_2 \cos 45^\circ = 39.8 \text{ ft/sec}$$

The Bernoulli equation becomes

$$\left[\frac{(8.73)^2}{2g} + \frac{p_1}{\gamma} + 0\right] - \left[\frac{(15.71)^2}{2g} - \frac{(7.33)^2}{2g}\right] - 4.00 = \left[\frac{(39.8)^2}{2g} + 1.0 + 0\right]$$

from which $p_1/\gamma = 31.4$ ft.

14.4. Determine the value of the head developed by the impeller of a pump.

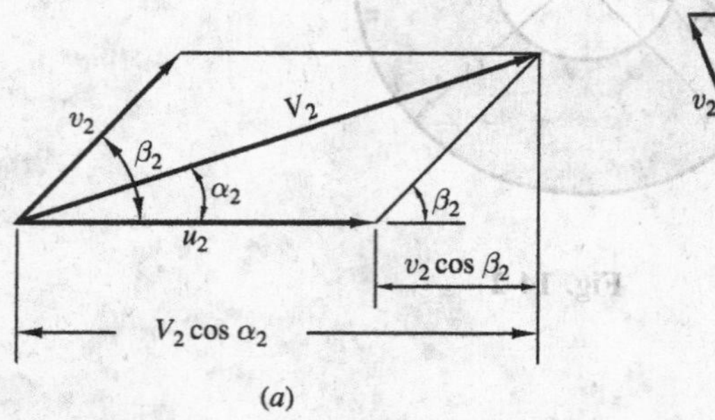

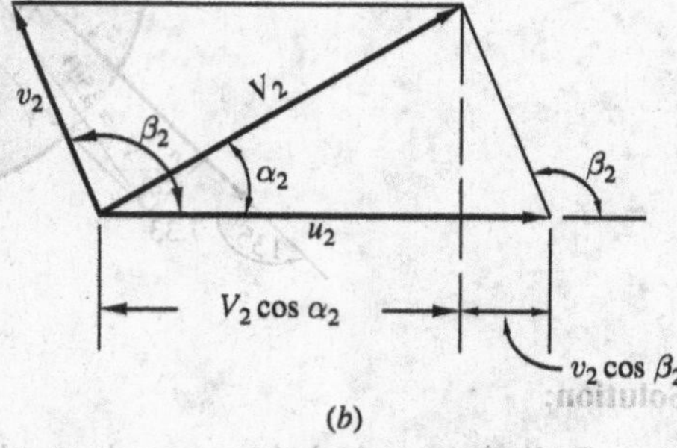

Fig. 14-3

Solution:

Expression (*1*) of Problem 14.1, applied in the direction of flow in a pump (where r_1 is the inner radius, etc.) becomes

$$\text{power input} = \frac{\gamma Q}{g}(u_2 V_2 \cos\alpha_2 - u_1 V_1 \cos\alpha_1)$$

and the head imparted by the impeller is obtained by dividing by γQ; thus,

$$\text{head } H' = \frac{1}{g}(u_2 V_2 \cos\alpha_2 - u_1 V_1 \cos\alpha_1)$$

In most pumps the flow at point 1 can be assumed to be radial and the value of the $u_1 V_1 \cos\alpha_1$ term is zero. The above equation then becomes

$$\text{head } H' = \frac{1}{g}(u_2 V_2 \cos\alpha_2) \tag{1}$$

It can be seen in Fig. 14-3(*a*) and (*b*) that $V_2 \cos\alpha_2$ can be expressed in terms of u_2 and v_2, thus,

$$V_2 \cos\alpha_2 = u_2 + v_2 \cos\beta_2$$

with due regard to the sign of $\cos\beta_2$. Then

$$\text{head } H' = \frac{u_2}{g}(u_2 + v_2 \cos\beta_2) \tag{2}$$

Furthermore, from the vector triangles,

$$V_2^2 = u_2^2 + v_2^2 - 2u_2 v_2 \cos(180^\circ - \beta_2)$$

from which we may write

$$u_2 v_2 \cos\beta_2 = \tfrac{1}{2}(V_2^2 - u_2^2 - v_2^2)$$

The head equation (2) becomes

$$\text{head } H' = \frac{u_2^2}{2g} + \frac{V_2^2}{2g} - \frac{v_2^2}{2g}$$

The head developed by the pump will be less than this amount by the loss in the impeller and by loss at exit. Then

$$\text{head developed } H = \left(\frac{u_2^2}{2g} + \frac{V_2^2}{2g} - \frac{v_2^2}{2g}\right) - \text{impeller loss} - \text{exit loss}$$

$$H = \left(\frac{u_2^2}{2g} + \frac{V_2^2}{2g} - \frac{v_2^2}{2g}\right) - K_i\frac{v_2^2}{2g} - K_e\frac{V_2^2}{2g}$$

14.5. Evaluate for pumps and turbines (*a*) the speed factor ϕ, (*b*) the unit speed N_u, (*c*) the unit flow Q_u, (*d*) the unit power P_u, and (*e*) the specific speed.

Solution:

(*a*) By definition, $\phi = \dfrac{u}{\sqrt{2gH}}$. But $u = r\omega = r\dfrac{2\pi N}{60} = \dfrac{\pi DN}{60} = \dfrac{\pi D_1 N}{720}$, where D_1 is the diameter in inches and N is the speed in revolutions per minute. Finally,

$$\phi = \frac{\pi D_1 N}{720} \times \frac{1}{\sqrt{2gH}} = \frac{D_1 N}{1840\sqrt{H}} \qquad \left(\text{for } g = 32.2 \text{ ft/sec}^2\right) \tag{1a}$$

(*b*) If $D_1 = 1$ in and $H = 1$ ft, we obtain from equation (*1a*) the unit speed N_u. Thus

$$N_u = 1840\phi \tag{1b}$$

which is constant for all wheels of like design if ϕ relates to best speed. Also, from (*1a*),

$$N_u = \frac{D_1 N}{\sqrt{H}} \quad \text{in rpm} \tag{2}$$

Thus, for homologous rotating elements, the best speed N varies inversely as the diameter and directly as the square root of H.

(*c*) For the tangential turbine, the flow Q through the unit can be expressed as

$$Q = cA\sqrt{2gH} = c\frac{\pi d_1^2}{4 \times 144}\sqrt{2gH} = \frac{c\pi\sqrt{2g}}{576}\left(\frac{d_1}{D_1}\right)^2 D_1^2\sqrt{H}$$

$$= (\text{factor})\ D_1^2\sqrt{H} = Q_u D_1^2\sqrt{H} \tag{3}$$

For $D_1 = 1$ in and $H = 1$ ft, the factor is defined as the unit flow Q_u.

For reaction turbines and pumps, flow Q can be expressed as the product

$$c \times A \times (\text{velocity component})$$

The velocity component depends upon the square root of H and the sine of angle α_1 (see Fig. 14-1). Thus flow Q can be written in the form of (*3*), above.

(*d*) Using expression (*3*),

$$\text{power } P = \frac{\gamma QH}{550} = \frac{\gamma\left(Q_u D_1^2\sqrt{H}\right)H}{550} \quad \text{hp}$$

For $D_1 = 1$ in and $H = 1$ ft, power $= \gamma Q_u/550 =$ (factor). When the efficiency is included in the power output for turbines and the water horsepower for pumps, the factor becomes the unit power P_u. Then

$$\text{power } P = P_u D_1^2 H^{3/2} \tag{4}$$

(*e*) In equation (*4*) we may substitute for D_1 its value in (*2*), obtaining

$$\text{power } P = P_u \frac{N_u^2 H}{N^2} H^{3/2}$$

Also,
$$P_u N_u^2 = \frac{PN^2}{H^{5/2}} \quad \text{or} \quad N_u\sqrt{P_u} = \frac{N\sqrt{P}}{H^{5/4}} \tag{5}$$

The term $N_u\sqrt{P_u}$ is called the specific speed N_S. Expression (*5*) then becomes

$$N_S = \frac{N\sqrt{P}}{H^{5/4}} \quad \text{(for turbines)} \tag{6}$$

If P is replaced by Q eliminating D_1 in equations (*2*) and (*3*), we obtain

$$N_u^2 Q_u = \frac{QN^2}{H^{3/2}}$$

and
$$N_S = \frac{N\sqrt{Q}}{H^{3/4}} \quad \text{(for pumps)} \tag{7}$$

where this specific speed refers to the speed at which 1 cfs would discharge against a 1-ft head.

These are the common expressions for pumps and water wheels. For homologous rotating elements in which different fluids may be used, see expressions (*9b*), (*11a*), and (*12a*) at the beginning of this chapter.

14.6. A tangential turbine develops 7200 hp at 200 rpm under a head of 790 ft at an efficiency of 82%. (*a*) If the speed factor is 0.46, compute the wheel diameter, flow, unit speed, unit power, unit flow, and specific speed. (*b*) For this turbine, what would be the speed, power, and flow under a head of 529 ft? (*c*) For a turbine having the same design, what size wheel should be used to develop 3800 hp under a 600-ft head, and what would be its speed and rate of discharge? Assume no change in efficiency.

Mathcad

Solution:

Referring to Problem 14.5 for the necessary formulas, we proceed as follows.

(*a*) Since $\phi = \dfrac{D_1 N}{1840\sqrt{H}}$, $D_1 = \dfrac{1840\sqrt{790} \times 0.46}{200} = 119$ in.

From horsepower output $= \dfrac{\gamma Q H e}{550}$, $Q = \dfrac{7200 \times 550}{62.4 \times 790 \times 0.82} = 98.0$ cfs.

$$N_u = \frac{ND_1}{\sqrt{H}} = \frac{200 \times 119}{\sqrt{790}} = 847 \text{ rpm}$$

$$P_u = \frac{P}{D_1^2 H^{3/2}} = \frac{7200}{(119)^2(790)^{3/2}} = 0.0000229 \text{ hp}$$

$$Q_u = \frac{Q}{D_1^2\sqrt{H}} = \frac{98.0}{(119)^2\sqrt{790}} = 0.000246 \text{ cfs}$$

$$N_S = \frac{N\sqrt{P}}{H^{5/4}} = \frac{200\sqrt{7200}}{(790)^{5/4}} = 4.05 \text{ rpm}$$

(*b*) Speed $N = \dfrac{N_u\sqrt{H}}{D_1} = \dfrac{847\sqrt{529}}{119} = 164$ rpm.

Power $P = P_u D_1^2 H^{3/2} = (0.0000229)(119)^2(529)^{3/2} = 3946$ hp.

Flow $Q = Q_u D_1^2\sqrt{H} = (0.000246)(119)^2\sqrt{529} = 80.1$ cfs.

The above three quantities might have been obtained by noting that, for the same turbine (D_1 unchanged), the speed varies as $H^{1/2}$, the power varies as $H^{3/2}$, and Q varies as $H^{1/2}$. Thus,

$$N = 200\sqrt{\frac{529}{790}} = 164 \text{ rpm}, \qquad P = (7200)\left(\frac{529}{790}\right)^{3/2} = 3945 \text{ hp}, \qquad Q = 98.0\sqrt{\frac{529}{790}} = 80.2 \text{ cfs}$$

(*c*) From $P = P_u D_1^2 H^{3/2}$ we obtain

$$3800 = 0.0000229\,(D_1)^2\,(600)^{3/2}, \qquad \text{from which} \qquad D_1 = 106 \text{ in}$$

$$N = \frac{N_u\sqrt{H}}{D_1} = \frac{847\sqrt{600}}{106} = 196 \text{ rpm}$$

$$Q = Q_u D_1^2 \sqrt{H} = (0.000246)(106^2)\sqrt{600} = 67.7 \text{ cfs}$$

14.7. A turbine develops 107 kW running at 100 rpm under a head of 7.62 m. (*a*) What power would be developed under a head of 10.97 m, assuming the same flow? (*b*) At what speed should the turbine run?

Solution:

(*a*) Power developed $= \gamma Q H e$, from which $\gamma Q e = P/H = 107/7.62$.

For the same flow (and efficiency), under the 10.97-m head we obtain

$$\gamma Q e = 107/7.62 = P/10.97 \qquad \text{or} \qquad P = 154 \text{ kW}$$

(*b*)
$$N_S = \frac{N\sqrt{P}}{H^{5/4}} = \frac{100\sqrt{107}}{(7.62)^{5/4}} = 81.7$$

Then
$$N = \frac{N_S H^{5/4}}{\sqrt{P}} = \frac{(81.7)(10.97)^{5/4}}{\sqrt{154}} = 131 \text{ rpm}$$

14.8. An impulse wheel at best speed produces 93 kW under a head of 64 m. (*a*) By what percent should the speed be increased for an 88-m head? (*b*) Assuming equal efficiencies, what power would result?

Solution:

(*a*) For the same wheel, the speed is proportional to the square root of the head. Thus

$$N_1/\sqrt{H_1} = N_2/\sqrt{H_2} \qquad \text{or} \qquad N_2 = N_1\sqrt{H_2/H_1} = N_1\sqrt{88/64} = 1.173 N_1$$

The speed should be increased 17.3%.

(*b*) The specific speed relation can be used to obtain the new horsepower produced.

From $N_S = \dfrac{N\sqrt{P}}{H^{5/4}}$ we have $\dfrac{N_1\sqrt{93}}{(64)^{5/4}} = \dfrac{N_2\sqrt{P_2}}{(88)^{5/4}}$.

Solving for the power produced, $P_2 = \left[\dfrac{N_1}{1.173N_1}\sqrt{93}(88/64)^{5/4}\right]^2 = 150 \text{ kW}$

The same value of horsepower can be obtained by noting that, for the same wheel, horsepower varies as $H^{3/2}$, giving $P = (93)(88/64)^{3/2} = 150$ kW.

14.9. Find the approximate diameter and angular velocity of a Pelton wheel, efficiency 85% and effective head 220 ft, when the flow is 0.95 cfs. Assume values of $\phi = 0.46$ and $c = 0.975$.

Solution:

For an impulse wheel, the general expression for power is

$$P = \frac{\gamma Q H e}{550} = \frac{(62.4)(cA\sqrt{2gH})He}{550} = \frac{62.4c\pi\sqrt{2g}\ e}{550 \times 4 \times 144} d^2 H^{3/2} = 0.00412 d^2 H^{3/2} \qquad (1)$$

where d = diameter of nozzle in inches and the values of c and e are 0.975 and 0.85, respectively. From the data we can also calculate the horsepower from

$$\text{power} = \frac{\gamma Q H e}{550} = \frac{62.4 \times 0.95 \times 220 \times 0.85}{550} = 20.2 \text{ hp}$$

Substituting this value in (*1*), we obtain $d = 1.23$ in. (This same value of diameter d can be calculated by using the equation $Q = cA\sqrt{2gH}$ from Chapter 12.)

Now the ratio of the diameter of nozzle to diameter of wheel will be established. This ratio will result from using the specific speed divided by the unit speed or

$$\frac{N_S}{N_u} = \frac{N\sqrt{P}/H^{5/4}}{ND_1/\sqrt{H}} = \frac{\sqrt{P} \times \sqrt{H}}{D_1 H^{5/4}}$$

Substituting the value of P from (*1*),

$$\frac{N_S}{N_u} = \frac{\sqrt{0.00412 d^2 H^{3/2}}\sqrt{H}}{D_1 H^{5/4}} = 0.0642 \frac{d}{D_1}$$

But $N_u = 1840\phi$ (see Problem 14.5). Then

$$N_S = (1840 \times 0.46)\left(0.0642 \frac{d}{D_1}\right) = 54.3 \frac{d}{D_1} \qquad (2)$$

It will be necessary to assume a value of N_S in (2). Using $N_S = 2.5$, we have

$$2.5 = \frac{N\sqrt{P}}{H^{5/4}} = \frac{N\sqrt{20.2}}{(220)^{5/4}} \qquad \text{or} \qquad N = 471 \text{ rpm}$$

The speed of an impulse wheel must synchronize with the speed of the generator. For a 60-cycle generator with 8 pairs of poles, speed $N = 7200/(2 \times 8) = 450$ rpm; and with 7 pairs, $N = 7200/(2 \times 7) = 514$ rpm. Using the 7-pair generator for illustration, recalculation produces

$$N_S = \frac{514\sqrt{20.2}}{(220)^{5/4}} = 2.73$$

Then from (*2*), $D_1 = 54.3d/N_S = (54.3)(1.23)/2.73 = 24.5$ in.

For the 7-pair generator, $N = 514$ rpm, from above.

14.10. The reaction turbines at the Hoover Dam installation have a rated capacity of 115,000 hp at 180 rpm under a head of 487 ft. The diameter of each turbine is 11 ft, and the discharge is 2350 cfs. Evaluate the speed factor, unit speed, unit discharge, unit power, and specific speed.

Solution:

Using equations (*4*) through (*11*) at the beginning of the chapter, we obtain the following.

$$\phi = \frac{D_1 N}{1840\sqrt{H}} = \frac{(11 \times 12)(180)}{1840\sqrt{487}} = 0.585$$

$$N_u = \frac{D_1 N}{\sqrt{H}} = \frac{(11 \times 12)(180)}{\sqrt{487}} = 1077 \text{ rpm}$$

$$Q_u = \frac{Q}{D_1^2\sqrt{H}} = \frac{2350}{(132)^2\sqrt{487}} = 0.00611 \text{ cfs}$$

$$P_u = \frac{\text{hp}}{D_1^2 H^{3/2}} = \frac{115{,}000}{(132)^2(487)^{3/2}} = 0.000614 \text{ hp}$$

$$N_S = N_u\sqrt{P_u} = 1077\sqrt{0.000614} = 26.7$$

14.11. An impulse wheel rotates at 400 rpm under an effective head of 196 ft and develops 90.0 bhp (brake hp). For values of $\phi = 0.46$, $c_v = 0.97$, and efficiency $e = 83\%$, determine (*a*) the diameter of the jet, (*b*) the flow in cfs, (*c*) the diameter of the wheel, and (*d*) the pressure head at the 8″-diameter base of the nozzle.

Solution:

(*a*) The velocity of the jet is $v = c_v\sqrt{2gh} = 0.97\sqrt{64.4 \times 196} = 109.0$ ft/sec.

The flow must be determined before the diameter of the jet can be calculated.

Horsepower developed $= \gamma QHe/550$, $90.0 = 62.4Q(196)(0.83)/550$ and $Q = 4.88$ cfs.

Then, area of jet $= Q/v = 4.88/109.0 = 0.0448 \text{ ft}^2$ and jet diameter $= 0.239 \text{ ft} = 2.87$ in.

(*b*) Solved under (*a*).

(*c*) $\phi = \dfrac{D_1 N}{1840\sqrt{H}}$, $0.46 = \dfrac{D_1(400)}{1840\sqrt{196}}$, and $D_1 = 29.6$ in.

(*d*) Effective head $h = (p/\gamma + V^2/2g)$, where p and V are average values of pressure and velocity measured at the base of the nozzle. The value of $V_8 = Q/A_8 = 4.88/0.349 = 13.98$ ft/sec.

$$\text{Then } \frac{p}{\gamma} = h - \frac{V_8^2}{2g} = 196 - \frac{(13.98)^2}{2g} = 193 \text{ ft.}$$

14.12. A Pelton wheel develops brake power of 4470 kW under a net head of 122 m at a speed of 200 rpm. Assuming $c_v = 0.98$, $\phi = 0.46$, efficiency = 88%, and jet-diameter-to-wheel-diameter ratio of 1/9, determine (*a*) the flow required, (*b*) the diameter of the wheel, and (*c*) the diameter and number of jets required.

Solution:

(*a*) Water power $= \gamma QH$ $4470/0.88 = 9.79Q(122)$ and $Q = 4.25 \text{ m}^3/\text{s}$.

(*b*) Jet velocity $v = c_v\sqrt{2gh} = 0.98\sqrt{(2 \times 9.81)(122)} = 47.9$ m/s.

Peripheral velocity $u = \phi\sqrt{2gh} = 0.46\sqrt{(2 \times 9.81)(122)} = 22.5$ m/s.

Then $u = r\omega = \pi DN/60$, $22.5 = \pi D(200/60)$, and $D = 2.15$ m.

(c) Since $d/D = 1/9$, $d = 2.15/9 = 0.239$ m diameter.

$$\text{Number of jets} = \frac{\text{flow } Q}{\text{flow per jet}} = \frac{Q}{A_{\text{jet}} v_{\text{jet}}} = \frac{4.25}{\frac{1}{4}\pi(0.239)^2(47.9)} = 1.98. \text{ Use two jets.}$$

14.13. At the Pickwick plant of TVA the propeller-type turbines are rated at 48,000 hp at 81.8 rpm under a 43-ft head. The discharge diameter is 292.3 in. For a geometrically similar turbine to develop 36,000 hp under a 36-ft head, what speed and diameter should be used? What percentage change in flow is probable?

Solution:

The specific speed of geometrically similar turbines can be expressed as $N_S = \dfrac{N\sqrt{P}}{H^{5/4}}$. Then

$$\frac{81.8\sqrt{48{,}000}}{(43)^{5/4}} = \frac{N\sqrt{36{,}000}}{(36)^{5/4}} \quad \text{and } N = 75.6 \text{ rpm}$$

The same result can be obtained by calculating N_u, then P_u and N_S. Apply these values to the turbine to be designed. Thus,

$$N_u = \frac{D_1 N}{\sqrt{H}} = \frac{(292.3)(81.8)}{\sqrt{43}} = 3646, \quad P_u = \frac{P}{D_1^2 H^{3/2}} = \frac{48{,}000}{(292.3)^2(43)^{3/2}} = 0.0020$$

$$N_s = N_u\sqrt{P_u} = 3646\sqrt{0.0020} = 163.1$$

and
$$N = \frac{N_S H^{5/4}}{\sqrt{P}} = \frac{(163.1)(36)^{5/4}}{\sqrt{36{,}000}} = 75.8 \text{ rpm, as above}$$

For the diameter of the new turbine, using $N_u = \dfrac{D_1 N}{\sqrt{H}}$, $D_1 = \dfrac{N_u\sqrt{H}}{N} = \dfrac{3646\sqrt{36}}{75.8} = 289$ in.

For the percentage change in flow Q, the flow relation for Pickwick and new turbines is

$$\text{new } \frac{Q}{D_1^2 H^{1/2}} = \text{Pickwick } \frac{Q}{D_1^2 H^{1/2}}, \qquad \frac{Q_{\text{Pick}}}{(292.3)^2(43)^{1/2}} = \frac{Q_{\text{new}}}{(289)^2(36)^{1/2}}$$

and new $Q = 0.894 Q_{\text{Pick}}$ or about 11% decrease in Q.

14.14. A model turbine, diameter 380 mm, develops 9 kW at a speed of 1500 rpm under a head of 7.6 m. A geometrically similar turbine 1.9 m in diameter will operate at the same efficiency under a 14.9-m head. What speed and power can be expected?

Solution:

From expression (5a) at the beginning of this chapter,

$$C'_N = \frac{ND}{\sqrt{gH}} = \text{constant for homologous turbines}$$

Thus,

$$\text{model } \frac{ND}{\sqrt{gH}} = \text{prototype} \frac{ND}{\sqrt{gH}}, \quad \frac{1500 \times 380}{\sqrt{g \times 7.6}} = \frac{N \times 1900}{\sqrt{g \times 14.9}}, \quad \text{and } N = 420 \text{ rpm.}$$

From expression (9a), $C_P = \dfrac{P}{D^2 H^{3/2}} = \text{constant}$. Thus,

$$\text{model } \frac{P}{D^2 H^{3/2}} = \text{prototype} \frac{P}{D^2 H^{3/2}}, \quad \frac{9}{(380)^2(7.6)^{3/2}} = \frac{P}{(1900)^2(14.9)^{3/2}}, \quad P = 618 \text{ kW}$$

14.15. A reaction turbine 500 mm in diameter, when running at 600 rpm, developed brake power of 195 kW when the flow was 0.74 m³/s. The pressure head at entrance to the turbine was 27.90 m, and the elevation of the turbine casing above tailwater level was 1.91 m. The water enters the turbine with a velocity of 3.66 m/s. Calculate (*a*) the effective head, (*b*) the efficiency, (*c*) the speed expected under a head of 68.6 m, and (*d*) the brake power and discharge under the 68.6-m head.

Solution:

(*a*) Effective head $H = \dfrac{p}{\gamma} + \dfrac{V^2}{2g} + z = 27.90 + \dfrac{(3.66)^2}{2g} + 1.91 = 30.5$ m.

(*b*) Power supplied by water $= \gamma QH = (9.79)(0.74)(30.5) = 221$ kW.

$$\text{Efficiency} = \frac{\text{power from shaft}}{\text{power supplied}} = \frac{195}{221} = 88.2\%$$

(*c*) For the same turbine, the $\dfrac{ND_1}{\sqrt{H}}$ ratio is constant. Then $\dfrac{N \times 500}{\sqrt{68.6}} = \dfrac{600 \times 500}{\sqrt{30.5}}$ or $N = 900$ rpm.

(*d*) For the same turbine, the $\dfrac{P}{D_1^2 H^{3/2}}$ and $\dfrac{Q}{D_1^2\sqrt{H}}$ ratios are also constant. Then

$$\frac{P}{(500)^2(68.6)^{3/2}} = \frac{195}{(500)^2(30.5)^{3/2}}, \; P = 658 \text{ kW} \quad \text{and} \quad \frac{Q}{(500)^2\sqrt{68.6}} = \frac{0.74}{(500)^2\sqrt{30.5}}, \; Q = 1.11\text{m}^3/\text{s}.$$

14.16. A pump impeller 12 inches in diameter discharges 5.25 cfs when running at 1200 rpm. The blade angle β_2 is 160°, and the exit area A_2 is 0.25 ft². Assuming losses of $2.8\left(v_2^2/2g\right)$ and $0.38\left(V_2^2/2g\right)$, compute the efficiency of the pump (exit area A_2 is measured normal to v_2).

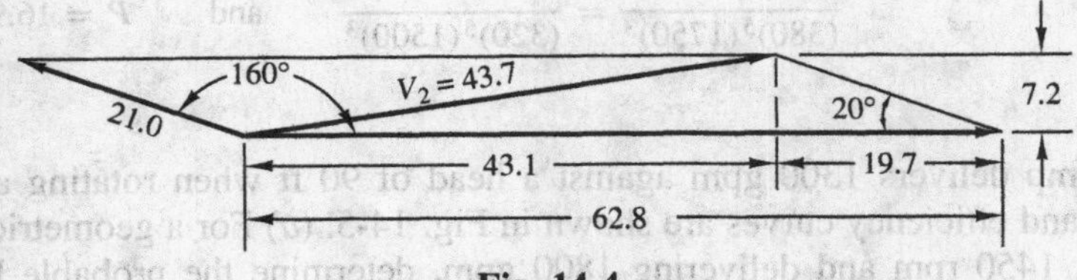

Fig. 14-4

Solution:

The absolute and relative velocities at exit must be calculated first. Velocities u_2 and v_2 are

$$u_2 = r_2\omega = (6/12)(2\pi \times 1200/60) = 62.8 \text{ ft/sec}, \qquad v_2 = Q/A_2 = 5.25/0.25 = 21.0 \text{ ft/sec}$$

From the vector diagram shown in Fig. 14-4, the value of the absolute velocity at exit is $V_2 = 43.7$ ft/sec. From Problem 14.4,

$$\text{head furnished by impeller, } H' = \frac{u_2^2}{2g} - \frac{v_2^2}{2g} + \frac{V_2^2}{2g} = \frac{(62.8)^2}{2g} - \frac{(21.0)^2}{2g} + \frac{(43.7)^2}{2g} = 84.0 \text{ ft}$$

$$\text{head delivered to water, } H = H' - \text{losses} = 84.0 - \left(2.8\frac{(21.0)^2}{2g} + 0.38\frac{(43.7)^2}{2g}\right) = 53.6 \text{ ft}$$

$$\text{efficiency } e = H/H' = 53.6/84.0 = 63.8\%$$

The value of H' might have been calculated by using the common expression

$$H' = \frac{u_2}{g}(u_2 + v_2\cos\beta_2) = \frac{62.8}{g}[62.8 + 21.0(-0.940)] = 84.0 \text{ ft}$$

14.17. A centrifugal pump discharged 0.019 m^3/s against a head of 16.8 m when the speed was 1500 rpm. The diameter of the impeller was 320 mm, and the brake power was 4.5 kW. A geometrically similar pump 380 mm in diameter is to run at 1750 rpm. Assuming equal efficiencies, (*a*) what head will be developed, (*b*) how much water will be pumped, and (*c*) what brake power will be developed?

Solution:

(*a*) From the speed relation, the $\dfrac{DN}{\sqrt{H}}$ ratios for model and prototype are equal. Then

$$\frac{320 \times 1500}{\sqrt{16.8}} = \frac{380 \times 1750}{\sqrt{H}} \quad \text{and} \quad H = 32.2 \text{ m}$$

(*b*) From the discharge relation, the $\dfrac{Q}{D^2\sqrt{H}}$ ratios are equal. Then

$$\frac{0.019}{(320)^2\sqrt{16.8}} = \frac{Q}{(380)^2\sqrt{32.2}}, \quad \text{and} \quad Q = 0.037 \text{ m}^3/\text{s}$$

Another useful discharge relation is $\dfrac{Q}{D^3N}$ = constant, from which

$$\frac{Q}{(380)^3(1750)} = \frac{0.019}{(320)^3(1500)} \quad \text{and} \quad Q = 0.037 \text{ m}^3/\text{s}$$

(*c*) The speed relation, $\dfrac{P}{D^5N^3}$ = constant, can be used for model and prototype. Then

$$\frac{P}{(380)^5(1750)^3} = \frac{4.5}{(320)^5(1500)^3} \quad \text{and} \quad P = 16.9 \text{ kW}$$

14.18. A 6-in pump delivers 1300 gpm against a head of 90 ft when rotating at 1750 rpm. The head discharge and efficiency curves are shown in Fig. 14-5. (*a*) For a geometrically similar 8-in pump running at 1450 rpm and delivering 1800 gpm, determine the probable head developed by the 8-in pump. (*b*) Assuming a similar efficiency curve for the 8-in pump, what power would be required to drive the pump at the 1800-gpm rate?

Solution:

(*a*) The homologous pumps will have identical characteristics at corresponding flows. Choose several rates of flow for the 6-in pump and read off the corresponding heads. Calculate the values of Q and H so that a curve for the 8-in pump can be plotted. One such calculation is detailed below and a table of values established by similar determinations.

Using the given 1300 gpm and the 90-ft head, we obtain from the speed relation

$$H_8 = (D_8/D_6)^2\,(N_8/N_6)^2\,H_6 = (8/6)^2(1450/1750)^2H_6 = 1.220H_6 = (1.220)(90) = 109.8 \text{ ft}$$

From the flow relation, $\dfrac{Q}{D^3N}$ = constant, we obtain

$$Q_8 = (D_8/D_6)^3\,(N_8/N_6)\,Q_6 = (8/6)^3(1450/1750)Q_6 = 1.964Q_6 = (1.964)(1300) = 2550 \text{ gpm}$$

Additional values, which have been plotted in Fig. 14-5, are as follows.

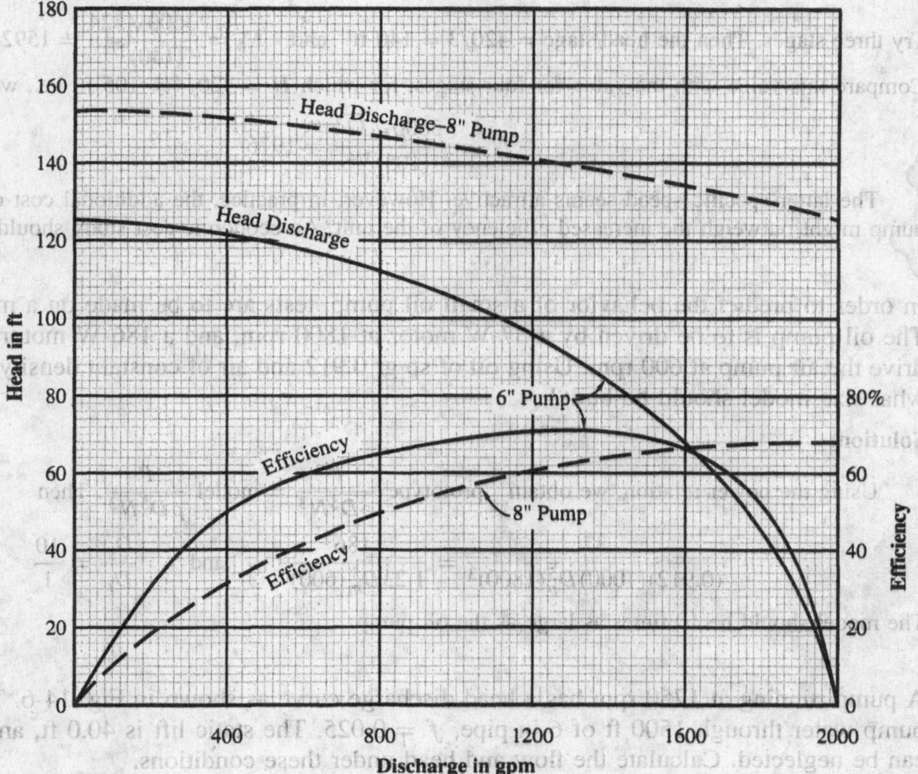

Fig. 14-5

For 6″ Pump at 1750 rpm			For 8″ Pump at 1450 rpm		
Q (gpm)	H (ft)	Efficiency	Q (gpm)	H (ft)	Efficiency
0	124	0	0	151.6	0
500	119	54%	980	145.5	54%
800	112	64%	1570	134.5	64%
1000	104	68%	1960	127.0	68%
1300	90	70%	2550	110.0	70%
1600	66	67%	3140	80.6	67%

From the head discharge curve, for $Q = 1800$ gpm the head is 130 ft.

(*b*) The efficiency of the 8″ pump would probably be somewhat higher than that of the 6″ pump at comparable rates of flow. For this case, the assumption is that the efficiency curves are the same at comparable rates of flow. The table above lists the values for the flows indicated. Figure 14-5 gives the efficiency curve for the 8″ pump, and for the 1800 gpm flow, the value is 67%. Then

$$P = \frac{\gamma QH}{550e} = \frac{(62.4)[1800/(60 \times 7.48)](130)}{(550)(0.67)} = 88.3 \text{ hp required}$$

14.19. It is required to deliver 324 gpm against a head of 420 ft at 3600 rpm. Assuming acceptable efficiency of pump at specific speeds of the impeller between 1200 and 4000 rpm when flow Q is in gpm, how many pumping stages should be used?

Solution:

For one stage, $N_S = \dfrac{N\sqrt{Q}}{H^{3/4}} = \dfrac{3600\sqrt{324}}{(420)^{3/4}} = 698$. This value is too low.

Try three stages. Then the head/stage $= 420/3 = 140$ ft and $N_S = \dfrac{3600\sqrt{324}}{(140)^{3/4}} = 1592.$

Compare this value with the value for four stages, for which $H = 420/4 = 105$ ft, i.e., with

$$N_S = \frac{3600\sqrt{324}}{(105)^{3/4}} = 1976$$

The latter specific speed seems attractive. However, in practice, the additional cost of the four-stage pump might outweigh the increased efficiency of the unit. An economic cost study should be made.

14.20. In order to predict the behavior of a small oil pump, tests are to be made on a model using air. The oil pump is to be driven by a 37-W motor at 1800 rpm, and a 186-W motor is available to drive the air pump at 600 rpm. Using oil of sp gr 0.912 and air of constant density at 1.23 kg/m^3, what size model should be built?

Solution:

Using the power relation, we obtain prototype $\dfrac{P}{\rho D^5 N^3}$ = model $\dfrac{P}{\rho D^5 N^3}$. Then

$$\frac{37}{(0.912)(1000)D_p^5(1800)^3} = \frac{186}{1.23D_m^5(600)^3} \quad \text{and} \quad \frac{D_m}{D_p} = \frac{10}{1}$$

The model should be 10 times as large as the oil pump.

14.21. A pump running at 1750 rpm has a head discharge curve as shown in Fig. 14-6. The pump is to pump water through 1500 ft of 6-in pipe, $f = 0.025$. The static lift is 40.0 ft, and minor losses can be neglected. Calculate the flow and head under these conditions.

Mathcad

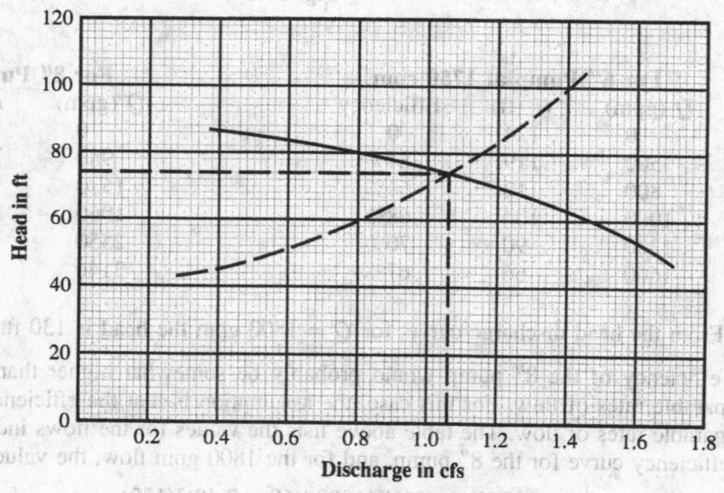

Fig. 14-6

Solution:

The loss of head through the pipeline will increase with increased flow. A curve can be drawn indicating the total head pumped against as a function of the flow (shown dashed). But

$$\text{head pumped against} = \text{lift} + \text{pipe loss}$$

$$= 40.0 + (0.025)\left(\frac{1500}{6/12}\right)\frac{V^2}{2g} = 40.0 + 75.0\frac{V^2}{2g}$$

We can evaluate this head as follows:

$Q =$	0.40	0.60	0.80	1.00	1.20	cfs
$V = Q/A =$	2.04	3.06	4.07	5.09	6.11	ft/sec
$75V^2/2g =$	4.8	10.9	19.3	30.2	43.5	ft lost
Total head =	44.8	50.9	59.3	70.2	83.5	ft

Figure 14-6 indicates that when the flow is 1.05 cfs the head developed by the pump will equal the total head being pumped against, i.e., 74 ft.

14.22. What is the power ratio of a pump and its 1/5 scale model if the ratio of the heads is 4 to 1?

Solution:

For geometrically similar pumps, $\dfrac{P}{D^2H^{3/2}}$ for pump $= \dfrac{P}{D^2H^{3/2}}$ for model. Then

$$\frac{P_p}{(5D)^2(4H)^{3/2}} = \frac{P_m}{D^2H^{3/2}} \quad \text{and} \quad P_p = (25)(4)^{3/2}P_m = 200P_m$$

14.23. Develop the expressions for thrust and power output of a propeller, the velocity through the propeller, and the efficiency of the propeller.

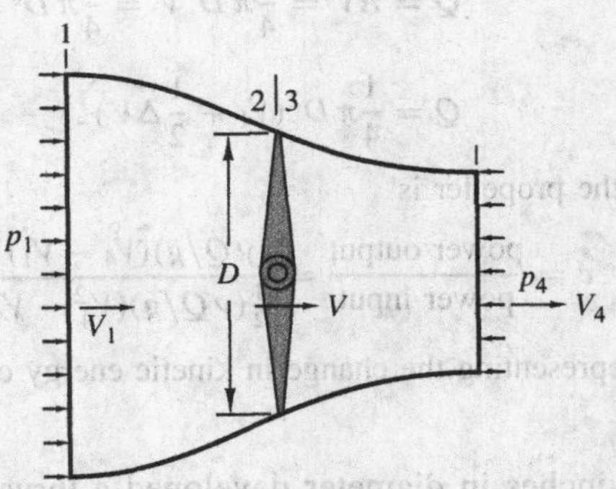

Fig. 14-7

Solution:

(*a*) Using the principle of impulse-momentum, the thrust F of the propeller changes the momentum of mass M of air in Fig 14-7. The propeller may be stationary in a fluid moving with velocity of approach V_1 or it may be moving to the left at velocity V_1 through quiescent fluid. Thus, neglecting vortices and friction,

$$\text{thrust } F = \frac{\gamma Q}{g}(\Delta V) = \frac{\gamma Q}{g}(V_4 - V_1) \tag{1a}$$

$$= \frac{\gamma}{g}\left(\tfrac{1}{4}\pi D^2 V\right)(V_4 - V_1) \tag{1b}$$

(*b*) The power output is simply

$$P = \text{thrust force} \times \text{velocity} = \frac{\gamma Q}{g}(V_4 - V_1)V_1 \tag{2}$$

(c) The thrust F is also equal to $(p_3 - p_2)\left(\frac{1}{4}\pi D^2\right)$. Therefore, from (1b),

$$p_3 - p_2 = \frac{\gamma}{g} V(V_4 - V_1) \tag{3}$$

Principles of work and kinetic energy, using a unit of 1 ft³ and assuming no lost head, produce the following:

$$\text{initial KE/ft}^3 + \text{work done/ft}^3 = \text{final KE/ft}^3$$

$$\tfrac{1}{2}(\gamma/g)V_1^2 + (p_3 - p_2) = \tfrac{1}{2}(\gamma/g)V_4^2$$

from which
$$p_3 - p_2 = \frac{\gamma}{g}\left(\frac{V_4^2 - V_1^2}{2}\right) \tag{4}$$

The same result can be obtained by applying the Bernoulli equation between 1 and 2 and between 3 and 4 and solving for $(p_3 - p_2)$. Note that $(p_3 - p_2)$ represents lb/ft² × ft/ft or ft-lb/ft³, or N/m² × m/m or N · m/m³.

Equating (3) and (4) yields

$$V = \frac{V_1 + V_4}{2} = \frac{V_1 + (V_1 + \Delta V)}{2} = V_1 + \frac{\Delta V}{2} \tag{5}$$

indicating that the velocity through the propeller is the average of the velocities ahead of and behind the propeller.

Flow of fluid Q can be stated in terms of this velocity V, as follows:

$$Q = AV = \frac{1}{4}\pi D^2 V = \frac{1}{4}\pi D^2\left(\frac{V_1 + V_4}{2}\right) \tag{6a}$$

or
$$Q = \frac{1}{4}\pi D^2 (V_1 + \frac{1}{2}\Delta V) \tag{6b}$$

(d) The efficiency of the propeller is

$$e = \frac{\text{power output}}{\text{power input}} = \frac{(\gamma Q/g)(V_4 - V_1)V_1}{\frac{1}{2}(\gamma Q/g)(V_4^2 - V_1^2)} = \frac{2V_1}{V_4 + V_1} = \frac{V_1}{V} \tag{7}$$

the denominator representing the change in kinetic energy created by the power input.

14.24. A propeller model 15 inches in diameter developed a thrust of 50 lb at a speed of 10 ft/sec in water. (a) What thrust should a similar 75-in propeller develop at the same velocity through the water? (b) At a velocity of 20 ft/sec? (c) What would be the slipstream velocity in (b)?

Solution:

(a) Linear velocity $V = r\omega$ or varies as DN. Then we may write

$$V_m \propto 15N_m \quad \text{and} \quad V_p \propto 75N_p$$

Since the velocities are equal, $15N_m = 75N_p$.

Using the thrust coefficient relation, equation (19) of the text, we obtain.

$$\frac{F}{\rho N^2 D^4}\text{ model} = \frac{F}{\rho N^2 D^4}\text{ prototype}, \qquad \frac{50}{\rho\left(\frac{75}{15}N_p\right)^2 (15)^4} = \frac{F}{\rho (N_p)^2 (75)^4}, \qquad F = 1250 \text{ lb}$$

In equation (19), diameter D is in ft and N is in revolutions/sec. However, when the ratios are equated to each other, as long as the same units are used for corresponding items (ft/ft, in/in, rpm/rpm), a correct solution results.

(*b*) Here $V_m \propto 15N_m$ and $(2V_m = V_p) \propto 75N_p$. These values give $30N_m = 75N_p$. Then

$$\frac{50}{\rho\left(\frac{75}{30}N_p\right)^2(15)^4} = \frac{F}{\rho N_p^2 (75)^4} \quad \text{and} \quad F = 5000 \text{ lb}$$

Note: The above linear velocity–angular velocity–diameter relationship can be written

$$\frac{V}{ND} \text{ for model} = \frac{V}{ND} \text{ for prototype} \tag{1}$$

This relation is called the advance–diameter ratio inasmuch as V/N is the distance the propeller advances in one revolution.

(*c*) The slipstream velocity (or velocity change) can be obtained by solving expression (*6b*) in Problem 14-23 for ΔV after substituting $\dfrac{F}{\rho \Delta V}$ for Q [from equation (*1a*)]. Then

$$\frac{F}{\rho \Delta V} = \left(\tfrac{1}{4}\pi D^2\right) V_1 + \left(\tfrac{1}{4}\pi D^2\right)\left(\tfrac{1}{2}\Delta V\right) \quad \text{and} \quad (\Delta V)^2 + 2V_1 \Delta V - \frac{8F}{\rho \pi D^2} = 0$$

Solving for ΔV gives, as the real root,

$$\Delta V = -V_1 + \sqrt{V_1^2 + \frac{8F}{\rho \pi D^2}} \tag{2}$$

From the values above, using D in ft,

$$\Delta V = -20.0 + \sqrt{(20.0)^2 + \frac{8 \times 5000}{1.94\pi(75/12)^2}} = 3.83 \text{ ft/sec} \quad \text{or} \quad V_4 = 23.83 \text{ ft/sec}$$

14.25. Determine the thrust coefficient of a propeller that is 100-mm in diameter, revolves at 1800 rpm, and develops a thrust of 11.1 N in fresh water.

Solution:

$$\text{Thrust coefficient} = \frac{F}{\rho N^2 D^4} = \frac{11.1}{(1000)(1800/60)^2(100/1000)^4} = 0.123.$$

The coefficient is dimensionless when F is in newtons, N in revolutions/s, and D in m.

14.26. The power and thrust coefficients of a 2.5-m-diameter propeller moving forward at 30 m/s at a rotational speed of 2400 rpm are 0.068 and 0.095, respectively. (*a*) Determine the power requirement and thrust in air ($\rho = 1.22$ kg/m^3). (*b*) If the advance diameter ratio for maximum efficiency is 0.70, what is the airspeed for the maximum efficiency?

Solution:

(*a*) Power $P = C_P \rho N^3 D^5 = (0.068)(1.22)(2400/60)^3(2.5)^5 = 518{,}000 \text{ W} = 518 \text{ kW}$.

Thrust $F = C_F \rho N^2 D^4 = (0.095)(1.22)(2400/60)^2(2.5)^4 = 7240 \text{ N} = 7.24 \text{ kN}$.

(*b*) Since $V/ND = 0.70$, $V = (0.70)(2400/60)(2.5) = 70.0$ m/s.

14.27. An airplane flies at 290 km/h in still air, $\gamma = 11.8$ N/m^3. The propeller is 1.68 m in diameter, and the velocity of the air through the propeller is 97.5 m/s. Determine (*a*) the slipstream velocity, (*b*) the thrust, (*c*) the horsepower input, (*d*) the horsepower output, (*e*) the efficiency, and (*f*) the pressure difference across the propeller.

Solution:

Using the expressions developed in Problem 14.23, we obtain, from (5),

(*a*) $V = \frac{1}{2}(V_1 + V_4)$, $97.5 = \frac{1}{2}[(290{,}000/3600) + V_4]$, $V_4 = 114$ m/s (relative to fuselage).

(*b*) $$\text{Thrust } F = \frac{\gamma}{g} Q(V_4 - V_1) = \frac{11.8}{9.81}\left[\frac{1}{4}\pi(1.68)^2(97.5)\right][114-(290{,}000/3600)] = 8690 \text{ N} = 8.69 \text{ kN}.$$

(*c*) Power input $P_i = FV = (8.69)(97.5) = 847$ kW.

(*d*) Power output $P_0 = FV_1 = (8.69)(290{,}000/3600) = 700$ kW.

(*e*) Efficiency $e = 700/847 = 82.6\%$.

(*f*) $$\text{Pressure difference} = \frac{\text{thrust } F}{\text{area } \left(\frac{1}{4}\pi D^2\right)} = \frac{8.69}{\frac{1}{4}\pi(1.68)^2} = 3.92 \text{ kPa}.$$

Supplementary Problems

14.28. An impulse wheel operates under an effective head of 190 m. The jet diameter is 100 mm. For values of $\phi = 0.45$, $c_v = 0.98$, $\beta = 160°$, and $v_2 = 0.85(V_1 - u)$, compute the power input to the shaft. *Ans.* 775 kW

14.29. An impulse wheel is to develop 2500 hp under an effective head of 900 ft. The nozzle diameter is 5 in, $c_v = 0.98$, $\phi = 0.46$, and $D/d = 10$. Calculate the efficiency and the speed of rotation. *Ans.* 76.3%, 508 rpm

14.30. A turbine model, built to a scale of 1:5, was found to develop 4.25 bhp at a speed of 400 rpm under a head of 6 ft. Assuming equivalent efficiencies, what speed and power of the full-size turbine can be expected under a head of 30 ft? *Ans.* 179 rpm, 1190 hp

14.31. From the following data, determine the diameter of the impulse wheel and its speed of rotation: $\phi = 0.46$, $e = 82\%$, $c_v = 0.98$, $D/d = 12$, head = 395 m, and power delivered = 3580 kW. *Ans.* 1.55 m, 500 rpm

14.32. A reaction turbine running at best speed produced 34.0 bhp at 620 rpm under a 100-ft head. If the efficiency proved to be 70.0% and the speed ratio $\phi = 0.75$, determine (*a*) the diameter of the wheel, (*b*) the flow in cfs, (*c*) the characteristic speed N_S, and (*d*) for a head of 196 ft, the brake horsepower and the discharge. *Ans.* (*a*) 22.3 in; (*b*) 4.28 cfs; (*c*) 11.42 rpm; (*d*) 93.3 hp, 6.00 cfs

14.33. Under conditions of maximum efficiency a 1.27-m turbine developed 224 kW under a head of 4.57 m and at 95 rpm. At what speed should a homologous turbine 0.64 m in diameter operate under a head of 7.62 m? What power should be developed? *Ans.* 245 rpm, 120 kW

14.34. A 60″ impulse wheel develops 625 brake horsepower when operating at 360 rpm under a head of 400 ft. (*a*) Under what head should a similar wheel operate at the same speed in order to develop 2500 bhp? (*b*) For the head just calculated, what diameter should be used? *Ans.* 697 ft, 79.2 in

14.35. The speed ratio ϕ of a turbine is 0.70 and the specific speed is 20.0. Determine the diameter of the runner in order that 1864 kW will be developed under a head of 98.8 m. *Ans.* 1.07 m

14.36. A turbine test produced the following data: brake horsepower = 22.5, head = 16.0 ft, N = 140 rpm, runner diameter = 36 in, and Q = 14.0 cfs. Calculate the power input, efficiency, speed ratio, and specific speed. *Ans.* 25.4 hp, 88.5%, 0.685, 20.8

14.37. A centrifugal pump rotates at 600 rpm. The following data are taken: $r_1 = 2''$, $r_2 = 8''$, radial $A_1 = 12\pi$ in^2, radial $A_2 = 30\pi$ in^2, $\beta_1 = 135°$, $\beta_2 = 120°$, radial flow at entrance to blades. Neglecting friction, calculate the relative velocities at entrance and exit and the power transmitted to the water.
Ans. 14.8 ft/sec, 4.83 ft/sec, 16.0 hp

14.38. What size centrifugal pump, running at 730 rpm, will pump 0.255 m^3/s against a head of 11 m, using a value of $C_N = 1450$? *Ans.* 305 mm

14.39. A centrifugal pump delivers 0.071 m^3/s against a 7.6-m head at 1450 rpm and requires 6.7 kW. If the speed is reduced to 1200 rpm, calculate the flow, head, and power, assuming the same efficiency.
Ans. 0.059 m^3/s, 5.2 m, 3.8 kW

14.40. An 80-in-diameter propeller rotates at 1200 rpm in an air stream moving at 132 ft/sec. Tests indicate the thrust of 720 lb and power absorbed of 220 hp. Calculate, for density of air of 0.00237 slug/ft^3, the thrust and power coefficients. *Ans.* 0.383, 0.483

14.41. A 1.5-m-diameter propeller moves through water at 9.1 m/s and develops a thrust of 15.6 kN. What is the increase in the velocity of the slipstream? *Ans.* 0.88 m/s

14.42. A 200-mm propeller developed a thrust of 71.2 N at 140 rpm and a water velocity of 3.66 m/s. For a similar propeller in a ship moving at 7.32 m/s, what size propeller would be required for a thrust of 178 kN? At what speed should it rotate? *Ans.* 5.08 m, 11.2 rpm

14.43. In a wind tunnel a fan produces an air velocity of 75 ft/sec when it rotates at 1200 rpm. (*a*) What velocity will be produced if the fan rotates at 1750 rpm? (*b*) If a 3.25-hp motor drives the fan at 1200 rpm, what size motor is required to drive the fan at 1750 rpm? *Ans.* 109.4 ft/sec, 10.1 hp

14.44. What size motor is required to supply 90,000 ft^3/min of air to a wind tunnel, if the losses in the tunnel are 5.67 in of water and the fan efficiency is 68%? (Use $\gamma_{\text{air}} = 0.0750$ lb/ft^3.) *Ans.* 118 hp

14.45. A 9-ft-diameter propeller moves through air, $\gamma = 0.0763$ lb/ft^3, at 300 ft/sec. If 1200 hp is delivered to the propeller, what thrust is developed and what is the efficiency of the propeller? *Ans.* 2030 lb, 92.2%

Appendix A

Tables and Diagrams

TABLE 1

(A) APPROXIMATE PROPERTIES OF SOME GASES
[at 68°F (20°C), 1 atm]

Gas	Specific Weight γ lb/ft³	Specific Weight γ N/m³	Gas Constant R ft/°R	Gas Constant R m/K	Adiabatic Exponent k	Kinematic Viscosity ν ft²/sec	Kinematic Viscosity ν m²/s
Air	0.0752	11.8	53.3	29.3	1.40	16.0×10^{-5}	14.9×10^{-6}
Ammonia	0.0448	7.0	89.5	49.2	1.32	16.5×10^{-5}	15.3×10^{-6}
Carbon dioxide	0.1146	18.0	34.9	19.2	1.30	9.1×10^{-5}	8.5×10^{-6}
Methane	0.0416	6.5	96.3	52.9	1.32	19.3×10^{-5}	17.9×10^{-6}
Nitrogen	0.0726	11.4	55.1	30.3	1.40	17.1×10^{-5}	15.9×10^{-6}
Oxygen	0.0830	13.0	48.3	26.6	1.40	17.1×10^{-5}	15.9×10^{-6}
Sulfur dioxide	0.1695	26.6	23.6	13.0	1.26	5.6×10^{-5}	5.2×10^{-6}

(B) SOME PROPERTIES OF AIR AT ATMOSPHERIC PRESSURE

Temperature °F	Density ρ slug/ft³	Specific Weight γ lb/ft³	Kinematic Viscosity ν ft²/sec	Dynamic Viscosity μ lb-sec/ft²
0	0.00268	0.0862	12.6×10^{-5}	3.28×10^{-7}
20	0.00257	0.0827	13.6×10^{-5}	3.50×10^{-7}
40	0.00247	0.0794	14.6×10^{-5}	3.62×10^{-7}
60	0.00237	0.0763	15.8×10^{-5}	3.74×10^{-7}
68	0.00233	0.0752	16.0×10^{-5}	3.75×10^{-7}
80	0.00228	0.0735	16.9×10^{-5}	3.85×10^{-7}
100	0.00220	0.0709	18.0×10^{-5}	3.96×10^{-7}
120	0.00215	0.0684	18.9×10^{-5}	4.07×10^{-7}

Temperature °C	Density ρ kg/m³	Specific Weight γ N/m³	Kinematic Viscosity ν m²/s	Dynamic Viscosity μ N · s/m²
0	1.29	12.7	13.3×10^{-6}	1.72×10^{-5}
10	1.25	12.2	14.2×10^{-6}	1.77×10^{-5}
20	1.20	11.8	15.1×10^{-6}	1.81×10^{-5}
30	1.16	11.4	16.0×10^{-6}	1.86×10^{-5}
40	1.13	11.0	16.9×10^{-6}	1.91×10^{-5}
50	1.09	10.7	17.9×10^{-6}	1.95×10^{-5}
60	1.06	10.4	18.9×10^{-6}	1.99×10^{-5}
70	1.03	10.1	19.9×10^{-6}	2.04×10^{-5}
80	1.00	9.80	20.9×10^{-6}	2.09×10^{-5}
90	0.972	9.53	21.9×10^{-6}	2.19×10^{-5}
100	0.946	9.28	23.0×10^{-6}	2.30×10^{-5}

(C) MECHANICAL PROPERTIES OF WATER AT ATMOSPHERIC PRESSURE

Temperature °F	Density slug/ft³	Specific Wt. lb/ft³	Dynamic Viscosity lb-sec/ft²	Surface Tension* lb/ft	Vapor Pressure lb/in² abs.	Elastic Modulus lb/in²
32	1.94	62.4	3.75×10^{-5}	0.00518	0.08	287000
40	1.94	62.4	3.24×10^{-5}	0.00514	0.12	296000
50	1.94	62.4	2.74×10^{-5}	0.00508	0.17	305000
60	1.94	62.4	2.36×10^{-5}	0.00504	0.26	313000
70	1.94	62.3	2.04×10^{-5}	0.00497	0.36	319000
80	1.93	62.2	1.80×10^{-5}	0.00492	0.51	325000
90	1.93	62.1	1.59×10^{-5}	0.00486	0.70	329000
100	1.93	62.0	1.42×10^{-5}	0.00479	0.96	331000
120	1.92	61.7	1.17×10^{-5}	0.00466	1.70	332000

Temperature °C	Density kg/m³	Specific Wt. kN/m³	Dynamic Viscosity N · s/m²	Surface Tension* N/m	Vapor Pressure kPa	Elastic Modulus GPa
0	1000	9.81	1.75×10^{-3}	0.0756	0.611	2.02
10	1000	9.81	1.30×10^{-3}	0.0742	1.23	2.10
20	998	9.79	1.02×10^{-3}	0.0728	2.34	2.18
30	996	9.77	8.00×10^{-4}	0.0712	4.24	2.25
40	992	9.73	6.51×10^{-4}	0.0696	7.38	2.28
50	988	9.69	5.41×10^{-4}	0.0679	12.3	2.29
60	984	9.65	4.60×10^{-4}	0.0662	19.9	2.28
70	978	9.59	4.02×10^{-4}	0.0644	31.2	2.25
80	971	9.53	3.50×10^{-4}	0.0626	47.4	2.20
90	965	9.47	3.11×10^{-4}	0.0608	70.1	2.14
100	958	9.40	2.82×10^{-4}	0.0589	101.3	2.07

* In contact with air.

TABLE 2

SPECIFIC GRAVITY AND KINEMATIC VISCOSITY OF CERTAIN LIQUIDS

(Kinematic Viscosity = tabular value $\times 10^{-5}$)

Note: $°C = (5/9)(°F - 32)$; $1\ ft^2/sec = 0.0929\ m^2/s$

	Water**		Commercial Solvent		Carbon Tetrachloride		Medium Lubricating Oil	
Temp. °F	Sp. Gr.	Kin. Visc. ft^2/sec	Sp. Gr.	Kin. Visc. ft^2/sec	Sp. Gr.	Kin. Visc. ft^2/sec	Sp. Gr.	Kin. Visc. ft^2/sec
40	1.000	1.664	0.728	1.61	1.621	0.810	0.905	477
50	1.000	1.410	0.725	1.48	1.608	0.750	0.900	280
60	0.999	1.217	0.721	1.37	1.595	0.700	0.896	188
70	0.998	1.059	0.717	1.26	1.582	0.650	0.891	125
80	0.997	0.930	0.713	1.17	1.569	0.607	0.888	94
90	0.995	0.826	0.709	1.10	1.555	0.560	0.885	69
100	0.993	0.739	0.705	1.03	1.542	0.530	0.882	49.2
110	0.991	0.667	0.702	0.96	1.520	0.500	0.874	37.5
120	0.990	0.610					0.866	29.3
150	0.980	0.475					0.865	16.1

	Dustproofing Oil*		Medium Fuel Oil*		Heavy Fuel Oil*		Regular Gasoline*	
Temp. °F	Sp. Gr.	Kin. Visc. ft^2/sec	Sp. Gr.	Kin. Visc. ft^2/sec	Sp. Gr.	Kin. Visc. ft^2/sec	Sp. Gr.	Kin. Visc. ft^2/sec
40	0.917	80.9	0.865	6.55	0.918	444	0.738	0.810
50	0.913	56.5	0.861	5.55	0.915	312	0.733	0.765
60	0.909	40.8	0.858	4.75	0.912	221	0.728	0.730
70	0.905	30.6	0.854	4.12	0.908	157	0.724	0.690
80	0.902	23.4	0.851	3.65	0.905	114	0.719	0.660
90	0.898	18.5	0.847	3.19	0.902	83.6	0.715	0.630
100	0.894	14.9	0.843	2.78	0.899	62.7	0.710	0.600
110	0.890	12.2	0.840	2.27	0.895	48.0	0.706	0.570

Some Other Liquids

Liquid and Temperature	Sp. Gr.	Kin. Visc. ft^2/sec
Turpentine at 68°F	0.862	1.86
Linseed oil at 86°F	0.925	38.6
Ethyl alcohol at 68°F	0.789	1.65
Benzene at 68°F	0.879	0.802
Glycerin at 68°F	1.262	711
Castor oil at 68°F	0.960	1110
Light machinery oil at 62°F	0.907	147

* Kessler & Lenz, University of Wisconsin, Madison.
** ASCE Manual 25.

TABLE 3

FRICTIONAL FACTORS f FOR WATER ONLY

(Temperature range about 50°F to 70°F or 10°C to 20°C)

For old pipe: approximate range ϵ from 0.004 ft to 0.020 ft
For average pipe: approximate range ϵ from 0.002 ft to 0.003 ft
For new pipe: approximate range ϵ from 0.0005 ft to 0.0010 ft

$(f = \text{tabular value} \times 10^{-4})$

Diameter and Type of Pipe		Velocity (ft/sec)										
		1	2	3	4	5	6	8	10	15	20	30
4″	Old, comm.	435	415	410	405	400	395	395	390	385	375	370
	Average, comm.	355	320	310	300	290	285	280	270	260	250	250
	New pipe	300	265	250	240	230	225	220	210	200	190	185
	Very smooth	240	205	190	180	170	165	155	150	140	130	120
6″	Old, comm.	425	410	405	400	395	395	390	385	380	375	365
	Average, comm.	335	310	300	285	280	275	265	260	250	240	235
	New pipe	275	250	240	225	220	210	205	200	190	180	175
	Very smooth	220	190	175	165	160	150	145	140	130	120	115
8″	Old, comm.	420	405	400	395	390	385	380	375	370	365	360
	Average, comm.	320	300	285	280	270	265	260	250	240	235	225
	New pipe	265	240	225	220	210	205	200	190	185	175	170
	Very smooth	205	180	165	155	150	140	135	130	120	115	110
10″	Old, comm.	415	405	400	395	390	385	380	375	370	365	360
	Average, comm.	315	295	280	270	265	260	255	245	240	230	225
	New pipe	260	230	220	210	205	200	190	185	180	170	165
	Very smooth	200	170	160	150	145	135	130	125	115	110	105
12″	Old, comm.	415	400	395	395	390	385	380	375	365	360	355
	Average, comm.	310	285	275	265	260	255	250	240	235	225	220
	New pipe	250	225	210	205	200	195	190	180	175	165	160
	Very smooth	190	165	150	140	140	135	125	120	115	110	105
16″	Old, comm.	405	395	390	385	380	375	370	365	360	350	350
	Average, comm.	300	280	265	260	255	250	240	235	225	215	210
	New pipe	240	220	205	200	195	190	180	175	170	160	155
	Very smooth	180	155	140	135	130	125	120	115	110	105	100
20″	Old, comm.	400	395	390	385	380	375	370	365	360	350	350
	Average, comm.	290	275	265	255	250	245	235	230	220	215	205
	New pipe	230	210	200	195	190	180	175	170	165	160	150
	Very smooth	170	150	135	130	125	120	115	110	105	100	95
24″	Old, comm.	400	395	385	380	375	370	365	360	355	350	345
	Average, comm.	285	265	255	250	245	240	230	225	220	210	200
	New pipe	225	200	195	190	185	180	175	170	165	155	150
	Very smooth	165	140	135	125	120	120	115	110	105	100	95
30″	Old, comm.	400	385	380	375	370	365	360	355	350	350	345
	Average, comm.	280	255	250	245	240	230	225	220	210	205	200
	New pipe	220	195	190	185	180	175	170	165	160	155	150
	Very smooth	160	135	130	120	115	115	110	110	105	100	95
36″	Old, comm.	395	385	375	370	365	360	355	355	350	345	340
	Average, comm.	275	255	245	240	235	230	225	220	210	200	195
	New pipe	215	195	185	180	175	170	165	160	155	150	145
	Very smooth	150	135	125	120	115	110	110	105	100	95	90
48″	Old, comm.	395	385	370	365	360	355	350	350	345	340	335
	Average, comm.	265	250	240	230	225	220	215	210	200	195	190
	New pipe	205	190	180	175	170	165	160	155	150	145	140
	Very smooth	140	125	120	115	110	110	105	100	95	90	90

Note: 1″ = 25.4 mm; 1 ft/sec = 0.3048 m/s.

TABLE 4

TYPICAL LOST HEAD ITEMS

(Subscript 1 = Upstream and Subscript 2 = Downstream)

Item	Average Lost Head
1. From Tank to Pipe, flush connection (entrance loss)	$0.50\dfrac{V_2^2}{2g}$
projecting connection	$1.00\dfrac{V_2^2}{2g}$
rounded connection	$0.05\dfrac{V_2^2}{2g}$
2. From Pipe to Tank (exit loss)	$1.00\dfrac{V_1^2}{2g}$
3. Sudden Enlargement	$\dfrac{(V_1 - V_2)^2}{2g}$
4. Gradual Enlargement (see Table 5)	$K\dfrac{(V_1 - V_2)^2}{2g}$
5. Venturi Meters, Nozzles, and Orifices	$\left(\dfrac{1}{c_v^2} - 1\right)\dfrac{V_2^2}{2g}$
6. Sudden Contraction (see Table 5)	$K_c\dfrac{V_2^2}{2g}$
7. Elbows, Fittings, Valves*	$K\dfrac{V^2}{2g}$
Some typical values of K are: 45° bend . . . . 0.35 to 0.45 90° bend . . . . 0.50 to 0.75 Tees . . . . 1.50 to 2.00 Gate valves (open) . . . . about 0.25 Check valves (open) . . . . . . about 3.0	

* See hydraulics handbooks for details.

TABLE 5

VALUES OF *K*
Contractions and Enlargements

Sudden Contraction		Gradual Enlargement for Total Angles of Cone						
d_1/d_2	K_c	4°	10°	15°	20°	30°	50°	60°
1.2	0.08	0.02	0.04	0.09	0.16	0.25	0.35	0.37
1.4	0.17	0.03	0.06	0.12	0.23	0.36	0.50	0.53
1.6	0.26	0.03	0.07	0.14	0.26	0.42	0.57	0.61
1.8	0.34	0.04	0.07	0.15	0.28	0.44	0.61	0.65
2.0	0.37	0.04	0.07	0.16	0.29	0.46	0.63	0.68
2.5	0.41	0.04	0.08	0.16	0.30	0.48	0.65	0.70
3.0	0.43	0.04	0.08	0.16	0.31	0.48	0.66	0.71
4.0	0.45	0.04	0.08	0.16	0.31	0.49	0.67	0.72
5.0	0.46	0.04	0.08	0.16	0.31	0.50	0.67	0.72

Source: Values from King's *Handbook of Hydraulics*, McGraw-Hill Book Company, 1954.

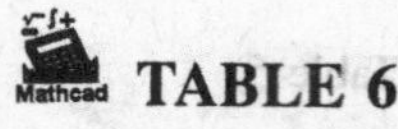

TABLE 6

SOME VALUES OF HAZEN-WILLIAMS COEFFICIENT *C*

Extremely smooth and straight pipes	140
New, smooth cast iron pipes	130
Average cast iron, new riveted steel pipes	110
Vitrified sewer pipes	110
Cast iron pipes, some years in service	100
Cast iron pipes, in bad condition	80

TABLE 7

DISCHARGE COEFFICIENTS FOR VERTICAL SHARP-EDGED CIRCULAR ORIFICES

For Water at 60°F (16°C) Discharging into Air at Same Temperature

Head in Feet	Orifice Diameter (in)					
	0.25	0.50	0.75	1.00	2.00	4.00
0.8	0.647	0.627	0.616	0.609	0.603	0.601
1.4	0.635	0.619	0.610	0.605	0.601	0.600
2.0	0.629	0.615	0.607	0.603	0.600	0.599
4.0	0.621	0.609	0.603	0.600	0.598	0.597
6.0	0.617	0.607	0.601	0.599	0.597	0.596
8.0	0.614	0.605	0.600	0.598	0.596	0.595
10.0	0.613	0.604	0.600	0.597	0.596	0.595
12.0	0.612	0.603	0.599	0.597	0.595	0.595
14.0	0.611	0.603	0.598	0.596	0.595	0.594
16.0	0.610	0.602	0.598	0.596	0.595	0.594
20.0	0.609	0.602	0.598	0.596	0.595	0.594
25.0	0.608	0.601	0.597	0.596	0.594	0.594
30.0	0.607	0.600	0.597	0.595	0.594	0.594
40.0	0.606	0.600	0.596	0.595	0.594	0.593
50.0	0.605	0.599	0.596	0.595	0.594	0.593
60.0	0.605	0.599	0.596	0.594	0.593	0.593

Note: 1″ = 25.4 mm; 1 ft = 0.3048 m.
Source: F. W. Medaugh and G. D. Johnson, *Civil Engr.*, July 1940, p. 424.

TABLE 8

SOME EXPANSION FACTORS *Y* FOR COMPRESSIBLE FLOW THROUGH FLOW NOZZLES AND VENTURI METERS

p_2/p_1	k	Ratio of Diameters (d_2/d_1)				
		0.30	0.40	0.50	0.60	0.70
	1.40	0.973	0.972	0.971	0.968	0.962
0.95	1.30	0.970	0.970	0.968	0.965	0.959
	1.20	0.968	0.967	0.966	0.963	0.956
	1.40	0.944	0.943	0.941	0.935	0.925
0.90	1.30	0.940	0.939	0.936	0.931	0.918
	1.20	0.935	0.933	0.931	0.925	0.912
	1.40	0.915	0.914	0.910	0.902	0.887
0.85	1.30	0.910	0.907	0.904	0.896	0.880
	1.20	0.902	0.900	0.896	0.887	0.870
	1.40	0.886	0.884	0.880	0.868	0.850
0.80	1.30	0.876	0.873	0.869	0.857	0.839
	1.20	0.866	0.864	0.859	0.848	0.829
	1.40	0.856	0.853	0.846	0.836	0.814
0.75	1.30	0.844	0.841	0.836	0.823	0.802
	1.20	0.820	0.818	0.812	0.798	0.776
	1.40	0.824	0.820	0.815	0.800	0.778
0.70	1.30	0.812	0.808	0.802	0.788	0.763
	1.20	0.794	0.791	0.784	0.770	0.745

For $p_2/p_1 = 1.00$, $Y = 1.00$.

TABLE 9

A FEW AVERAGE VALUES OF *n* FOR USE IN THE KUTTER AND MANNING FORMULAS AND *m* IN THE BAZIN FORMULA

Type of Open Channel	n	m
Smooth cement lining, best planed timber	0.010	0.11
Planed timber, new wood-stave flumes, lined cast iron	0.012	0.20
Good vitrified sewer pipe, good brickwork, average concrete pipe, unplaned timber, smooth metal flumes	0.013	0.29
Average clay sewer pipe and cast iron pipe, average cement lining	0.015	0.40
Earth canals, straight and well maintained	0.023	1.54
Dredged earth canals, average condition	0.027	2.36
Canals cut in rock	0.040	3.50
Rivers in good condition	0.030	3.00

TABLE 10

VALUES OF *C* FROM THE KUTTER FORMULA

Slope *S*	*n*	Hydraulic Radius *R* (ft) 0.2	0.3	0.4	0.6	0.8	1.0	1.5	2.0	2.5	3.0	4.0	6.0	8.0	10.0	15.0
0.00005	0.010	87	98	109	123	133	140	154	164	172	177	187	199	207	213	220
	0.012	68	78	88	98	107	113	126	135	142	148	157	168	176	182	189
	0.015	52	58	66	76	83	89	99	107	113	118	126	138	145	150	159
	0.017	43	50	57	65	72	77	86	93	98	103	112	122	129	134	142
	0.020	35	41	45	53	59	64	72	80	84	88	95	105	111	116	125
	0.025	26	30	35	41	45	49	57	62	66	70	78	85	92	96	104
	0.030	22	25	28	33	37	40	47	51	55	58	65	74	78	84	90
0.0001	0.010	98	108	118	131	140	147	158	167	173	178	186	196	202	206	212
	0.012	76	86	95	105	113	119	130	138	144	148	155	165	170	174	180
	0.015	57	64	72	81	88	92	103	109	114	118	125	134	140	143	150
	0.017	48	55	62	70	75	80	88	95	99	104	111	118	125	128	135
	0.020	38	45	50	57	63	67	75	81	85	88	95	102	107	111	118
	0.025	28	34	38	43	48	51	59	64	67	70	77	84	89	93	98
	0.030	23	27	30	35	39	42	48	52	55	59	64	72	75	80	85
0.0002	0.010	105	115	125	137	145	150	162	169	174	178	185	193	198	202	206
	0.012	83	92	100	110	117	123	133	139	144	148	154	162	167	170	175
	0.015	61	69	76	84	91	96	105	110	114	118	124	132	137	140	145
	0.017	52	59	65	73	78	83	90	97	100	104	110	117	122	125	130
	0.020	42	48	53	60	65	68	76	82	85	88	94	100	105	108	113
	0.025	30	35	40	45	50	54	60	65	68	70	76	83	86	90	95
	0.030	25	28	32	37	40	43	49	53	56	59	63	69	74	77	82
0.0004	0.010	110	121	128	140	148	153	164	171	174	178	184	192	197	199	203
	0.012	87	95	103	113	120	125	134	141	145	149	153	161	165	168	172
	0.015	64	73	78	87	93	98	106	112	115	118	123	130	134	137	142
	0.017	54	62	68	75	80	84	92	98	101	104	110	116	120	123	128
	0.020	43	50	55	61	67	70	77	83	86	88	94	99	104	106	110
	0.025	32	37	42	47	51	55	60	65	68	70	75	82	85	88	92
	0.030	26	30	33	38	41	44	50	54	57	59	63	68	73	75	80
0.001	0.010	113	124	132	143	150	155	165	172	175	178	184	190	195	197	201
	0.012	88	97	105	115	121	127	135	142	145	149	154	160	164	167	171
	0.015	66	75	80	88	94	98	107	112	116	119	123	130	133	135	141
	0.017	55	63	68	76	81	85	92	98	102	105	110	115	119	122	127
	0.020	45	51	56	62	68	71	78	84	87	89	93	98	103	105	109
	0.025	33	38	43	48	52	55	61	65	68	70	75	81	84	87	91
	0.030	27	30	34	38	42	45	50	54	57	59	63	68	72	74	78
0.01	0.010	114	125	133	143	151	156	165	172	175	178	184	190	194	196	200
	0.012	89	99	106	116	122	128	136	142	145	149	154	159	163	166	170
	0.015	67	76	81	89	95	99	107	113	116	119	123	129	133	135	140
	0.017	56	64	69	77	82	86	93	99	103	105	109	115	118	121	126
	0.020	46	52	57	63	68	72	78	84	87	89	93	98	102	105	108
	0.025	34	39	44	49	52	56	62	65	68	70	75	80	83	86	90
	0.030	27	31	35	39	43	45	51	55	58	59	63	67	71	73	77

TABLE 11

VALUES OF DISCHARGE FACTOR K in $Q = (K/n)y^{8/3}S^{1/2}$ FOR TRAPEZOIDAL CHANNELS*

(y = depth of flow, b = bottom width of channel)

Side Slopes of Channel Section (horizontal to vertical)

y/b	Vertical	$\frac{1}{4}:1$	$\frac{1}{2}:1$	$\frac{3}{4}:1$	$1:1$	$1\frac{1}{2}:1$	$2:1$	$2\frac{1}{2}:1$	$3:1$	$4:1$
0.01	146.7	147.2	147.6	148.0	148.3	148.8	149.2	149.5	149.9	150.5
0.02	72.4	72.9	73.4	73.7	74.0	74.5	74.9	75.3	75.6	76.3
0.03	47.6	48.2	48.6	49.0	49.3	49.8	50.2	50.6	50.9	51.6
0.04	35.3	35.8	36.3	36.6	36.9	37.4	37.8	38.2	38.6	39.3
0.05	27.9	28.4	28.9	29.2	29.5	30.0	30.5	30.9	31.2	32.0
0.06	23.0	23.5	23.9	24.3	24.6	25.1	25.5	26.0	26.3	27.1
0.07	19.5	20.0	20.4	20.8	21.1	21.6	22.0	22.4	22.8	23.6
0.08	16.8	17.3	17.8	18.1	18.4	18.9	19.4	19.8	20.2	21.0
0.09	14.8	15.3	15.7	16.1	16.4	16.9	17.4	17.8	18.2	19.0
0.10	13.2	13.7	14.1	14.4	14.8	15.3	15.7	16.2	16.6	17.4
0.11	11.83	12.33	12.76	13.11	13.42	13.9	14.4	14.9	15.3	16.1
0.12	10.73	11.23	11.65	12.00	12.31	12.8	13.3	13.8	14.2	15.0
0.13	9.80	10.29	10.71	11.06	11.37	11.9	12.4	12.8	13.3	14.1
0.14	9.00	9.49	9.91	10.26	10.57	11.1	11.6	12.0	12.5	13.4
0.15	8.32	8.80	9.22	9.57	9.88	10.4	10.9	11.4	11.8	12.7
0.16	7.72	8.20	8.61	8.96	9.27	9.81	10.29	10.75	11.20	12.1
0.17	7.19	7.67	8.08	8.43	8.74	9.28	9.77	10.23	10.68	11.6
0.18	6.73	7.20	7.61	7.96	8.27	8.81	9.30	9.76	10.21	11.1
0.19	6.31	6.78	7.19	7.54	7.85	8.39	8.88	9.34	9.80	10.7
0.20	5.94	6.40	6.81	7.16	7.47	8.01	8.50	8.97	9.43	10.3
0.22	5.30	5.76	6.16	6.51	6.82	7.36	7.86	8.33	8.79	9.70
0.24	4.77	5.22	5.62	5.96	6.27	6.82	7.32	7.79	8.26	9.18
0.26	4.32	4.77	5.16	5.51	5.82	6.37	6.87	7.35	7.81	8.74
0.28	3.95	4.38	4.77	5.12	5.43	5.98	6.48	6.96	7.43	8.36
0.30	3.62	4.05	4.44	4.78	5.09	5.64	6.15	6.63	7.10	8.04
0.32	3.34	3.77	4.15	4.49	4.80	5.35	5.86	6.34	6.82	7.75
0.34	3.09	3.51	3.89	4.23	4.54	5.10	5.60	6.09	6.56	7.50
0.36	2.88	3.29	3.67	4.01	4.31	4.87	5.38	5.86	6.34	7.28
0.38	2.68	3.09	3.47	3.81	4.11	4.67	5.17	5.66	6.14	7.09
0.40	2.51	2.92	3.29	3.62	3.93	4.48	4.99	5.48	5.96	6.91
0.42	2.36	2.76	3.13	3.46	3.77	4.32	4.83	5.32	5.80	6.75
0.44	2.22	2.61	2.98	3.31	3.62	4.17	4.68	5.17	5.66	6.60
0.46	2.09	2.48	2.85	3.18	3.48	4.04	4.55	5.04	5.52	6.47
0.48	1.98	2.36	2.72	3.06	3.36	3.91	4.43	4.92	5.40	6.35
0.50	1.87	2.26	2.61	2.94	3.25	3.80	4.31	4.81	5.29	6.24
0.55	1.65	2.02	2.37	2.70	3.00	3.55	4.07	4.56	5.05	6.00
0.60	1.46	1.83	2.17	2.50	2.80	3.35	3.86	4.36	4.84	5.80
0.70	1.18	1.53	1.87	2.19	2.48	3.03	3.55	4.04	4.53	5.49
0.80	0.982	1.31	1.64	1.95	2.25	2.80	3.31	3.81	4.30	5.26
0.90	0.831	1.15	1.47	1.78	2.07	2.62	3.13	3.63	4.12	5.08
1.00	0.714	1.02	1.33	1.64	1.93	2.47	2.99	3.48	3.97	4.93
1.20	0.548	0.836	1.14	1.43	1.72	2.26	2.77	3.27	3.76	4.72
1.40	0.436	0.708	0.998	1.29	1.57	2.11	2.62	3.12	3.60	4.57
1.60	0.357	0.616	0.897	1.18	1.46	2.00	2.51	3.00	3.49	4.45
1.80	0.298	0.546	0.820	1.10	1.38	1.91	2.42	2.91	3.40	4.36
2.00	0.254	0.491	0.760	1.04	1.31	1.84	2.35	2.84	3.33	4.29
2.25	0.212	0.439	0.700	0.973	1.24	1.77	2.28	2.77	3.26	4.22

* Tabular values are for British Engineering System units; for SI units, multiply tabular values by 1.486.

Source: Values from King's *Handbook of Hydraulics*, 4th ed., McGraw-Hill Co., 1954.

TABLE 12

VALUES OF DISCHARGE FACTOR K' in $Q = (K'/n)b^{8/3}S^{1/2}$ FOR TRAPEZOIDAL CHANNELS*

(y = depth of flow, b = bottom width of channel)

Side Slopes of Channel Section (horizontal to vertical)

y/b	Vertical	$\frac{1}{4}:1$	$\frac{1}{2}:1$	$\frac{3}{4}:1$	1 : 1	$1\frac{1}{2}:1$	2 : 1	$2\frac{1}{2}:1$	3 : 1	4 : 1
0.01	0.00068	0.00068	0.00069	0.00069	0.00069	0.00069	0.00069	0.00069	0.00070	0.00070
0.02	0.00213	0.00215	0.00216	0.00217	0.00218	0.00220	0.00221	0.00222	0.00223	0.00225
0.03	0.00414	0.00419	0.00423	0.00426	0.00428	0.00433	0.00436	0.00439	0.00443	0.00449
0.04	0.00660	0.00670	0.00679	0.00685	0.00691	0.00700	0.00708	0.00716	0.00723	0.00736
0.05	0.00946	0.00964	0.00979	0.00991	0.01002	0.01019	0.01033	0.01047	0.01060	0.01086
0.06	0.0127	0.0130	0.0132	0.0134	0.0136	0.0138	0.0141	0.0143	0.0145	0.0150
0.07	0.0162	0.0166	0.0170	0.0173	0.0175	0.0180	0.0183	0.0187	0.0190	0.0197
0.08	0.0200	0.0206	0.0211	0.0215	0.0219	0.0225	0.0231	0.0236	0.0240	0.0250
0.09	0.0241	0.0249	0.0256	0.0262	0.0267	0.0275	0.0282	0.0289	0.0296	0.0310
0.10	0.0284	0.0294	0.0304	0.0311	0.0318	0.0329	0.0339	0.0348	0.0358	0.0376
0.11	0.0329	0.0343	0.0354	0.0364	0.0373	0.0387	0.0400	0.0413	0.0424	0.0448
0.12	0.0376	0.0393	0.0408	0.0420	0.0431	0.0450	0.0466	0.0482	0.0497	0.0527
0.13	0.0425	0.0446	0.0464	0.0480	0.0493	0.0516	0.0537	0.0556	0.0575	0.0613
0.14	0.0476	0.0502	0.0524	0.0542	0.0559	0.0587	0.0612	0.0636	0.0659	0.0706
0.15	0.0528	0.0559	0.0585	0.0608	0.0627	0.0662	0.0692	0.0721	0.0749	0.0805
0.16	0.0582	0.0619	0.0650	0.0676	0.0700	0.0740	0.0777	0.0811	0.0845	0.0912
0.17	0.0638	0.0680	0.0716	0.0748	0.0775	0.0823	0.0866	0.0907	0.0947	0.1026
0.18	0.0695	0.0744	0.0786	0.0822	0.0854	0.0910	0.0960	0.1008	0.1055	0.1148
0.19	0.0753	0.0809	0.0857	0.0899	0.0936	0.1001	0.1059	0.1115	0.1169	0.1277
0.20	0.0812	0.0876	0.0931	0.0979	0.1021	0.1096	0.1163	0.1227	0.1290	0.1414
0.22	0.0934	0.1015	0.109	0.115	0.120	0.130	0.139	0.147	0.155	0.171
0.24	0.1061	0.1161	0.125	0.133	0.140	0.152	0.163	0.173	0.184	0.204
0.26	0.119	0.131	0.142	0.152	0.160	0.175	0.189	0.202	0.215	0.241
0.28	0.132	0.147	0.160	0.172	0.182	0.201	0.217	0.234	0.249	0.281
0.30	0.146	0.163	0.179	0.193	0.205	0.228	0.248	0.267	0.287	0.324
0.32	0.160	0.180	0.199	0.215	0.230	0.256	0.281	0.304	0.327	0.371
0.34	0.174	0.198	0.219	0.238	0.256	0.287	0.316	0.343	0.370	0.423
0.36	0.189	0.216	0.241	0.263	0.283	0.319	0.353	0.385	0.416	0.478
0.38	0.203	0.234	0.263	0.288	0.312	0.353	0.392	0.429	0.465	0.537
0.40	0.218	0.253	0.286	0.315	0.341	0.389	0.434	0.476	0.518	0.600
0.42	0.233	0.273	0.309	0.342	0.373	0.427	0.478	0.526	0.574	0.668
0.44	0.248	0.293	0.334	0.371	0.405	0.467	0.525	0.580	0.633	0.740
0.46	0.264	0.313	0.359	0.401	0.439	0.509	0.574	0.636	0.696	0.816
0.48	0.279	0.334	0.385	0.432	0.474	0.553	0.625	0.695	0.763	0.897
0.50	0.295	0.355	0.412	0.463	0.511	0.598	0.679	0.757	0.833	0.983
0.55	0.335	0.410	0.482	0.548	0.609	0.722	0.826	0.926	1.025	1.22
0.60	0.375	0.468	0.557	0.640	0.717	0.858	0.990	1.117	1.24	1.49
0.70	0.457	0.592	0.722	0.844	0.959	1.17	1.37	1.56	1.75	2.12
0.80	0.542	0.725	0.906	1.078	1.24	1.54	1.83	2.10	2.37	2.90
0.90	0.628	0.869	1.11	1.34	1.56	1.98	2.36	2.74	3.11	3.83
1.00	0.714	1.022	1.33	1.64	1.93	2.47	2.99	3.48	3.97	4.93
1.20	0.891	1.36	1.85	2.33	2.79	3.67	4.51	5.32	6.11	7.67
1.40	1.07	1.74	2.45	3.16	3.85	5.17	6.42	7.64	8.84	11.2
1.60	1.25	2.16	3.14	4.14	5.12	6.99	8.78	10.52	12.2	15.6
1.80	1.43	2.62	3.93	5.28	6.60	9.15	11.6	14.0	16.3	20.9
2.00	1.61	3.12	4.82	6.58	8.32	11.7	14.9	18.1	21.2	27.3
2.25	1.84	3.81	6.09	8.46	10.8	15.4	19.8	24.1	28.4	36.7

* Tabular values are for British Engineering System units; for SI units, multiply tabular values by 1.486.

Source: Values from King's *Handbook of Hydraulics*, 4th ed., McGraw-Hill Co., 1954.

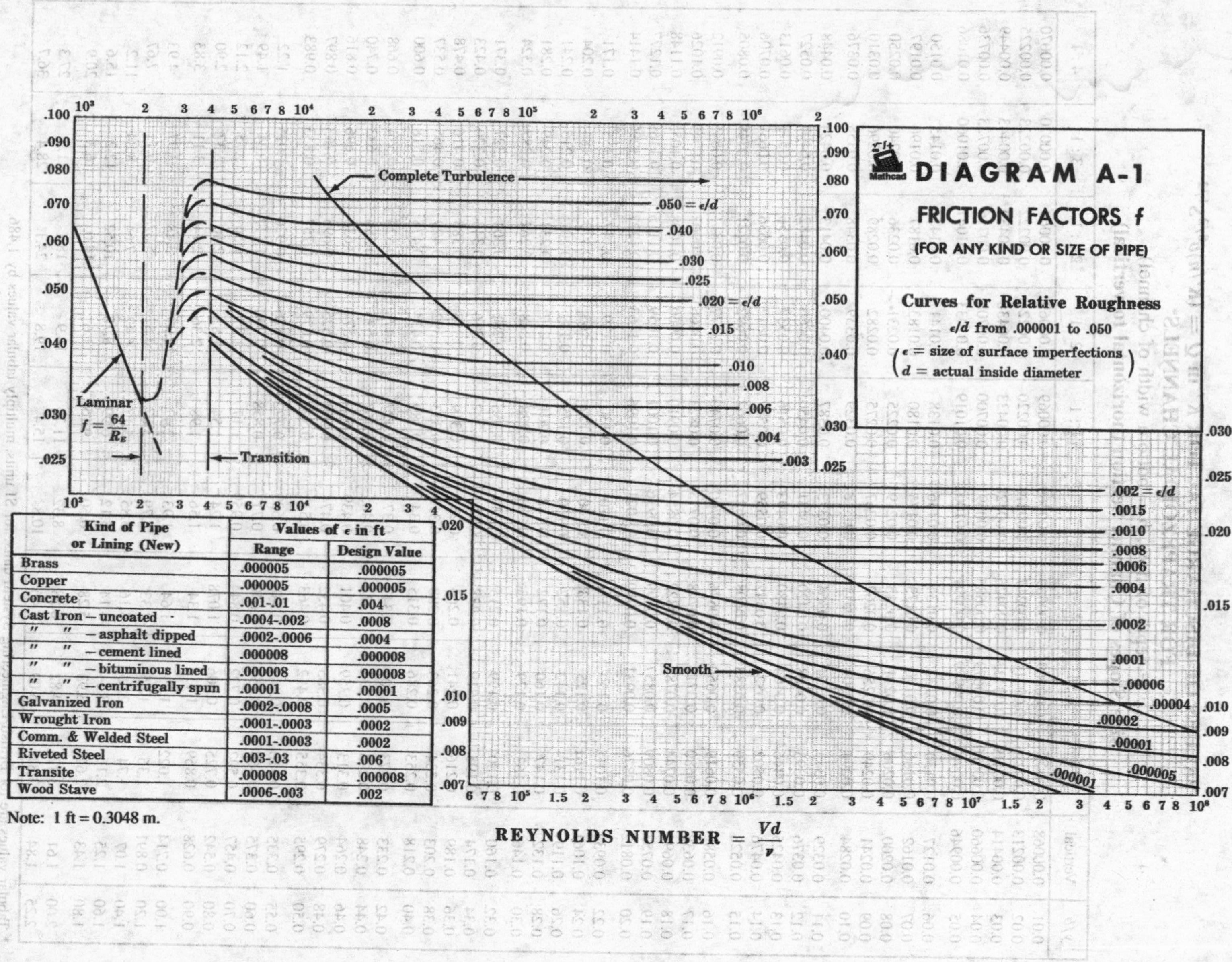

Kind of Pipe or Lining (New)	Values of ε in ft Range	Design Value
Brass	.000005	.000005
Copper	.000005	.000005
Concrete	.001-.01	.004
Cast Iron – uncoated	.0004-.002	.0008
" " – asphalt dipped	.0002-.0006	.0004
" " – cement lined	.000008	.000008
" " – bituminous lined	.000008	.000008
" " – centrifugally spun	.00001	.00001
Galvanized Iron	.0002-.0008	.0005
Wrought Iron	.0001-.0003	.0002
Comm. & Welded Steel	.0001-.0003	.0002
Riveted Steel	.003-.03	.006
Transite	.000008	.000008
Wood Stave	.0006-.003	.002

Note: 1 ft = 0.3048 m.

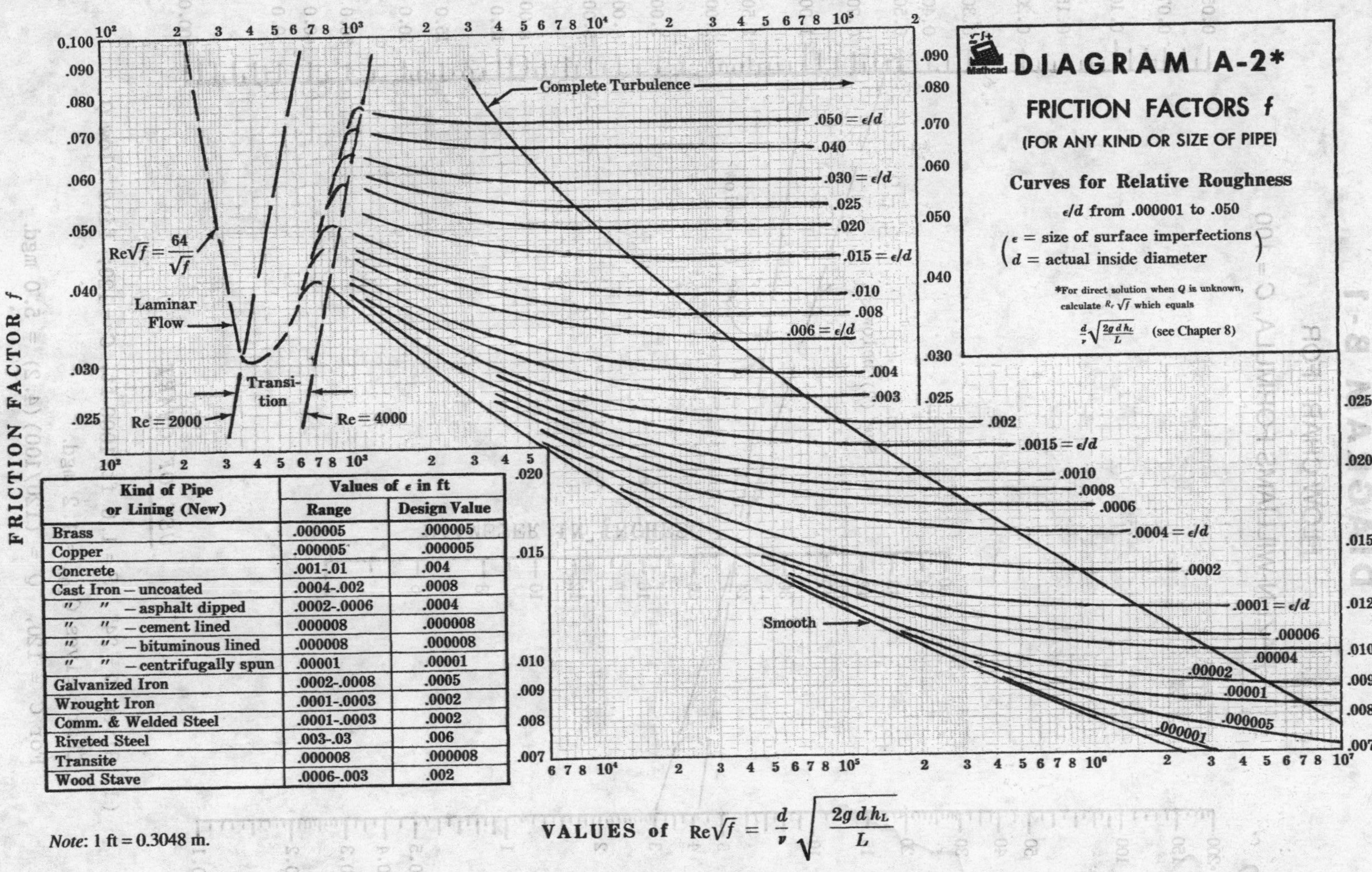

Kind of Pipe or Lining (New)	Values of ϵ in ft: Range	Design Value
Brass	.000005	.000005
Copper	.000005	.000005
Concrete	.001-.01	.004
Cast Iron – uncoated	.0004-.002	.0008
" " – asphalt dipped	.0002-.0006	.0004
" " – cement lined	.000008	.000008
" " – bituminous lined	.000008	.000008
" " – centrifugally spun	.00001	.00001
Galvanized Iron	.0002-.0008	.0005
Wrought Iron	.0001-.0003	.0002
Comm. & Welded Steel	.0001-.0003	.0002
Riveted Steel	.003-.03	.006
Transite	.000008	.000008
Wood Stave	.0006-.003	.002

Note: 1 ft = 0.3048 m.

DIAGRAM B-1

FLOW CHART FOR HAZEN-WILLIAMS FORMULA, $C = 100$

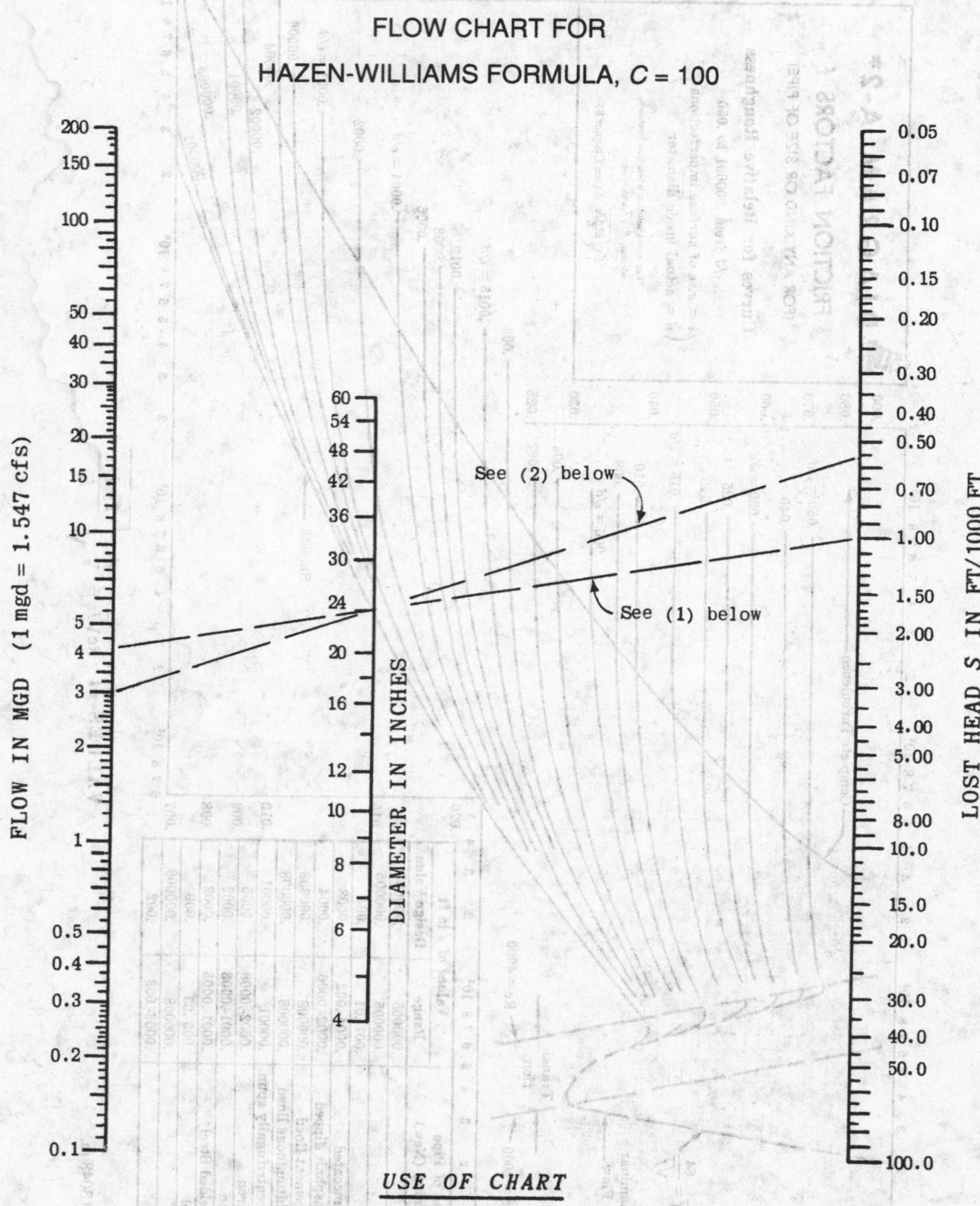

USE OF CHART

(1) Given $D = 24''$, $S = 1.0$ ft/1000 ft, $C = 120$; find flow Q.
Chart gives $Q_{100} = 4.2$ mgd.
For $C = 120$, $Q = (120/100)(4.2) = 5.0$ mgd.

(2) Given $Q = 3.6$ mgd, $D = 24''$, $C = 120$; find Lost Head.
Change Q_{120} to Q_{100}: $Q_{100} = (100/120)(3.6) = 3.0$ mgd.
Chart gives $S = 0.55$ ft/1000 ft.

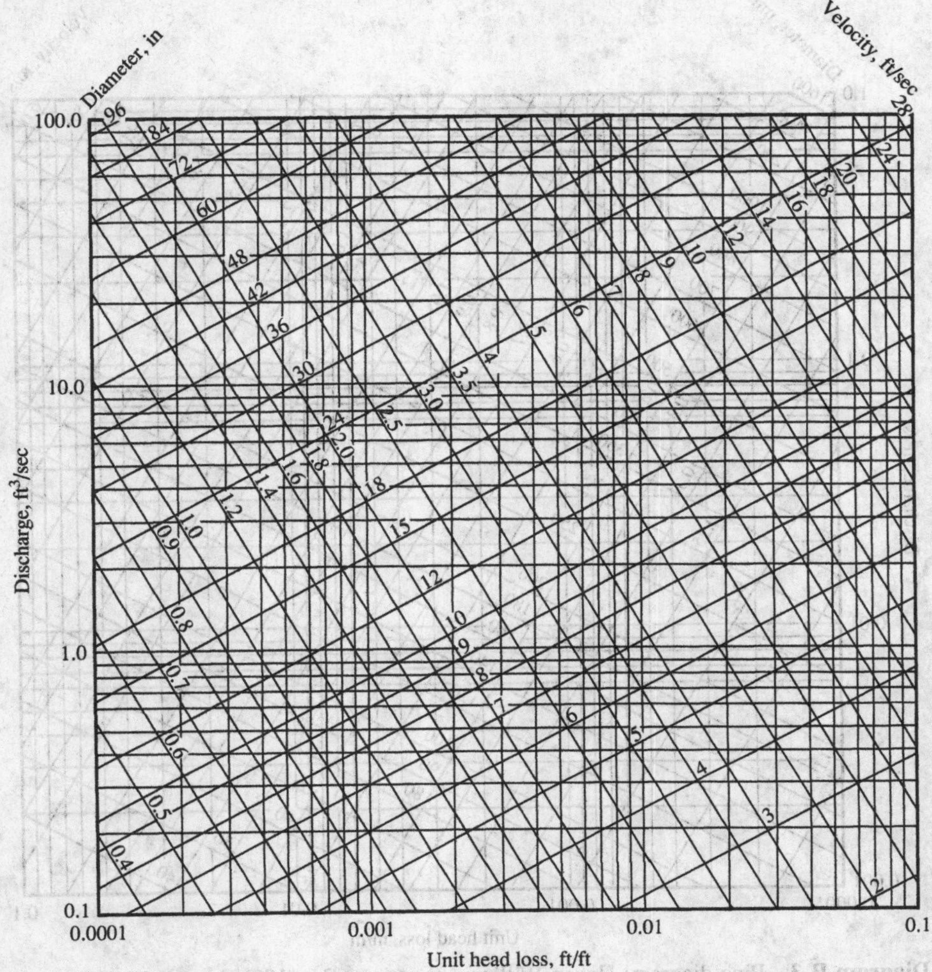

Diagram B-2 Pipe diagram: Hazen-Williams equation ($C = 120$), British Engineering System.

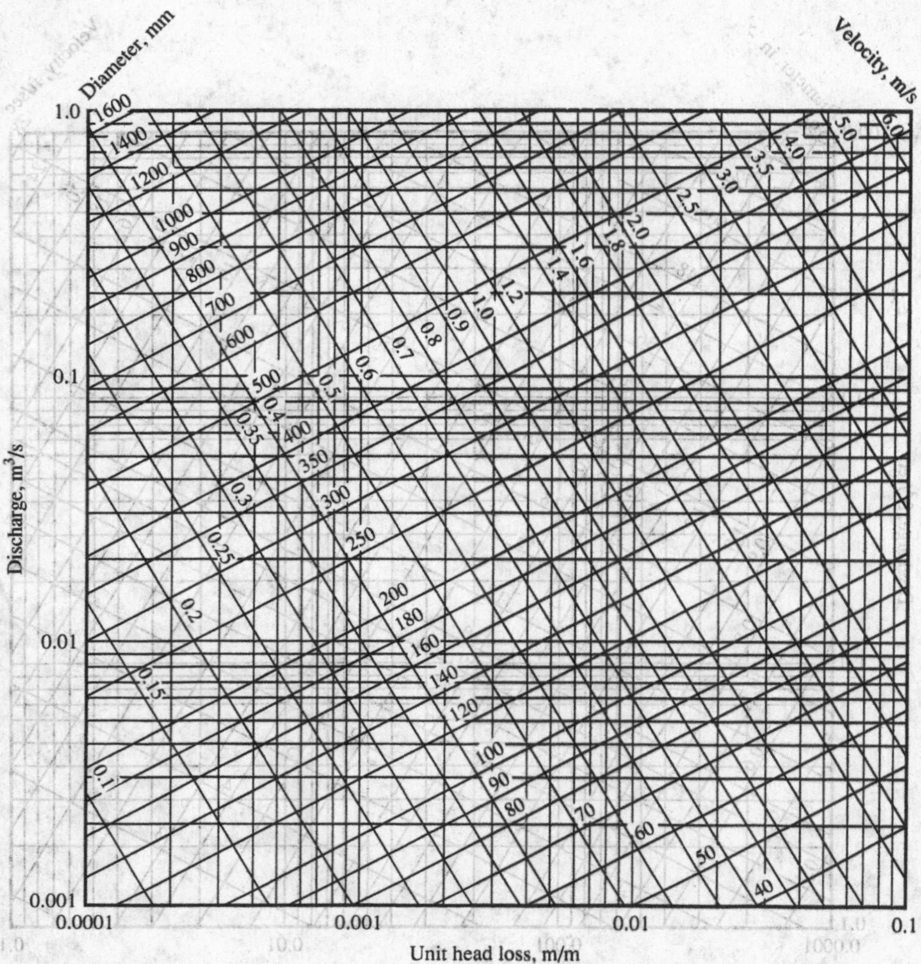

Diagram B-3 Pipe diagram: Hazen-Williams equation ($C = 120$), International System.

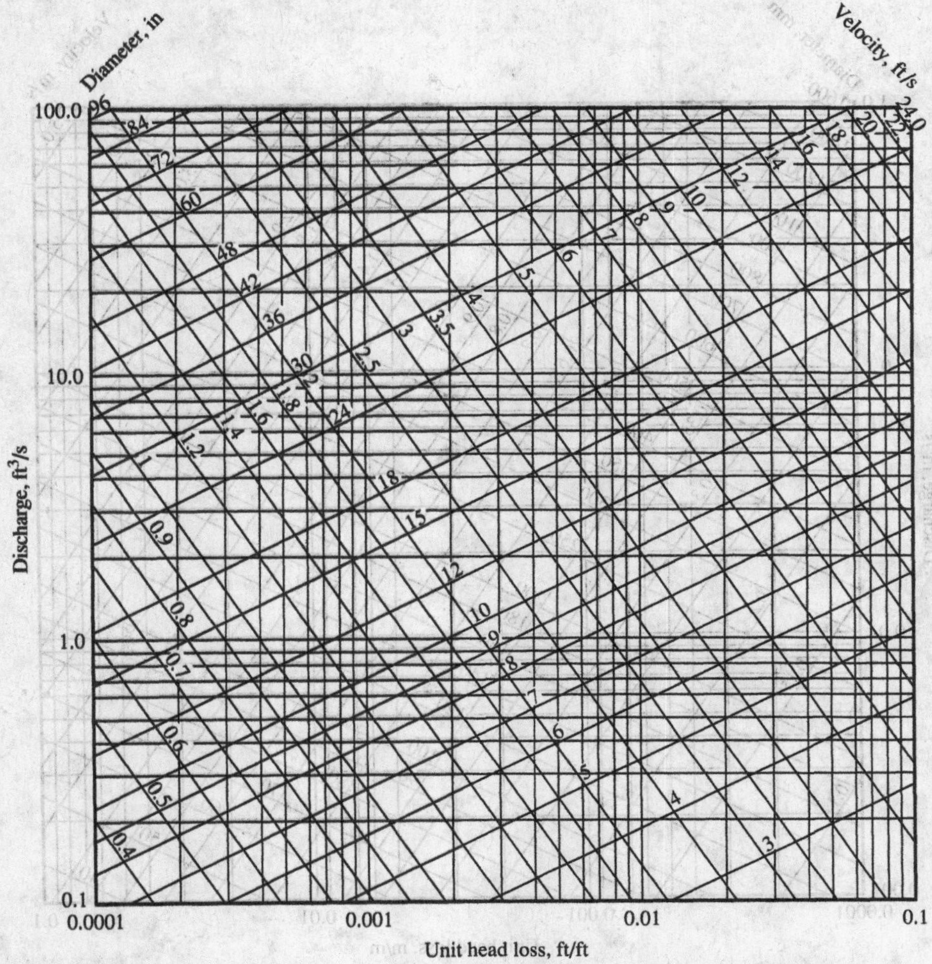

Diagram B-4 Pipe diagram: Manning equation ($n = 0.013$), British Engineering System.

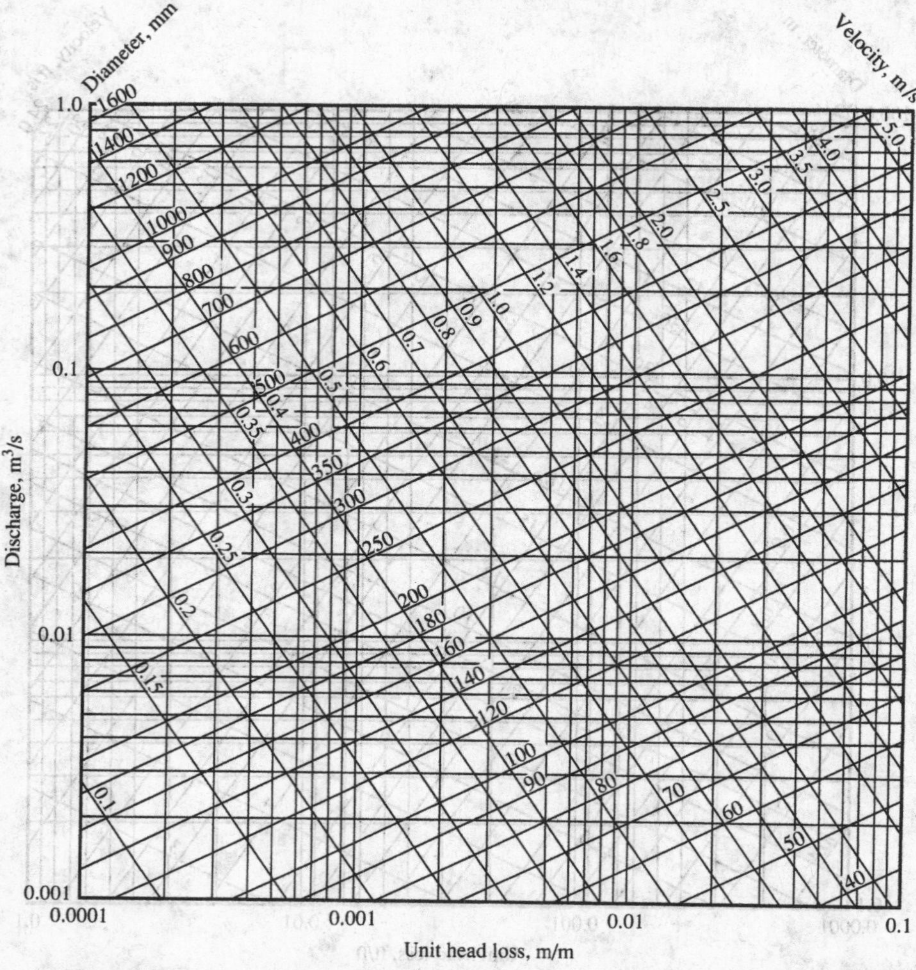

Diagram B-5 Pipe diagram: Manning equation ($n = 0.013$), International System.

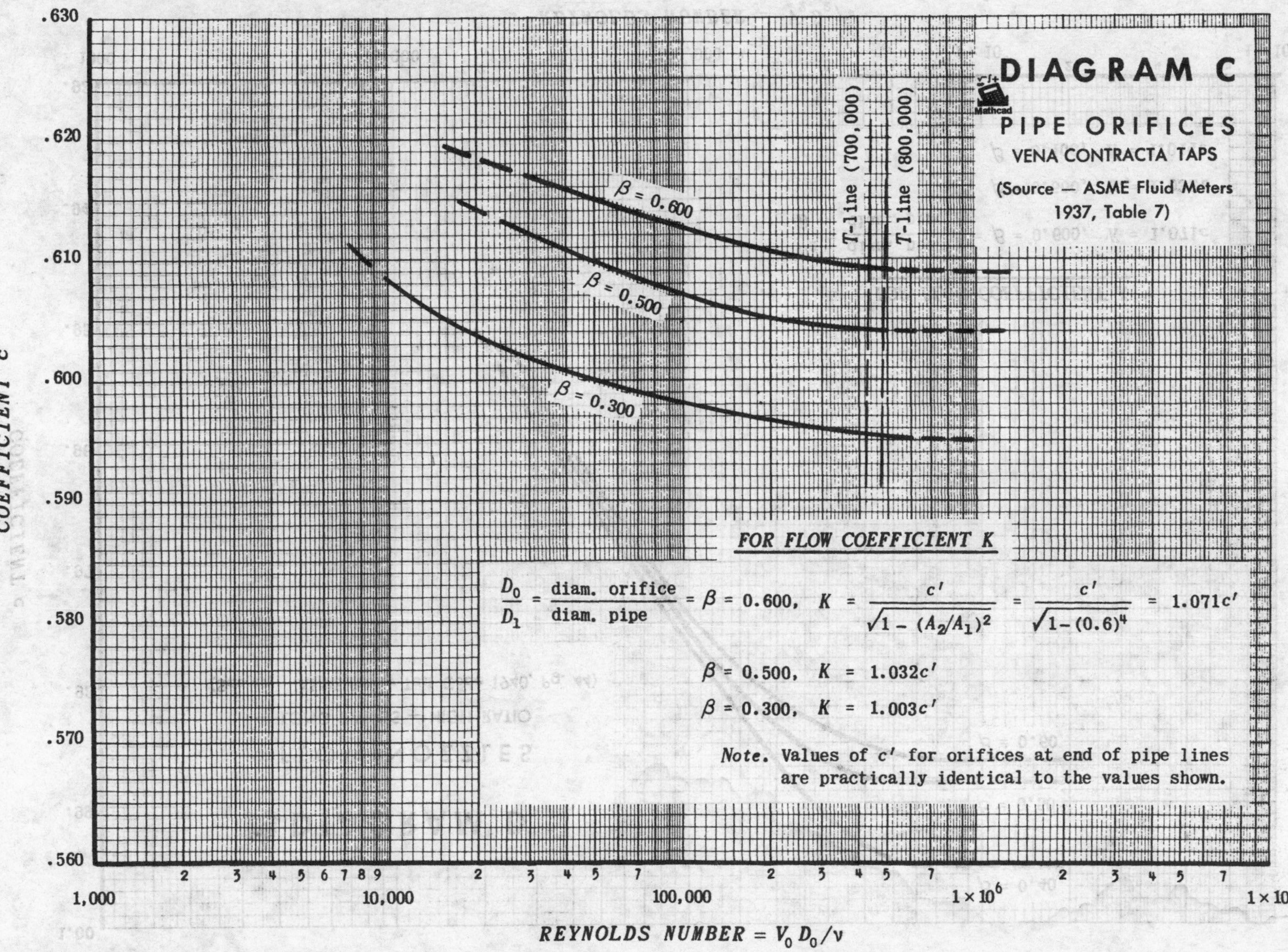
DIAGRAM C
PIPE ORIFICES
VENA CONTRACTA TAPS
(Source — ASME Fluid Meters 1937, Table 7)
T-line (700,000)
T-line (800,000)
β = 0.600
β = 0.500
β = 0.300
FOR FLOW COEFFICIENT K
$\frac{D_0}{D_1} = \frac{\text{diam. orifice}}{\text{diam. pipe}} = \beta = 0.600,\ K = \frac{c'}{\sqrt{1-(A_2/A_1)^2}} = \frac{c'}{\sqrt{1-(0.6)^4}} = 1.071c'$
$\beta = 0.500,\ K = 1.032c'$
$\beta = 0.300,\ K = 1.003c'$
Note. Values of c′ for orifices at end of pipe lines are practically identical to the values shown.
COEFFICIENT c′
.630
.620
.610
.600
.590
.580
.570
.560
1,000
10,000
100,000
1×10^6
1×10^7
REYNOLDS NUMBER = $V_0 D_0/\nu$

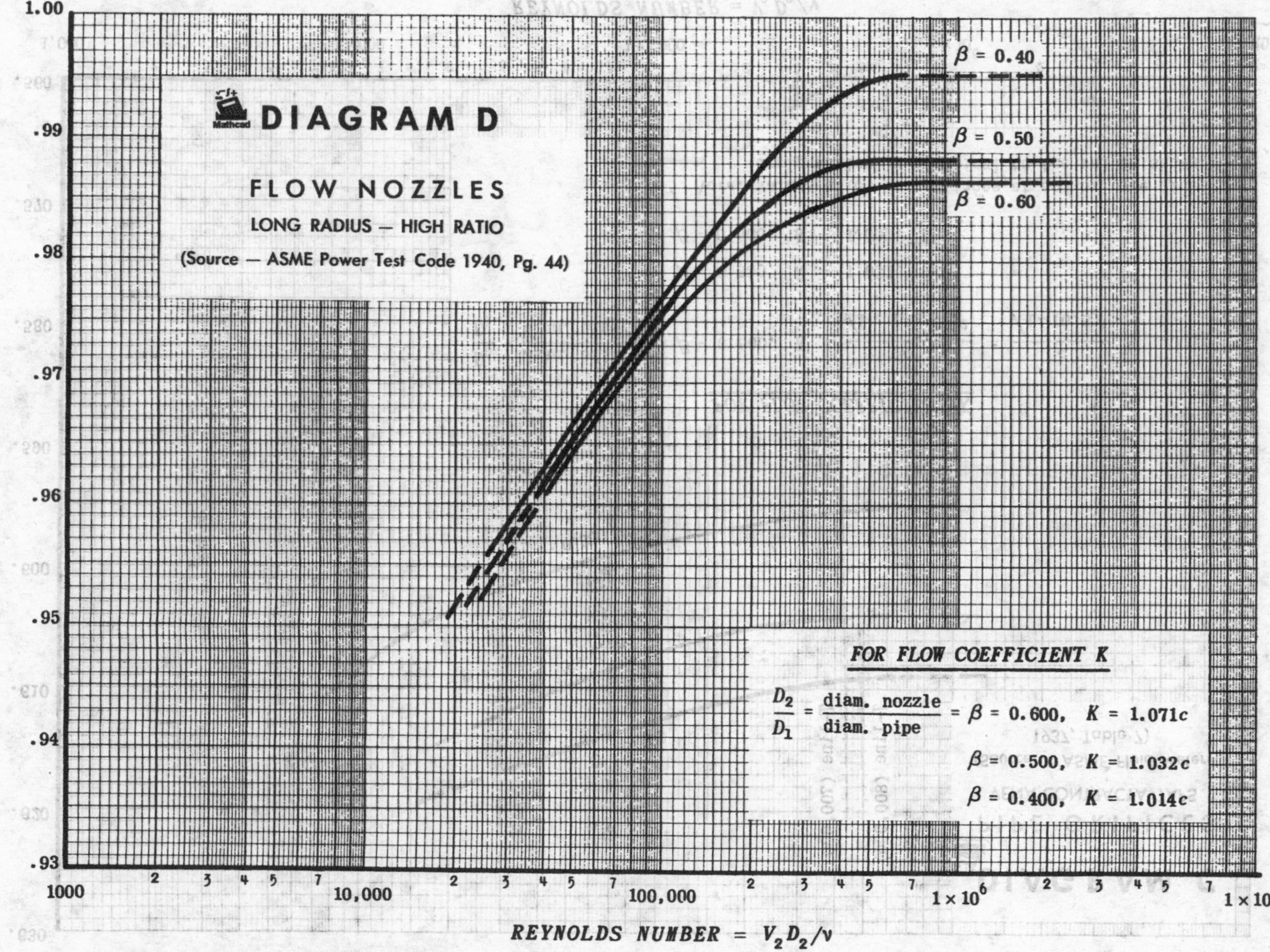
DIAGRAM D
FLOW NOZZLES
LONG RADIUS — HIGH RATIO
(Source — ASME Power Test Code 1940, Pg. 44)
β = 0.40
β = 0.50
β = 0.60
FOR FLOW COEFFICIENT K
$\frac{D_2}{D_1} = \frac{\text{diam. nozzle}}{\text{diam. pipe}} = \beta = 0.600, \quad K = 1.071c$
$\beta = 0.500, \quad K = 1.032c$
$\beta = 0.400, \quad K = 1.014c$
1.00
.99
.98
.97
.96
.95
.94
.93
COEFFICIENT c
1000
10,000
100,000
1×10^6
1×10^7
REYNOLDS NUMBER = $V_2 D_2/\nu$

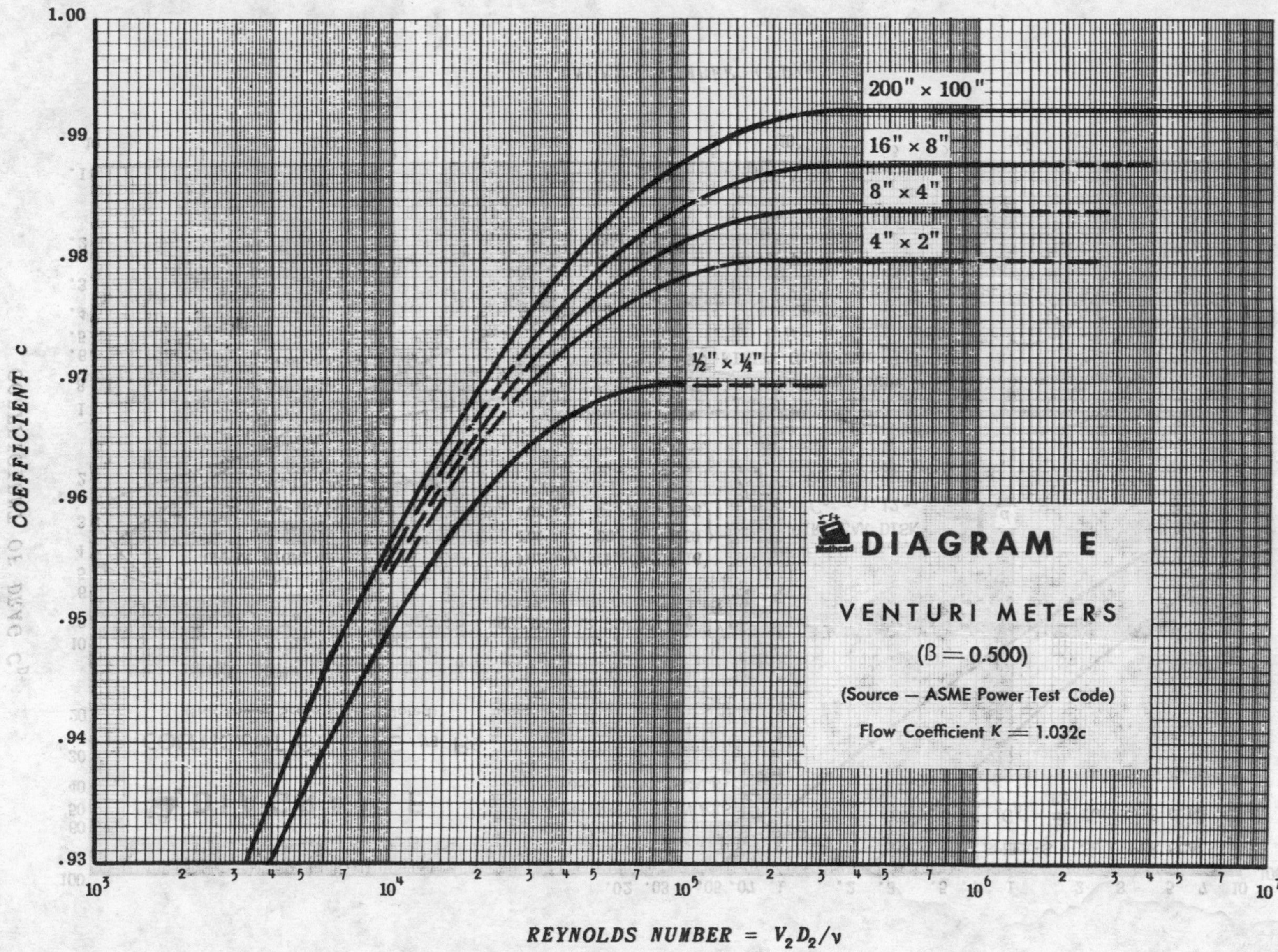
DIAGRAM E
VENTURI METERS
(β = 0.500)
(Source — ASME Power Test Code)
Flow Coefficient K = 1.032c
200" × 100"
16" × 8"
8" × 4"
4" × 2"
½" × ¼"
1.00
.99
.98
.97
.96
.95
.94
.93
COEFFICIENT c
10³
10⁴
10⁵
10⁶
10⁷
REYNOLDS NUMBER = $V_2 D_2/\nu$

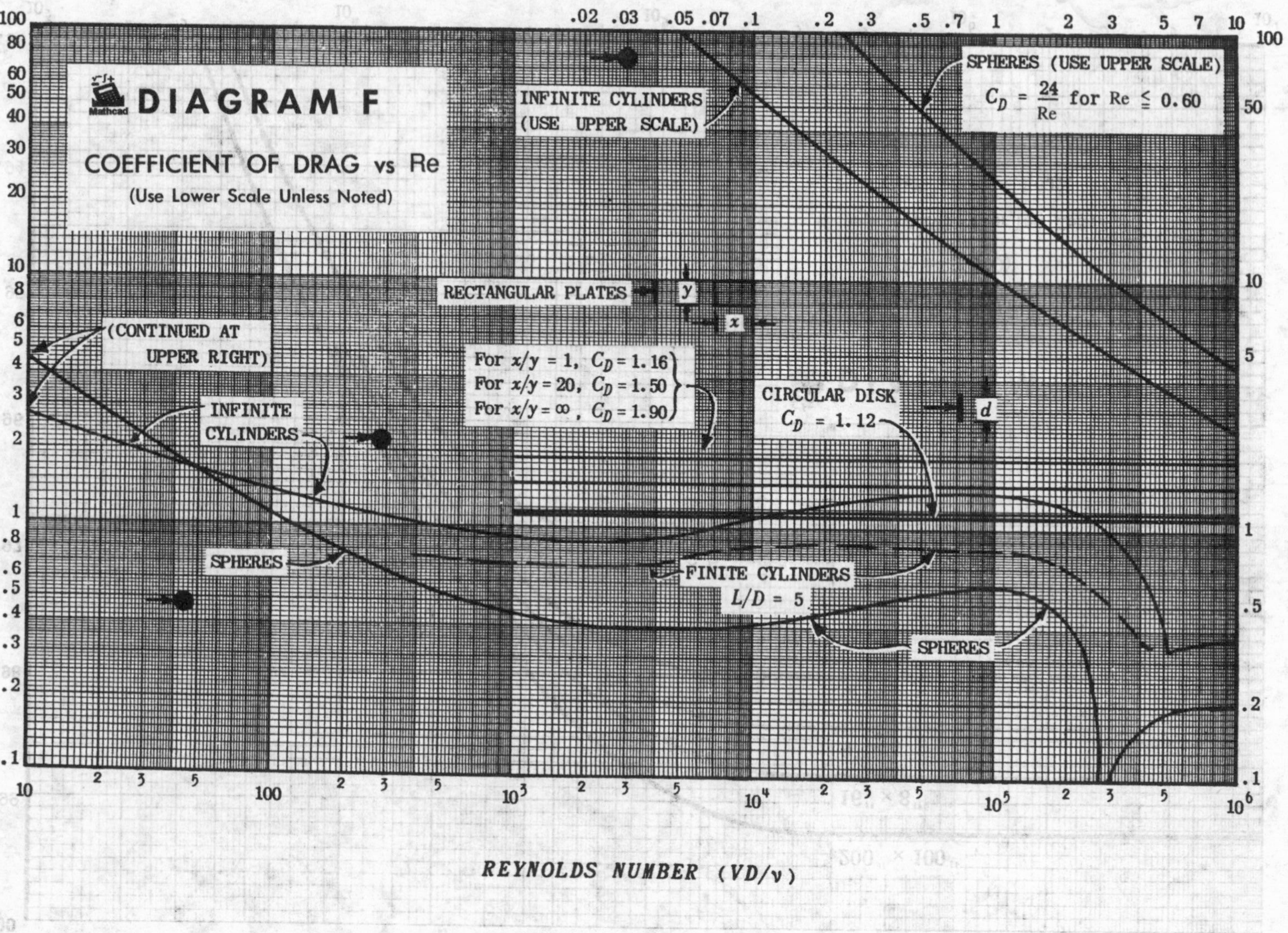
DIAGRAM F
COEFFICIENT OF DRAG vs Re
(Use Lower Scale Unless Noted)
INFINITE CYLINDERS
(USE UPPER SCALE)
SPHERES (USE UPPER SCALE)
$C_D = \frac{24}{Re}$ for Re ≦ 0.60
RECTANGULAR PLATES
(CONTINUED AT
UPPER RIGHT)
For x/y = 1, C_D = 1.16
For x/y = 20, C_D = 1.50
For x/y = ∞ , C_D = 1.90
CIRCULAR DISK
C_D = 1.12
INFINITE
CYLINDERS
SPHERES
FINITE CYLINDERS
L/D = 5
SPHERES
COEFFICIENT OF DRAG C_D
REYNOLDS NUMBER (VD/ν)

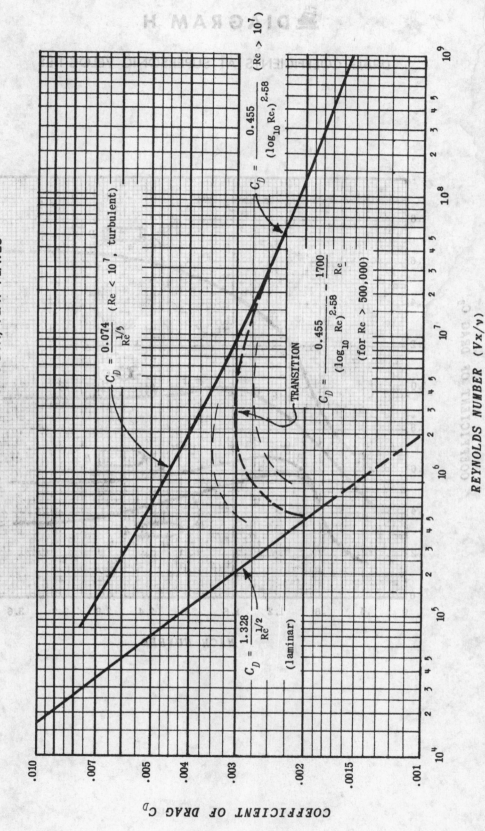

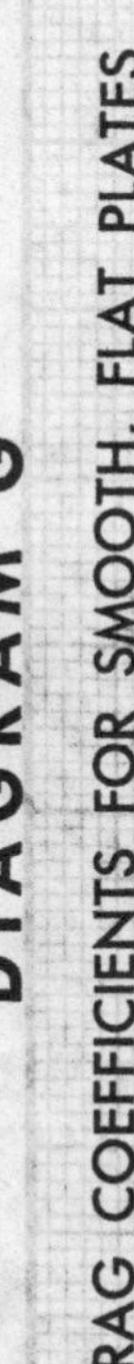
DIAGRAM G
DRAG COEFFICIENTS FOR SMOOTH, FLAT PLATES
$C_D = \frac{0.074}{Re^{1/5}}$ (Re < 10^7, turbulent)
$C_D = \frac{0.455}{(\log_{10} Re)^{2.58}}$ (Re > 10^7)
TRANSITION
$C_D = \frac{0.455}{(\log_{10} Re)^{2.58}} - \frac{1700}{Re}$
(for Re > 500,000)
$C_D = \frac{1.328}{Re^{1/2}}$
(laminar)
COEFFICIENT OF DRAG C_D
REYNOLDS NUMBER (Vx/ν)

DIAGRAM H

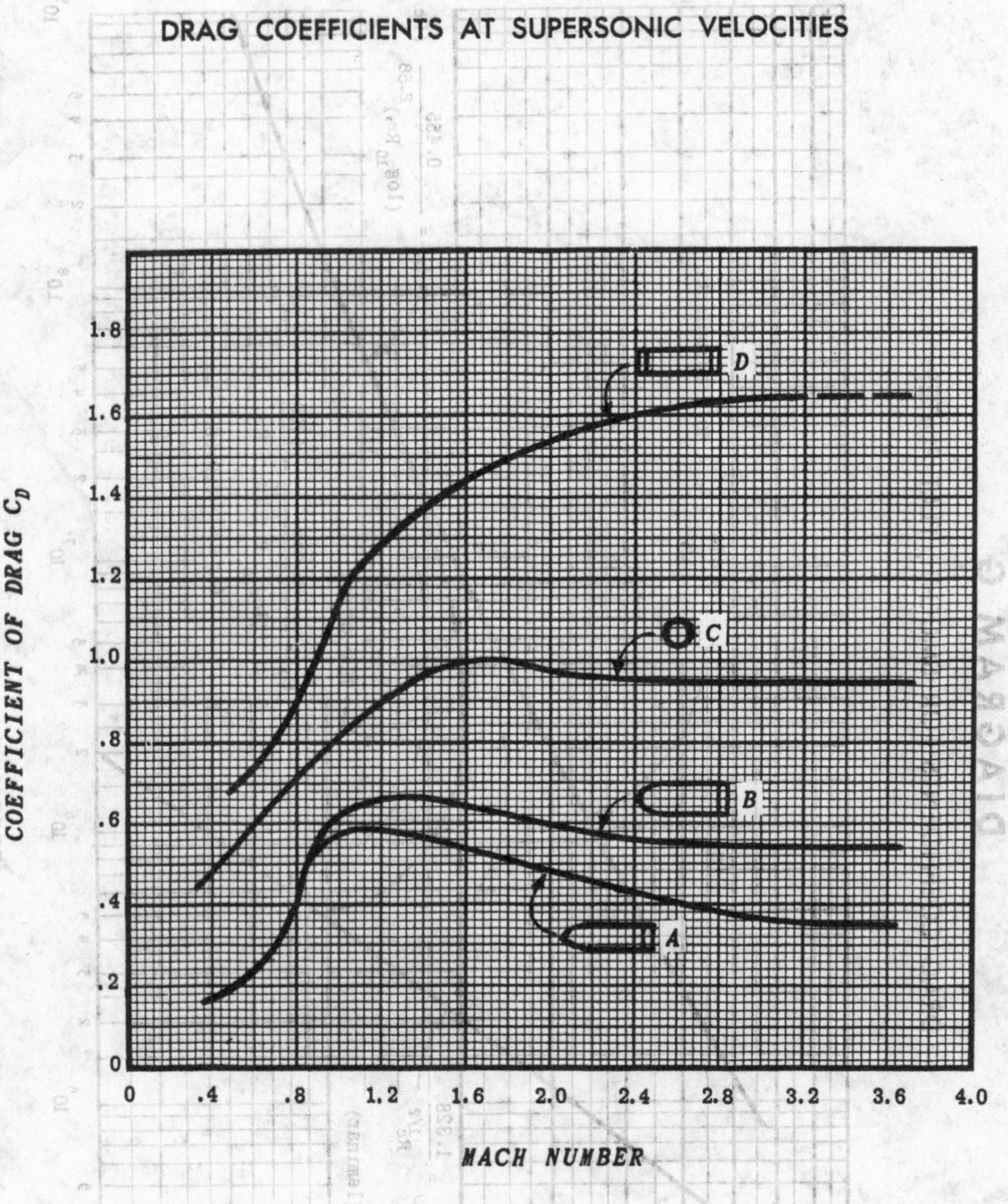

COEFFICIENT OF DRAG C_D

1.8
1.6
1.4
1.2
1.0
.8
.6
.4
.2
0

D
C
B
A

0 .4 .8 1.2 1.6 2.0 2.4 2.8 3.2 3.6 4.0

MACH NUMBER

Index